OPTICAL AND MOLECULAR PHYSICS

Theoretical Principles and
Experimental Methods

OPTICAL AND MOLECULAR PHYSICS

Theoretical Principles and Experimental Methods

Edited by
Miguel A. Esteso, PhD
Ana Cristina Faria Ribeiro, PhD
Soney C. George, PhD
Ann Rose Abraham, PhD
A. K. Haghi, PhD

AAP | APPLE ACADEMIC PRESS

First edition published 2022

Apple Academic Press Inc.
1265 Goldenrod Circle, NE,
Palm Bay, FL 32905 USA

4164 Lakeshore Road, Burlington,
ON, L7L 1A4 Canada

CRC Press
6000 Broken Sound Parkway NW,
Suite 300, Boca Raton, FL 33487-2742 USA

2 Park Square, Milton Park,
Abingdon, Oxon, OX14 4RN UK

Library and Archives Canada Cataloguing in Publication

Title: Optical and molecular physics : theoretical principles and experimental methods / edited by Miguel A. Esteso, PhD, Ana Cristina Faria Ribeiro, PhD, Soney C. George, PhD, Ann Rose Abraham, PhD, A.K. Haghi, PhD.

Names: Esteso, Miguel A., editor. | Ribeiro, Ana Cristina Faria, editor. | George, Soney C., editor. | Abraham, Ann Rose, editor. | Haghi, A. K., editor.

Description: First edition. | Includes bibliographical references and index.

Identifiers: Canadiana (print) 20210190035 | Canadiana (ebook) 20210190124 | ISBN 9781771889834 (hardcover) | ISBN 9781774639405 (softcover) | ISBN 9781003150053 (ebook)

Subjects: LCSH: Optical materials. | LCSH: Plasmonics. | LCSH: Carbon. | LCSH: Diffusion. | LCSH: Nanoparticles. | LCSH: Macromolecules.

Classification: LCC QC374 .O68 2021 | DDC 621.36—dc23

Library of Congress Cataloging-in-Publication Data

....

CIP data on file with US Library of Congress

....

ISBN: 978-1-77188-983-4 (hbk)
ISBN: 978-1-77463-940-5 (pbk)
ISBN: 978-1-00315-005-3 (ebk)

About the Editors

Miguel A. Esteso, PhD
Emeritus Professor, Department of Analytical Chemistry,
Physical Chemistry and Chemical Engineering, University of Alcalá, Spain

Miguel A. Esteso, PhD, is Emeritus Professor in Physical Chemistry at the University of Alcalá, Spain. He is the author of more than 300 published papers in various journals and conference proceedings, as well as chapters in specialized books. He has supervised several PhD and master's degree theses. He is a member of the International Society of Electrochemistry (ISE), the Ibero-American Society of Electrochemistry (SIBAE), the Portuguese Society of Electrochemistry (SPE), and the Biophysics Spanish Society (SEB). He is on the editorial boards of various international journals. His research activity is focused on electrochemical thermodynamics. He developed his postdoctoral work at the Imperial College of London (United Kingdom) and has made several research-stays at different universities and research centers, including the University of Regensburg, Germany; CITEFA (Institute of Scientific and Technical Research for the Armed Forces), Argentina; Theoretical and Applied Physicochemical Research Institute, La Plata, Argentina; Pontifical Catholic University of Chile, Chile; University of Zulia, Venezuela; University of Antioquia, Colombia; National University of Colombia, Bogota, Colombia; University of Coimbra, Portugal; and University of Ljubljana, Slovenia.

Ana Cristina Faria Ribeiro, PhD
Researcher, Department of Chemistry, University of Coimbra, Portugal

Ana Cristina Faria Ribeiro, PhD, is a researcher in the Department of Chemistry at the University of Coimbra, Portugal. Her area of scientific activity is physical chemistry and electrochemistry. Her main areas of research interest are transport properties of ionic and nonionic components in aqueous solutions. Dr Ribeiro has supervised master's degree theses as well as some PhD theses, and has been a theses jury member. She has been a referee for various journals as well an expert evaluator of some of the research programs funded by the Romanian government through the National Council for Scientific Research. She has been a member of the organizing committee of scientific conferences, and she is an editorial member of several journals. She is a member of the Research Chemistry Centre, Coimbra, Portugal.

Soney C. George, PhD
Dean (Research) and Director Amal Jyothi Centre for Nanoscience and Technology, Kerala, India
Email: soneygeo@gmail.com; soneygeo@yahoo.co.in

Soney C. George, PhD, is the Dean (research) and Director of the Amal Jyothi Centre for Nanoscience and Technology, Kerala, India. He is a fellow of the Royal Society of Chemistry (FRSC), London and a recipient of the best researcher of the year award 2018 of APJ Abdul Kalam Technological University. He has also received a best faculty award from the Indian Society for Technical Education, best citation award from the *International Journal of Hydrogen Energy*, fast-track award of young scientist by DST, India, and Indian Young Scientist award instituted by the Indian Science Congress Association. He has done postdoctoral studies from the University of Blaise Pascal, France, and Inha University, South Korea. He has 190 publications in journal and conferences. His major research fields are polymer nanocomposites, polymer membranes, polymer tribology, pervaporation, and supercapacitors. He has guided eight PhD scholars and 95 student projects.

Ann Rose Abraham, PhD
Assistant Professor, Department of Basic Sciences,
Amal Jyothi College of Engineering, Kanjirappally, Kerala, India
Email: annroseabraham86@gmail.com

Ann Rose Abraham, PhD, is currently an Assistant Professor in the Department of Basic Sciences, Amal Jyothi College of Engineering, Kanjirappally, Kerala, India. Dr. Abraham was involved in teaching and also served as the examiner for valuation of answer scripts of Engineering Physics at APJ Abdul Kalam Kerala Technological University. She received her MSc, MPhil, and PhD degrees in Physics from the School of Pure and Applied Physics, Mahatma Gandhi University, Kerala, India. Her PhD thesis was on the title "Development of Hybrid Multiferroic Materials for Tailored Applications." She has expertise in the field of materials science, nanomagnetic materials, multiferroics, and polymeric nanocomposites, and so forth. She has research experience at various national institutes including Bose Institute, SAHA Institute of Nuclear Physics, and UGC-DAE CSR Centre, Kolkata and collaborated with various international laboratories including the University of Johannesburg, South Africa, Institute of Physics Belgrade, and others. She is a recipient of Young Researcher award in the area of Physics, a prestigious forum to showcase intellectual capability. She has delivered invited lectures and sessions at national and international conferences. During her tenure as a doctoral fellow, she taught students at the postgraduate level at the International and Inter University Centre for Nanoscience and Nanotechnology and mentored postgraduate students in their research projects. She has coauthored many book chapters and coedited many books. She has a good number of publications to her credit in many peer-reviewed journals of international repute.

A. K. Haghi, PhD

Editor in Chief, International Journal of Chemoinformatics and Chemical Engineering and Polymers Research Journal; Member, Canadian Research and Development Center of Sciences and Cultures (CRDCSC), Canada Email: AKHaghi@gmail.com

A. K. Haghi, PhD, is the author and editor of over 200 books, as well as 1000 published papers in various journals and conference proceedings. Dr. Haghi has received several grants, consulted for a number of major corporations, and is a frequent speaker to national and international audiences. Since 1983, he served as a professor at several universities. He served as editor in chief of the *International Journal of Chemoinformatics and Chemical Engineering* and *Polymers Research Journal* and is on the editorial boards of many international journals. He is also a member of the Canadian Research and Development Center of Sciences and Cultures (CRDCSC), Montreal, Quebec, Canada. He holds a BSc in urban and environmental engineering from the University of North Carolina (United States), an MSc in mechanical engineering from North Carolina A&T State University (United States), a DEA in applied mechanics, acoustics and materials from the Université de Technologie de Compiègne (France), and a PhD in engineering sciences from Université de Franche-Comté (France).

Contents

Contributors

Dinu Alexander
Department of Physics, Newman College, Thodupuzha, Kerala, India

S. C. G. S. Andrade
Department of Chemistry, University of Coimbra, 3004-535 Coimbra, Portugal

Marisa C. F. Barros
Department of Chemistry, University of Coimbra, 3004535 Coimbra, Portugal

P. R. Biju
School of Pure and Applied Physics, Mahatma Gandhi University, Kottayam 686560, Kerala, India

A. M. T. D. P. V. Cabral
Faculty of Pharmacy, University of Coimbra, 3000-295 Coimbra, Portugal

M. A. Esteso
U.D. Physical Chemistry, University of Alcala, 28805 Alcala de Henares, Madrid, Spain
Catholic University of "Santa Teresa de Jesus de Avila," Los Canteros Street, 05005 Avila, Spain

Robson Fernandes De Farias
The Federal University of Rio Grande do Norte, 59078-970 Natal, RN, Brazil

Sijo Francis
Department of Chemistry, St. Joseph's College, Moolamattom, Kerala, India

Rani George
Department of Physics, St. Aloysius College, Edathua, Alappuzha 689573, Kerala, India
School of Pure and Applied Physics, Mahatma Gandhi University, Kottayam 686560, Kerala, India

Nazmul Islam
Ramgarh Engineering College, Murubanda, Ramgarh, Jharkhand 825101, India

Cyriac Joseph
School of Pure and Applied Physics, Mahatma Gandhi University, Kottayam, Kerala, India

Savaş Kaya
Department of Chemistry, Faculty of Science, Cumhuriyet University, Sivas 58140, Turkey

M. Kailasnath
International School of Photonics, Cochin University of Science and Technology, Kochi 682022, Kerala, India

Ebey P. Koshy
Department of Chemistry, St. Joseph's College, Moolamattom, Kerala, India

Shrikaant Kulkarni
Department of Chemical Engineering, Vishwakarma Institute of Technology, Pune 411037, India

K. V. Arun Kumar
Department of Physics, CMS College (Autonomous), Kottayam 686001, Kerala, India
Nanotechnology and Advanced Materials Research Centre, CMS College (Autonomous), Kottayam 686001, Kerala, India

Kamal P. Mani
International School of Photonics, Cochin University of Science and Technology, Kochi 682022, Kerala, India

K. A. Ann Mary
Department of Physics, St. Thomas College, Thrissur 680001, Kerala, India

Beena Mathew
School of Chemical Sciences, Mahatma Gandhi University, Kottayam, India

Johns Naduvath
Department of Physics, St. Thomas College, Thrissur 680001, Kerala, India

S. Prasanth
Centre for Excellence in Advanced Materials, Cochin University of Science and Technology, Cochin, India

Mamatha Susan Punnoose
School of Chemical Sciences, Mahatma Gandhi University, Kottayam, India

Aparna Raj
School of Pure & Applied Physics, Mahatma Gandhi University, Kottayam, Kerala, India

Ana C. F. Ribeiro
Department of Chemistry, University of Coimbra, 3004-535 Coimbra, Portugal

Cecília I. A. V. Santos
Department of Chemistry, University of Coimbra, 3004535 Coimbra, Portugal

C. Sudarsanakumar
School of Pure and Applied Physics, Mahatma Gandhi University, Kottayam 686560, Kerala, India

A. P. Sunitha
Department of Physics, Government Victoria College, Palakkad 678001, Kerala, India

Sunil Thomas
Department of Physics, United Arab Emirates University, P.O. Box 15551, Al Ain, United Arab Emirates

Riju K. Thomas
Bharata Mata College, Thrikkakara, Ernakulam, Kerala, India

N. V. Unnikrishnan
School of Pure and Applied Physics, Mahatma Gandhi University, Kottayam, Kerala, India

Gianluca Utzeri
Coimbra Chemistry Centre, Department of Chemistry, University of Coimbra, 3004-535 Coimbra, Portugal

Artur J. M. Valente
Department of Chemistry, University of Coimbra, 3004-535 Coimbra, Portugal

Viji Vidyadharan
Department of Optoelectronics, University of Kerala, Thiruvananthapuram 695581, Kerala, India

Luis M. P. Verissimo
Department of Chemistry, University of Coimbra, 3004-535 Coimbra, Portugal

Remya Vijayan
School of Chemical Sciences, Mahatma Gandhi University, Kottayam, Kerala, India

Abbreviations

THN	1,2,3,4-tetrahydronaphthalene
TNT	2,4,6-trinitrotoluene
MBT	2-mercaptobenzothiazole
AA	ascorbic acid
AS	asymmetry ratio
ALD	atomic layer deposition
BRET	bioluminescence resonance energy transfer
PDns	bis-phosphonate dendrons
BP	black phosphorous
CDs	carbon dots
CNT	carbon nanotube
CQDs	carbon quantum dots
CRT	cathode ray tube
CAE	cathodic arc evaporation
CVD	chemical vapor deposition
CL	chemiluminescence
CRI	color rendering index
CIE	Commission Internationale de l'Eclairage
DMF	dimethylformamide
DB	Doxorubicin
DSSC	dye-sensitized solar cells
ECL	electroluminescence
ECHA	European Chemical Agency
EFSA	European Food Safety Agency
EU	European Union
FRET	fluorescence resonance energy transfer
FTO	fluorine tin oxide
FAO	Food and Agriculture Organization
FTIR	Fourier transform infrared
FJO	Furry–Jones–Onsager
GI	graded-index
GQDS	graphene quantum dots
HOMO	highest occupied molecular orbital
IFE	Inner Filter Effect

ISS   International Space Station
IAD   ion-assisted deposition
IBB   isobutyl benzene
JO   Judd–Ofelt
LEDs   light-emitting diodes
LOD   limit of detection
LSPR   localized surface plasmon resonance
MRI   magnetic resonance imaging
MPUs   Manitowoc Public Utilities
MRLs   maximum residual limits
MNPs   metal nanoparticles
MOFs   metal organic frameworks
MO   methyl orange
MB   methylene blue
MWCNTs   multiwalled carbon nanotubes
NAC   *N*-acetyl-l-cysteine
NCs   nanocrystals
NPs   nanoparticles
NSL   nanosphere lithography
NIR   near-infrared
NB   nitrobenzene
NF-RGO   noncovalent functionalized reduced graphene oxide
NLO   nonlinear optical
OCP   organochlorine pesticides
POPs   persistent organic pollutants
pc-WLEDs   phosphor-converted white LEDs
PECs   photoelectrochemical cells
PET   photoinduced electron transfer
PL   photoluminescence
PTT   photothermal therapy
PPPs   plant production products
PMMA   poly(methyl methacrylate)
PEG   polyethylene glycol
PDs   polymer dots
PNCs   polymer nanocomposites
POFs   polymer optical fibers
PSCs   polymer solar cells
PCE   power conversion efficiency
PLD   pulsed laser deposition

QDs	quantum dots
QY	quantum yields
RE	rare earth
RIU	refractive index unit
REACH	Registration, Evaluation, and Restriction of Chemicals
RSA	reverse saturable absorption
RhB	rhodamine B
SA	saturable absorption
SHG	second-harmonic generation
SPIONs	super-paramagnetic iron oxide NPs
SPP	surface plasmon polariton
SERS	surface-enhanced Raman spectroscopy
TEOS	tetraethylorthosilicate
TGC	thermogravitational column
TIP	titanium isopropoxide
TEM	transmission electron microscopy
UCPL	upconversion photoluminescence
VFD	vacuum fluorescent display
WHO	World Health Organization
XPS	X-ray photoelectron spectroscopy

Preface

This book is an invaluable ready reference for many scientists. It offers broad information on key ideas, formulae, techniques, and results in optical and molecular physics. The book is divided into five parts to ensure that this ambitious publication covers the most important areas for scientists today.

The first part concentrates on the plasmonics and carbon dots physics with applications.

The second part is devoted entirely to optical films, fibers, and materials.

The third part treats new developments in optical properties of advanced materials.

The fourth part presents molecular physics and diffusion

Finally macromolecular physics is presented in the last and fifth part of this book.

Optical and molecular physics is a mature field but is still an actively developing field today.

Understanding the basic properties of optical and molecular physics and how these properties can be modified and adjusted to meet specific functional requirements is needed to meet the increasing demands.

Chapter 1 is devoted to plasmonic sensing by green synthesized silver nanoparticles.

In Chapter 2, plasmonic silver nanoparticles successfully synthesized one of the low-cost and easy processing methods, like sol–gel technique, which finds several advantages over other methods, like high purity, low-processing temperatures, ultrahomogeneity, and most significantly the possibility of making glasses of new compositions.

To designate the interactions of light with nanostructured metals, the research community has created the new term "plasmonics." The field of plasmonics has blossomed over the last few years. To make out the gap between conventional optics and highly integrated nanophotonics, plasmonic nanoparticles has most severely studied. The importance of the plasmonic nanoparticles (NPs) over other NPs in a variety of applications like catalysis, electronics, magnetic, optical, optoelectronic, materials for solar cell, medical, bioimaging, and the diagnosis was established on the account of their properties such as surface-enhanced plasmonic character. The optical properties, mainly localized surface plasmon resonance (LSPR) present in gold and silver nanoparticles, play a significant role in the development of these diverse fields. The versatility of the plasmonic NPs led us to review their most salient physicochemical properties and potential applications. In Chapter 3, we present a general idea about the LSPR-based applications of gold and silver NPs.

Green fluorescent carbon dots (CDs) have gained ample interest due to their less toxicity and low cost. They constitute a new class of nanocarbon materials prepared by simple green synthetic routes. The optical properties of the prepared CDs deeply depend on the surface passivation and functionalization provided by the green carbon sources on the surface of CDs. CDs have found profound applications in various fields such as sensing, bioimaging, catalysis, and drug delivery. In Chapter 4, we discuss briefly the synthesis of CDs by means of different green sources, their optical properties, principles of sensing, heavy metal pollution, and also the application of CDs as fluorescent probes for the detection of different heavy metal ions.

Carbon quantum dots (CQDs) are fluorescent nanomaterials intensively investigated and a hot topic of research for the past few years. CQDs with tunable emissions and high quantum yield can be synthesized using low-cost methods and from a wide variety of raw materials. CQDs have emerged into as a new class of photoactive nanomaterials with excitation dependent fluorescence, biocompatibility, aqueous solubility, high photostability, and low toxicity. These fascinating qualities made CQDs a potential alternative to organic dyes, polymer dots, and semiconductors quantum dots and promising materials for photovoltaic cells, photocatalytic reactions, bioimaging, and chemical sensing applications. Chapter 5 includes a brief description on certain remarkable properties of CQDs and their potential applications.

Chapter 6 takes a review of the enormously broad range of types of coatings and the evolution in optical coatings technology as well as merits and limitations of different kinds of coatings.

Chapter 7 reviews applications of polymer fiber optics in various forms and further discusses the possible science-fiction-like applications by virtue of various phenomena involved in it.

In Chapter 8, different types of luminescence quenching mechanisms and the importance of quenching free luminescent materials are discussed. The advantages of the oxalate host and the spectroscopic properties of efficient green-emitting terbium oxalate in the platform of single crystal structure is detailed. Moreover, the structural specificity leading to the quenching free nature of fully concentrated terbium oxalate is described.

In Chapter 9, we discuss some optical properties of semiconductor NPs and also their biological applications like optical sensing, imaging, and some therapeutic applications.

In Chapter 10, we report the optical properties of sol–gel synthesized Pr^{3+}-doped $PVDF–TiO_2$ hybrid. The optical bandgap, bonding nature, absorption, emission, and colorimetric analysis of $PVDF–TiO_2:Pr^{3+}$ are described.

CDs are a new class of the nanocarbon materials displaying a wide range of significant advantages such as low toxicity, chemical inertness, good water solubility, and physicochemical properties. Chapter 11 demonstrates the use of various natural sources as carbon precursor for the synthesis of CDs. The structural and optical properties of green CDs are also discussed.

In Chapter 12, it is shown that nonlinear optical materials have a wide range of applications in industry and electronics because of their versatile mechanical properties. Inorganic polymers are macromolecules having noncarbon skeletal structure and inorganic or organic side chain. Coordination polymers, cross-linked polymers, metal-organic frameworks, thin films, hybrid materials, and so forth have additional optoelectronic qualities.

Nonlinear optics was originated shortly after the discovery of the laser in 1960, and its research field has been constantly emerging. A nonlinear optics study includes several types of nonlinear effects in the interactions between a laser and matter. To date, nonlinear optics gained a wide range of applications. In Chapter 13, we have discussed the nonlinear properties of some metal nanoparticles and their application. Also, we have briefly explained some important methods for the preparation of different metal nanoparticles and nanocomposites with their nonlinear optical behavior. The investigation on nonlinear optical properties of metal nanoparticles is very essential for the invention of new optoelectronic elements and device.

In Chapter 14, we discuss the phosphors, various luminescence and their basic mechanisms, and various phosphor synthesis methods, and their advantages are explained with various applications of phosphor materials and important optical properties of some of the rare-earth-doped oxide phosphors are also included.

The last two sections' chapters of this book are devoted to the presentation of new insights in molecular and macromolecular physics along with diffusion.

—Miguel A. Esteso, PhD
Ana Cristina Faria Ribeiro, PhD
Soney C George, PhD
Ann Rose Abraham, PhD
A. K. Haghi, PhD

PART I

Plasmonics and Carbon Dots Physics with Applications

Plasmonic Sensing by Green Synthesized Silver Nanoparticles

MAMATHA SUSAN PUNNOOSE and BEENA MATHEW*

School of Chemical Sciences, Mahatma Gandhi University, Kottayam, India

Corresponding author. E-mail: beenamscs@gmail.com

ABSTRACT

Recent advancement in nanotechnology has allowed the development of green strategies for the synthesis of silver metal nanoparticles. These green silver nanoparticles have emerged as a powerful tool in the field of sensing due to their unique optical properties. In this chapter, we discuss the green approach toward silver nanoparticles synthesis, their surface plasmon resonance property, and their application for the detection of different environmental pollutants including heavy metal ions and several organic moieties.

1.1 INTRODUCTION

Metal nanoparticles is a fastly emerging current field of nanotechnology having applications in the fields of medicine, materials, energy, catalysts, sensors, water treatment, and antimicrobials [1–3]. Nanoscience deals with materials of the order of nanometer or 10^{-9} m. Unlike bulk materials, particles at the nanoscale regime show unique characteristic optical, electrical, chemical, biological, catalytic, and magnetic properties [4, 5]. These significant novel properties are attributed to its stability, large surface to volume ratio, and quantum size effects.

Properties like high extinction coefficient, a distinct color variation upon dispersion, and aggregation have contributed much toward the use of metallic nanoparticles as promising analytical colorimetric probes. Recently noble metal nanoparticles have drawn special interest on the selective sensing of a wide variety of molecules. These sensors find applicability in the detection of pollutants in industrial effluents, trace elements in water samples, and monitoring of potable water.

Noble metal nanoparticles are synthesized by various methods such as electrochemical, sonochemical, chemical reduction, and microwave irradiations [6–10]. Most of these physicochemical routes for the preparation of metal nanoparticles involve use of environmentally hazardous and potent toxic organic solvents and reducing agents like hydrazine, thiophenol, sodium borohydride, citrate, and mercapto acetate [11, 12]. They are potentially hazardous to the environment and human health. The use of dangerous chemicals, high energy reaction requirements, time consumption, and the limitation of effective waste disposal are the major drawbacks of wet chemical methods [13]. In order to overcome these flaws, an ecofriendly or green chemistry-based metal nanoparticle synthesis is often desirable.

1.2   GREEN NANOTECHNOLOGY

Green chemistry is a set of principles and procedures in order to avoid or minimize the use of toxic chemicals in reaction products and process. Green nanotechnology utilizes renewable biomolecules of natural origin such as algae, plant extracts, microorganisms, seed extracts, and enzymes as the reducing and stabilizing agents for the formation of well-dispersed nanoparticles [1, 14]. Green synthetic methods for nanoparticles have several advantages over conventional synthetic schemes such as simplicity, clean, cost-effectiveness, and safety in handling [15]. The approach of green synthesis has emerged as one of the active areas of research due to the wide availability of natural resources, the use of environmentally benign solvents, low toxicity, and biocompatibility of the products.

Several biosynthetic methods for the silver metal nanoparticles using neem, *Synedrella nodiflora, Dillenia indica, Polygonum hydropiper, Biophytum sensitivum, Indigofera tinctorial,* and algae are reported in the literature [16–22]. Coffee and green tea extracts were used for the synthesis of stable nanoparticles of various sizes and morphologies [23]. Citrus fruit extracts of orange and lemon have been explored for the green synthesis of silver nanoparticles (AgNPs) [24, 25].

1.3 SURFACE PLASMON RESONANCE (SPR)

Nanoparticles of noble metal silver exhibit characteristic optical properties that differentiate them from their bulk counterparts. The phenomenon of surface plasmon resonance (SPR) involves the exhibition of the peculiar strong absorption band in the visible region while monitored using UV–visible spectroscopy. The absorption and scattering properties of AgNPs are highly efficient [26]. SPR is the collective oscillation of conduction electrons on the metal surface upon excitation with the oscillating electromagnetic field of the incident light. The optical resonance related to these oscillations at a particular frequency of light is called localized SPR. The optical properties of metal nanoparticles were first proposed by Mie. According to his theory, reverberation condition is accomplished only when the real part of the dielectric elements of the metal and surrounding match each other.

Silver nanoparticles (AgNPs) offer simple, rapid, and easy selective sensing toward different moieties by means of their highly efficient SPR, which can be easily tuned to different wavelengths by means of variation in size and shape of particles, interparticle distance, refractive index, and dielectric environment [28]. The unique optical properties inherited from SPR are often advantageous toward the sensing applications.

The literature describes the various powerful techniques like atomic absorption or emission spectrometry [29], inductively coupled plasma mass spectrometry [30], atomic fluorescence spectrometry [31], high-performance liquid chromatography [32], ion-selective electrode, and flame photometry [33], for the detection of various heavy metals. Most of these conventional methods, are limited with respect to sensitivity, selectivity, need for high energy requirements, sophisticated instrumentation, and high time requiring procedures. The use of metal nanoparticles particularly AgNPs as a sensor is of great interest due to the several advantages like high molar extinction coefficient, high sensitivity to the physicochemical surrounding, practical implementation, and economical operation compared to other metal nanoparticle-based sensors.

1.4 PLASMONIC SENSING BY GREEN SILVER NANOPARTICLES (AGNPS)

The very small size, high surface-to-volume ratios, and stability of AgNPs make the interaction of them with any chemical species to be easier, which results in changes of their SPR properties. Both tunability and sensitivity of SPR empower the silver nanoparticles to be the best contender for optical detection applications. Following is a brief account on the plasmonic sensing of different metal ions and organic ions by the various green synthesized AgNPs.

1.4.1 SENSING OF HEAVY METAL IONS

1.4.1.1   MERCURY

Mercury is a highly persistent pollutant that can damage the brain, heart, kidney, stomach, and intestine even at their very low concentrations. The maximum permissible limit of Hg^{2+} ions in drinking water is limited to 30 nm as per World Health Organization.

Syzygium aqueum fruit extract mediated AgNPs show an SPR at 420 nm and is used as a selective plasmon sensor for Hg^{2+} ions [36]. Upon successive addition of mercuric ions, the absorbance of SPR peak at 420 nm gradually decreases with a slight blueshift to 380 nm. The mechanism proposes the oxidation of Ag(0) to Ag(I) by Hg(II) caused the blueshift of AgNP absorption peak. This Mie blueshift phenomenon may occur due to the close approach of the AgNPs upon mercuric ions reduction [37]. The limit of detection is estimated to be 8.5×10^{-7} M over a linear range of 5–100 ppm. These AgNPs show a precision better than 5% for the determination of Hg^{2+} in real water samples. Over the linear addition of 10–90 µM Hg^{2+}, characteristic SPR of *Agaricus bisporus*-AgNPs at 412 nm diminished without any shift [38]. The maximum effectual complexation of 90 µM Hg^{2+} with AgNP resulted in the vanishing of SPR band.

Farhadi et al. reported the synthesis of green AgNPs using the extracts of soap-root and manna of Hedysarum plants as the stabilizing and reducing agents, respectively [39]. In the absence of mercuric ions, yellowish-brown AgNPs show an SPR band around 408 nm. Upon addition of 10–100 µM Hg^{2+}, the original color of AgNPs gradually faded away accompanied by the broadening and blueshifting of SPR peak (Figure 1.1). The changes in intensity and shift of SPR were monitored by UV–visible spectroscopy. The plausible mechanism for plasmonic sensing involves a redox reaction between Ag^0 and Hg^{2+} ions, by the effectual removal of stabilizing components from the silver surface by means of mercuric addition. The vanishing of spherical AgNPs upon Hg^{2+} addition is clearly depicted by the SEM images of freeze-dried samples of AgNPs in the absence and presence of Hg^{2+} ions, which is in agreement with the proposed mechanism. The mercury atoms formed as the resultant of redox reaction could strongly bind to the silver surface is accompanied by a blueshift of the plasmon absorption band of AgNPs [40]. The limit of detection was calculated to be 2.2×10^{-6} M. Upon increase of Hg^{2+} concentration a decrease in SPR absorption peak along with broadening and blueshifting is also observed with AgNPs stabilized using fructose [41], and *Cucurma longa* rhizome [42], extract. Complete decolorization of *Citrus limon*-AgNPs occurred along with SPR broadening and blueshift over a pH range of 3.2–8.5 [43].

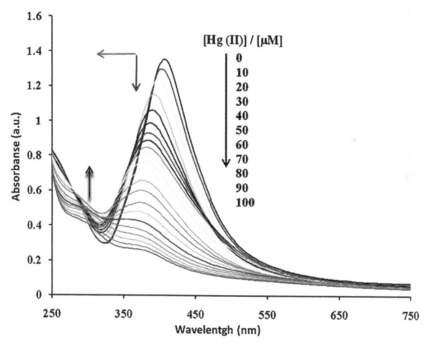

FIGURE 1.1 UV–visible absorption spectra of AgNPs upon addition of 10–100 µM of Hg^{2+} ions. (Reproduced from Farhadi et al. [39] with permission from Elsevier).

In the absence of Hg^{2+} ions, *Epilobium parviflorum* green tea extract derived AgNPs show an SPR band at 414 nm [44]. In the presence of mercuric ions, the SPR band disappeared along with the remarkable visual color change. A detection limit of 58.11 nM was calculated for the successive addition of Hg^{2+} ions over the 0.5–235 µM range. Polyphenols like tannins, gallic acid, and ellagic acid present in the extract are used for the reduction and stabilization of AgNPs. The presence of these phytochemicals on the AgNPs surface facilitates the bonding with Hg^{2+} ions, thus favors the redox reaction between zero-valent silver and mercuric ions. The Hg^{2+} act as a better oxidizing agent than Ag^+, which resulted in the easy formation of Hg–Ag amalgam. Moreover, this amalgamation process led to the alteration of AgNP coloration. A similar mechanism is observed in the case of plasmonic sensing by blood-red colored AgNPs synthesized using *Opuntia ficus indica* [45]. Upon the increase of Hg^{2+} concentration from 10^{-8} to 10^{-3} M, SPR band at 441 nm decreased in intensity with a slight redshift and is accompanied by a progressive color change to white.

The phytoconstituents stabilizing the *Derris trifoliate* derived AgNPs efficiently reduce Hg(II) to Hg(0) [46]. Thus effective electrostatic binding of these biological constituents with the AgNP get destabilized and redox reaction leads to amalgam formation. Amalgamated AgNPs attain stability via aggregation formation into larger particles that results in the quenching of SPR band intensity with a blueshift. The DLS measurements showed that the average size of nanoparticles increased with the addition of Hg^{2+} ions. The mechanism involving the redox reaction and amalgamation of elemental mercury with silver particles is observed in Onion [47] and *Soymida febrifuga* [48] derived AgNPs.

The selectivity and sensitivity of *Hibiscus Sabdariffa* derived AgNPs depend on the pH and plant parts used for the synthesis [49]. At pH 7, *H. sabdariffa* leaves derived AgNPs show selectivity toward Hg^{2+}, Cd^{2+}, and Pb^{2+}. The concentration-dependent sensing of Cd^{2+} and Pb^{2+} involves a bathochromic shift of absorption peak because of the interaction of nanosurface functionalities that lead to the aggregation of AgNPs. When metal nanoparticles are brought in closer proximity, the plasmon oscillation of them couple with each other [27]. The respective additions of 300, 420, and 260 µL of Hg^{2+}, Cd^{2+}, and Pb^{2+} into the AgNPs caused a complete reduction of SPR band. At pH 7, *H. Sabdariffa* stem mediated AgNPs show selective decolorization toward Hg^{2+}. The mechanism of Hg^{2+} sensing could be explained on the basis of electrochemical differences between Ag^+ and Hg^{2+} ions.

1.4.1.2   COPPER

Copper is the main constituent of the metabolism system. The increase in the Cu^{2+} level in the biological cells leads to kidney-related and neurodegenerative diseases and the free Cu^{2+} bring out toxicity to cells as they generate hydroxyl radicals that cause apoptosis [50]. *Callistemon viminalis* leaves functionalized AgNPs show selective plasmonic sensing toward Cu^{2+} ions [51]. This is visually identified by the color change from yellow to colorless. This visual assay is accompanied by the broadening and shifting of SPR at 450 nm while monitored using UV–visible spectroscopy. TEM analysis proves the aggregation of AgNPs in the presence of Cu^{2+} ions. Aggregation is attributed to the host–guest and Л–Л interactions of AgNPs with Cu^{2+} ions, which resulted in the color change. Strong binding of Cu^{4+} with AgNPs derived from *Moringa oleifera* flower causes etching of antioxidants from the nanosurface [52]. Upon increase of Cu^{4+} concentration, diminishment of SPR peak occurred along with broadening due to the electron surface scattering that may be enhanced for very small clusters [39]. Sensitivity is found to be 0.249 mM. *M. oleifera* bark extract mediated AgNPs show complexation with Cu^{2+} ions [53]. The intensity of SPR at 410 nm quenched with an increase of Cu^{2+} concentration from 10 to 90 µM. Rate of aggregation of silver particles that leads to complexation with Cu^{2+} ions linearly increased with the concentration of Cu^{2+}. *Azadirachta indica*-AgNPs-Rh6G act as a probe for the detection of Cu^{2+} [54]. The mechanism involves the fixation of Rh6G over the surface of AgNPs and replacement of these dye moieties by the cupric ions. The original purple color of the dye is not retained on noncovalent interaction with nanosurface. Upon addition of Cu^{2+} ions, Rh6G moieties are replaced from the AgNPs that shows a purple color. The *A. indica*-AgNPs show an SPR at 400 nm. On bonding with Rh6G, the absorbance of the AgNPs peak decreased with the emergence of an additional adsorption band. The adsorption values of the new peak decreased on high amalgamation for complex formation with Cu^{2+} ions. Spherical monodispersed reddish-brown AgNPs are derived from *Ananas comosus* [55]. On increasing the Cu^{4+} concentration between 2 and 7 mM, the intensity of SPR at 478 nm gradually decreased with broadening [56] and a new SPR band at 780 nm increased with intensity. New SPR peak originates due to the adsorption of entities causing variations in the dielectric environment around AgNPs. The biomolecules present on the AgNPs surface that emerged from the plant extract act as a link for bonding with Cu^{4+} ions. Elimination of these biomolecules resulted in the aggregation of nanoparticles that was evident from the color change to light blue. Sensitivity is estimated to be 0.605/mM. On varying the concentration of Zn^{2+} between 3×10^{-4} and 12×10^{-4} M, the optical absorbance decreases with broadening and blueshift due to the removal of capping biomolecules from the nanosurface. The reddish-brown color of AgNPs became colorless upon maximum sensing with Zn^{2+} ions. Sensitivity is found to be 0.036×10^{-4}/M.

1.4.1.3   Other Metal Ions

The negative charge over the *Calotropis procera*-AgNP surface is responsible for the electrostatic attraction between the oppositely charged Fe^{2+} ions [57]. Upon successive addition of Fe^{2+} ions in the linear range of 2–40 μg/mL, AgNPs show a slight blueshift in association with the increase in absorbance. The drastic color change from yellow to brown indicates the strong complexation between them. Aravind et al. reported the optical selectivity of microwave synthesized *Allium sativum*-AgNPs toward Cd^{2+} ions [58]. On increasing the concentration of Cd^{2+} ions from 10 to 90 μM, the SPR peak at 403 nm decreases with broadening. Upon maximum complexation of AgNPs with 90 μM Cd^{2+} ions, absorbance peak completely vanished and is visually marked by a color change from brown to golden yellow. According to Pearson soft and hard acid–base concept, soft acid cadmium can complex with the soft sulfur atoms of the ligands present in *A. sativum* extract. TEM analysis shows the increment in the size of AgNPs upon complexation with Cd^{2+} from 19.7 to 26.82 nm, which proved the aggregation of AgNPs. *Lycopersicon esculentum* modified AgNPs of brown color fade to feeble pink upon selectivity sensing of Cr^{3+} [59]. The Cr^{3+} can easily chelate with citrate ions present on AgNPs surface, which induces the easy aggregation between the nanoparticles [60]. Upon the linear addition of Cr^{3+} in the range of 10–90 μM, the characteristic absorption maxima at 413 nm decreased with a slight blueshift. Microwave-assisted *Brassica oleracea var. italica* mediated brown AgNPs show a characteristic SPR peak at 402 nm [61]. Selective complexation with Ni^{2+} ions is visually indicated by the color change to orange. In order to study the effect of concentration of Ni^{2+} ions the concentration is varied between 10 and 90 μM. As the concentration of Ni^{2+} increases the SPR absorbance decreases and disappears accordingly. As a result of the effective complexation between Ni^{2+} and AgNPs a new peak at 290 nm arises. Aggregation of particles due to the effective complexation is evident from the TEM and EDX results.

1.4.2   ORGANIC MOIETIES

1.4.2.1 H_2O_2

H_2O_2 possess strong oxidizing property. It possesses great relevance in various fields such as food, pharmaceutical, cosmetics, environmental, wood, and pulp industries [62,63]. It causes various health issues like cellular damage

and environmental hazards even at their very low concentrations [64]. Therefore it is very essential to develop simple and accurate green methods to
detect H_2O_2 in spite of the sophisticated electrochemical and amperometric
techniques. A concentration-dependent degradation of *Sargassum boveanum*
mediated AgNPs by H_2O_2 is reported [65]. Changes in the SPR before and
after addition of H_2O_2 are monitored using UV–visible spectra. SPR at
417 nm progressively decreased with increment of H_2O_2 concentration in
the 1–120 μM range. It easily oxidizes AgNPs from Ag^0 to Ag^+ state. The
maximum degradation of yellowish AgNP occurred at 120 μM addition of
the analyte solution that was primarily evident from the colorless appearance of the solution mixture and was further confirmed by the decrease in
nanoparticle size from TEM analysis (Figure 1.2).

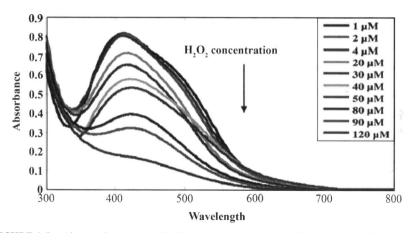

FIGURE 1.2   Plasmonic sensing of H_2O_2 using *S. boveanum* mediated AgNPs. (Reproduced
from Farrokhnia et al. [65] with permission from Elsevier).

After the addition of 1 mL 20 mM H_2O_2 into the *Calliandra haema-*
tocephala [66] leaf extract mediated AgNPs, UV–visible spectrum was
noted at regular intervals of time. The fading of brownish-yellow color of
AgNPs to colorless is associated with a decrease of SPR absorbance at 414
nm that eventually dies out. Addition of H_2O_2 to AgNPs resulted in a free
radical generation that oxidizes Ag^0 to Ag^+. This initiated the degradation
of AgNPs and diminishment of absorbance maximum. Upon addition of
10^{-6}–1 M concentrations of H_2O_2, SPR band of chitosan stabilized AgNPs
decreased with a redshift from 415 to 423 nm [67]. This is explained by the
agglomeration of AgNPs with the chitosan layer of molecules. The fading of

the yellowish color of chitosan stabilized AgNPs to colorless with the linear increase of H_2O_2 could be due to the oxidation of Ag^0 to Ag^+. Subramanian et al. reported the decrease in SPR at 427 nm of *Mangifera indica*-AgNP upon addition of 1 μM–1 M H_2O_2 [68].

1.4.2.2   OTHER ORGANIC MOIETIES

Addition of ammonia caused a blueshift in SPR *Durenta erecta*-AgNPs from 451 to 429 nm [60]. The surface charge on AgNPs surface increased due to the formation of $Ag(NH_3)_2^+$ coordination complex between silver ions and ammonia. The complexation resulted in a decrease of AgNPs amount that eventually led to the nanoparticle decolorization. Increased addition of 10–100 ppm Cr^{6+} into AgNPs also shows a shift in SPR. The strong covalent binding of dichromate anions with AgNPs resulted by the removal of capping agents from the nanosurface was evident from XRD and FE-SEM analyses.

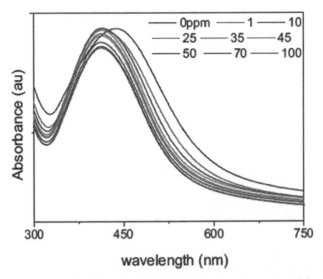

FIGURE 1.3   Changes in the absorption spectrum of *Cyamopsis tetragonaloba*-AgNPs with various ammonia concentrations. (Reproduced from Pandey et al. [69] with permission from Elsevier).

SPR peak at 440 nm of *Cyamopsis tetragonaloba*-AgNPs decreased with the increase of aqueous ammonia concentration (Figure 1.3) [69]. As ammonia content increased, peak slightly shifted to 413 nm due to the variations in the interparticle distance and the dielectric constant [70]. Ammonia concentration within 1–50 ppm range is linear with absorbance having a R^2 value of 0.966. Tea leaf extract mediated AgNPs show high colorimetric and optical sensitivity toward an amino acid, cysteine [71]. Addition of L-cysteine into the aqueous AgNPs solution caused a decrease and shift in SPR band at 406 nm along with the appearance of a new peak at 506 nm. Addition of cysteine caused a distinct color change to the nanosolution from yellow to pink. The remarkable color change and shift in wavelength are attributed to the thiol groups of cysteine-mediated aggregation of AgNPs via hydrogen bonding and electrostatic interactions. These strong interactions resulted in the formation of immobilized Ag–S bonds on the surface of AgNPs [72]. The absorbance of new peak shows linear dependence on the concentration of cysteine added. The detection limit was found to be 50 µM from the colorimetric assay. Selective sensing of microwave-assisted *Coscinium fenestratum*-AgNPs toward the dithiocarbamate fungicide thiram is reported [73]. The SPR peak at 423 nm decreased upon addition of thiram from 10 to 90 µM. This minimization is ascribed to the intermolecular H-bonding interactions between thiram and the functional groups present on capping molecules of AgNPs. Silver nanoparticles embedded in bacterial cellulose nanopaper act as bionanocomposite plasmonic sensor for cyanide and 2-mercaptobenzothiazole (MBT) in water samples [74]. For the synthesis of bionanocomposite, hydroxyl groups present in the cellulose nanofibers act as the reducing agent for silver ions adsorbed on the bacterial cellulose nanopaper. Proportional to the MBT concentration plasmonic adsorption band of bionanocomposite decreased in intensity and redshifted to higher wavelengths. FE-SEM results proved the increase in the size of AgNPs after the addition of MBT. While on increasing the CN⁻ concentration, intensity of the absorption band decreased and resulted in a blueshift. Upon addition of CN⁻, FE-SEM shows a decrease in the size of the nanoparticles. Table 1.1 provides an overall view of plasmonic sensing by green synthesized silver nanoparticles.

TABLE 1.1 Overview of Plasmonic Sensing by Green Synthesized Silver Nanoparticles

Green Precursor	Analyte	Linear Range	Detection Limit	Reference
Agaricus bisporus	Hg²⁺	0–90 µM	–	38
Cucurma longa	Hg²⁺	0.01–0.2 µM	8.2 nM	42
Citrus limon	Hg²⁺	20–440 µL	–	43

TABLE 1.1   *(Continued)*

Green Precursor	Analyte	Linear Range	Detection Limit	Reference
Derris trifoliate	Hg^{2+}	0–13 µM	1.55 µM	46
Epilobium parviflorum	Hg^{2+}	0.5–235 µM	58.11 nM	44
Fructose	Hg^{2+}	100–1000 µL	–	41
Hedysarum	Hg^{2+}	10–100 µM	2.2 µM	39
Onion	Hg^{2+}	0–1400 µL	–	47
Opuntia ficus indica	Hg^{2+}	10^{-8}–10^{-3} M	–	45
Syzygium aqueum	Hg^{2+}	5–100 ppm	8.5×10^{-7} M	36
Soymida febrifuga	Hg^{2+}	0.001–1000 µM	13.32 µM	48
Callistemon viminalis	Cu^{2+}	10–400 µM	10 µM	51
Moringa oleifera	Cu^{2+}	10–90 µM	–	53
M. oleifera	Cu^{4+}	1–12 mM	–	52
Ananas comosus	Cu^{4+}	2–7 mM	–	55
	Zn^{2+}	3×10^{-4}–12×10^{-4} M	–	
Allium sativum	Cd^{2+}	10–90 µM	–	58
Brassica oleracea	Ni^{2+}	10–90 µM	3.302 µM	61
Calotropis procera	Fe^{2+}	2–40 µg/mL	–	57
Lycopersicon esculentum	Cr^{3+}	10–90 µM	–	59
Chitosan	H_2O_2	10^{-6}–1M	–	67
M. indica	H_2O_2	1 µM–1 M	–	68
S. boveanum	H_2O_2	1–120 µM	8.64 nM	65
Cyamopsis tetragonaloba	NH_3	1–100 ppm	–	69
Durenta erecta	NH_3	5–100 ppm	0.5 ppm	60
	Cr^{6+}	10–100 ppm	–	
Tea leaf	ʟ-cysteine	0–60 µM	–	71
Coscinium fenestratum	Thiram	0–90 µM	–	73

TABLE 1.1 *(Continued)*

Green Precursor	Analyte	Linear Range	Detection Limit	Reference
Bacterial cellulose	CN⁻	0.2–2.5 µg mL⁻¹	0.012 µg mL⁻¹	74
	MBT	2–110 µg mL⁻¹	1.37 µg mL⁻¹	

1.4 CONCLUSION

The silver nanoparticles were developed by green approach utilizing different natural sources. The as-prepared AgNPs show high stability and unique SPR. These AgNPs act as a potential optical sensor for the selective recognition and detection of heavy metal ions and different organic moieties of high environmental toxicology. Thus a simple, cost-effective, and unmodified AgNPs prepared via green techniques offer a new approach for the selective sensing of aqueous environmental pollutants.

KEYWORDS

- **nanotechnology**
- **green strategies**
- **silver nanoparticles**
- **surface plasmon resonance**
- **sensing**
- **optical property**
- **environmental pollutant**
- **heavy metal**

REFERENCES

1. Sharma, V. K., Yngard, R. A., Lin, Y. Silver nanoparticles: green synthesis and their antimicrobial activities. *Advances in Colloid and Interface Science.* 2009, vol. 145(1–2), 83–96.
2. Annavaram, V., Posa, V. R., Uppara, V. G., Jorepalli, S., Somala, A. R. Facile green synthesis of silver nanoparticles using *Limonia acidissima* leaf extract and its antibacterial activity. *BioNanoScience.* 2015, vol. 5(2), 97–103.

3. Du, J., Zhao, M., Huang, W., Deng, Y., He, Y. Visual colorimetric detection of tin(II) and nitrite using a molybdenum oxide nanomaterial-based three-input logic gate. *Analytical and Bioanalytical Chemistry*. 2018, vol. 410(18), 4519–4526.

4. Krasteva, N., Besnard, I., Guse, B., Bauer, R. E., Müllen, K., Yasuda, A., Vossmeyer, T. Self-assembled gold nanoparticle/dendrimer composite films for vapor sensing applications. *Nano Letters*. 2002, vol. 2(5), 551–555.

5. Prow, T. W., Salazar, J. H., Rose, W. A., Smith, J. N., Reece, L., Fontenot, A. A., Wang, N. A., Lloyd, R. S., Leary, J. F. Nanomedicine: nanoparticles, molecular biosensors, and targeted gene/drug delivery for combined single-cell diagnostics and therapeutics. *Advanced Biomedical and Clinical Diagnostic Systems II*. 2004.

6. Jana, N. R., Gearheart, L., Murphy, C. J. Wet chemical synthesis of silver nanorods and nanowires of controllable aspect ratio. *Chemical Communications*. 2001, vol. 7, 617–618.

7. Yamauchi, Y., Tonegawa, A., Komatsu, M., Wang, H., Wang, L., Nemoto, Y., Suzuki, N., Kuroda, K. Electrochemical synthesis of mesoporous Pt–Au binary alloys with tunable compositions for enhancement of electrochemical performance. *Journal of the American Chemical Society*. 2012, vol. 134(11), 5100–5109.

8. Makarov, V. V., Love, A. J., Sinitsyna, O. V., Makarova, S. S., Yaminsky, I. V., Taliansky, M. E., Kalinina, N. O. Green nanotechnologies: synthesis of metal nanoparticles using plants. *Acta Naturae*. 2014, vol. 6(1), 35–44.

9. Joseph, S., Mathew, B. Microwave-assisted facile synthesis of silver nanoparticles in aqueous medium and investigation of their catalytic and antibacterial activities. *Journal of Molecular Liquids*. 2014, vol. 197, 346–352.

10. Joseph, S., Mathew, B. Microwave-assisted facile green synthesis of silver nanoparticles and spectroscopic investigation of the catalytic activity. *Bulletin of Materials Science*. 2015, vol. 38(3), 659–666.

11. Sener, G., Uzun, L., Denizli, A. Lysine-promoted colorimetric response of gold nanoparticles: a simple assay for ultrasensitive mercury(II) detection. *Analytical Chemistry*. 2013, vol. 86(1), 514–520.

12. Chen, Z., Zhang, X., Cao, H., Huang, Y. Chitosan-capped silver nanoparticles as a highly selective colorimetric probe for visual detection of aromatic ortho-trihydroxy phenols. *The Analyst*. 2013, vol. 138(8), 2343.

13. Abdel-Halim, E. S., El-Rafie, M. H., Al-Deyab, S. S. Polyacrylamide/guar gum graft copolymer for preparation of silver nanoparticles. *Carbohydrate Polymers*. 2011, vol. 85(3), 692–697.

14. Prasad, K., Jha, A. K., Kulkarni, A. R. Lactobacillus assisted synthesis of titanium nanoparticles. *Nanoscale Research Letters*. 2007, vol. 2(5), 248–250.

15. Punnoose, M. S., Mathew, B. Treatment of water effluents using silver nanoparticles. *Material Science & Engineering International Journal*. 2018, vol. 2(5).

16. Shankar, S. S., Rai, A., Ahmad, A., Sastry, M. Rapid synthesis of Au, Ag, and bimetallic Au core–Ag shell nanoparticles using Neem (*Azadirachta indica*) leaf broth. *Journal of Colloid and Interface Science*. 2004, vol. 275(2), 496–502.

17. Vijayan, R., Joseph, S., Mathew, B. Eco-friendly synthesis of silver and gold nanoparticles with enhanced antimicrobial, antioxidant, and catalytic activities. *IET Nanobiotechnology*. 2018, vol. 12(6), 850–856.

18. Mohanty, A. S., Jena, B. S. Innate catalytic and free radical scavenging activities of silver nanoparticles synthesized using *Dillenia indica* bark extract. *Journal of Colloid and Interface Science.* 2017, vol. 496, 513–521.

19. Bonnia, N. N., Kamaruddin, M. S., Nawawi, M. H., Ratim, S., Azlina, H. N., Ali, E. S. Green biosynthesis of silver nanoparticles using '*Polygonum hydropiper*' and study its catalytic degradation of methylene blue. *Procedia Chemistry.* 2016, vol. 19, 594–602.

20. Joseph, S., Mathew, B. Microwave-assisted green synthesis of silver nanoparticles and the study on catalytic activity in the degradation of dyes. *Journal of Molecular Liquids.* 2015, vol. 204, 184–191.

21. Vijayan, R., Joseph, S., Mathew, B. *Indigofera tinctoria* leaf extract mediated green synthesis of silver and gold nanoparticles and assessment of their anticancer, antimicrobial, antioxidant and catalytic properties. *Artificial Cells, Nanomedicine, and Biotechnology.* 2017, vol. 46(4), 861–871.

22. Dhas, T. S., Ganesh Kumar, V., Karthick, V., Angel, K. J., Govindaraju, K. Facile synthesis of silver chloride nanoparticles using marine alga and its antibacterial efficacy. *Spectrochimica Acta Part A: Molecular and Biomolecular Spectroscopy.* 2014, vol. 120, 416–420.

23. Nadagouda, M. N., Varma, R. S. Green synthesis of silver and palladium nanoparticles at room temperature using coffee and tea extract. *Green Chemistry.* 2008, vol. 10(8), 859.

24. Cruz, D., Falé, P. L., Mourato, A., Vaz, P. D., Serralheiro, M. L., Lino, A. R. L. Preparation and physicochemical characterization of Ag nanoparticles biosynthesized by *Lippia citriodora* (Lemon Verbena). *Colloids and Surfaces B: Biointerfaces.* 2010, vol. 81(1), 67–73.

25. Jiang, X. C., Chen, C. Y., Chen, W. M., Yu, A. B. Role of citric acid in the formation of silver nanoplates through a synergistic reduction approach. *Langmuir.* 2010, vol. 26(6), 4400–4408.

26. Jensen, T., Schatz, G., Lazarides, A., Kelly, K. L. Modeling metal nanoparticle optical properties. *Metal Nanoparticles.* 2001, 89–118.

27.

28. Mie, G. Contributions to the optics of turbid media, especially colloidal metal solutions. *Annals of Physics.* 1908, vol. 25, 377–445.

29. Aslan, K., Lakowicz, J. R., Geddes, C. D. Nanogold-plasmon-resonance-based glucose sensing. *Analytical Biochemistry.* 2004, vol. 330(1), 145–155.

30. Yang, Q., Tan, Q., Zhou, K., Xu, K., Hou, X. Direct detection of mercury in vapor and aerosol from chemical atomization and nebulization at ambient temperature: exploiting the flame atomic absorption spectrometer. *Journal of Analytical Atomic Spectrometry.* 2005, vol. 20(8), 760.

31. Karunasagar, D., Arunachalam, J., Gangadharan, S. Development of a 'collect and punch' cold vapour inductively coupled plasma mass spectrometric method for the direct determination of mercury at nanograms per litre levels. *Journal of Analytical Atomic Spectrometry.* 1998, vol. 13(7), 679–682.

32. Nevado, J. J. B., Martín-Doimeadios, R. C. R., Bernardo, F. J. G., Moreno, M. J. Determination of mercury species in fish reference materials by gas chromatography-atomic fluorescence detection after closed-vessel microwave assisted extraction. *Journal of Chromatography A.* 2005, vol. 1093(1–2), 21–28.

33. Ichinoki, S., Kitahata, N., Fujii, Y. Selective determination of mercury(II) ion in water by solvent extraction followed by reversed-phase HPLC. *Journal of Liquid Chromatography & Related Technologies*. 2004, vol. 27(11), 1785–1798.
34. Kuswandi, B., Nuriman, Dam, H. H., Reinhoudt, D. N., Verboom, W. Development of a disposable mercury ion-selective optode based on trityl-picolinamide as ionophore. *Analytica Chimica Acta*. 2007, vol. 591(2), 208–213.
35. Clarkson, T. W., Magos, L., Myers, G. J. The toxicology of mercury—current exposures and clinical manifestations. *New England Journal of Medicine*. 2003, vol. 349(18), 1731–1737.
36. Driscoll, C. T., Mason, R. P., Chan, H. M., Jacob, D. J., Pirrone, N. Mercury as a global pollutant: sources, pathways, and effects. *Environmental Science & Technology*. 2013, vol. 47(10), 4967–4983.
37. Firdaus, M. L., Fitriani, I., Wyantuti, S., Hartati, Y. W., Khaydarov, R., McAlister, J. A., Obata, H., Gamo, T. Colorimetric detection of mercury(II) ion in aqueous solution using silver nanoparticles. *Analytical Sciences*. 2017, vol. 33(7), 831–837.
38. Tiggesbäumker, J., Köller, L., Meiwes-Broer, K. H., Liebsch, A. Blue shift of the Mie plasma frequency in Ag clusters and particles. *Physical Review A*. 1993, vol. 48(3), R1749–R1752.
39. Sebastian, M., Aravind, A., Mathew, B. Green silver-nanoparticle-based dual sensor for toxic Hg(II) ions. *Nanotechnology*. 2018, vol. 29 (35), 355502.
40. Farhadi, K., Forough, M., Molaei, R., Hajizadeh, S., Rafipour, A. Highly selective Hg^{2+} colorimetric sensor using green synthesized and unmodified silver nanoparticles. *Sensors and Actuators B: Chemical*. 2012, vol. 161(1), 880–885.
41. Katsikas, L., Gutiérrez, M., Henglein, A. Bimetallic colloids: silver and mercury. *The Journal of Physical Chemistry*. 1996, vol. 100(27), 11203–11206.
42. Mehtab, S., Zaidi, M. G. H., Siddiqi, T. I. Designing fructose stabilized silver nanoparticles for mercury (II) detection and potential antibacterial agents. *Material Science Research India*. 2018, vol. 15(3), 241–249.
43. George, J. M., Mathew, B. Curcuma longa rhizome extract mediated unmodified silver nanoparticles as multisensing probe for Hg(II) ions. *Materials Research Express*. 2019, vol. 6(11), 1150h5.
44. Ravi, S. S., Christena, L. R., SaiSubramanian, N., Anthony, S. P. Green synthesized silver nanoparticles for selective colorimetric sensing of Hg^{2+} in aqueous solution at wide pH range. *The Analyst*. 2013, vol. 138(15), 4370.
45. Ertürk, A. S. Biosynthesis of silver nanoparticles using *Epilobium parviflorum* green tea extract: analytical applications to colorimetric detection of Hg^{2+} ions and reduction of hazardous organic dyes. *Journal of Cluster Science*. 2019, vol. 30(5), 1363–1373.
46. Kalam, A., Al-Sehemi, A. G., Alrumman, S. A., Assiri, M. A., Moustafa, M. F., Pannipara, M. In vitro antimicrobial activity and metal ion sensing by green synthesized silver nanoparticles from fruits of *Opuntia ficus indica* grown in the Abha region, Saudi Arabia. *Arabian Journal for Science and Engineering*. 2018, vol. 44(1), 43–49.
47. Cyril, N., George, J. B., Joseph, L., Sylas, V. P. Catalytic degradation of methyl orange and selective sensing of mercury ion in aqueous solutions using green synthesized silver nanoparticles from the seeds of *Derris trifoliate*. *Journal of Cluster Science*. 2019, vol. 30(2), 459–468.

48. Alzahrani, E. Colorimetric detection based on localized surface plasmon resonance optical characteristics for sensing of mercury using green-synthesized silver nanoparticles. *Journal of Analytical Methods in Chemistry.* 2020, vol. 2020, 1–14.

49. Sowmyya, T., Lakshmi, G. V. Soymida febrifuga aqueous root extract maneuvered silver nanoparticles as mercury nanosensor and potential microbicide. *World Scientific News.* 2018, 114, 84–105.

50. Vinod Kumar, V., Anbarasan, S., Christena, L. R., SaiSubramanian, N., Anthony, S. P. Bio-functionalized silver nanoparticles for selective colorimetric sensing of toxic metal ions and antimicrobial studies. *Spectrochimica Acta Part A: Molecular and Biomolecular Spectroscopy.* 2014, vol. 129, 35–42.

51. Durgadas, C. V., Sharma, C. P., Sreenivasan, K. Fluorescent gold clusters as nanosensors for copper ions in live cells. *The Analyst.* 2011, vol. 136(5), 933–940.

52. Aki, A., Rahisuddin. Biomediated unmodified silver nanoparticles as a green probe for Cu^{2+} ion detection. *Sensor Letters.* 2015, vol. 13(11), 953–960.

53. Bindhu, M. R., Umadevi, M., Esmail, G. A., Al-Dhabi, N. A., Arasu, M. V. Green synthesis and characterization of silver nanoparticles from *Moringa oleifera* flower and assessment of antimicrobial and sensing properties. *Journal of Photochemistry and Photobiology B: Biology.* 2020, vol. 205, 111836.

54. Sebastian, M., Aravind, A., Mathew, B. Green silver nanoparticles based multi-technique sensor for environmental hazardous Cu(II) ion. *BioNanoScience.* 2019, vol. 9(2), 373–385.

55. Kirubaharan, C. J., Kalpana, D., Lee, Y. S., Kim, A. R., Yoo, D. J., Nahm, K. S., Kumar, G. G. Biomediated silver nanoparticles for the highly selective copper(II) ion sensor applications. *Industrial & Engineering Chemistry Research.* 2012, vol. 51(21), 7441–7446.

56. Bindhu, M. R., Umadevi, M. Surface plasmon resonance optical sensor and antibacterial activities of biosynthesized silver nanoparticles. *Spectrochimica Acta Part A: Molecular and Biomolecular Spectroscopy.* 2014, vol. 121, 596–604.

57. Link, S., El-Sayed, M. A. Shape and size dependence of radiative, non-radiative and photothermal properties of gold nanocrystals. *International Reviews in Physical Chemistry.* 2000, vol. 19(3), 409–453.

58. Nipane, S. V., Mahajan, P. G., Gokavi, G. S. Green synthesis of silver nanoparticle in *Calotropis procera* flower extract and its application for Fe^{2+} sensing in aqueous solution. *International Journal of Recent and Innovation Trends in Computing and Communication.* 2016, vol. 4(10), 98–107.

59. Aravind, A., Sebastian, M., Mathew, B. Green silver nanoparticles as a multifunctional sensor for toxic Cd(II) ions. *New Journal of Chemistry.* 2018, vol. 42(18), 15022–15031.

60. Aravind, A., Sebastian, M., Mathew, B. Green synthesized unmodified silver nanoparticles as a multi-sensor for Cr(III) ions. *Environmental Science: Water Research & Technology.* 2018, vol. 4(10), 1531–1542.

61. Ismail, M., Khan, M. I., Akhtar, K., Khan, M. A., Asiri, A. M., Khan, S. B. Biosynthesis of silver nanoparticles: a colorimetric optical sensor for detection of hexavalent chromium and ammonia in aqueous solution. *Physica E: Low-dimensional Systems and Nanostructures.* 2018, vol. 103, 367–376.

62. Aravind, A., Sebastian, M., Mathew, B. Unmodified silver nanoparticles based multisensor for Ni(II) ions in real samples. *International Journal of Environmental Analytical Chemistry.* 2019, vol. 99(4), 380–395.

63. Lei, W., Durkop, A., Lin, Z., Wu, M., Wolfbeis, O. S. Detection of hydrogen peroxide in river water via a microplate luminescence assay with time-resolved (gated) detection. *Microchimica Acta.* 2003, vol. 143(4), 269–274.

64. Wang, J., Lin, Y., Chen, L. Organic-phase biosensors for monitoring phenol and hydrogen peroxide in pharmaceutical antibacterial products. *The Analyst.* 1993, vol. 118(3), 277.

65. Tagad, C. K., Kim, H. U., Aiyer, R. C., More, P., Kim, T., Moh, S. H., Kulkarni, A., Sabharwal, S. G. A sensitive hydrogen peroxide optical sensor based on polysaccharide stabilized silver nanoparticles. *RSC Advances.* 2013, vol. 3(45), 22940.

66. Farrokhnia, M., Karimi, S., Momeni, S., Khalililaghab, S. Colorimetric sensor assay for detection of hydrogen peroxide using green synthesis of silver chloride nanoparticles: experimental and theoretical evidence. *Sensors and Actuators B: Chemical.* 2017, vol. 246, 979–987.

67. Raja, S., Ramesh, V., Thivaharan, V. Green biosynthesis of silver nanoparticles using *Calliandra haematocephala* leaf extract, their antibacterial activity and hydrogen peroxide sensing capability. *Arabian Journal of Chemistry.* 2017, vol. 10(2), 253–261.

68. Noghabi, M. P., Parizadeh, M. R., Ghayour-Mobarhan, M., Taherzadeh, D., Hosseini, H. A., Darroudi, M. Green synthesis of silver nanoparticles and investigation of their colorimetric sensing and cytotoxicity effects. *Journal of Molecular Structure.* 2017, vol. 1146, 499–503.

69. Subramanian, L., Thomas, S., Koshy, O. Green synthesis of silver nanoparticles using aqueous plant extracts and its application as optical sensor. *International Journal of Biosensors & Bioelectronics.* 2017, vol. 2(3).

70. Pandey, S., Goswami, G. K., Nanda, K. K. Green synthesis of biopolymer–silver nanoparticle nanocomposite: an optical sensor for ammonia detection. *International Journal of Biological Macromolecules.* 2012, vol. 51(4), 583–589.

71. Dubas, S. T., Pimpan, V. Green synthesis of silver nanoparticles for ammonia sensing. *Talanta.* 2008, vol. 76(1), 29–33.

72. Babu, S., Claville, M. O., Ghebreyessus, K. Rapid synthesis of highly stable silver nanoparticles and its application for colourimetric sensing of cysteine. *Journal of Experimental Nanoscience.* 2015, vol. 10(16), 1242–1255.

73. Csapó, E., Patakfalvi, R., Hornok, V., Tóth, L. T., Sipos, Á., Szalai, A., Csete, M., Dékány, I. Effect of pH on stability and plasmonic properties of cysteine-functionalized silver nanoparticle dispersion. *Colloids and Surfaces B: Biointerfaces.* 2012, vol. 98, 43–49.

74. Ragam, P. N., Mathew, B. Unmodified silver nanoparticles for dual detection of dithiocarbamate fungicide and rapid degradation of water pollutants. *International Journal of Environmental Science and Technology.* 2019, vol. 17(3), 1739–1752.

75. Pourreza, N., Golmohammadi, H., Naghdi, T., Yousefi, H. Green in-situ synthesized silver nanoparticles embedded in bacterial cellulose nanopaper as a bionanocomposite plasmonic sensor. *Biosensors and Bioelectronics.* 2015, vol. 74, 353–359.

CHAPTER 2

Surface Plasmon Response and Applications of Silver Nanoparticles Synthesized via Sol–Gel Route

K. V. ARUN KUMAR[1, 2*] and N. V. UNNIKRISHNAN[3]

1Department of Physics, CMS College (Autonomous), Kottayam 686001, Kerala, India

2Nanotechnology and Advanced Materials Research Centre, CMS College (Autonomous), Kottayam 686001, Kerala, India

3School of Pure and Applied Physics, Mahatma Gandhi University, Kottayam, Kerala, India

*Corresponding author. E-mail: arunkumar@cmscollege.ac.in

ABSTRACT

The surface plasmon resonance response of noble metallic nanoparticles is a fascinating research field in the future due to a wide range of applications. The advances in the synthesis and characterization methods of metal nanostructures open a new pathway to different fields of applications in optics, electronics, biological sensing, and detection. An accurate and precise synthesis of plasmon nanoparticles and their characteristics is one of the challenging tasks due to multiple factors. The synthesis method has a definite role in uniform nanostructure formation and repeatability. We herein summarize the method, characterization advantages, and applications of sol–gel synthesized plasmonic nanoparticles. Here we used silver nanoparticles synthesis in silica via the hydrolytic method and titanosilicate via the nonhydrolytic sol–gel method through a low-cost technique. We analyzed in detail the effect of plasmonics in phosphor materials, dielectric materials, and solar cells. For sol–gel synthesis, we used materials like titanium isopropoxide, tetraethylorthosilicate (TEOS) ethanol, and europium nitrate. For plasmonic nanoparticles, silver nitrate was used. We analyzed the optical and structural features of plasmonic nanoparticles by different techniques like absorption spectroscopy, X-ray diffraction, and transmission electron microscopy, and impedance spectroscopy.

2.1 INTRODUCTION

The scientific world looking for materials with unpredictable properties so nanoparticles are much interesting due to their properties like high reactivity and large surface area to volume ratio. Some of the noble metal nanoparticles like platinum, gold, silver, and copper show a broad absorption peak in the visible region of the electromagnetic. It has a special attraction to the research world. Fields like electronic, magnetic, optical, optoelectronic, mechanic, solar cell and fuel cell materials, medical, bioimaging, cosmetic, data communication, and optical data storage, the use of metal nanoparticles are increasing due to improved application. Nanoparticles are the strong bridge between the bulk materials and atomic or molecular structures. Nanomaterials show a different physical, chemical, and biological properties from bulk material counterpart due to the size, morphological structure of the substance, and shape. But for a bulk material always show a constant physical behavior regardless of their size and shape. The novel properties of nanoobjects are due to the changes in size and scale. The size and shape of the nanoparticles depend on the surface area to volume ratio of the particle, if it is very small in at least limited to one dimension. Quantum tunneling, random motion of the small particles, the uncertainty of the matter, discreteness of energy, duality nature of the mass, and energy for wave particles, and others are some of the extra phenomena exhibited in the nanoparticles. Moreover in nanomaterials at the nanoscale, the van der Waals force became exceptionally strong but the gravity force becomes markedly less. The size of the nanomaterials is one of the important factors for the optical properties of materials. Electrons at nanoscale restricted and cannot move freely, these confinement causes them to react light differently. For example, gold appears golden at the bulk material, but the nanosized gold particles show red color. Thus the size of the particles decreases creating different colors through which quantum dots change in their optical properties [1–3]. If the noble metal nanoparticle size is comparable to that of light wavelength and interacts with the sunlight, the free electrons of the nanoparticles of metals are integrated with the photon energy that produces subwaves and such a collective oscillations of conducting electrons named as the localized surface plasmonic resonance of the material. Metal nanoparticles like Cu, Ag, Au, Pt, and Pd are a few examples of the noble metals with surface plasmonic resonance. Nanoparticles of noble metals can trap, scatter, and concentrate light and in that way enhance the number of active sites by forming the electron–hole pairs, which enables a fast charge transfer mechanism. The novel physical–chemical properties of metal nanoparticles have been making use of potential applications in medicine, optics, cosmetics, renewable energies, microelectronics, medical imaging, and biomedical devices. Besides, many of the bactericidal and fungicidal activities are controlled by silver nanoparticles. The market value of the consumer goods can be enhanced by the use of MNPs and the product become remarkably trendy for the market especially in plastics, soaps, pastes, food, textiles, and others. Due to their unique physical, chemical, and biological properties silver nanoparticles are the most popular compared to other metal nanoparticles. Special advantages of silver nanoparticles over the other noble metals are low cost than gold and platinum, stability at ambient conditions, nontoxicity, during the surface-plasmon propagation small loss in the optical frequency, high electrical conductivity, high thermal conductivity, wide absorption region in the visible and far IR part, high-primitive nature, enhanced surface Raman scattering, nonlinear optical behavior, chemical stability, and catalytic activity. Scientists and technologists with much interest to develop nanosilver-based disinfectant products due to their high antimicrobial activity (bactericidal and fungicidal activity). Silver nanoparticles is considered as one of the most vital and charming nanomaterials due to increased biomedical applications. The role of silver nanoparticles is most relevent in cancer diagnosis and therapy, so nanoparticles in nanoscience and nanotechnology, particularly in nanomedicine, are waiting for many future applications. Their function is beyond imagination. Due to the unusual properties of silver nanoparticles, they have been used in many fields like industrial, medical device coatings, healthcare-related products, optical sensors, cosmetics, pharmaceutical industry, food industry, diagnostics, drug delivery, anticancer agents, tumor killing, and anticancer drugs. Recently, AgNPs have been frequently used in many textiles, keyboards, wound dressings, and biomedical devices [4–18]. Sol–gel method is one of the techniques used for silver nanoparticles synthesis. It has several advantages over other methods like meltquenching, ion implantation, ion exchange, and high-temperature glass fusion. High purity, low-temperature processing, ultrahomogeneity, and most significantly the possibility of making glasses of new compositions are some of the added advantages of sol–gel method. Sol–gel process is identified as one of the productive techniques for incorporating different types of metal dopants into different matrixes. Molecular scale homogeneity of the starting solutions was ensured by the sol–gel method. The sol–gel method can easily incorporate nanoparticles into the inorganic matrix, through which silver nanoparticles are incorporated and used for various applications. Other properties like transparency in the UV–visible range, chemical inertness, good mechanical strength, and large 3D porous network are added advantages of the sol–gel method. Low-temperature processing of homogeneous multicomponent oxide films, good control of composition, and optical properties of the final material the sol–gel method is still a challenging technique and is useful in many future applications of coatings and films [19–23].

2.2 EXPERIMENTAL PART

2.2.1 SOL–GEL PROCESS

Sol–gel technique was used in the early to mid-1800s for processing materials like ceramic and glass. Higher homogeneity, purity, and processing temperature at the lower level is some of the primary motivations for choosing sol–gel processing as compared with traditional glass or ceramic processing methods. Sol–gel monoliths are prepared by the method of hydrolysis and polycondensation of alkoxide precursors followed by aging and drying under ambient atmospheres. Solid particles of diameter ranging from 1 to 100 nm are the colloids particles. Dispersions of colloidal particles in a liquid are called sols. The interconnected, rigid network with pores of submicrometer dimensions and polymeric chains with average length is greater than a micrometer is called as gel. The processing steps involved in making sol–gel derived monoliths mainly are (1) sol preparation: formation of a colloidal suspension from the dispersion of solid nanoscale particles, derived from a starting material, within a solvent; (2) gelation (transition from sol to gel state): with the addition of an acid or a base catalyst cross-linking and branching of particles occur that initiates polymerization, forming an interconnected chain structure; (3) aging of the gel: increase in the mechanical strength of the gel; (4) drying of the gel: removal of the solvent from the pores of the gel in a manner to avoid gel fracture. Solid nanoparticles materials are dispersed within a solution of reactants and solvents that is the starting stage of the formation of a colloidal suspension. The added catalyst promotes polymerization, which includes hydrolysis and polycondensation reactions. The development of a 3D porous network within a wet, gel-like structure occurred due to the cross-linking and branching between polymeric species. Various materials such as oxides, like silica, metal oxides, organic materials, such as polymers like cellulose and carbon materials can be produced from sol–gel-derived materials. The composite materials strengthen the material properties of the gel which have significant applications in different fields. The different stages involved in sol–gel processing are summarized in Figure 2.1 [19–23].

Precursor Solution   Sol   Gel formation   Xero gel   Dense Sample

FIGURE 2.1   Sol preparation and gel transformation.

2.2.2   SOL–GEL SILVER NANOPARTICLE SYNTHESIS

The precursor materials used for preparing sol–gel silica sample are tetraethyl orthosilicate, $(Si(OC2H_5)_4)$, ethanol (C_2H_5OH), and silver nitrate $(AgNO_3)$. Highly dispersed silver nanoparticles were successfully incorporated into the matrix by a simple and conventional hydrolytic sol–gel technique. For the synthesis of silica matrix via the sol–gel process, tetraethylorthosilicate (TEOS) was used as a precursor for SiO_2. Ethanol as the solvent, distilled water for hydrolysis, and nitric acid (HNO_3) added as a catalyst. Magnetic stirring process was used for ensuring the homogeneity of the solution. For making the sample the required amount of the dopant (silver nitrate) is weighed and mixed with ethanol (C_2H_5OH) and distilled water. A measured volume of TEOS taken in a beaker and mixed with ethanol and incorporating the mixture of silver nitrate and ethanol. This solution is subjected to continuous stirring with the help of magnetic stirrer. During this time a few drops of HNO_3 were added that act as a catalyst for the reaction. The stirring process was continued until the solution became clear that indicates that the solution is homogeneously mixed. The prepared homogeneous solutions of the mixture are poured into small polypropylene containers and sealed with a parafilm. These sealed samples were placed without disturbing for a month and for the gel formation. For the densification of samples, they were heat-treated at high temperatures.

For titanosilicate samples prepared by a simple and conventional nonhydrolytic sol–gel technique. For titanosilicate matrix synthesis, we used precursor materials such as titanium isopropoxide, tetraethylorthosilicate, ethanol, and silver nitrate. For the synthesis of titanosilicate matrix via a sol–gel process we used a measured volume of titanium isopropoxide, tetraethylorthosilicate, ethanol, nitric acid, and most importantly silver nitrate. The samples were undergone continuous magnetic stirring of the solution for ensuring the homogeneity of the final solution. Figure 2.2 shows the mixing of two solutions for preparing titanosilicate sample. The first solution is a mixture of tetraethylorthosilicate and ethanol. The second one is a mixture of titanium isopropoxide and ethanol. These solutions were separately prepared and later was added to a third beaker and stirred well using a magnetic stirrer. During stirring a few drops of HNO_3 were added that acts as a catalyst for improving the reaction rate. A weighed amount of silver nitrate was added to the above mixture. The stirring process was continued until a clear and homogeneous solution of the mixture was obtained. The homogenous solutions were poured into small polypropylene containers and tightly sealed with a parafilm and kept in a dry place without disturbing the sample for a month and allowed to form the gel. The densified samples are characterized by different techniques. Figure 2.2 shows the scheme of the experimental procedure for sol–gel (a) silica (b) titanosilicate samples [24–28].

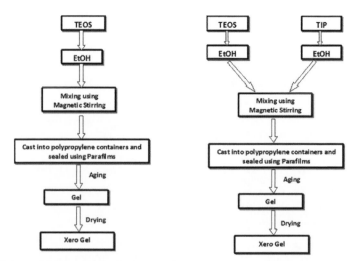

FIGURE 2.2   Scheme of the experimental procedure for sol–gel (a) silica (b) titanosilicate samples.

2.3   RESULTS AND DISCUSSION

2.3.1   *SILVER NANOPARTICLE ABSORPTION STUDIES*

The surface plasmon resonance (SPR) mechanism of metal nanoparticles is one of the important properties due to the collective oscillation of free electrons. The noble metal nanoparticles exhibit a strong optical extinction (absorption + scattering) cross-section due to this collective oscillation. When electromagnetic waves interact with nanoparticles, the surface plasmons become excited and begin to resonate. The presence of the electric field the charge separation mechanism creates forming dipole oscillation, which is represented in Figure 2.3. The SPR allows a strong absorption of the incident electromagnetic wave due to frequency matching, while also allowing some scattering of light. A UV–vis spectrometer can be used for measuring the SPR band. The tuning of surface plasmon resonance bandwidth, intensity, and wavelength mainly dependent on the size, shape, structure, type of metal particle used, and dielectric material surrounding the medium. The noble metals showing the strongest band intensity is Ag and Au. For the strongest bands, sharpest peaks, and low cost commonly Ag is used for applications in different fields.

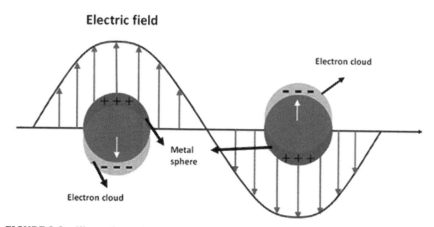

FIGURE 2.3   Illustrations of surface plasmon resonance (SPR).

The optical extinction studies were completed by using Shimadzu UVPC 2401 spectrometer. Figure 2.4 is the optical spectra of silver nanoparticles in silica and titanosilicate matrix. Figure 2.4a shows the optical extinction spectra Ag-doped SiO_2 matrix and Figure 2.4b is of Ag in $SiO_2–TiO_2$ matrix heat treated at 800 °C. The sum of scattering and absorption is named as the optical extinction spectrum when an electromagnetic wave going through a material. The broadband around 350–450 nm is due to the surface plasmon resonance of the silver nanoparticles. The broad surface plasmon peak is due to the light-induced collective oscillation of conducting electrons on the metal surface. The electromagnetic radiation excites the metallic nanoparticles resulting in extraordinary scattering in which light is trapped to confine light produces a strong electromagnetic field that strengthens the absorption. The absorption wavelength depends on several factors of metal nanoparticles, in which some of the prominent factors are particle size and shape, as well as the nature of the surrounding medium. So changing the size, shape, and dielectric medium of the nanoparticle, the surface plasmonic resonance wavelength can be tuned to the desired region. The shift in the optical extinction spectra of the Ag nanoparticle in SiO_2 and SiO_2–TiO_2 is due to the change of the dielectric medium. Comparing the surrounding medium of silver nanoparticle we can tune the surface plasmon wavelength in the desired position and used for different types of applications [27–34].

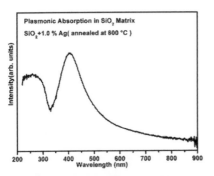 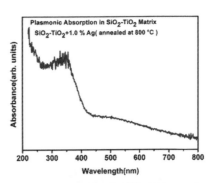

FIGURE 2.4   Optical extinction spectra Ag nanoparticles (a) SiO_2 samples (b) SiO_2–TiO_2 sample.

2.3.2   STRUCTURAL STUDIES OF SILVER NANOPARTICLES

2.3.2.1   X-RAY DIFFRACTION (XRD)

The structural features of the silver nanoparticle were analyzed by X-ray diffraction method. PANalytical X'Pert instrument was used for measuring XRD. Figure 2.5 shows the XRD pattern of Ag nanoparticles in (a) sol–gel SiO_2 Matrix (b) sol–gel SiO_2–TiO_2 matrix. The crystalline phase of silver nanoparticles was confirmed from XRD measurements. The broad peak in the diffraction measurements cantered at 2θ ~23° is due to the characteristic of amorphous SiO$_2$. Figure 2.5(a) shows the XRD of Ag nanoparticle in silica sample and the peaks at 38°, 44°, 64°, and 77° correspond to [111], [200], [220], and [311] planes of silver (ICDD file 04-0783), respectively. The silver nanoparticles in the titanosilicate matrix is shown in Figure 2.5(b). The diffraction pattern centered at 2θ ~ 23° is the amorphous nature of silica. The narrow peak corresponding at 25°, 48°, 54°, and 62° is due to [101], [200], [211], and [204] planes of TiO$_2$ and the values at 38°, 44°, 64°, 78s° correspond to [111], [200], [220], and [311] planes of silver.

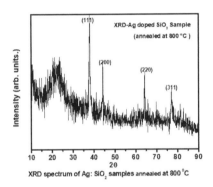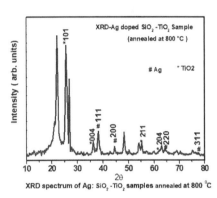

FIGURE 2.5 XRD of Ag nanoparticles in (a) sol–gel SiO$_2$ matrix (b) sol–gel SiO$_2$–TiO$_2$ matrix.

The particle size was analyzed by Debye–Scherrer formula 0.9 $\lambda/\beta\cos\theta$, where β is the measured FWHM value, 2θ is the Bragg diffraction angle peak, and λ is the X-ray wavelength. The XRD shows a major peak of silver particles at 2θ~38°. The calculated size of silver nanoparticle in sol–gel SiO$_2$ matrix is 18 nm and in sol–gel SiO$_2$–TiO$_2$ matrix it is 17.5 nm [27, 28, 35].

2.3.2.2 *TRANSMISSION ELECTRON MICROSCOPY (TEM)*

The transmission electron microscopy (TEM) studies reveal the size distribution, size, and shape of the nanoparticles. Measurements were done using Joel/JEM2100 TEM facility. Figure 2.6a and b shows silver nanoparticles in sol–gel SiO$_2$ matrix and sol–gel SiO$_2$–TiO$_2$ matrix. Figure 2.6c is the higher magnification images of Ag nanoparticles in silica matrix. HRTEM images show the spherical morphology of the particles. The size of the particle is measured using Image J program and the average size is found to be approximately 20 nm in sol–gel synthesized SiO_2 matrix and 16 nm in SiO_2–TiO_2 matrix [27, 28].

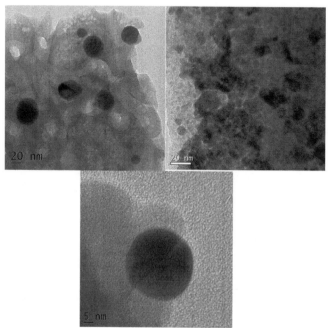

FIGURE 2.6   TEM images of Ag nanoparticles in (a) sol–gel SiO_2 matrix, (b) sol–gel SiO_2-TiO_2 matrix, (c) HRTEM of Ag nanoparticle.

2.2.4   SILVER NANOPARTICLES APPLICATIONS

2.2.4.1   LUMINESCENCE ENHANCEMENT DUE TO THE PLASMONIC EFFECT

Researchers are trying to improve the fluorescence efficiency of phosphor materials. Fluorescence enhancement properties of rare-earth ions in glassy media codoped with metal nanoparticles are employing applications in many fields. The chemical detection and biomolecular analysis were done using optical phenomena of the plasmonic coupling of nanoparticles with an electromagnetic wave. The observed luminescence enhancement is due to the energy transfer mechanism between the noble metal particles and europium ions. The possible energy transfer process from silver to Eu^{3+} ion is schematically represented in Figure 2.7. The enhancement mechanism is mainly due to the metal particle–luminophore distance and particle size. Metal particles like Ag and Au with larger than 5 nm show strong plasmon absorption. The fluorescence band of silver is in resonance with the absorption band of Eu^{3+} ions that is a favorable condition for energy transfer and hence luminescence enhancement that is shown in Figure 2.8. The absorption of the exciting light by the Ag particles and their emitted energy transferred to the Eu^{3+} ions in the ground state is the process involves in the enhancement of mechanism. Thus the additional Eu^{3+} ions are excited to the $|3>$ state leading to an enhancement of luminescence. The important emission bands of Eu^{3+} ions are 577, 589, 612, and 649 nm which is due to transitions $^5D_0 \rightarrow {}^7F_0$, $^5D_0 \rightarrow {}^7F_1$, $^5D_0 \rightarrow {}^7F_2$, and $^5D_0 \rightarrow {}^7F_3$, respectively. The magnetic dipole transition intensity of the $^5D_0 \rightarrow {}^7F_1$ hardly varies with the bonding environment of the Eu^{3+} ions, but the electronic dipole allowed transition $^5D_0 \rightarrow {}^7F_2$ is hypersensitive to the coordination environment of the Eu^{3+} ions. The site symmetries of Eu3+ ions can be estimated by the intensity ratio of $^5D_0 \rightarrow {}^7F_2$ to $^5D_0 \rightarrow {}^7F_1$. One of the possible reasons for the luminescence intensity variation with the addition of silver nanoparticles is due to the variation in the asymmetry ratio (AS). The ratio of the magnetic dipole transition ($^5D_0 \rightarrow {}^7F_1$) and the electric dipole transition ($^5D_0 \rightarrow {}^7F_2$) is defined as the AS. The intensity transition ratio of $^5D_0 \rightarrow {}^7F_2 / {}^5D_0 \rightarrow {}^7F_1$ in luminescence indicates the degree of asymmetry in the vicinity of Eu^{3+} ions and Eu–O covalency. The calculated value of the AS for the europium and europium codoped silver samples is 2.6274 and 3.6148, respectively. The addition of nanoparticles the asymmetry value variation can be described by two mechanisms: one is the refractive index around the europium ions and the other is the modification of the ligand field. Interaction with Ag nanoparticles the local field effect which modifies the refractive index and also increases the distortion of the ligand field, which increases the AS value. The emission intensity of Eu^{3+} ions increased with the addition of silver, due to an efficient energy transfer from silver particles to Eu^{3+} ions. The structural features, size, shape, and concentration of the metallic particles also influence the fluorescence enhancement. The structural features of the particle are directly dependent on the dielectric function of the composite medium. The electromagnetic interaction between the europium ion and the SPR of the silver particles can also cause enhancement. Many investigations show that rare-earth ions in sol–gel silica glasses tend to form clusters. In most of the optical materials, the clusters of rare-earth ions are undesirable, which enhances concentration quenching. The incorporation of TiO_2 in the matrix results in the inhibition of clustering that in turn favors the dispersion of silver nanoparticles. Titania also plays a crucial role in the propagation of plasmonic waves and enhance the fluorescence of europium ions more relative to silica glasses. The presence of titania has a crucial role for the easy propagation of plasmonic waves through the dielectric medium. In the interaction between Eu^{3+} and silver on the pore surface of the host matrix, silver particles strongly influence Eu^{3+} ions emission due to the intensified electromagnetic field around them, resulting in luminescence enhancement [26, 36–39].

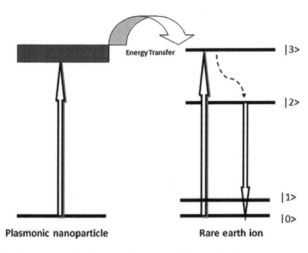

FIGURE 2.7   Energy transfer diagram of rare earth ion and plasmonic nanoparticle particle.

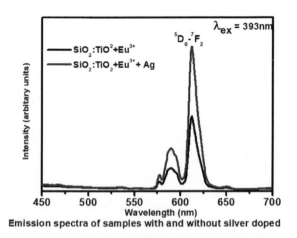

FIGURE 2.8   Plasmonic effects on luminescence enhancement.

2.4.2   RC EQUIVALENT CIRCUIT TUNING DUE TO THE PLASMONIC EFFECT

Impedance spectroscopy is one of the powerful tools for the electrical or dielectric characterization of solid materials. With the help of impedance spectroscopy, it is possible to tune the properties of the materials thereby construct the equivalent circuits for the devices. Many parameters like AC conductivity, complex impedance, and dielectric relaxation can be measured by the dielectric spectroscopy. Materials' insulating character, conduction behavior and the structural aspects can be analyzed by studying dielectric parameters of dielectric constant, dielectric loss, and AC conductivity over a wide range of frequencies.

The variation in the real and imaginary part of dielectric constant values above 100 kHz is a constant. The variation dielectric constant with frequency is due to the mechanism of the simultaneous presence of space charge, dipolar, and ionic, electronic polarization. The silver doped sample shows low dielectric values compared to the undoped sample that may be due to the interfacial charge transfer mechanism. It is one of the possible reasons for low dielectric values at a lower frequency due to the induced dipoles in the silver nanoparticles. Nanomaterial exhibit different polarization mechanism like space charge polarization, hopping exchange of charge carriers between the localized states and the resultant displacement of dipoles for the applied field. The possible reason for the changes in the dielectric constant values with Ag substitution in the low-frequency region observed due to the effects of grain boundaries is prominent at lower frequencies. The silver nanoparticles thicken the grain boundary and can cause a decrease in the polarization value and hence the dielectric constant decreases. At lower frequencies, the conductivity value is same for both the samples but at higher frequencies the value of conductivity goes on increasing. This may be due to large surface scattering that results in a decrease in conductivity and also the short-range intrawell hopping of charge carriers between localized states. The decrease in conductivity values with the addition of silver nanoparticles may be due to the effective number of charge carriers involved. Cole–Cole plot is a representation of the real and imaginary part of the impedance values. The Cole–Cole plot (Figure 2.9) gives the value of R_b, C_b, and τ—known as Cole–Cole parameters. The observed values of R_b, C_b, and τ for undoped sample is $R_b = 2.42 \times 10^6$, $C_b = 7.66 \times 10^{-11}$, and $\tau = 1.85 \times 10^{-4}$ and that of silver doped sample is $R_b = 11.81 \times 10^6$, $C_b = 3.16 \times 10^{-11}$, and $\tau = 3.73 \times 10^{-4}$. Cole–Cole plot giving single semicircles are observed for all the samples that can be modelled by an equivalent parallel RC circuit. The peaks of the semicircles are used to determine the relaxation time (τ) [40–42].

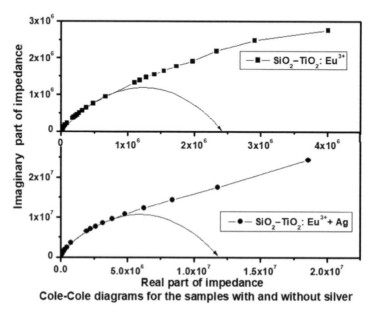

Cole-Cole diagrams for the samples with and without silver

FIGURE 2.9   Cole–Cole plot for samples with and without silver nanoparticle.

2.4.3   PLASMONIC EFFECTS ON DYE-SENSITIZED SOLAR CELLS

Currently, increasing energy consumption has forced researchers to search for new inexpensive sources of renewable energy. Sunlight is one of the plentiful, inexhaustible, and eco-friendly sources of energy, and trapping of solar energy in an effective way is one of the interesting research areas. Dye-sensitized solar cells (DSSCs) are one of the third generation solar cells, with a different kind of mesoporous solar device having moderate efficiency and low cost than the other two kinds of solar cells. One of the methods for improving the efficiency of a DSSC solar cell is by introducing the plasmonic structure in the solar cell. Plasmonic structure excites more electrons due to increased light absorption and scattering by the plasmon–polaritons trapped in more incident light within the solar cell. The plasmonic DSSCs are more efficient compared to normal DSSCs due to the above-said facts. The plasmonic nanoparticles (NPs) employed in energy harvesting in two different ways (1) adopting light trapping and (2) spectral modification processes to shift frequencies of the solar spectrum. In both aspects the NPs boost, by scattering and concentrating the electromagnetic field into the active region of the device. The broad optical absorption due to the plasmonic effect is useful for solar energy harvesting and it improves the efficiency of solar cells

2.4.3.1 DYE-SENSITIZED SOLAR CELLS FABRICATION

For DSSC fabrication, sol–gel synthesized samples of composition SiO_2–TiO_2 and SiO_2–TiO_2+Ag3% were used. Schematic diagram of fabricated DSSC shown in Figure 2.10a and b represents the measured output. Synthesized sol was coated on the conducting side of the fluorine tin oxide (FTO) using a glass rod. The FTO coated film was placed in an oven for 30 min at 75 °C and then the film was allowed to cool at room temperature. Then a few drops of amaranthus red (natural dye) are incorporated into the film. After incorporating the dye the films were annealed at 450 °C. Other part of FTO glass plate act as a counter electrode coated with a graphite material. For secured combination these two plates were sandwiched together using a binder clip. A digital multimeter was used for measuring the output voltage of the DSSC. The output voltage measured value of dye-sensitized solar cells is 87 mV, whereas the output voltage measured value of plasmonic dye-sensitized solar cells is 160 mV. The measurement shows that the plasmonic effect of silver nanoparticles produce an appreciable amount of the output voltage, which means efficiency improvement possible with plasmonic nanoparticle [28, 43–45].

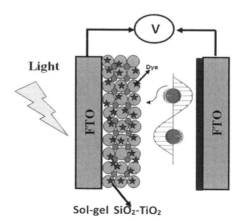

FIGURE 2.10A   Schematic diagram of plasmonic DSSC.

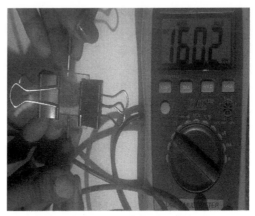

FIGURE 2.10B   DSSC with SiO_2-TiO_2: Ag film composition.

2.4   PLASMONIC NANOPARTICLES IN BIOLOGICAL APPLICATIONS

The unique physical and chemical properties of silver nanoparticles are very useful in different applications in the fields of optical, electrical, thermal, high electrical conductivity, and biological properties. Figure 2.11 shows the various kinds of applications of Ag nanoparticles. Silver nanoparticles are used for many biological applications such as antibacterial agents, medical device coatings, healthcare-related products, in the pharmaceutical industry, drug delivery, as anticancer agents, tumor-killing effects of anticancer drugs, in diagnostics, household, in consumer products, optical sensors, cosmetics, and the food industry. Silver nanoparticles are also used in many advanced applications in textiles, keyboards, wound dressings, and biomedical devices. Some of the factors on which the biological activity of silver nanoparticles depends are surface chemistry, size distribution, shape, particle morphology, particle composition, coating/capping, agglomeration, particle reactivity in solution, type of reducing agents, the efficiency of ion release, and others. For various biomedical applications, silver nanoparticles with controlled structures of uniform size, morphology, and functionality are essentially used. For treating diseases like cancer, which has the characteristic feature of the uncontrolled growth and spread of abnormal cells caused by several factors and it is treated by various treatments, therefore the challenge is to identify effective, cost-effective, and sensitive lead molecules that have cell-targeted specificity and increase the sensitivity. Due to the therapeutic applications in cancer as anticancer agents, AgNPs have been shown much interest because in diagnostics, and in probing. Sol–gel synthesized silica glass containing silver is considered as a candidate of antibacterial material for medical applications. A sol−gel prepared thin film embedded a porous silver nanoparticle biosensor used as matrix-free laser desorption ionization mass spectrometry of analytes of various chemical classes. The AgNP-impregnated thin film absorbs incoming photons, resulting in desorption and ionization of analytes [18, 46–48].

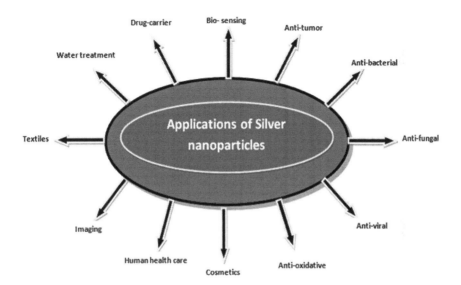

FIGURE 2.11   Biological applications of plasmonic nanoparticles.

2.5   CONCLUSIONS

Plasmonic silver nanoparticles successfully synthesized one of the low cost and easy processing methods like sol−gel technique, which finds several advantages over other methods like high purity, low processing tempera-tures, ultrahomogeneity, and most significantly the possibility of making glasses of new compositions. The hydrolytic sol−gel method used for plas-monic silver nanoparticles synthesis in the silica matrix and nonhydrolytic sol−gel method in titanosilicate matrix. The UV−vis absorption spectra confirmed the presence of plasmonic nanoparticle through the broadband around 350–450 nm, which is due to the plasmon resonance of the silver nanoparticles and it is due to the collective oscillation of free electrons in the noble metal nanoparticles. The structural features of plasmonic silver nanoparticles were characterized and confirmed by the X-ray diffraction and TEM techniques. The XRD analysis confirmed the nanocrystalline nature of the particle and TEM analysis confirmed the physical morphology and size of the nanoparticle. The sol–gel synthesized nanoparticles successfully employed different fields of applications like luminescence enhancement of rare earth elements, RC equivalent circuit tuning of dielectric materials, efficiency improvement in solar cells, and biomedical applications. The energy transfer mechanism from plasmonic nanoparticles is the possible reason for the luminescence enhancement. RC equivalent circuit tuning property of the dielectric material is due to phenomena like simultaneous presence of space charge, dipolar and ionic electronic polarization exhibit in the metal nanoparticle. The plasmonic nanostructure trap more incident light within the material improves the efficiency of the DSSC solar cell. Silver nanoparticles are more effective in many fields of biological applications. Finally, we summarize that the plasmonic nanoparticles are useful in every research field and find applications in different aspects of life.

KEYWORDS

- **sol–gel**
- **nanoparticles**
- **plasmonics**
- **luminescence**
- **impedance spectroscopy**
- **dye-sensitized solar cell**

REFERENCES

1. Khan Maaz, Silver Nanoparticles: Fabrication, Characterization and Applications, DOI: 10.5772/intechopen.71247.
2. Shawn Y. Stevens, LeeAnn M. Sutherland, Joseph S. Krajcik, The Big Ideas of Nanoscale Science and Engineering, Princeton University Press, 2009, ISBN 978-1-935155-07-2.
3. Earl Boysen and Nancy Boysen Nanotechnology For Dummies, 2nd Edition, Wiley Publishing, Inc, 2011, ISBN: 978-1-118-13688-1 (ebk).
4. Stefan A. Maier, Plasmonics: Fundamentals and Applications, Springer, e-ISBN 978-0387-37825-1.
5. Starowicz Z, Lipinski M, Socha R P, Berent K, Kulesza G, Ozga P, Photochemical silver nanoparticles deposition on sol–gel TiO_2 for plasmonic properties utilization, J. Sol-Gel Sci. Technol. 73, 2015, 563–571.

6. Lal S, Link S, and Halas N. Nano-optics from sensing to waveguiding. Nat. Photon., 2007, 641–648.

7. McConnell Wyatt P, Novak James P, Brousseau Louis C, Fuierer Ryan R., Tenent Robert C, and, Feldheim Daniel L, Electronic and Optical Properties of Chemically Modified Metal Nanoparticles and Molecularly Bridged Nanoparticle Arrays, J. Phys. Chem. B, 2000, 104(38) 8925–8930.

8. Mrinmoy De, Ghosh Partha S, Rotello Vincent M, Applications of nanoparticles in biology, Adv. Mater. 2008, 20, 4225–4241.

9. An-Hui Lu, Salabas E L, Ferdi Schüth, Magnetic nanoparticles: synthesis, protection, functionalization, and application, Angew. Chem. Int. Ed., 2007, 46, 1222–1244.

10. Rajib Ghosh Chaudhuri, and Santanu Paria, Core/shell nanoparticles: classes, properties, synthesis mechanisms, characterization, and applications, Chem. Rev. 2012, 112, 4, 2373–2433.

11. Douglas Roberto Monteiro, Luiz Fernando Gorup, Aline Satie Takamiya, Adhemar CollaRuvollo-Filho, Emerson Rodrigues deCamargo, Debora Barros Barbosa, The growing importance of materials that prevent microbial adhesion: antimicrobial effect of medical devices containing silver, Int. J. Antimicrob. Agents, 2009, 34,(2), 103–110.

12. Maqusood Ahamed, Mohamad S.AlSalhi, M.K.J.Siddiqui, Silver nanoparticle applications and human health, Clin. Chim. Acta, 2010, 411 (23) 1841–1848.

13. Hossain M K, Drmosh Q A, Yamani Z H, and Tabet N. Silver nanoparticles on Zinc Oxide thin film: an insight in fabrication and characterization, 2014 IOP Conf. Ser.: Mater. Sci. Eng. 64, 012018 .

14. Julia Fabrega, Samuel N Luoma, Charles R. Tyler, Tamara S. Galloway, Jamie R. Lead, Silver nanoparticles: behaviour and effects in the aquatic environment, Environ. Int. 2011, 37, 517–531.

15. Panagiotis Dallas a, Virender K. Sharma, Radek Zboril, Silver polymeric nanocomposites as advanced antimicrobial agents: classification, synthetic paths, applications, and erspectives, Adv. Colloid Interface Sci. 2011, 166, 119–135.

16. Hong-Mei Gong, Li Zhou, Xiong-Rui Su, Si Xiao, Shao-Ding Liu, and Qu-Quan Wang, Illuminating dark plasmons of silver nanoantenna rings to enhance exciton–plasmon interactions , Adv. Funct. Mater. 2009, 19, 298–303.

17. Yu A Krutyakov, Kudrinskiy A, Olenin A Yu, Lisichkin G V, Synthesis and properties of silver nanoparticles: advances and prospects, Russ. Chem. Rev. 2008, 77 (3) 233–257.

18. Xi-Feng Zhang, Zhi-Guo Liu, Wei Shen and Sangiliyandi Gurunathan, Silver nanoparticles: synthesis, characterization, properties, applications, and therapeutic approaches, , Int. J. Mol. Sci. 2016, 17, 1534.

19. Brinker George Scherer, C. Sol-Gel Science, The Physics and Chemistry of Sol-Gel Processing, Academic Press, ISBN: 9780121349707

20. Larry L. Hench, Jon K. West, The sol–gel process, Chem. Rev, 1990, 90, 33–72.

21. Suresh C. Pillai , Sarah Hehir. Sol–gel materials for energy, environment and electronic applications, DOI 10.1007/978-3-319-50144-4, ISBN 978-3-319-50142-0

22. De ay, G., Licciulli, A., Massaro, C., Tapfer, L. Catalano, M., Battaglin, G., Meneghini C., Mazzoldi, P. Silver nanocrystals in silica by sol–gel processing, J. Non-Cryst. Solids, 1996, 194, 225–234.

23. Mauro Epifani, Cinzia Giannini, Leander Tapfer, Lorenzo Vasanelli, Sol–gel synthesis and characterization of Ag and Au nanoparticles in SiO_2, TiO_2, and ZrO_2 thin films, J. Am. Ceram. Soc. 2000, 83, 2385–2393.

24. Toney Fernandez, Gijo Jose, Siby Mathew, Rejikumar PR, Unnikrishnan NV, An ultra-low hydrolysis sol–gel route for titanosilicate xerogels and their characterization, J. Sol-Gel Sci. Technol., 2007, 41, 163–168.

25. Arun Kumar K V, Sajna M S, Vinoy Thomas, Cyriac Joseph, Unnikrishnan N V, Plasmonic and energy studies of Ag nanoparticles in silica-titania hosts, Plasmonics, 2014, 9, 631–636.

26. Arun Kumar K V, Revathy K P, Prathibha Vasudevan, Sunil Thomas, Biju P R, Unnikrishnan N V, Structural and luminescence enhancement properties of Eu^{3+}/Ag nanocrystallites doped SiO_2 TiO_2 matrices, J. Rare Earths, 2013, 31, 441–448.

27. K.V. Arun Kumar, Jini John, T.R. Sooraj, Shedhal Anu Raj, N.V. Unnikrishnan, Nivas Babu Selvaraj , Surface plasmon response of silver nanoparticles doped silica synthesised via sol–gel route, Appl. Surf. Sci., 2019, 472, 40–45.

28. Arun Kumar K V , Arya Balu, Athira Ramachandran, Unnikrishnan N V, Nivas Babu Selvaraj, Sol–gel synthesized plasmonic nanoparticles and their integration into dye sensitized solar cells, Appl. Surf. Sci., 2019, 491, 670–674.

29. Kreibig U, Vollmer M, Optical Properties of Metal Clusters, Springer, Berlin, 1995

30. Huijuan Bi, Weiping Cai, Caixia Kan, Lide Zhang, D. Martin, F. Trager, Optical study of redox process of Ag nanoparticles at high temperature, J. Appl. Phys., 2002, 92, 7491–7497.

31. Moores A, Goettmann F. The plasmon band in noble metal nanoparticles: an introduction to theory and applications, New J. Chem., 2006, 30, 1121

32. Pinchuk A, Kreibig U, Interface decay channel of particle surface plasmon resonance, New J. Phys., 2003, 5, 151.1–151.15.

33. De G, Licciulli A, Massaro C., Tapfer L, Catalano M, Battaglin G, Meneghini C, Mazzoldi P, Silver nanocrystals in silica by sol–gel processing, J. Non-Crystall. Solids, 1996, 194, 225–234.

34. Toshiharu Teranishi, Miharu Eguchi, Masayuki Kanehara and Shangjr Gwo, Controlled localized surface plasmon resonance wavelength for conductive nanoparticles over the ultraviolet to near-infrared region, J. Mater. Chem., 2011, 21, 10238.

35. Kim Y H, Kang Y S, Jo B G., Preparation and characterization of $Ag-TiO_2$ core–shell type nanoparticles, J. Ind. Eng. Chem., 2004, 10, 739.

36. Gijo Jose, Gin Jose, Thomas V, Joseph C, Ittyachen M A, Unnikrishnan N V., Fluorescence enhancement from Eu^{3+} ions in CdSe nanocrystal containing silica matrix hosts. Mater. Lett., 2003, 57, 1051.

37. Michael J Lochhead, Kevin L Bray. Rare-earth clustering and aluminum codoping in sol–gel silica: investigation using europium(III) fluorescence spectroscopy, Chem. Mater., 1995, 7, 572.

38. Eichelbaum M, Rademann K. Plasmonic enhancement or energy transfer on the luminescence of gold, silver, and lanthanide doped silicate glasses and its potential for light-emitting devices, Adv. Funct. Mater., 2009, 19, 2045.

39. Nabika H, Deki S. Enhancing and quenching functions of silver nanoparticles on the luminescent properties of europium complex in the solution phase, J. Phys. Chem. B, 2003, 107, 9161.

40. Abdul Gafoor, A.K., Musthafa, M.M., Pradeep Kumar, K., Pradyumnan, P.P. Effect of Ag doping on structural, electrical and dielectric properties of TiO_2 nanoparticles synthesized by a low temperature hydrothermal method, J. Mater. Sci. Mater. Electron., 2012, 23, 2011–2016.

41. Chung-Chia Chen, Bo-Chao Huang, Ming-Shiang Lin, Yin-Jui Lu, Ting-Yi Cho, Chih Hao Chang, Kun-Cheng Tien, Su-Hao Liu, Tung-Hui Ke, Chung-Chih Wu, Impedance spectroscopy and equivalent circuits of conductively doped organic hole-transport materials, Org. Electron., 2010, 11, 1901–1908.

42. Arun Kumar K V, Sunil Thomas, Manju Gopinath , Biju P R, Unnikrishnan N V, Structural and dielectric studies of Eu^{3+}/Ag nanocrystallites: SiO_2–TiO_2 matrices, J. Mater. Sci. Mater. Electron, 2013, 24, 1727–1733.

43. Prabhakar Rai, Plasmonic noble metal@metal oxide core–shell nanoparticles for dye-sensitized solar cell applications, Sustain. Energy Fuels, 2019, 3, 63–91.

44. Md Ashraf Hossain, Jieun Park, Dayoung Yoo, Youn-kyoung Baek, Yangdo Kim, Soo Hyung Kim, Dongyun Lee, Surface plasmonic effects on dye-sensitized solar cells by SiO_2-encapsulated Ag nanoparticles, Curr. Appl. Phys. 2016, 16, 397–403.

45. Holly F. Zarick, William R. Erwin, Abdelaziz Boulesbaa, Olivia K. Hurd, Joseph A. Webb, Alexander A. Puretzky, David B. Geohegan, Rizia Bardhan, Improving light harvesting in dye-sensitized solar cells using hybrid bimetallic nanostructures, ACS Photon, 2016, 3, 385–394.

46. Kawashita M, Tsuneyama S, Miyaji F, Kokubo T, Kozuka H, Yamamoto K, Antibacterial silver-containing silica glass prepared by sol–gel method, Biomaterials, 2000, 21, 393–398.

47. Roberto C. Gamez, Edward T. Castellana, and David H. Russell, Sol–gel-derived silver-nanoparticle-embedded thin film for mass spectrometry-based biosensing, Langmuir, 2013, 29 (21), 6502–6507.

48. Essraa A Hussein, Moustafa M Zagho, Gheyath K Nasrallah, and Ahmed A Elzatahry, Recent advances in functional nanostructures as cancer photothermal therapy, Int. J. Nanomed., 2018, 13, 2897–2906.

CHAPTER 3

Localized Surface Plasmon Resonance (LSPR) Applications of Gold (Au) and Silver (Ag) Nanoparticles

APARNA RAJ[1] and RIJU K. THOMAS[2]

[1]*School of Pure & Applied Physics, Mahatma Gandhi University, Kottayam, Kerala, India*

[2]*Bharata Mata College, Thrikkakara, Ernakulam, Kerala, India*

Corresponding author. E-mail: rijukthomas@gmail.com

ABSTRACT

To designate the interactions of light with nanostructured metals, the research community has created the new term "plasmonics." The field of plasmonics has blossomed over the last few years. To make out the gap between conventional optics and highly integrated nanophotonics, plasmonic nanoparticles (NPs) had most severely studied. The significance of the plasmonic NPs over other NPs in diverse applications like catalysis, electronics, magnetic, optical, optoelectronic, materials for solar cell, medical, bioimaging, and the diagnosis was established because of their characteristics such as surface-enhanced plasmonic behavior. The optical properties of gold and silver NPs, mainly localized surface plasmon resonance (LSPR) play a significant role in the development of these diverse fields. The versatility of the plasmonic NPs led us to review their most salient physicochemical properties and potential applications. Here we are presenting a general idea about the LSPR-based applications of gold and silver NPs.

3.1 INTRODUCTION

Half a century ago, Richard Feynman invited his listeners to uncloak a new field of physics [1]. His lecture named "There's Plenty of Room at the Bottom: An Invitation to Enter a New Field of Physics" elucidated the possibility of direct manipulation of individual atoms. In the 1970s the term *nanotechnology* was formed and the remarkable furtherance in the field of nanoscience had blossomed over the last centuries [2]. The modern era of technology is categorized by the varied applications of nanomaterials in almost all biomedical and industrial sectors. Particles having one structural dimension less than 100 nm are defined as NPs [3]. Prodigiousness in the designing and fabrication approaches of nanosized materials in the last decades allowed as going into the fascinating world at the length scale of subcellular structures including DNA strands [4]. NPs often have idiomatic physical and chemical behaviors. While comparing the electronic, thermal, physical, optical, and chemical properties of NPs with bulk materials, it is found that totally unlike in nature. In the nanometer range, materials behave very disparately compared to voluminous scales and it is still very difficult to forecast the chemical and physical properties of particles of very small size. The quantum confinement effect is the reason behind the unique properties of nanomaterials [5–7]. Quantum confinement can be observed when the particle dimension is in the order of de Broglie wavelength. The surface-to-volume ratio increases with the reduction in the particle size; as a result, there is an increase in the dangling bonds (the free radicle which exists in an immobilized environment) that in turn contributes higher surface free energy [8].

NPs can be categorized according to their size, morphology, chemical, and physical properties [9]. They are carbon-based NPs, ceramic NPs (inorganic nonmetallic solids), metal NPs, semiconductor NPs, polymeric NPs (organic based), and lipid-based NPs (contain lipid moieties). Other properties like strength, degree of activity were also pendent on their unique size, shape, and structure. Due to these characteristics, they are competent candidates for a variety of corporate, marketable, and domestic applications. Nowadays, NPs provide solutions to various biological and environmental challenges in many areas like solar energy conversion, medical application, catalysis, water treatment, energy-based research, and ultrasensitive molecular detectors. NPs extensive applications in biomarkers, diagnostics, biological imaging, biosensors, cell labeling, antimicrobial agents, and drug delivery systems make them potential successor of the next-generation medicine

3.2 METAL NANOPARTICLES

Metal NPs are chastely made up of the metals harbingers. These metal NPs possess peculiar optoelectrical properties due to the noted LSPR [10, 11]. Noble metal NPs and alkali metal NPs, that is, Au, Ag, and Cu NPs produce a strong UV–visible extinction band that is not seen in the spectrum of the bulk metal. This excitation band results from the stimulation effect caused by incident light on the charge density oscillations of conduction electrons at the metal-dielectric interface. Surface plasmon oscillations experience a resonant condition when excited with corresponding identical wavelength known as surface plasmon resonance (SPR). When the surface plasmons of conductive nanostructures confined to a particle size that is comparatively smaller to the wavelength of the incident light, it is called LSPR. The research society has introduced the term "plasmonics" to denominate the interactions of light with nanostructured metals. By confining the light on subwavelength volumes, plasmonics build a bridge between two different length scales. Metal NPs and colloids are the building blocks of that bridge. The LSPR can be tuned based on the properties of capping materials, particle shape, refractive index, surface modification, surface charge and interparticle interactions.Recently, heterostructured metallic nanostructures are receiving attention. They are nanoshelled particles containing a shell and a core and each of them made up of different materials. When the shell material is metallic, there will be dramatic changes in the SPR. Resonances wavelengths are typically shifted to much longer wavelengths comparing with corresponding metal NPs.

Due to their advanced optical properties, metal NPs find applications in many research areas. Gold NPs coating is widely used for the sampling of SEM to enhance the electronic stream in order to produce high-quality SEM images, which is a simple example for the application. The potentiality to manipulate and control light on the nanometer scale as results in miscellaneous applications such as quantum information processing, data storage, data processing, and optoelectronic devices [12]. Metallic NPs are very much important in biomedical sciences due to their wide range of physical, chemical, and biological properties. The optical and biological properties of the NPs will be assorted on account of their surface modifications and functionalization with suitable functional groups, which may contribute stable interactions with ligands, antibodies, and drugs, thereby opening a wide range of potential applications in biotechnology and drug delivery systems. To extrapolate the fundamental mechanisms of transport and interaction of NPs in specific biological systems high level of control is available by adjusting the shape, size, and surface of capped metal NPs can be employed. The silver and gold NPs play a major part in the curative field of research due to their medicinal applications against various diseases and antimicrobial activities. The characteristic color of metal NPs is due to their SPR, which has been widely applicable to the colorimetric recognition of biomolecules.

Nowadays, it is well probed that numerous biological molecules subsuming algae, bacteria, plants, fungi, and human cells are used to synthesize metal nanostructures, through the reductive conditioning of the phytochemicals, proteins, and metabolites in these organisms. Biological applications demand nontoxicity, but various physical and chemical procedures are currently used for the synthesis of metal NPs are very expensive and are potentially hazardous to the living organisms and environment. Therefore it is necessary to explore an eco-friendly and cost-effective alternative method for NPs synthesis. Recently, the synthesis of metal NPs using biological molecules (Green synthesis) has become a significant focus for researchers due to their simplicity of procedures, stability, and potential applications in the pharmaceutical industry (manufacturing of pharmaceutical), biomedical field, and others. Figure 3.1 discusses biological synthesis and applications of metal NPs in biomedical and environmental fields.

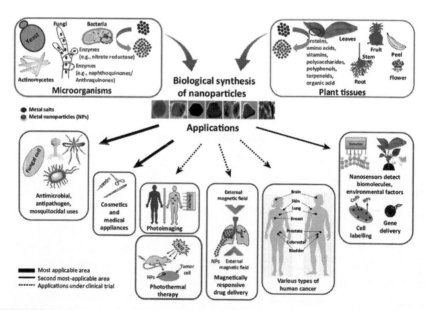

FIGURE 3.1   Green synthesis and employment of metal nanomaterials in environmental and biomedical fields.

Source: Reproduced with permission from Ref [13]. © 2016 Elsevier.

3.2.1 PROPERTIES OF METAL NANOPARTICLES

3.2.1.1 THERMAL PROPERTIES

Thermal conductivity of metal NPs decreases with decreasing particle size. Owning to the low melting point electronic wiring can be made with NPs whose diameter is less than 10 nm. NPs of this much low dimension will have a melting point lower than its bulk form [14].

3.2.1.2 ELECTRICAL PROPERTIES

Metal NPs behave in an electronically distinct manner from the corresponding bulk metals because of the restriction of their dimensions in three, two, or one direction (size quantization). The number of electron wave modes propping to the electrical conductivity is getting smaller according to the decrease of the diameter of the particle. The electronic properties depend resolutely on the particles' size and shape. It can be used to make high-temperature superconductivity material. The size-dependent change of the electronic properties has consequences on chemical behavior [15].

3.2.1.3 MECHANICAL PROPERTIES

Mechanical properties (stiffness, adhesion, elastic modulus, etc.) of metallic NPs are the reason behind many applications of metal NPs in various fields like surface engineering and nanofabrication. For the modification and manipulation of materials at the atomic level, understanding of the armature and mechanical properties of nanomaterials is needed.From the recent studies about the mechanical strength of gold NPs, a significant result was found that the bond strength of the atoms in the nanostructure was double compared with the bond strength in the bulk. Apparently, the effective hardness of a nanomaterial hangs on the perfect configuration of the base atoms [16].

3.2.1.4 MAGNETIC PROPERTIES

Even though metals like gold and platinum are nonmagnetic materials in the bulk form, they exhibit magnetic property at the nano size level. Yoshiyuki

Yamamoto et al. reported that platinum NPs having the diameter below 3.8 nm exhibited superparamagnetic behavior by the magnetization process, that is,comparing with the paramagnetic moment of bulk state, gold NPs shows an enhanced magnetization. In this study, the magnetic moment increases with the reduction in size and reaches $5.0\mu_B$/Pt particle at 2.3 nm [17]. Magnetic properties of nanoparticles can be modified by varying the size, shape, and capping agent of nanoparticles.

3.2.1.5   OPTICAL PROPERTIES

Noble metal NPs, particularly Ag and Au, have been attracting a great deal of engrossment due to their size and shape-dependent optical properties, which emanate from their SPR property [18–20]. In semiconductor NPs, the reason behind the major modifications of the optical spectra is the quantization of the electron and hole energy states. While in metal NPs, the unique optical properties of metal NPs evolved from the nanoscale optical interactions and nanoscale confinement of electronic interactions [21].

Nanoscale electronic interactions are mainly classified into three different types. They are (1) quantum confinement effects: localization of electronic motion to modify optical interactions by producing new optical transitions. (2) Nanoscopic interactions to control dynamics: manipulate radiative and nonradiative transitions by controlling of local interactions and phonon density of states. (3) New cooperative optical transitions: this is again classified into three types of interactions, that is, new cooperative absorption (molecules within nanometer distances absorbing collectively), Nanoscale electronic energy transfer (fluorescence resonance energy transfer [22] and exciton transfer), cooperative emission [23] (emission by an ion pair through a virtual state).

To induce optical interactions on the nanoscale, the electric field associated with a photon can be confined by using a number of geometries. The nanoscale optical field can be localized both axially and laterally. Methods for nanoscale localization of electromagnetic (EM) fields are shown in Figure 3.2.

3.2.1.5.1   *What Are Polaritons and Surface Plasmons?*

When an electromagnetic field travels through a bulk material, it will cause the charged particles on the surface to move. Obviously, the moving charges

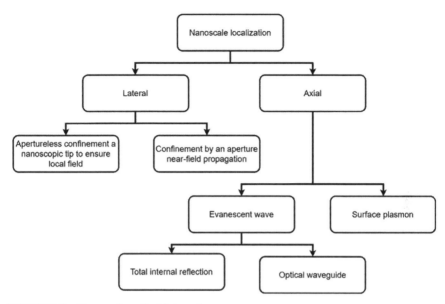

FIGURE 3.2 Nanoscale optical interactions.

radiate. When the resonance occurs and the oscillatory mechanical motion of charged particles will be coupled to the electromagnetic field oscillations. Polariton is the name coined for these coupled mechanical–electromagnetic waves. If the bulk material is an ionic crystal, then the polaritons are called phonon–polaritons. In the case of a metal, a plasmon–polariton is formed when the electromagnetic field coupled to a collective motion of the valence free electrons. This prodigy is called plasmon resonance. From another approach, it can be said that the whole system acts as an oscillator whose natural frequencies results to the formation of a plasmon–polariton. At the metal–dielectric interface longitudinal motion of the electrons will correspond to surface plasmon polariton (SPP) [24]. SPPs are surface-bound electromagnetic waves propagates at the interface. SPP is another version of evanescent wave interaction when the waveguide is replaced with the boundary of negative and positive permittivity material materials (metal–dielectric interface). Like evanescent waves, the intensity of wave will be maximum at the interface and exponentially decreases when enter into the metal. SPPs in metal NPs are enslaved to the NP contour, that is, they cannot freely propagate, so the resonance conditions depend on the size, exact shape, and the polarization of the field with respect to the NPs. These plasmon resonances are called LSPRs [25]. Each plasmon has a feature electric field distribution around the NP since every plasmon will produce a distinct charge distribution on the NP surface. This new type of resonance localized near the boundary between the metal nanostructure and the surrounding dielectric called SPR, which also produces an enhanced electromagnetic field at the boundary. This enhanced field has enormous opportunities in this modern era that it may be employed for boundary-supersensitive optical interactions. That in turn forms a powerful basis for nanoscale localized optical imaging as well as for optical sensing. Apertureless near-field imaging is also a wonderful application of the above case. Six basic fingerprints of LSPRs for their applications:

Electric field distribution surrounding of nanoparticle: Each plasmon mode generates a specific electric field around the NP with two important characteristics. Their intensity is maximum in the NP interface and exponentially decays toward surrounding. First is decay length equals to half of the incident light wavelength. Moreover, the second one is several hot spot fields are present there. Hot spot fields are the regions where intensity is very stronger than the incident fields.

Spectral response: Each LSPR mode has its own specific frequency, that is, each mode will interact with a given color of light. The plasmon resonance spectra hanging on the size, shape, dielectric properties of the immediate surroundings and the material of the NP.

Enhanced absorption: In resonant condition, metallic NPs are efficient light absorbers. The absorption cross-section may be several times higher than the actual physical cross-section of the NPs at the LSPR frequency.

Enhanced scattering: Scattering cross-section of metal NPs will be enriched at LSPR frequency. Scattering or absorption may dominate depending on the NP size. For NPs having sizes below 20 nm absorption takes over scattering, and the diametric occurs for larger NPs.

Directionality: There will be a particular angular pattern for the interaction of light with each plasmon. Depending on the direction, the light will be absorbed or scattered with differing efficiencies.

Heat generation: Metallic NPs are not efficient light emitters. For the reason, almost all absorbed energy may turns into heat. It is possible to selectively heat different NPs by tuning main LSPR modes and enhanced resonant absorption.

3.2.1.5.2   *Theory of Surface Plasmon Resonance*

The principal photonic applications of the metal nanostructures are derived from the local field enhancement of various light-induced linear and nonlinear optical behaviors within the nanoscopic volume of the media surrounding the metal nanostructures. When the metal–dielectric interface is irradiated by light, it will travel as an electromagnetic wave in a direction parallel to the boundary of metal–dielectric and hence the electron cloud around the metal is displaced slightly relative to the nuclei. There occurs a columbic attraction between them, that is, electron cloud gets polarized. This results in oscillation in the electron cloud. The shape and size of the charge distribution, the effective electron mass, and the density of electrons are the determining factors of the oscillation frequency.

Even there are many proposed theoretical models, the classical Mie theory is often used for the explanation of metal NPs' optical properties. Spherical and spheroid particles best explained using Mie theory. There are extensions of Mie theory for aggregates of particles. Particles having a size less than the wavelength of light can be explained using dipole or quasi-static approximation also known as Rayleigh approximation. Multi-mode expansion of the field will help us to explain the SPR phenomenon. For small size particles, plasmon oscillations beset by the electromagnetic field of light will produce oscillating dipoles along the field direction. The EM field is uniform across a particle such that all the conduction electrons move in-phase produce only dipole-type oscillations when the size of the NPs are small comparing with the wavelength of incident light, that is, for smaller particles dipole approximation is conspicuous. The spectrum will be of slender width. However, higher modes of plasmon excitation can also occur, such as the quadrapole and octapole mode. For larger size NPs (e.g., >25 nm for gold particles), higher-order (such as quadrapolar) charge cloud distortion of conduction electrons becomes important. The field across the particle becomes nonuniform. As the particle size increased, the higher-order distortions induce a pronounced shift in the plasmon resonance condition. The spectrum broadens and excites higher multipole resonances.

3.2.2 LOCALIZED SURFACE PLASMON RESONANCE APPLICATIONS OF SILVER AND GOLD NANOPARTICLES

Silver shows the highest efficacy of plasmon excitation among the other metals NPs (Au, Ag, and Cu) that shows visible plasmon resonances spectrum. The particular frequency in the incident light will invoke collective oscillation of the surrounding electrons of silver NPs. The peculiar frequency of the LSPR is dependent not only on the size and shape of silver NP but also on the agglomeration state. Most applications of silver NPs are conferred by this LSPR effect. Ag NPs have potential than any other particle of the same dimension composed of any known organic or inorganic chromophore to interact with light. The geometric cross-section of silver NPs is nearly 10-fold diminutive than the light-interaction cross-slice or cross-section, which foretells that the particles capture much more light than physically incident on them. Plasmon resonance of the silver NPs can be tuned to any wavelength in the visible spectrum. An idiomatic characteristic of spherical AgNPs is that by altering the particle shape, size, and the refractive index of the surrounding of particles outer layer, the SPR peak wavelength can vary from 400 nm (violet light) to region 530 nm (green light). Aggregation state of the particles will also influence the optical properties. Figure 3.3 explains the effect of size on LSPR peak.

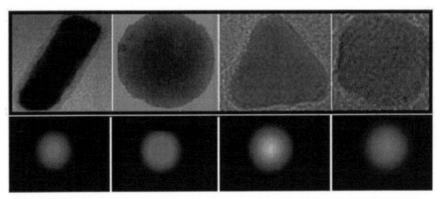

FIGURE 3.3   LSPR with the shape of silver NPs. Reproduced with permission from [26].
Soruce: Reproduced with permission from Ref. [26]. © 2013 Elsevier.

The AuNPs display awesome bio congeniality with biological-macromolecules and exposed promising optoelectronic, magnetic, and medicinal applications in the field of scientific research. By their unique SPR, plasmonic NPs (noble metals) distinguish themselves from other nanoplatforms such as semiconductor quantum dots, polymeric and magnetic NPs. The optoelectronic properties of AuNPs are tunable by varying its surface chemistry, aggregation state, size, and shape. Besides unique optoelectrical properties, gold is one of the best weather-resistant metals. Even a single-layer gold oxide trace was not detected yet on the surface exposed to the air. The bioinert (stable) obsolete of gold NPs makes the AuNPs as excellent candidates for the potential biomedical applications. Figure 3.4 explains the biomedical applications of gold nanoparticles.

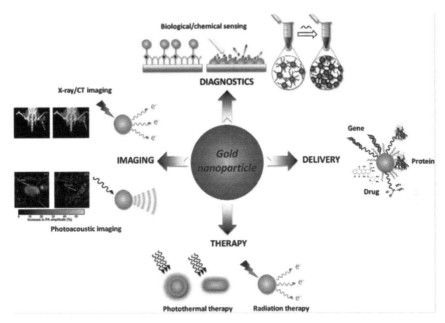

FIGURE 3.4 Biomedical applications of AuNPs. Owing to their unique physiochemical, optical and electronic properties, AuNPs have been used for a variety of applications in biomedical fields.

Soruce: Reproduced with permission from [27]. © 2017 Elsevier.

The AuNPs are also being explored widely in diverse areas of sensory probes, electronic conductors, therapeutic agents, photothermal therapy, drug-gene therapy, organic photovoltaic, and others. Duo to the awesome biocompatibility seemed capable enough for clinic settings as well as firmly tunable and highlighted optical properties, gold (Au) NPs have been revived to the leading edge of cancer research. Furthermore, AuNPs are also used as biomarkers in the diagnosis of cardiovascular diseases and in thermal ablation of tumors. For monodispersed small-sized AuNPs, the SPR bring on the absorption of light in the blue-green region of the spectrum (~450 nm) while red portion (~700 nm) is reflected, yielding an intense wine red color. SPR absorption wavelength has a redshift to higher wavelength regions, when particle size progressing.

By taking advantage of the desirable optical, conductive, and antibacterial properties of the Ag NPs, they are having wide areas of applications such as photonics, biosensing, antimicrobial applications and electronics. Most applications in biosensing and detection exploit the optical properties of silver

NPs. In diagnosis application, make use of silver nanomaterials as biological tags for quantitative spotting. The antibacterial properties of silver NPs are employed for the manufacturing of paints, footwear, apparel, wound dressings, appliances, cosmetics, and plastics. Silver NPs are accustomed in conductive inks due to their enhanced thermal and electrical conductivity. Even if silver NPs has this much applications, it is highly essential to study their potential risk of being used in real-life applications of environment and human health. For the purpose toxicological effects have to be studied and accordingly the size, shape, and functionalization should be precisely controlled. El et al. reported the curcumin conjugated AgNPs seemed to exhibit antimicrobial effect comparable to that of Penicillin and Amoxicillin along with its nucleic acid sensing mechanism in the concentration range 100–1000 ng/mL.

There are four morphological structures of gold nanostructures that are most extensively used, especially in biomedical applications. They are nanospheres and nanorods, nanoshells (silica balls covered with a polycrystalline gold layer), and nanocells. Except nanospheres the others have firm SPR spectra in the near-infrared zone. Gold nanospheres are the simplest species for synthesis and may be obtained with various sizes of highly uniform particles. The properties of these obtained nanostructures can be tuned according to their size. Figure 3.5 explains the relation of size with the properties of NPs.

Gold NPs are widely used in many applications. Computation of spectra of gold NPs (resonance wavelength and absorption efficacy) can be done according to Mie theory and discrete-dipole approach. In general, optical properties of nanostructures depend on size, shape, and core/shell structure to an extent. From different studies, it is agreed upon that gold nanomaterials are the utmost suitable NPs for studies in a living body. In the near-infrared region (in which the body tissue is transparent to) the SPR peak of gold NPs indicates itself. Even gold NPs possess dominated absorption and scattering characteristics than (even five orders in magnitude) different organic dye molecules. The thermal properties of gold NPs under light beams are of great interest. Compared to other metal NPs, gold has high chemical stability. They are biocompatible too [29]. Depending on the structure of the particle, the absorption wavelength can be departing from 500 to 1000 nm. Irradiation in the plasmon range of gold NPs will result in an increase of temperature from 4 to 40 °C. By using a femtosecond laser, it is possible a more strongly marked temperature increase near the gold surface. These photothermal effects are able to kill bacteria to cancer cells. Modern biomedical techniques can triumphantly treat cancer through plasmon-mediated photothermal therapy, in which metal nanoprobes perform as intense heaters to kill cancer cells [25, 28, 30]. Metal nanomaterials have their use from cancer therapy to third-generation solar battery cells. Figure 3.6 unveils the clinical trials [a tryout or experiment to test quality, value, or usefulness] for cancer immunotherapy applications.

A. gold nanoshell
B. gold nanocage
20nm
10nm
7nm
5nm
500 600 700 800 900 1000 1100 1200
Absorbance (a.u.)
0
400 600 800 1000
Wavelength (nm)

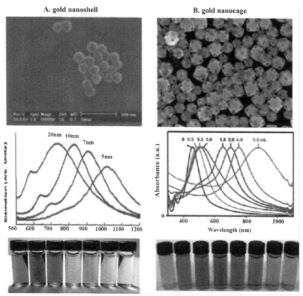

FIGURE 3.5 Tunable properties of gold nanocages and nanoshells with size.

Source: Reproduced with permission from Ref. [28]. © 2010 Elsevier.

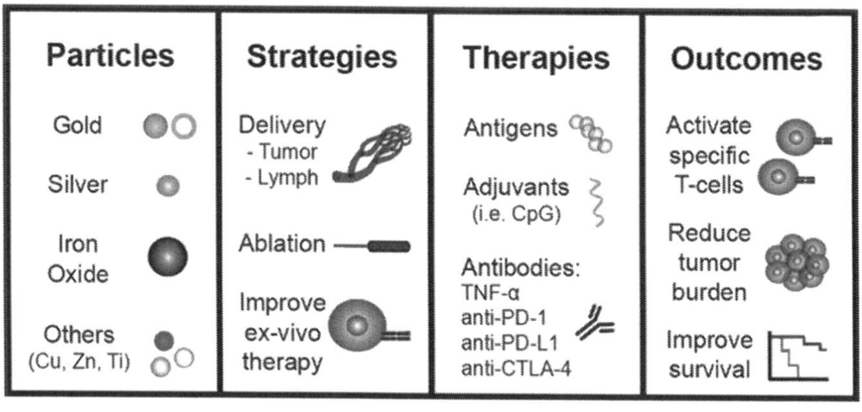

Particles
Gold
Silver
Iron Oxide
Others (Cu, Zn, Ti)
Strategies
Delivery
- Tumor
- Lymph
Ablation
Improve ex-vivo therapy
Therapies
Antigens
Adjuvants (i.e. CpG)
Antibodies:
TNF-α
anti-PD-1
anti-PD-L1
anti-CTLA-4
Outcomes
Activate specific T-cells
Reduce tumor burden
Improve survival

FIGURE 3.6 Cancer immunology.

Source: Reproduced with permission from Ref. [31]. © 2018 Elsevier.

Gold material in a bulk order is stable and chemically inert, but the small cluster formations of gold NPs possess catalytic activity. This can be explained by the specific electron structure, size, and the oxidation state of NPs. The surface occupied by nanoclusters may also exert an influence on their catalytic activity. One can enhance the photocatalytic activity of nonmetal NPs by applying metal NPs such as gold and silver because of their plasmonic properties. Garniture of metal NPs on the top of the semiconductor can increase the charge-separation efficacy; accordingly, result in the better photocatalytic activity. But the synthesis is not that much easy since metal NPs will vigorously aggregate in high temperature and in contrary nonmetal NPs need a high temperature for doping. To overcome these synthetic challenges, many researchers synthesized sandwich-structured metal-non-metal photocatalyst. Let us take an instance, gold NPs sediment uniformly on a silica sphere and covered with a titanium layer. The prepared composites will show great photocatalytic efficiency under both direct sunlight and visible light.

Our society also has an increased interest in the development of alternative (green) energy sources [32]. The plasmonic nanostructures have an important role in improving the performance and feasibility of photovoltaic devices. A solar cell is a prime example for the tunability of plasmonic resonance. An efficient solar device must for example have a very broad absorption spectrum for all parts of sunlight. So it is very important to maximum overlap between the light-absorption spectra of solar cells and plasmonic peaks of the metal NPs. New possibilities for the design of solar-cells were open up by the combination metal NPs and semiconductor materials that allow a considerable reduction in the physical thickness of absorber layers. Performance of solar cells can be improved by changing the arrangement of metal NPs, composition or mixing two different types of metal NPs. Temple and Bagnall reported that spectra of gold and aluminum metal NPs can be extended across the solar spectrum by tuning the lateral size and shape [33]. Planar metal NPs offer several advantages over spherical particles for PV applications. From their work, it is confirmed that the peak position of planar metal NPs can be tuned across the entire solar spectrum without increasing the height of NP.

To resolve the low efficacy problem of dye-sensitized solar cells (DSSC), Nikhil et al. used the plasmonic NPs. To synthesize highly efficient DSSC device, the gold NPs embedded in a 3D TiO_2 mesoporous matrix [34]. Gold NPs of various size (5–85 nm) synthesized by Turkevich method were subjected to optical and electrical measurements. They studied the influence of size and concentration of spherical GNPs on the overall device performance of plasmonic DSSC. A maximum enhancement in DSSC performance was observed for the gold NPs of 15–40 nm size and concentration of 0.1%–0.25% with respect to the matrix in the device due to the enriched near field excitation of dye molecules along with the incident light far-fields. Since the outsized Au NPs own high scattering efficiency and hence little or no enhancement in photocurrent, they suggest the choice of mid-size Au NPs along with a 13 μm mesoporous TiO_2 layer and 5 μm of commercial titania paste layer.

The comparatively poor light absorption efficiency of thin-film crystalline silicon (c-Si) solar cells has always made researchers think of a systematic design that could improve its absorption capacity. Shuyuan Zhang et al. presented a systematic design and study of thin-film c-Si solar cells garnished with a bilayer AgNPs array of dissimilar particle size [35]. Employing mathematical simulations a comparison of performance with its optimized counterpart decorated with uniform Ag NPs resulting a 9.97% increase in short circuit current density and a 9.94% increase in intergraded quantum efficiency across the solar spectrum. Light absorption at shorter wavelengths was enhanced by smaller particles arranged in the upper layer and plasmonic enhancement at longer wavelengths was provided by larger particles arranged in the lower layer. Backscattering of double-layer Ag NPs with different particle size enhanced the diffused scattering light and reduced the light transmission. This in turn enhances the light confining interior part of the solar cells.

Nanostructure morphology plays a key role in determining the photoconversion efficiency of solar cells. Ganesha et al. reported that platinum (Pt) counter electrodes were embedded with the plasmonic effects of silver nanostructures (Ag rods, sphere, and prism) [36]. The observed PCEs are 8.10% for nanospheres, 8.55% for nanoprism, and 8.68% for nanorods devices. But PCEs for reference device without Ag was only 7.60% only. Owing to the longitudinal localized surface plasmon resonance (L-LSPR), nanorods showed greater improvement to the device efficiency. It showed greater scattering effects to the light in longer wavelength resulting in the highest enhancement in short circuit photocurrent density (J_{sc}).

The metallic nanostructures adsorbed on metal surfaces will ameliorate the Raman scattering of molecules owing to the plasmonic characteristics. This is the so-called surface-enhanced Raman spectroscopy (SERS). With metal nanostructures, it will be a powerful vibrational spectroscopy and the enhancement comparing with plasmonic NPs will be 14 to 15 orders of magnitude. This phenomenon is indifferent from SPR or SPP. Intelligent design and fabrication of different type of plasmonic nanostructures can be utilized in rough metal surfaces to upgrade the application SERS in disparate fields. It has been well substantiated that the SERS performance of plasmonic NPs can be drastically enriched by using anisotropic nanostructures, like nanostars, nanoplates, and nanostructures with small gaps. For example, gold nanohedge with small interparticle gaps has shown great SERS performance.SERS signal of the Au@void@Au nanorattles could be enriched especially by tuning the shape of AuNPs [37]. Comparing the same-sized spherical gold NPs and highly asymmetric nanocrystals, it is found that the asymmetric nanocrystals show stronger field localizations (nearly 100-fold) at the vertexes.

Gold NPs have a wide variety of usages in parasitology and entomology. AuNPs offer high toxicity toward insect vectors, cestodes, and trematodes. They are toxic to various parasites [38]. Figure 3.7 shows detail about the toxicity of AuNPs toward various parasites and insects.

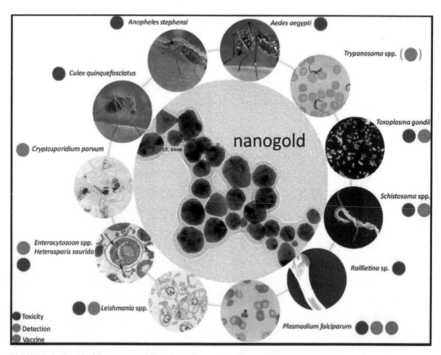

FIGURE 3.7   Gold nanoparticles—against parasites and insect vectors.

Source: Reproduced with permission from Ref. [38]. © 2018 Elsevier.

One of the major applications of plasmonic NPs is biosensing.Comparing with the other metal NPs, gold (Au) and silver (Ag) nanostructures show the most suitable physical properties for biosensing. Because of their high chemical stability and biocompatibility AuNPs are the highly considered one for the biosensing applications. But in terms of sensitivity silver (Ag) NPs offer better results. LSPR is the most significant property of metallic NP [25, 29, 30, 39]. Since AuNPs and AgNPs have strong interactions with light LSPR is responsible for the bright color of the colloid suspensions of NPs. As already discussed this property of LSPR can be tuned by domineered the parameters like size, uniformity, shape, and surface morphology and this is enormously employed for diverse applications in various fields such as biomedical and biochemistry. Monodispersed tiny spherical AgNPs (10 nm diameter) are yellow in color and has an absorption band at 380 nm, while similar AuNPs are red and absorb around 520 nm. Since the absorption bands of gold and silver NPs are in between the visible spectrum of the electromagnetic spectrum, they permit a colorimetric detection of different biomolecules by causing changes in the intensity and wavelength of the LSPR band. Modifying the size, shape, uniformity, composition (ratio Au: Ag), dispersion, and dielectric constant of the surrounding medium induces wavelength shift since the LSPR band has dependent on these properties. For example, LSPR will be shifted to a higher wavelength, that is, redshifted while the size of the NPs increases. The shift in wavelength of the LSPR peak is measured in units of nanometer per refractive index unit (RIU) [40]. When the shift is high, the refractive index variation is low and in turn, the sensitivity of the biosensor will be high. From various studies, it is confirmed that the refractive index sensitivity for AuNPs and AgNPs in the size range from 5 to 50 nm will increase from 153 to 265 nm/RIU and 128 to 233 nm/RIU, respectively. Among each other AgNPs are more sensitive, but AuNPs are more biocompatible [41]. However, nanostructures involving both the metals are very consequential and will result in a long range of possibilities. Noble metal NPs have been used to detect a variety of biomolecules such as proteins, disease biomarkers, enzymes, DNA, and even drugs. There are limitations like poor stability, toxicity, and their surface chemistry for bare Ag and Au NPs for the biosensing application. In order to overcome these limitations, these NPs were coated with a variety of compounds both organic and inorganic. This coating process reduces the toxicity, improves the stability, and avoids agglomeration. The coating also improves electrostatic, steric, and electrosteric stabilization of AgNPs. Figure 3.8 shows AgNPs functionalized with a variety of biomolecules.

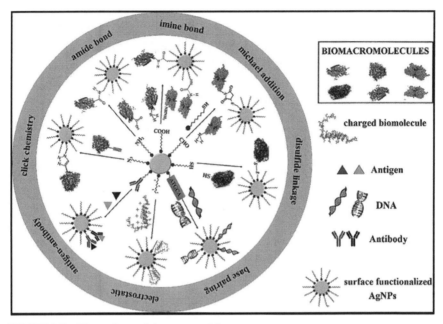

FIGURE 3.8   Bio conjugated Ag nanoparticles.

Source: Reproduced with permission from Ref. [42]. © 2020 Elsevier.

The nanomaterial-based colorimetric and fiber-optic sensors are excellent choices to effectively ease the quantification of contaminants in the food industry for safety and quality control monitoring. AgNPs and AuNPs are distinctive and attracted much interest among other nanomaterials for both biological and nonbiological sensors owing to their optical properties (which is capable of being tuned), stability, biocompatibility, antimicrobial and various therapeutic activities [43]. The gold and silver NPs show a marked LSPR absorption since these are highly sensitive to structure and plasmon coupling (interparticle) on the analyte-induced aggregation [44]. Fiber optic sensors and colorimetric sensor working on the basis of LSPR property have many benefits over conventional sensors such as resistance to electromagnetic interference, compact size, sensitivity, selectivity, remote sensing, and ability to be multiplexed and embedded into various textiles [45]. Fiber optic sensors display an elevated selectivity and sensitivity toward the health hazardous food preservatives/additives and various toxic metals [46]. The aggregation/reduction normally results in a color change for AuNPs and AgNPs, this can be observed by the naked eye. Figure 3.9 displays the UV–Vis spectral evolution of silver triangular plates in the presence of KCl as the etchant [47]. The original silver colloid was blue in color with an intense LSPR peak located at around 700 nm. The edge length of the silver triangles was about 70 nm. Upon the addition of KCl, a gradual change in the color of the silver colloid was observed. The color changed gradually from blue into bluish violet, violet, amaranth, orange red, and finally orange. Figure 3.9 is the corresponding UV–visible spectrum. The spectrum shows a quenching of the LSPR peaks around 700 nm, concomitant with the appearance of the other set of peaks around 480 nm that is in an opposite manner of the former set. From the TEM images in the figure, it is found that AgNPs are disc shape, with a diameter of about 39 nm. Thus the silver triangles were transformed into discs by chemical etching of KCl.

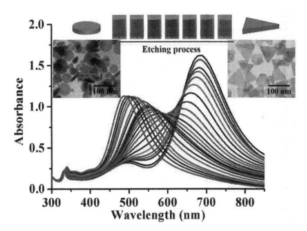

FIGURE 3.9 LSPR spectrum of silver nanoparticles [synthesis method: photochemical regrowth process].

Source: Reproduced with permission from Ref. [47]. © 2016 Elsevier.

In the case of LSPR-based fiber optic sensors [48–50], the spectral alterations in the resonance wavelength possibly occur due to the interaction between toxicants with NPs. LSPR can be tuned based on the properties of capping material, appearance, the refractive index of the surroundings, surface charge and interparticle interaction [44], and different preparation techniques such as etching and photochemical regrowth [44]. Earlier reports inferred that Ag decorated SnO_2 NPs exhibits the high response to formaldehyde but also has better stability and selectivity at an operating temperature of 125°C with 0.53 ppm limit of detection [51]. Silica nanostructure capped with thiol-reactive polymer units and polar polyamines were also used for the selective colorimetric detection of formaldehyde expressed a limit of detection (LOD) of 36 ppb in water [52]. Lakade et al. reported an effective gold NP-based method for detection of CaC_2 in artificially ripened mangoes. From the Mass spectroscopy analysis of artificially ripened mangoes, they have found that a higher concentration of arsenic is present in the sample. AuNPs functionalized with lauryl sulphate aggregate in the presence of ions of arsenic, resulting in a color change from red to purple [53]. Kashani et al. reported the development of a colorimetric sensor array consisting of citrate-capped 13 nm AuNPs for the detection and discrimination of five organophosphate pesticides, at concentration ranges of 120–400 ng/mL [54]. Rithesh et al. reported the SPR based fiber optic sensor fabricated using green synthesized NPs as sensing material for mercury and ammonia in the solution with a detection limit of 1×10^{-6}, 0.0077, and 2×10^{-7} M with minimal response time [45, 55].

The AuNPs have been largely employed for the colorimetric sensing of the degree of genetic similarity between pools of DNA sequences (DNA-hybridization). These techniques are enthused by the agglomeration or aggregation-based wine red-to-blue color transition, which is due to a redshifting and dampening of the SPR band of the AuNPs. The plasmonic metal NPs have a great capacity for the evolution of impressive chemical and biological sensors. For example, there are reports for highly selective colorimetric recognition of Cr (III) and Cr (VI) ions using gallic acid capped AuNPs. Also, the diosmin (a flavonoid) capped AgNPs (DM-AgNPs) shows good sensing capacity toward cysteine with the LOD value of 0.0077 M. The diosmin capped gold nanoparticles (DM-AuNPs) in the presence of an increasing concentration of ctDNA (0–15 µM) will decrease the intensity of LSPR peak at 537 nm along with a bathochromic shift (4 nm). This indicating a significant modification happened to the dielectric microenvironment of DM-AuNPs due to their complex formation with ctDNA. Agglomeration of AuNPs in the presence of ctDNA results in the hypochromic effects in the LSPR peak [56] (Figure 3.10).

Similarly, there are reports about silver NPs functionalized with chlorogenic acid (CA-AgNPs). In a higher value concentration of ctDNA (18–20 µM) a significant redshift was observed in the LSPR peak of CA-AgNPs accompanied by the formation of a double peak at the higher wavelength region (~600 nm). The colorimetric alterations from yellowish to brown color along with the intense hypochromic effect in the LSPR peak of CA-AgNPs appeared due to the agglomeration of AgNPs in the presence of ctDNA [57]. Figure 3.11 shows the sensing of Ct-DNA by Ag NPs. From Figure 3.10, it is identified that quenching of spectrum with the increasing concentrations of Ct-DNA.

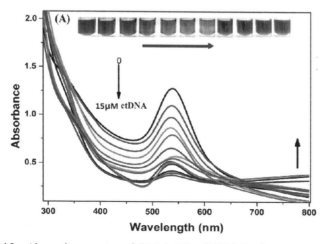

FIGURE 3.10 Absorption spectra of DM-AuNPs (LSPR) in the presence of different concentrations of ctDNA (0–15 μM).

Source: Reproduced with permission from Ref. [56]. © 2019 Elsevier.

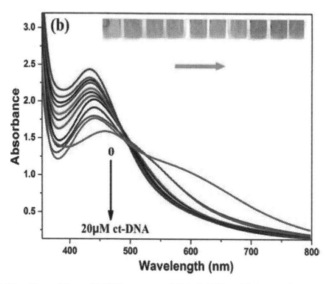

FIGURE 3.11 Quenching of LSPR spectra of CA-AgNPs with increasing concentration of ctDNA.

Source: Reproduced with permission from Ref. [57]. © 2019 Elsevier.

Another consequent application of plasmonic NPs is in the diagnosis of health issues [58]. Metal NPs have a significant part in the progression of new imaging techniques. Because of their effective light-scattering properties the noble metal NPs act as contrast agents for optical imaging. For example, targeted gold NPs can be used to deliver high-contrast images of cancer cells using diverse optical techniques, including one- and two-photon fluorescence imaging, dark-field microscopy, reflectance confocal microscopy, and others.

Plasmonic NPs (Ag and AuNPs) became an essential part of targeted drug delivery application in biomedical fields [58, 59]. Enhanced permeability and retention is a process by which NP accumulates passively at tumor sites. This will provide the possibility of targeted drug delivery or photo thermally destroy the tumor cells when the target is reached. Also, this NP can prevent the degradation of drugs inside the body before they reached the target (therapeutic timings control). Also, functionalization of the metal NPs with suitable material will increase the efficiency of targeted drug delivery. In order to enhance the quality of the treatment and narrow down the side effects, the drug release can be controlled by external stimuli such as light. For this technique light-triggered nanocarriers have developed and this can be used analogous to photothermal therapy [28]. Gold and silver nanostructures are developed as nanocarriers and sometimes the NPs will act as active material that helps the bound drug to release upon irradiated with a proper laser beam.

Radiative decay engineering is the field having a close connection with controlling the radiative decay properties [60, 61]. Plasmonics NPs may be used in this field of radiative decay engineering. For achieving the enhanced quantum yield and increased quenching of fluorescence, the distance of the fluorophore relative to the surface can be changed and different sized particles can be used. The intrinsic radiative decay rate of the fluorophore can be increased by the interaction between a fluorophore and a metal nanostructure. When the distance between the metal surface and fluorophore is <5 nm, the metal-induced quenching dominates. But if the distances increased from 5 to 20 nm, either field enhancement dominates or radiative rate. Silver films containing silver metal of circular domains (approx. 2 nm diameters) are repeatedly employed to examine metal-induced enhancement of fluorescence and for SERS. When deposited on silver island films, the fluorescence intensity of the chelate increased while the lifetime decreased. Photostability of dyes can be improved by using the application of plasmonic-induced lifetime shortening. If a molecule spends more time in the excited state, there is a possibility of photothermal destruction of the molecule. The major problem in fluorescence-based sensing and imaging is the photo thermal destruction of dyes. But when the lifetime shortens, this possibility photo bleaching of dyes will be significantly reduced. This can be achieved by using plasmonics. Hence a significant reduction in the lifetime along with a quantum yield enhancement can highly reduce photobleaching together with a great increase in the detection limit.

3.3 FUTURE PERSPECTIVE

The optical properties of noble metal NPs create exciting opportunities for new technologies. We had presented here a general readership on optical properties of noble metal NPs and its rich varieties of applications. Plasmonics is a well-established and increasingly acceptable research field that developed from the study of optical phenomena related to the electromagnetic response of metals. The field plasmonics builds a bridge between two different length scales by converging light into subwavelength volumes. The building bricks of this arch are metallic NPs. Metal NPs have distinctively characteristic electrical and optical properties. The possibility of tailoring the optical properties of metal NPs has made it a promising candidate in large verities of applications such as pharmaceutical, biomedical fields, and photovoltaic devices. The metal NPs may elucidate the future of tomorrow through many fields as distinct as, new possibilities to treat cancer, prediction of new superfast computer chips, ultrasensitive molecular detectors or the ability to make things invisible with negative-refraction materials. Also, the potentiality to manipulate and control light on the nanometer scale opens up a new world of possible applications such as quantum optics, data storage, optoelectronics, optical data processing, quantum information processing, and photovoltaics. Confidingly metal NPs could open a new province of research and application.

KEYWORDS

- silver nanoparticles
- gold nanoparticles
- SPR
- LSPR
- LSPR applications
- metal NPs

REFERENCES

1. Feynman, R. P. There's Plenty of Room at the Bottom. *Micromechanics MEMS Class. Semin. Pap. to 1990* **1997**, 2–9. https://doi.org/10.1109/9780470545263.sect1.
2. Heiligtag, F. J.; Niederberger, M. The Fascinating World of Nanoparticle Research. *Mater. Today* **2013**, *16* (7–8), 262–271. https://doi.org/10.1016/j.mattod.2013.07.004.
3. Jeevanandam, J.; Barhoum, A.; Chan, Y. S.; Dufresne, A.; Danquah, M. K. Review on Nanoparticles and Nanostructured Materials: History, Sources, Toxicity and Regulations. *Beilstein J. Nanotechnol.* **2018**, *9* (1), 1050–1074. https://doi.org/10.3762/bjnano.9.98.
4. Shnoudeh, A. J.; Hamad, I.; Abdo, R. W.; Qadumii, L.; Jaber, A. Y.; Surchi, H. S.; Alkelany, S. Z. Synthesis, Characterization, and Applications of Metal Nanoparticles. **2019**. https://doi.org/10.1016/B978-0-12-814427-5.00015-9.
5. Cid, A.; Simal-Gandara, J. Synthesis, Characterization, and Potential Applications of Transition Metal Nanoparticles. *J. Inorg. Organomet. Polym. Mater.* **2020**, *30* (4), 1011–1032. https://doi.org/10.1007/s10904-019-01331-9.
6. Chouhan, N. Silver Nanoparticles: Synthesis, Characterization and Applications. *Silver NanoparticlesFabr. Charact. Appl.* **2018**. https://doi.org/10.5772/intechopen.75611.
7. Hawaiz, F. E.; Raheem, D. J.; Samad, M. K. Synthesis and Characterization of Some New Azo-Imine Dyes and Their Applications. *J. Zankoy Sulaimani A* **2016**, *18* (3), 25–36. https://doi.org/10.17656/jzs.10532.
8. Xuan, T.; Technology, P.; Xuan, T. Metallic Nanoparticles: Synthesis, Characterisation and Application Nguyen Hoang Luong, Nguyen Ngoc Long , Le Van Vu and Nguyen Hoang Hai Phan Tuan Nghia and Nguyen Thi Van Anh. **2011**, *8*, 227–240.
9. Khan, I.; Saeed, K.; Khan, I. Nanoparticles: Properties, Applications and Toxicities. *Arab. J. Chem.* **2017**. https://doi.org/10.1016/j.arabjc.2017.05.011.
10. Luong, N. H.; Long, N. N.; Vu, L. Van; Hai, N. H.; Nghia, P. T.; Van Anh, N. T. Metallic Nanoparticles: Synthesis, Characterisation and Application. *Int. J. Nanotechnol.* **2011**, *8* (3–5), 227–240. https://doi.org/10.1504/IJNT.2011.038201.
11. Zhang, X. F.; Liu, Z. G.; Shen, W.; Gurunathan, S. Silver Nanoparticles: Synthesis, Characterization, Properties, Applications, and Therapeutic Approaches. *Int. J. Mol. Sci.* **2016**, *17* (9). https://doi.org/10.3390/ijms17091534.
12. Venkatesh, N. Metallic Nanoparticle: A Review. *Biomed. J. Sci. Tech. Res.* **2018**, *4* (2), 3765–3775. https://doi.org/10.26717/bjstr.2018.04.0001011.
13. Singh, P.; Kim, Y. J.; Zhang, D.; Yang, D. C. Biological Synthesis of Nanoparticles from Plants and Microorganisms. *Trends Biotechnol.* **2016**, *34* (7), 588–599. https://doi.org/10.1016/j.tibtech.2016.02.006.
14. Warrier, P.; Teja, A. Effect of Particle Size on the Thermal Conductivity of Nanofluids Containing Metallic Nanoparticles. *Nanoscale Res. Lett.* **2011**, *6* (1), 1–6. https://doi.org/10.1186/1556-276X-6-247.
15. Schmid, G. Metal Nanoparticles: Electronic Properties, Bioresponse, and Synthesis Update Based on the Original Article by Günter Schmid, Encyclopedia of Inorganic Chemistry Second Edition, 2005, John Wiley & Sons, Ltd . *Encycl. Inorg. Bioinorg. Chem.* **2012**. https://doi.org/10.1002/9781119951438.eibc0284.pub2.
16. Guo, D.; Xie, G.; Luo, J. Mechanical Properties of Nanoparticles: Basics and Applications. *J. Phys. D. Appl. Phys.* **2014**, *47* (1). https://doi.org/10.1088/0022-3727/47/1/013001.

17. Yamamoto, Y.; Miura, T.; Nakae, Y.; Teranishi, T.; Miyake, M.; Hori, H. Magnetic Properties of the Noble Metal Nanoparticles Protected by Polymer. *Phys. B Condens. Matter* **2003**, *329–333* (II), 1183–1184. https://doi.org/10.1016/S0921-4526(02)02102-6.

18. Republic, C. Surface Plasmon Resonance (SPR) Sensors. **2006**, No. July, 45–67.

19. Homola, J.; Yee, S. S.; Gauglitz, G. Surface Plasmon Resonance Sensors: Review. *Sensors Actuators, B Chem.* **1999**, *54* (1), 3–15. https://doi.org/10.1016/S0925-4005(98)00321-9.

20. Kim, J.; Globerman, S. Physical Distance vs. Clustering as Influences on Contracting Complexity for Biopharmaceutical Alliances. *Ind. Innov.* **2017**, *4*, 1–28. https://doi.org/10.1080/13662716.2017.1395730.

21. Lienau, C.; Noginov, M. A.; Lončar, M. Light-Matter Interactions at the Nanoscale. *J. Opt.* **2014**, *16* (11), 13–15. https://doi.org/10.1088/2040-8978/16/11/110201.

22. Sekar, R. B.; Periasamy, A. Fluorescence Resonance Energy Transfer (FRET) Microscopy Imaging of Live Cell Protein Localizations. *J. Cell Biol.* **2003**, *160* (5), 629–633. https://doi.org/10.1083/jcb.200210140.

23. Wannemacher, R.; Heber, J. Cooperative Emission of Photons by Weakly Coupled Chromium Ions in Al2O3. *Zeitschrift für Phys. B Condens. Matter* **1987**, *65* (4), 491–501. https://doi.org/10.1007/BF01303771.

24. Derrien, T. J. Y.; Itina, T. E.; Torres, R.; Sarnet, T.; Sentis, M. Possible Surface Plasmon Polariton Excitation under Femtosecond Laser Irradiation of Silicon. *J. Appl. Phys.* **2013**, *114* (8). https://doi.org/10.1063/1.4818433.

25. Coronado, E. A.; Encina, E. R.; Stefani, F. D. Optical Properties of Metallic Nanoparticles: Manipulating Light, Heat and Forces at the Nanoscale. *Nanoscale* **2011**, *3* (10), 4042–4059. https://doi.org/10.1039/c1nr10788g.

26. Al-Ghamdi, H. S.; Mahmoud, W. E. One Pot Synthesis of Multi-Plasmonic Shapes of Silver Nanoparticles. Mater. Lett. 2013, 105, 62–64. https://doi.org/10.1016/j.matlet.2013.04.086.

27. Her, S.; Jaffray, D. A.; Allen, C. Gold Nanoparticles for Applications in Cancer Radiotherapy: Mechanisms and Recent Advancements. Adv. Drug Deliv. Rev. 2017, 109, 84–101. https://doi.org/10.1016/j.addr.2015.12.012.

28. Huang, X.; El-Sayed, M. A. Gold Nanoparticles: Optical Properties and Implementations in Cancer Diagnosis and Photothermal Therapy. J. Adv. Res. 2010, 1 (1), 13–28. https://doi.org/10.1016/j.jare.2010.02.002.

29. Ogarev, V. A.; Rudoi, V. M.; Dement'eva, O. V. Gold Nanoparticles: Synthesis, Optical Properties, and Application. *Inorg. Mater. Appl. Res.* **2018**, *9* (1), 134–140. https://doi.org/10.1134/S2075113318010197.

30. Carrillo-Cazares, A.; Jiménez-Mancilla, N. P.; Luna-Gutiérrez, M. A.; Isaac-Olivé, K.; Camacho-López, M. A. Study of the Optical Properties of Functionalized Gold Nanoparticles in Different Tissues and Their Correlation with the Temperature Increase. *J. Nanomater.* **2017**, *2017* (3). https://doi.org/10.1155/2017/3628970.

31. Evans, E. R.; Bugga, P.; Asthana, V.; Drezek, R. Metallic Nanoparticles for Cancer Immunotherapy. Mater. Today 2018, 21 (6), 673–685. https://doi.org/10.1016/j.mattod.2017.11.022.

32. de Aberasturi, D. J.; Serrano-Montes, A. B.; Liz-Marzán, L. M. Modern Applications of Plasmonic Nanoparticles: From Energy to Health. *Adv. Opt. Mater.* **2015**, *3* (5), 602–617. https://doi.org/10.1002/adom.201500053.

33. Temple, T. L.; Bagnall, D. M. Optical Properties of Gold and Aluminium Nanoparticles for Silicon Solar Cell Applications. *J. Appl. Phys.* **2011**, *109* (8), 1–13. https://doi.org/10.1063/1.3574657.

34. Chander, N.; Khan, A. F.; Thouti, E.; Sardana, S. K.; Chandrasekhar, P. S.; Dutta, V.; Komarala, V. K. Size and Concentration Effects of Gold Nanoparticles on Optical and Electrical Properties of Plasmonic Dye Sensitized Solar Cells. *Sol. Energy* **2014**, *109*, 11–23. https://doi.org/10.1016/j.solener.2014.08.011.

35. Communications, P.; Manuscript, A. Ce Pte an Us Absorption Enhancement in Thin Film Solar Cells With. **2018**.

36. Ganeshan, D.; Xie, F.; Sun, Q.; Li, Y.; Wei, M. Plasmonic Effects of Silver Nanoparticles Embedded in the Counter Electrode on the Enhanced Performance of Dye-Sensitized Solar Cells. *Langmuir* **2018**, *34* (19), 5367–5373. https://doi.org/10.1021/acs.langmuir.7b03086.

37. Fainstein, Alejandro & Calvo, Ernesto & Abdelsalam, Mamdouh & Bartlett, Philip. Incident Wavelength Resolved Resonant SERS on Au Sphere Segment Void (SSV) Arrays. J. Phys. Chem. C. **2012**, *116*, 3414–3420. 10.1021/jp211049u.

38. Benelli, G. Gold Nanoparticles—against Parasites and Insect Vectors. Acta Trop. 2018, 178, 73–80. https://doi.org/10.1016/j.actatropica.2017.10.021.

39. Kelly, K. L.; Coronado, E.; Zhao, L. L.; Schatz, G. C. The Optical Properties of Metal Nanoparticles: The Influence of Size, Shape, and Dielectric Environment. *J. Phys. Chem. B* **2003**, *107* (3), 668–677. https://doi.org/10.1021/jp026731y.

40. Cennamo, N.; Zeni, L.; Catalano, E.; Arcadio, F.; Minardo, A. Refractive Index Sensing through Surface Plasmon Resonance in Light-Diffusing Fibers. *Appl. Sci.* **2018**, *8* (7). https://doi.org/10.3390/app8071172.

41. Tuersun, P.; Yusufu, T.; Yimiti, A.; Sidike, A. Refractive Index Sensitivity Analysis of Gold Nanoparticles. *Optik* **2017**, *149* (September), 384–390. https://doi.org/10.1016/j.ijleo.2017.09.058.

42. Prasher, P.; Sharma, M.; Mudila, H.; Gupta, G.; Sharma, A. K.; Kumar, D.; Bakshi, H. A.; Negi, P.; Kapoor, D. N.; Chellappan, D. K.; et al. Emerging Trends in Clinical Implications of Bio-Conjugated Silver Nanoparticles in Drug Delivery. *Colloids Interface Sci. Commun.* **2020**, *35* (December 2019), 100244. https://doi.org/10.1016/j.colcom.2020.100244.

43. Majdalawieh, A., Kanan, M. C., El-Kadri, O., and Kanan, S. M. Recent Advances in Gold and Silver Nanoparticles: Synthesis and Applications. J. Nanosci. Nanotechnol. **2014**, *14*(7), 4757–4780.

44. Lee, K. S., andEl-Sayed, M. A. Gold and Silver Nanoparticles in Sensing and Imaging: Sensitivity of Plasmon Response to Size, Shape, and Metal Composition. J. Phys. Chem. B **2006**, *110*(39), 19220–19225.

45. Raj, D. R., Prasanth, S., Vineeshkumar, T. V., and Sudarsanakumar, C. mmonia sensing properties of tapered plastic optical fiber coated with silver nanoparticles/PVP/PVA hybrid. Opt. Commun. **2015**, *340*, 86–92.

46. Bülbül, G.; Hayat, A.; Andreescu, S. Portable Nanoparticle-Based Sensors for Food Safety Assessment. *Sensors* **2015**, *15* (12), 30736–30758. https://doi.org/10.3390/s151229826.

47. Zheng, X.; Peng, Y.; Cui, X.; Zheng, W. Modulation of the Shape and Localized Surface Plasmon Resonance of Silver Nanoparticles via Halide Ion Etching and Photochemical Regrowth. *Mater. Lett.* **2016**, *173*, 88–90. https://doi.org/10.1016/j.matlet.2016.02.120.

48. Tambe, A.; Kumbhaj, S.; Lalla, N. P.; Sen, P. LSPR Based Fiber Optic Sensor for Fluoride Impurity Sensing in Potable Water. *J. Phys. Conf. Ser.* **2016**, *755* (1). https://doi.org/10.1088/1742-6596/755/1/012058.

49. Willets, K. A.; Van Duyne, R. P. Localized Surface Plasmon Resonance Spectroscopy and Sensing. *Annu. Rev. Phys. Chem.* **2007**, *58* (1), 267–297. https://doi.org/10.1146/annurev.physchem.58.032806.104607.

50. Gouvêa, P. M. P.; Carvalho, I. C. S.; Jang, H.; Cremona, M.; Braga, A. M. B.; Fokine, M. Characterization of a Fiber Optic Sensor Based on LSPR and Specular Reflection. *Opt. InfoBase Conf. Pap.* **2010**, 4–6. https://doi.org/10.1364/sensors.2010.stua4.

51. Tian, S., Ding, X., Zeng, D., Wu, J., Zhang, S., and; Xie, C. A low temperature gas sensor based on Pd-functionalized mesoporous SnO_2 fibers for detecting trace formaldehyde. RSC Adv. **2013**, *3*(29), 11823–11831.

52. El Sayed, S., Pascual, L., Licchelli, M., Martínez-Máñez, R., Gil, S., Costero, A. M., & Sancenón, F. Chromogenic detection of aqueous formaldehyde using functionalized silica nanoparticles. ACS Appl. Mater. Interfaces, **2016** *8*(23), 14318–14322. https://doi.org/10.1021/acsami.6b03224

53. Lakade, A. J., Sundar, K., & Shetty, P. H. Gold Nanoparticle-Based Method for Detection of Calcium Carbide in Artificially Ripened Mangoes (*Magnifera indica*). Food Additiv. Contam. A **2018**, *35*(6), 1078–1084.

54. Fahimi-Kashani, N., & Hormozi-Nezhad, M. R. Gold-Nanoparticle-Based Colorimetric Sensor Array for Discrimination of Organophosphate Pesticides. Anal. Chem. **2016**, *88*(16), 8099–8106.

55. Raj, D. R., Prasanth, S., Vineeshkumar, T. V., and Sudarsanakumar, C. Surface plasmon resonance based fiber optic sensor for mercury detection using gold nanoparticles PVA hybrid. Opt. Commun. **2016**, *367*, 102–107.

56. Thomas, R. K.; Sukumaran, S.; Prasanth, S.; Sudarsanakumar, C. Revealing the Interaction Strategy of Diosmin Functionalized Gold Nanoparticles with CtDNA: Multi-Spectroscopic, Calorimetric and Thermodynamic Approach. J. Lumin. **2019**, *205*. https://doi.org/10.1016/j.jlumin.2018.09.004.

57. Thomas, R. K.; Sukumaran, S.; Sudarsanakumar, C. An Insight into the Comparative Binding Affinities of Chlorogenic Acid Functionalized Gold and Silver Nanoparticles with CtDNA along with Its Cytotoxicity Analysis. J. Mol. Liq. **2019**, 287, 110911. https://doi.org/10.1016/j.molliq.2019.110911.

58. Doane, T. L., & Burda, C. The Unique Role of Nanoparticles in Nanomedicine: Imaging, Drug Delivery and Therapy. Chem. Soc. Rev. **2012**, *41*(7), 2885. doi:10.1039/c2cs15260f

59. Parveen, S., Misra, R., and Sahoo, S. K. Nanoparticles: A Boon to Drug Delivery, Therapeutics, Diagnostics and Imaging. Nanomed. Nanotechnol. Biol. Med. **2012**, *8*(2), 147–166. doi:10.1016/j.nano.2011.05.016

60. Badugu, R., Nowaczyk, K., Descrovi, E., and Lakowicz, J. R. Radiative Decay Engineering 6: Fluorescence on One-Dimensional Photonic Crystals. Anal. Biochem. **2013**, *442*(1), 83–96. doi:10.1016/j.ab.2013.07.021

61. Lakowicz, J. R. Radiative Decay Engineering: Biophysical and Biomedical Applications. Analytical Biochemistry, **2001**, *298*(1), 1–24. doi:10.1006/abio.2001.5377

CHAPTER 4

Green Carbon Dots as Optical Sensors for Metal Ions

MAMATHA SUSAN PUNNOOSE and BEENA MATHEW*

School of Chemical Sciences, Mahatma Gandhi University, Kottayam, India

*Corresponding author. E-mail: beenamscs@gmail.com

ABSTRACT

Green fluorescent carbon dots (CDs) have gained ample interest due to their less toxicity and low cost. They constitute a new class of nanocarbon materials prepared by simple green synthetic routes. The optical properties of the prepared CDs deeply depend on the surface passivation and functionalization provided by the green carbon sources on the surface of CDs. CDs have found profound applications in various fields such as sensing, bioimaging, catalysis, and drug delivery. In this chapter, we discuss briefly on the synthesis of CDs by means of different green sources, their optical properties, principles of sensing, heavy metal pollution, and also the application of CDs as fluorescent probes for the detection of different heavy metal ions.

4.1 INTRODUCTION

The advancement in carbon nanomaterials such as carbon tubes, carbon fibers, graphene, and fullerenes have now become essential ingredients of material science due to their wide optical and electrical properties, high strength, stability, low weight, and flexibility. Carbon dots (CDs) are the newest member to the carbon family. They are fluorescent nanocarbon materials with a diameter between 1 and 10 nm. They were accidentally discovered during gel electrophoresis purification of single-walled CNTs in 2004 [1]. CDs found paramount use as biosensors [2], gene transmitters [3], drug carriers [4], and bioimaging probes [5] due to their distinct optical properties, good biocompatibility, low toxicity and facile synthesis using versatile sources. Their excellent fluorescent properties exhibited potential applications in the analytical fields like sensing, photodynamic therapy, catalysis, and bioimaging [6, 7].

Various strategies for the synthesis of fluorescent CDs involve hydrothermal, solvothermal, ultrasonication, electrochemical, microwave, pyrolysis, simple heating, laser ablation, arc discharge, and chemical oxidation [8–13]. The various chemical precursors used were citric acid, ammonium citrate, ethylene glycol, benzene, phenylenediamine, phytic acid, EDTA, glucose, sucrose, folic acid, ascorbic acid, lactic acid, carbon soot, or activated carbon [14–24].

4.2   GREEN PRECURSORS

Most of the conventional synthetic approaches of CDs involve multistep operations, strong organic chemicals or solvents for posttreatment with surface passivating the nanosurface in order to improve water solubility, quantum yield, and luminescence properties. Recently, self-passivated CDs are reported by means of microwave, hydrothermal, and carbonization of green carbon precursors. Therefore one-pot green synthesis of CDs without any hazardous chemicals has attracted the attention of young scientists. The term "Green" CDs refers to the carbon quantum dots synthesized using "green precursors" as the source of carbon. And "green precursors" are substances that are either natural or are derivatives of renewable natural products or processes [25]. Green chemistry approach attempts to achieve a one-pot, cost-effective, environmental friendly method for the large scale production of CDs by the use of inexpensive renewable precursors and abundant biomass. The deficiency in control of lateral dimensions, surface chemistry, low quantum yield, and novel applications was overcome by the green CDs.

In comparison with toxic organic molecules, the use of green precursors is advantageous that they do not require any further reactants for the purpose of doping or surface passivation. The various carbohydrates, proteins, and biomolecules present in the natural source provide self-passivation to green CDs. Phytoconstituents in the green source is not only responsible for the availability of functional groups on the nanocarbon surface but also their reactivity. CDs synthesis, functionalization, and passivation without the addition of any chemicals can be achieved by the use of these green carbon sources. Different fruits, vegetables, spices, plants, biowaste materials, and human derivatives act as the carbon source for the synthesis of CDs. *Musa acuminate* [26] fruit is used for the synthesis of CDs with an 8.95% quantum yield. CDs derived by hydrothermal carbonization of apple at 150 °C, contained surface functional groups such as hydroxyl, amino, keto, and carboxylic acid with a quantum yield of 4.27% [27]. Papaya [28], peach [29], lemon peel [30] were found to be efficient green precursors. The mixing of ammonia as a nitrogen source with fruit extract of *Hylocereus undatus* resulted in CDs with 5.82% nitrogen content [31]. CDs obtained by the hydrothermal treatment of crushed cabbage followed by carbonization at 140 °C for 5 h show a quantum yield of 16.5% [32]. A high quantum yield of 19.3% was obtained for sweet red pepper derived CDs via the hydrothermal method [33]. Willow leaves [34] and rose petals [35] are reported plant sources for CDs preparation.

In recent research, tomatoes [36], peanut shells [37], citrus peel [38], biowaste [39], cellulose [39], egg yolk [40], alovera [41], potatoes [42], garlic [43], shrimps [44], hair [45], cucumber [46], and wheat straw [47] have been employed as a carbon source for the preparation of CDs. The degree of surface functionalization and variations in optical properties of CDs depends on the green precursor used. This highlights the importance of exploring various natural precursors for green CDs synthesis.

4.3 OPTICAL PROPERTIES

4.3.1 ABSORBANCE

The UV–visible absorption spectra of green CDs showed a prominent peak in the UV region with a tail extending into the visible region. The absorption band located at around 270 nm is associated with Л–Л* transitions of C=C in the carbon core and the shoulder peak around 300–330 nm is attributed to *n*–Л* transitions of carbonyl groups on the surface of CDs (Figure 4.1). These functional groups are available on the CDs surface because the green sources mainly consist of carbon and oxygen-abundant materials [48–50].

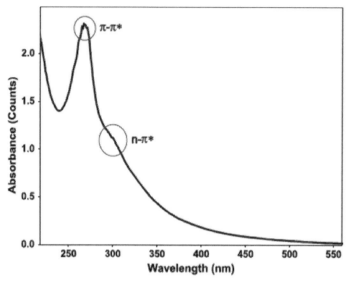

FIGURE 4.1   The absorption spectrum of CDs.

Source: Reproduced from Atchudan et al. [50] with permission from Elsevier. © 2016.

4.3.2 FLUORESCENCE

Fluorescence is the most interesting optical property of green CDs. The different precursors cause the synthesis of CDs with different sizes that leads to variations in the photoluminescence (PL) emission spectra. CDs often display excitation dependent emission spectra. The emission peak of *Prunus persica* [50] derived CDs shifted from 400 to 426 nm with the variation of excitation wavelength from 310 to 350 nm (Figure 4.2). And a maximum fluorescence was obtained at excitation at 325 nm. Hydrothermally synthesized CDs using *Phyllanthus emblica* [51] showed a bathochromic shift of the emission peaks upon variation of excitation wavelength from 320 to 400 nm. The luminescent CDs derived from coriander leaves showed an excitation-dependent emission character [52]. With an increase of excitation wavelength from 320 to 480 nm resulted in a redshift of fluorescence peak from 400 to 510 nm with a gradual decrease in emission intensity. CDs derived from tea shows two peaks in PL spectra [53]. Both peaks show an excitation dependent emission upon excitation from 320 to 400 nm. Papaya-based CDs exhibited wavelength tuned emission [28]. Excitation-independent emission was observed in the case of lemon juice derived CDs [54].

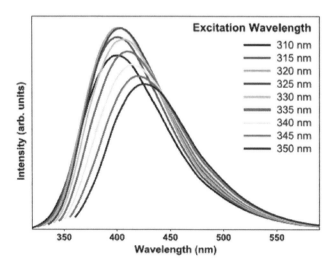

FIGURE 4.2   Excitation-dependent PL spectra of *Prunus persica* derived CDs.

Source: Reproduced from Atchudan et al. [50] with permission from Elsevier. © 2016.

4.3.3   DOPING

The fluorescence ability can be improved by doping of electron-rich hetero-atoms like nitrogen, boron, phosphorus, and sulfur on the CDs surface. Doping of heteroatoms on CDs improves active sites, fluorescence quantum yield, chemical, optical, and electronic properties due to the availability of extra electrons on the heteroatom. Thus heteroatom-doped CDs become significant for broad applications including heavy metal sensing, catalysis, optoelectronic, and energy storage devices [55, 56]. Nitrogen-doped carbon dots (N-CDs) has gained much attention as a potential method to improve quantum yield and emission properties. Aqueous ammonia is used as the nitrogen dopant for the *Prunus avium*-derived CDs to obtain a quantum yield of 13% [57]. Increment in quantum yield of N-CDs derived from peach increased from 5.31% to 28.46% by the surface passivation using ethylenediamine as the dopant [29]. Enhancement of the optical properties of green CDs by self-passivation and heteroatom doping led researchers to attempt codoping of multiple elements in CDs. In order to provide magnetic properties to carb shell derived CDs, metal ions say Mn^{2+}, Eu^{3+}, and Gd^{3+} were used as dopants possessing five, six, and seven unpaired electrons, respectively [58]. A quantum yield of 19.84% was obtained for Gd^{3+}- doped, followed by Eu^{3+}- (14.97%) and Mn^{2+}-doped (12.86%) CDs. Synthesis of heteroatom-doped CDs by a microwave-assisted method using feathers as the source has been reported [59]. The elemental composition of as-synthesized CDs showed the presence of 48.4% carbon, 16.3% nitrogen, 1.90% sulfur, and 33.3% oxygen. The quantum yield was found to be 17.1%. S and N codoped CDs were obtained by heating of water chestnut and onion together in an autoclave at 180 °C [60]. Sun et al. used hair fibers for the synthesis of nitrogen and sulfur co-doped CDs [61]. Garlic acts as the source of carbon, nitrogen, and sulfur for synthesis of N, S co-doped CDs, containing 6.9% nitrogen and 1% sulfur [43].

4.4   HEAVY METAL POLLUTION

The term heavy metal is used to refer to any metallic elements that have a relatively high density and is toxic at low concentrations. Toxic heavy metal ions are serious environmental pollutants due to their latent toxic effects, persistence in the environment, and bioaccumulative nature. Natural phenomena such as weathering and volcanic eruptions have significantly contributed to heavy metal pollution [62]. Most environmental contamination and human exposure result from anthropogenic activities such as mining and smelting operations; domestic and agricultural use of metals and metal-containing compounds [63]. Industrial sources include electroplating, textile dyeing, petroleum combustion, wood preservation, paper processing, and leather tanning [64]. Environmental contamination can also occur through metal corrosion, atmospheric deposition, soil erosion of metal ions, leaching of heavy metals, sediment resuspension, and metal evaporation from water resources to soil and groundwater [65]. Contamination of aquatic and terrestrial ecosystems with heavy metal ions is posing environmental and health risks. Being persistent pollutants, metal ions accumulate in the environment and consequently can reach the higher organism level by means of the food chain contamination. Accumulation of potentially toxic heavy metals like arsenic, cadmium, chromium, lead, and mercury rank high among the priority metals that are of a potential health threat to both humans and animals [66–70]. Therefore, the development of versatile systems for the continuous monitoring of trace heavy metal ions is very crucial in the current society.

Several analytical techniques such as inductively coupled plasma mass spectrometry, atomic absorption spectrometry, voltammetry, spectrophotometry, and high-performance liquid chromatography have been demonstrated for the quantification of metal ions in various aqueous samples [71–75].

Although these techniques exhibit good sensitivity, unfortunately these often required sophisticated operations, extra processing in terms of surface modification, and tedious sample preparation. Recently, biocompatible carbon nanomaterials-based fluorescent sensors have opened up a new avenue for the detection of heavy metal ions. CDs are potable detectors because of their large abundance, stability, low toxicity, and relatively low cost. Fluorescence-based green sensor has gained increasing attention due to its simplicity, good sensitivity with high selectivity, rapid detection, minimal interference effect, and inexpensive nature [76].

4.5 SENSING PRINCIPLES

Following are the different mechanisms responsible for the fluorescence changes of green CDs, which have been utilized in sensing applications.

4.5.1 PHOTO-INDUCED ELECTRON TRANSFER (PET)

PET is one of the most commonly used mechanisms to describe quenching phenomena because functional groups on the surface of CDs allow their interaction with target analytes. It involves a formation of a complex between electron donor and electron acceptor. The excited complex can return to the ground state without the release of photons. PET processes can be divided into two, oxidative PET and reductive PET. In reductive PET, CDs act as an electron acceptor for the target analyte. Electrons are transferred from the highest occupied molecular orbital (HOMO) of the target analytes to the HOMO of CDs. In contrast, CDs serve as the electron donor in oxidative PET processes. The excited electrons are transferred from the lowest unoccupied molecular orbitals of CDs to the HOMO of the electron-acceptor analyte.

4.5.2 FORSTER RESONANCE ENERGY TRANSFER (FRET)

FRET is another mechanism commonly used in CD-based fluorescence sensors. FRET is a nonradiative energy transfer process originating from dipole–dipole interactions between donor and acceptor molecules [77]. In a FRET process, an initially light excited donor molecule returns to the ground state by simultaneous transferring of the energy to excite the acceptor molecule. Rate of energy transfer in FRET process depends on the essential overlap of the donor emission spectrum with the acceptor absorption spectrum. The FRET efficiency is inversely proportional to the sixth power of the separation distance between the donor and acceptor. Hence, the donor and acceptor should be in close proximity. Dynamic quenching dominates in the quenching process, where the fluorescence lifetime decreases with increasing concentration of quencher.

4.5.3   INNER FILTER EFFECT (IFE)

IFE is a typical nonirradiation energy conversion model in spectrofluorometry, which only requires a good spectral overlap between the absorption band of the absorber and the excitation/emission band of the fluorophore. The observed fluorescence intensity is proportional to the intensity of the exciting light, and the quantum yield is lowest for an infinitely dilute solution. This is called an inner filter effect. Attenuation due to absorption light or absorption of the emitted light is sometimes called the primary or secondary inner filter effects. In both cases, IFE leads to a reduction in the PL intensity without affecting the decay time.

4.6   METAL-ION SENSORS BASED ON GREEN CARBON DOTS

The very finite size, high surface to volume ratios, and facile surface functionalization make CDs very reactive. Their excellent PL properties, water solubility, high fluorescence quantum yields, and availability of functional groups for interaction with analytes have rendered green CDs as an excellent candidate for the detection of heavy metals. The binding and interaction of CDs with the metal ions via surface bonding cause variations in fluorescence intensity, lifetime, and provides a measurable response signal corresponding to the sensitivity of analytes [79]. The following section outlines recent studies related to different green CDs that facilitate the heavy metal detection.

4.6.1   FERRIC IONS

Fe^{3+} ions are indispensable for a large number of living systems and play an important role in many physiological and pathological processes including cellular metabolism, enzyme catalysis, oxygen transport, and RNA synthesis [80]. The excessive intake of iron can lead to diseases such as hemochromatosis, Alzheimer, and Parkinson. The development of rapid and efficient detection for Fe^{3+} ions is thus valuable. A novel fluorescent CDs probe for sensing of Fe^{3+} ions was hydrothermally synthesized using Rose-heart radish as the carbon source [81]. The synthesized CDs are considered to be composed of an amorphous or crystalline core of chiefly sp^2 carbon and an oxidized carbon surface consisting of carboxyl and hydroxyl groups. A maximum strong emission at a wavelength of 420 nm was obtained at an excitation wavelength of 330 nm. With the increase of Fe^{3+} concentration, a progressive decrease in the blue emission corresponding to quenching of CDs fluorescence was obtained. The quenching resulted from the strong interactions of Fe^{3+} ions with the phenolic hydroxyl groups on the surface of CDs, which aids the transfer of photoelectrons from the excited state of CDs to the vacant d orbitals of Fe^{3+} ions [15]. LOD was determined to be as 0.13 µM, over a linear concentration range between 0.02 and 40 µM. These CDs are practically used for the environmental water sample analysis that resulted in satisfactory recoveries. CDs prepared by the hydrothermal carbonization of *Malpighia emarginata* fruit juice at 180 °C for 18 h exhibited a green fluorescence of emission wavelength 459 nm [82]. The as-prepared CDs were used for detection of Fe^{3+} ions in the range of 0.001–0.012 mol L^{-1}. The fabricated CDs demonstrated quenching of PL intensity upon the increase of Fe^{3+} ions concentration via efficient charge and energy transfer process. Fluorescence quenching may contribute to nonradiative electron transfer that involves partial transfer for an electron in the excited state to the d orbital of Fe^{3+}. Ag^+ ions also showed a quenching effect but much lesser than ferric ions. The aqueous solution of CDs derived from coriander leaves without using any passivating agents showed a bright green fluorescence [52]. The two peaks at 273 and 320 nm corresponding to $\pi–\pi^*$ transition of C=C bonds and $n–\pi^*$ transition of C=O bonds were observed in the absorption spectra. Out of the 12 different metal ions studied, Ag^+, Cu^{2+}, Hg^{2+}, and Fe^{2+} ions show a slight reduction in the fluorescence of CDs. Strongest quenching was obtained in the presence of ferric ions that shows the high selectivity of CDs toward Fe^{3+} ions. Quenching was due to exceptional coordination between Fe^{3+} ions and hydroxyl groups of CDs. The sensitivity study of CDs toward Fe^{3+} ions was carried out in the concentration range of 0–60 µM and the detection limit was determined to be 0.4 µM. This value is much lower than the permissible level (5.36 µM) of ferric ions in drinking water as per World Health Organization (WHO) [83]. The decrease of fluorescence intensity of honey-derived CDs [84] permits the ferric ions detection over a linear range of 5.0×10^{-9}–1.0×10^{-4} mol L^{-1}, with a detection limit of 1.7×10^{-9} mol L^{-1}. CDs synthesized from onion waste [85] could detect Fe^{3+} in the linear range of 0–20 µM with a LOD of 0.31 µM. Tea waste-derived CDs [86] show much more sensitivity toward ferric ions with a detection limit of 0.15 μM. Significant fluorescent quenching was observed for orient plane leaves-derived CDs [87] toward ferric ions in the dynamic range of 0–100 μM. Water dispersible CDs from *Lycii fructus* prepared via hydrothermal treatment showed a captivating sensitivity toward ferric ions over a linear range of 0–30 μM [79]. The estimated limit of detection value is 21 nM, which is lower than the maximum permissible limit of Fe^{3+} in drinking water. Quenching studies were done by PL emission spectra with an excitation wavelength of 350 nm. Fluorescence quenching of CDs induced by Fe^{3+} occurs due to the inner filter effect. Complex formation due to the coordination of vacant orbit of Fe^{3+} with hydroxyl groups on the CDs may accelerate the nonradiative electron/hole recombination through effective photoelectron or energy transfer, which results in the quenching effect. These CDs were also employed as fluorescent probes for the determination of Fe^{3+} in the urine samples, the water samples from the Yellow River and living Henrietta Lacks cells [80].

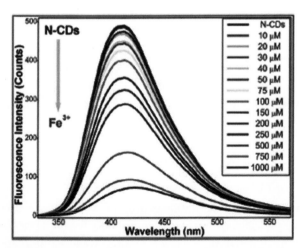

FIGURE 4.3   Fluorescence spectra of Prunus avium-derived N-CDs in the presence of various concentrations of ferric ions.

Source: Reproduced from Edison et al. [57] with permission from Elsevier. © 2016.

N-CDs synthesized from *Chionanthus retusus* [88] fruit using a one-pot hydrothermal carbonization method. On completion of the hydrothermal process, solution color changed from pale yellowish brown to dark brown due to the formation of N-CDs. The synthesized N-CDs showed high selectivity toward Fe^{3+}, over a linear concentration range between 0 and 2 μM and having a detection limit of 70 μM. The fluorescence quenching mechanism of N-CDs in the presence of Fe^{3+} involves higher affinity of ferric ions for nitrogen and oxygen on the surface of N-CDs and nonradiative electron transfer process between them. The oxygen abundant hydroxyl and carboxyl functional groups present on the N-CDs surface contribute to the water solubility and strong interaction with metal ions. Thus they can act as a fluorescent probe for the selective detection of Fe^{3+} ions. The high sensitivity of N-CDs is attributed to the formation of complexes due to the coordination between Fe^{3+} and the phenolic hydroxyl and amine groups on carbon nanostructure. After quenching, N-CDs remain aggregated due to effective complexation within ferric ions and N-CDs. *Magnolia liliiflora*-derived N-CDs shows a detection limit of 1.2 μM over a linear range of 1–1000 μM ferric ions [89]. Complete quenching resulted from the strong adsorption affinity and highly efficient nonradiative energy transfer during the complex formation between N-CDs and Fe^{3+} ions. Upon addition of ferric ions with *P. avium* fruit extract-derived N-CDs, nonradiative electron transfer from the N-CDs surface to metal ion causes the continuous quenching of fluorescence (Figure 4.3) [57]. The mechanism involves the formation of phenolic hydroxyl group complexation with ferric ions. And these N-CDs can sense ferric ions of concentrations even up to 0.96 μM (Figure 4.4). To date, iron is the most widely sensed metal ion by green CDs including papaya, [28] banana, [90] and sweet potatoes [91] having a detection limit of 0.29 μmol L^{-1}, 6.5×10^{-9} M, and 0.32 μM, respectively.

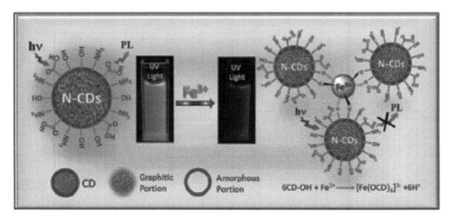

FIGURE 4.4   Plausible mechanism for the fluorescence quenching of Prunus avium-derived N-CDs by ferric ions.

Source: Reproduced from Edison et al. [57] with permission from Elsevier. © 2016.

4.6.2   MERCURY IONS

One of the most toxic heavy metal is mercury that causes brain damage, kidney problems, and deformity of limbs even at their very low concentration. Its inorganic form is highly toxic due to its water solubility. These mercury ions can be absorbed by the human body through the skin, respiratory, and gastrointestinal tissues. The US Environmental Protection Agency (EPA) has set a maximum allowable level of Hg^{2+} in drinking water to be 10 nM. Hair-derived [92] CDs show sensitivity toward Hg^{2+} with a linear response in the range of 0–75 µM showing a limit of detection of 10 nM. With an increase of Hg^{2+} concentration, CDs showed a dynamic quenching with an increase of fluorescence lifetime from 5.15 to 6.22 ns. Disappearance of the absorption spectra of CDs upon addition of mercuric ions indicates the existence of a static mechanism of quenching. Thus this system proves the coexistence of both static and dynamic quenching mechanisms. Yeng et al. used tea [53] as the carbon source for the synthesis of CDs for selective sensing of Hg^{2+} ions of LOD 1 nM and they showed real water sample applicability. *Hongcaitai*-derived CDs shows quenching of fluorescence by Hg^{2+} in PBS at a pH of 3 [93]. Mercuric ions interaction with functional groups on the surface of CDs offers nonradiative electron transfer between them. CDs show LOD of 1 nM. This sensor found practical applicability for the accurate detection of mercuric ions in river water. CDs prepared from *Tamarindus indica* [94] leaves are used as a sensitive probe for the turn-off sensing of Hg^{2+} with a detection limit of 6 nM in the dynamic range from 0 to 0.1 mM.

Lotus root-derived N-CDs show selectivity toward Hg^{2+} with a linear range from 0.1 to 60.0 µM and a detection limit of 18.7 nM [95]. High affinity and higher chelating kinetics of mercuric ions toward the surface function-alities of CDs contributes to static quenching. The reduction potentials of Hg^{2+} is in between the conduction band and valence band of CDs, induces photo-induced electron transfer from the conduction band to the mercuric ions. Hydrothermally prepared N-CDs using folic acid as both carbon and nitrogen source is used for the detection of Hg^{2+} [96]. Effective quenching of N-CDs was obtained with the gradual increase of Hg^{2+} concentration. The quenching mechanism involves nonradiative electron transfer interactions of N-CDs with Hg^{2+}. Mercuric ions persuade the conversion of a closed ring –CONH– functional group to an open ring that effactually contributes to fluorescence quenching. LOD was calculated to be 0.23 µM over a linear range of 0–25 µM. And the system is found to be successful for the determi-nation of real water samples. Feng's group [97] achieved a detection limit of 3 nM using strawberry-derived N-CDs for Hg^{2+} in a linear range from 10 nM to 50 μM. *Jinhua bergamot* [98] derived CDs acts as a dual sensor for Hg^{2+} and Fe^{3+} ions. Static quenching of CDs is observed for both ions. The detection limit was 5.5 nM for mercuric ions over the linear range between 0.01 and 100 mM, whereas for Fe^{3+} the LOD was 0.075 mM and the linear range was between 0.025 and 100 mM.

4.6.3 COPPER IONS

Exposure to Cu^{2+} beyond the tolerable limit is toxic to human cells and could cause many severe disorders to the kidney, atherosclerosis, infant lever damage, and Wilson's disease. Ultrasonic dispersion after the pyrolysis of *Eleusine coracana* [99] at 300 °C for 3 h in distilled water resulted in the formation of homogenous dark brown CDs solution. The fluorescence of resultant CDs was quenched by cupric ions in the linear range of 0 to 100 μM. The detection limit is estimated to be 10 nM. Upon quenching, fluorescence color of the solution was changed from dark blue to light blue. The chelation interaction of Cu^{2+} ions with the oxygen abundant carboxyl and hydroxyl groups at the surface of the CDs might have attributed to the quenching of the fluorescence through inner filter effect. This allows the CDs to serve as a fluorescent probe with excellent sensitivity and selectivity for Cu^{2+} through the formation of a complex. Cu^{2+} is a paramagnetic ion with d^9 system having one unoccupied d shell. Therefore Cu^{2+} act as an electron acceptor that can be easily adsorbed onto the surface of CDs that leads to the effective quenching of the fluorescence [100]. Ultrasmall fluorescent CDs are derived from green precursor *Acacia concinna* seeds via microwave heating for 2 min at 800 W [101]. Ultrasmall CDs act as turn off fluorescent probe for Cu^{2+}. With increasing concentration of cupric ions, bright blue emission of CDs was quenched to colorless. The formation of strong coordination complexes of Cu^{2+} ions with CDs causes to quench the fluorescence of CDs via photo-induced electron transfer mechanism. The detection limit is estimated to be 4.3 nM. Liu et al. developed a novel green hydrothermal method to prepare branced polyethylenimine (BPEI) capped CDs using bamboo leaves as the carbon precursor [102]. BPEI capped CDs showed a reliable and sensitive detection of Cu^{2+} with a limit of detection of 115 nM over a linear range of 0.333–66.6 μM. LOD is much lower than the maximum level of 20 μM of Cu^{2+} in drinking water permitted by the US EPA. The amine rich BPEI introduces a high density of positive charges on CDs' surface that makes them highly stable in water and various buffers.

The abundant nitrogen present on BPEI act as the effectual binding sites for cupric ions. Upon cupric ion addition, electrostatically capped BPEI on CDs surface is detached because of the strong binding between Cu^{2+} and BPEI. Polyamine desorption from the CDs surface resulted in the gradual blueshift in emission peaks. The quenching effect is attributed to the inner filter effect of cupric ion complexation with the functionalized CDs. Resultant CDs of hydrothermal synthesis of pear juice showed a quenching in fluorescence by cupric ions in the linear concentration range of 0.1–50 mg L^{-1} [103]. Also CDs from waste polyolefins residue showed a LOD of 6.33 nM over a linear range of 1–8 μM concentration of Cu^{2+} [104].

N-CDs derived from the one-pot hydrothermal treatment of natural green pakchoi without any additional solvents showed a quantum yield of 37.5% [105]. The high quantum yield is probably due to the existence of nitrogen-containing functional groups, which are excellent auxochromes. The sensitivity of the synthesized CD toward Cu^{2+} was revealed from the progressive decrease of fluorescence intensity with increasing concentration of Cu^{2+} between 0 and 100 nM. The LOD is found to be 9.98 nM. N-CDs prepared using lemon extract and L-arginine through thermal coupling shows a PL strong emission peak at 485 nm when excited at 340 nm [106]. N-CDs show a limit of detection of 0.047 μM over a linear range from 0 to 15 μM of Cu^{2+} ions. A continuous decrease of PL peak was observed upon the increase of Cu^{2+} concentration. Chelation of Cu^{2+} with the nitrogen at the CDs surface results in the formation of cupric ammine complex resulting in quenching of fluorescence via an inner filter effect. The complex formation can bring all reactants in a closer distance resulting in a decrease of PL luminescence. Paramagnetic quenching mechanism involved by the Cu^{2+} ions can effectively quench the fluorescence of the N-CDs by energy or electron transfer. CDs developed from prawn shells are used for fluorometric detection of cupric ions with a LOD of 5 nM. The prawn shells [100] were deproteinized and demineralized with sodium hydroxide and hydrochloric acid respectively to produce chitin. Chitosan was produced from chitin by deacetylation process involving sodium hydroxide. Chitosan upon hydrothermal treatment produced fluorescent CDs with a quantum yield of 9%.

4.6.4   LEAD IONS

Lead ions can cause serious problems such as brain damage, neurological and cardiovascular disorders, and mental retardation. A novel fluorescent

CD using the leaves *Ocimum sanctum* is synthesized through a hydrothermal reaction [107]. A fluorometric detection toward Pb^{2+} ions with a detection limit of 0.59 nM over a range of 0.01–1.0 µM was obtained by these CDs. The fluorescence intensity of CDs continuously decreased with increase in the concentration of Pb^{2+} ions accompanied by a color change of bright green to light green (Figure 4.5). A nonradiative type electron transfer from the nitrogen atoms on the nanosurface to the vacant d-orbital of Pb^{2+} ions is the reason for high selectivity and sensitivity of CDs toward Pb^{2+}ions. Surface complexation arises due to the high binding affinity of Pb^{2+} ions toward the amine group available on the surface of CDs.

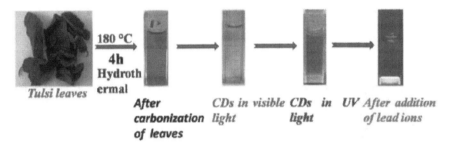

FIGURE 4.5   Schematic representation for the formation of CDs from *Ocimum sanctum* and its effective fluorescent sensing of lead ions.

Source: Reproduced from Kumar et al. [107] with permission from Elsevier. © 2018.)

Sago waste-derived CDs [108] by the method of thermal pyrolysis act as an excellent probe for detection of Cu^{2+} and Pb^{2+} ions. Detection limits for the sensing of $Cu^{2+}$and Pb^{2+} were calculated to be 7. 78 µM and 7.49 µM, respectively, over a concentration range of 0–47.62 µM. Quenching mechanism involves the strong affinity of the paramagnetic metal ions toward the carbon nanostructure. The high absorption affinity causes closer proximal approach of metal ions with CDs, which may obstruct the initial electronic conversion process within the nanostructure to a nonradiative pathway.

4.6.5   CHROMIUM IONS

Contrast to other valence states of chromium, Cr^{6+} is highly toxic and carcinogenic in nature due to its higher oxidation potential, smaller size, and greater mobility. Also, it has the ability to produce reactive oxygen species which ultimately results in the generation of other highly genotoxic by-products.

CDs from the Tulsi leaves [109] show potential sensing selectivity toward Cr^{6+} ions over 1.6 to 50 µM. The limit of detection is found to be 1.6 µM, which is much lower than the permissible limits of EPA. The excitation and emission spectra of N-CDs were shown to have effective overlapping with absorption spectra of Cr^{6+}, leading to quenching inner filter effect. Compared to absorption spectra of only CDs and Cr^{6+}, a redshift occurred upon addition of Cr^{6+} into CDs. The diminishment of zeta potential values with an increment of Cr^{6+}ions confirms the effective interaction of CDs surface with metal ions. Hydrothermally prepared CDs from lemon peel waste show a sensitivity toward Cr^{6+} ions with a LOD of 73 nM [30]. A linear correlation of fluorescence quenching with Cr^{6+} concentration in the range of 2.5–50 mM was observed. As per WHO, Cr^{6+} concentrations lower than 900 nM are acceptable in drinking water. The presence of vacant d orbitals and low lying d–d transition state makes it convenient for Cr^{6+} ions to take part in nonradiative electron transfer process resulting in fluorescence quenching of CDs [80].

4.6.6   OTHER METAL IONS

Biocompatible CDs synthesized from Broccoli juice [110] through a single-step hydrothermal reaction at 190 °C for 6 h. The developed CDs showed a strong emission at 450 nm and exhibited a selective detection of Ag^+ ions via photoluminescence quenching (Figure 4.6). The quenching of CDs upon addition of various concentrations of Ag^+ from 0 to 600 µM involves the formation of chelate complexes due to the energy transfer between the Ag^+ and oxygen functional groups on the surface of the CDs. The as-synthesized CDs were able to detect Ag^+ concentration as low as 0.5 µM.

CDs derived from *Pyrus pyrifolia* fruit [111] acts as the first "turn on" probe for the detection of Al^{3+} ions. Al^{3+} plays the role of a "hard" acid that preferably binds with "hard" electron-rich groups like –COOH, –CHO, –CO and –NH$_2$ available on the surface of CDs. This facilitates the formation of Al^{3+}–CDs complexes which leads to a high degree increase in luminescence of CDs via CHEF mechanism. The detection limit was found to be 2.5 nM over 0.005–100 µM concentration range. WHO has set the maximum permissible limit of Al^{3+} ion as 3–10 mg day^{-1} based on body weight [112]. Pig skin-derived N-CDs is used for the detection of Co^{2+} ions over 1×10^{-6} to 3×10^{-4} M with a detection limit of 6.8×10^{-7} M [113]. The essential criterion for the inner filter effect is the precise overlap between the absorption band of Co^{2+} with both excitation and emission spectra of the fluorophore is satis-fied by N-CDs. It is the mechanism for fluorescence quenching. LOD of Peach gum-derived [29] NCDs as a gold ion sensor is 6.4×10^{-8} M. Sensing response was attributed toward the fluorescence resonance energy transfer between Au nanoparticles and NCDs.

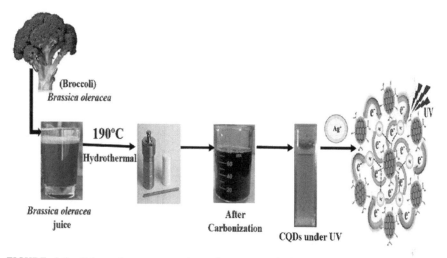

FIGURE 4.6   Schematic representation of green synthetic procedure of CDs for Ag^+ detection.

Source: Reproduced from Arumugam et al. [110] with permission from Elsevier. © 2018.

4.7   CONCLUSION

Recent research in carbon-based CDs has reported the synthesis and char-acterization of green CDs by making use of natural or renewable sources. It offers an environmentally benign approach for the minimization of chemical usage and accident prevention. The use of natural renewable sources for the CDs synthesis is cost-effective, and find applications in the field of bioimaging, metal ion sensing, and catalysis. Although some heavy metals are essential for the human body in very trace amounts, the excess dosage of them cause diseases like Minamata, Itai-Itai, Parkinson, and problems related to heart, lungs, and kidneys. Recent advancement of fluorescent CDs has opened the possibility of using them as portable detectors for toxic heavy metals. This chapter has outlined recent research related to the progress of biocompatible fluorescent CDs as optical sensors for different metal ions. The green CDs provide an easy and harmless onsite interpretation via fluorescent color detection. Moreover, relatively high water solubility and cost-effective nature of CDs offer new avenues of applications having a practical use.

KEYWORDS

- **carbon dots**
- **green precursors**
- **optical properties**
- **photoluminescence**
- **quenching**
- **fluorescence**
- **sensing**

REFERENCES

1. Xu, X., Ray, R., Gu, Y., Ploehn, H. J., Gearheart, L., Raker, K., Scrivens, W. A. electrophoretic analysis and purification of fluorescent single-walled carbon nanotube fragments. *Journal of the American Chemical Society*. 2004, vol. 126(40), 12736–12737.
2. Lin, L., Rong, M., Luo, F., Chen, D., Wang, Y., Chen, X. Luminescent graphene quantum dots as new fluorescent materials for environmental and biological applications. *Trends in Analytical Chemistry*. 2014, vol. 54, 83–102.
3. Liu, C., Zhang, P., Zhai, X., Tian, F., Li, W., Yang, J., Liu, Y., Wang, H., Wang, W., Liu, W. Nano-carrier for gene delivery and bioimaging based on carbon dots with PEI-passivation enhanced fluorescence. *Biomaterials*. 2012, vol. 33(13), 3604–3613.
4. Wang, Q., Huang, X., Long, Y., Wang, X., Zhang, H., Zhu, R., Liang, L., Teng, P., Zheng, H. Hollow luminescent carbon dots for drug delivery. *Carbon*. 2013, vol. 59, 192–199.
5. Lai, C. W., Hsiao, Y. H, Peng, Y. K., Chou, P. T. Facile synthesis of highly emissive carbon dots from pyrolysis of glycerol; gram scale production of carbon dots/mSiO2 for cell imaging and drug release. *Journal of Materials Chemistry*. 2012, vol. 22(29), 14403.
6. Sarkar, S., Banerjee, D., Ghorai, U. K., Das, N. S., Chattopadhyay, K. K. Size dependent photoluminescence property of hydrothermally synthesized crystalline carbon quantum dots. *Journal of Luminescence*. 2016, vol. 178, 314–323.
7. Niino, S., Takeshita, S., Iso, Y., Isobe, T. Influence of chemical states of doped nitrogen on photoluminescence intensity of hydrothermally synthesized carbon dots. *Journal of Luminescence*. 2016, vol. 180, 123–131.
8. Sharma, V., Saini, A. K., Mobin, S. M. Multicolour fluorescent carbon nanoparticle probes for live cell imaging and dual palladium and mercury sensors. *Journal of Materials Chemistry* B. 2016, vol. 4(14), 2466–2476.

9. Qian, Z., Ma, J., Shan, X., Feng, H., Shao, L., Chen, J. Highly luminescent N-doped carbon quantum dots as an effective multifunctional fluorescence sensing platform. *Chemistry A European Journal.* 2014, vol. 20(8), 2254–2263.

10. Ruan, S., Qian, J., Shen, S., Zhu, J., Jiang, X., He, Q., Gao, H. A simple one-step method to prepare fluorescent carbon dots and their potential application in non-invasive glioma imaging. *Nanoscale.* 2014, vol. 6(17), 10040.

11. Wang, X., Qu, K., Xu, B., Ren, J., Qu, X. Microwave assisted one-step green synthesis of cell-permeable multicolor photoluminescent carbon dots without surface passivation reagents. *Journal of Materials Chemistry.* 2011, vol. 21(8), 2445.

12. Yang, P., Zhao, J., Zhang, L., Li, L., Zhu, Z. Intramolecular hydrogen bonds quench photoluminescence and enhance photocatalytic activity of carbon Nnnodots. *Chemistry—A European Journal.* 2015, vol. 21(23), 8561–8568.

13. Qiao, Z. A., Wang, Y., Gao, Y., Li, H., Dai, T., Liu, Y., Huo, Q. Commercially activated carbon as the source for producing multicolor photoluminescent carbon dots by chemical oxidation. *Chemical Communications*, 2010, vol. 46(46), 8812.

14. Zhu, S., Meng, Q., Wang, L., Zhang, J., Song, Y., Jin, H., Zhang, K., Sun, H., Wang, H., Yang, B. Highly photoluminescent carbon dots for multicolor patterning, sensors, and bioimaging. *Angewandte Chemie International Edition.* 2013, vol. 52(14), 3953–3957.

15. Zhai, X., Zhang, P., Liu, C., Bai, T., Li, W., Dai, L., Liu, W. Highly luminescent carbon nanodots by microwave-assisted pyrolysis. *Chemical Communications.* 2012, vol. 48(64), 7955–7957.

16. Ju, E., Liu, Z., Du, Y., Tao, Y., Ren, J., Qu, X. Heterogeneous assembled nanocomplexes for ratiometric detection of highly reactive oxygen species in vitro and in vivo. *ACS Nano.* 2014, vol. 8(6), 6014–6023.

17. Yang, Z., Xu, M., Liu, Y., He, F., Gao, F., Su, Y., Wei, H., Zhang, Y. Nitrogen-doped, carbon-rich, highly photoluminescent carbon dots from ammonium citrate. *Nanoscale.* 2014, vol. 6(3), 1890–1895.

18. Jaiswal, A., Ghosh, S. S., Chattopadhyay, A. One step synthesis of C-dots by microwave mediated caramelization of poly(ethylene glycol). *Chemical Communications.* 2012, vol. 48(3), 407–409.

19. Jiang, H., Chen, F., Lagally, M. G., Denes, F. S. New Strategy for synthesis and functionalization of carbon nanoparticles. *Langmuir.* 2010, vol. 26(3), 1991–1995.

20. Vedamalai, M., Periasamy, A. P., Wang, C. W., Tseng, Y. T., Ho, L. C., Shih,C. C., Chang, H. T. Carbon nanodots prepared from o-phenylenediamine for sensing of Cu^{2+}ions in cells," *Nanoscale.* 2014, vol. 6(21), 13119–13125.

21. Wang, W., Li, Y., Cheng, L., Cao, Z., Liu, W. Water-soluble and phosphorus-containing carbon dots with strong green fluorescence for cell labeling. *Journal of Materials Chemistry* B. 2014, vol. 2(1), 46–48.

22. Zhou, L., Li, Z., Liu, Z., Ren, J., Qu, X. Luminescent carbon dot-gated nanovehicles for pH-triggered intracellular controlled release and imaging. *Langmuir.* 2013, vol. 29(21), 6396–6403.

23. Zhou, L., Lin, Y., Huang, Z., Ren, J., Qu, X. Carbon nanodots as fluorescence probes for rapid, sensitive, and label-free detection of $Hg^{2+}$and biothiols in complex matrices. *Chemical Communications.* 2012, vol. 48(8), 1147–1149.

24. Wang, L., Bi, Y., Gao, J., Li, Y., Ding, H., Ding, L. Carbon dots based turn-on fluorescent probes for the sensitive determination of glyphosate in environmental water samples. *RSC Advances.* 2016, vol. 6(89), 85820–85828.

25. Sharma, V., Tiwari, P., Mobin, S. M. Sustainable carbon-dots: recent advances in green carbon dots for sensing and bioimaging. *Journal of Materials Chemistry* B. 2017, vol. 5(45), 8904–8924.

26. De, B., Karak, N. A green and facile approach for the synthesis of water soluble fluorescent carbon dots from banana juice. *RSC Advances*. 2013, vol. 3(22), 8286.

27. Mehta, V. N., Jha, S., Basu, H., Singhal, R. K., Kailasa, S. K. One-step hydrothermal approach to fabricate carbon dots from apple juice for imaging of mycobacterium and fungal cells. *Sensors and Actuators B: Chemical*. 2015, vol. 213, 434–443.

28. Wang, N., Wang, Y., Guo, T., Yang, T., Chen, M., Wang, J. Green preparation of carbon dots with papaya as carbon source for effective fluorescent sensing of Iron (III) and Escherichia coli. *Biosensors and Bioelectronics*. 2016, vol. 85, 68–75.

29. Liao, J., Cheng, Z., Zhou, L. Nitrogen-doping enhanced fluorescent carbon dots: green synthesis and their applications for bioimaging and label-free detection of Au3+ ions. *ACS Sustainable Chemistry & Engineering*. 2016, vol. 4(6), 3053–3061.

30. Tyagi, A., Tripathi, K. M., Singh, N., Choudhary, S., Gupta, R. K. Green synthesis of carbon quantum dots from lemon peel waste: applications in sensing and photocatalysis. *RSC Advances*. 2016, vol. 6(76), 72423–72432.

31. Arul, V., Edison, T. N. J. I., Lee, Y. R., Sethuraman, M. G. Biological and catalytic applications of green synthesized fluorescent N-doped carbon dots using *Hylocereus undatus*. *Journal of Photochemistry and Photobiology B: Biology*. 2017, vol. 168, 142–148.

32. Alam, A. M., Park, B. Y., Ghouri, Z. K., Park, M., Kim, H. Y. Synthesis of carbon quantum dots from cabbage with down- and up-conversion photoluminescence properties: excellent imaging agent for biomedical applications. *Green Chemistry*. 2015, vol. 17(7), 3791–3797.

33. Yin, B., Deng, J., Peng, X., Long, Q., Zhao, J., Lu, Q., Chen, Q., Li, H., Tang, H., Zhang, Y., Yao, S. Green synthesis of carbon dots with down- and up-conversion fluorescent properties for sensitive detection of hypochlorite with a dual-readout assay. *The Analyst*. 2013, vol. 138(21), 6551.

34. Gao, S., Chen, Y., Fan, H., Wei, X., Hu, C., Wang, L., Qu, L. A green one-arrow-two-hawks strategy for nitrogen-doped carbon dots as fluorescent ink and oxygen reduction electrocatalysts. *Journal of Materials Chemistry* A. 2014, vol. 2(18), 6320.

35. Feng, Y., Zhong, D., Miao, H., Yang, X. Carbon dots derived from rose flowers for tetracycline sensing. *Talanta*. 2015, vol. 140,128–133.

36. Liu, W., Li, C., Sun, X., Pan, W., Yu, G., Wang, J. Highly crystalline carbon dots from fresh tomato: UV emission and quantum confinement. *Nanotechnology*. 2017, vol. 28(48), 485705.

37. Ma, X., Dong, Y., Sun, H., Chen, N. Highly fluorescent carbon dots from peanut shells as potential probes for copper ion: the optimization and analysis of the synthetic process. *Materials Today Chemistry*. 2017, vol. 5, 1–10

38. Benelli, G. Gold Nanoparticles—against Parasites and Insect Vectors. Acta Trop. 2018, 178, 73–80. https://doi.org/10.1016/j.actatropica.2017.10.021.

39. Pramanik, A., Biswas, S., Kumbhakar, P. Solvatochromism in highly luminescent environmental friendly carbon quantum dots for sensing applications: conversion of bio-waste into bio-asset. *Spectrochimica Acta Part A: Molecular and Biomolecular Spectroscopy*. 2018, vol. 191, 498–512.

40. Zhao, Y., Zhang, Y., Liu,X., Kong, H., Wang, Y., Qin, G., Cao, P., Song, X., Yan, X., Wang, Q., Qu, H. Novel carbon quantum dots from egg yolk oil and their haemostatic effects. *Scientific Reports*. 2017, vol. 7(1),

41. Xu, H., Yang, X., Li, G., Zhao, C., Liao, X. Green synthesis of fluorescent carbon dots for selective detection of tartrazine in food samples. *Journal of Agricultural and Food Chemistry*. 2015, vol. 63(30), 6707–6714.

42. Lu, W., Qin, X., Asiri, A. M., Al-Youbi, A. O., Sun, X. Green synthesis of carbon nanodots as an effective fluorescent probe for sensitive and selective detection of mercury(II) ions. *Journal of Nanoparticle Research*. 2012, vol. 15(1).

43. Zhao, M. Lan, X. Zhu, H. Xue, T.-W. Ng, X. Meng, C.-S. Lee, P., Zhang, W. Green synthesis of bifunctional fluorescent carbon dots from garlic for cellular imaging and free radical scavenging. *ACS Applied Materials & Interfaces*. 2015, vol. 7(31), 17054–17060.

44. D'souza, S. L., Deshmukh, B., Bhamore, J. R., Rawat, K. A., Lenka, N., Kailasa, S. K. Synthesis of fluorescent nitrogen-doped carbon dots from dried shrimps for cell imaging and boldine drug delivery system. *RSC Advances*. 2016, vol. 6(15), 12169–12179.

45. Liu, S. S., Wang, C. F., Li, C. X., Wang, J., Mao, L. H., Chen, S. Hair-derived carbon dots toward versatile multidimensional fluorescent materials. *Journal of Materials Chemistry* C. 2014, vol. 2(32), 6477–6483.

46. Wang, C., Sun, D., Zhuo, K., Zhang, H., Wang, J. Simple and green synthesis of nitrogen-, sulfur-, and phosphorus-co-doped carbon dots with tunable luminescence properties and sensing application. *RSC Advances*. 2014, vol. 4(96), 54060–54065.

47. Yuan, M., Zhong, R., Gao, H., Li, W., Yun, X., Liu, J., Zhao, X., Zhao, G., Zhang, F. "One-step, green, and economic synthesis of water-soluble photoluminescent carbon dots by hydrothermal treatment of wheat straw, and their bio-applications in labeling, imaging, and sensing. *Applied Surface Science*. 2015, vol. 355, 1136–1144.

48. Ramanan, V., Thiyagarajan, S. K., Raji, K., Suresh, R., Sekar, R., Ramamurthy, P. Outright green synthesis of fluorescent carbon dots from eutrophic algal blooms for in vitro imaging. *ACS Sustainable Chemistry & Engineering*. 2016, vol. 4(9), 4724–4731.

49. Wang, W. J., Xia, J. M., Feng, J., He, M. Q., Chen, M. L., Wang, J. H. Green preparation of carbon dots for intracellular pH sensing and multicolor live cell imaging. *Journal of Materials Chemistry* B. 2016, vol. 4(44), 7130–7137.

50. Atchudan, R., Edison, T. N. J. I., Lee, Y. R. Nitrogen-doped carbon dots originating from unripe peach for fluorescent bioimaging and electrocatalytic oxygen reduction reaction. *Journal of Colloid and Interface Science*. 2016, vol. 482, 8–18.

51. Arul, V., Sethuraman, M. G. Hydrothermally green synthesized nitrogen-doped carbon dots from *Phyllanthus emblica* and their catalytic ability in the detoxification of textile effluents. *ACS Omega*. 2019, vol. 4(2), 3449–3457.

52. Sachdev, A., Gopinath, P. Green synthesis of multifunctional carbon dots from coriander leaves and their potential application as antioxidants, sensors and bioimaging agents. *The Analyst*. 2015, vol. 140(12), 4260–4269.

53. Wei, J., Liu, B.,Yin, P. Dual functional carbonaceous nanodots exist in a cup of tea. *RSC Advance*. 2014, vol. 4(108), 63414–63419.

54. Ding, H., Ji, Y., Wei, J. S., Gao, Q. Y., Zhou, Z. Y., Xiong, H. M. Facile synthesis of red-emitting carbon dots from pulp-free lemon juice for bioimaging. *Journal of Materials Chemistry* B. 2017, vol. 5(26), 5272–5277.

55. Guo, Y., Zhang, L., Zhang, S., Yang, Y., Chen, X., Zhang, M. Fluorescent carbon nanoparticles for the fluorescent detection of metal ions. *Biosensors and Bioelectronics.* 2015, vol. 63, 61–71.

56. Mondal, P., Ghosal, K., Bhattacharyya, S. K., Das, M., Bera, A., Ganguly, D., Kumar, P., Dwivedi, J., Gupta, R. K., Martí, A. A., Gupta, B. K., Maiti, S. Formation of a gold–carbon dot nanocomposite with superior catalytic ability for the reduction of aromatic nitro groups in water. *RSC Advances.* 2014, vol. 4(49), 25863–25866.

57. Edison, T. N. J. I., Atchudan, R., Shim, J.-J., Kalimuthu, S., Ahn, B. C., Lee, Y. R. Turn-off fluorescence sensor for the detection of ferric ion in water using green synthesized N-doped carbon dots and its bio-imaging. *Journal of Photochemistry and Photobiology B: Biology.* 2016, vol. 158, 235–242.

58. Yao, Y. Y., Gedda, G., Girma, W. M., Yen, C. L., Ling, Y. C., Chang, J. Y. Magnetofluorescent carbon dots derived from crab shell for targeted dual-modality bioimaging and drug delivery. *ACS Applied Materials & Interfaces.* 2017, vol. 9(16), 13887–13899.

59. Liu, R., Zhang, J., Gao, M., Li, Z., Chen, J., Wu, D., Liu, P. A facile microwave-hydrothermal approach towards highly photoluminescent carbon dots from goose feathers. *RSC Advances.* 2015, vol. 5(6), 4428–4433.

60. Hu, Y., Zhang, L., Li, X., Liu, R., Lin, L., Zhao, S. Green preparation of S and N co-doped carbon dots from water chestnut and onion as well as their use as an off–on fluorescent probe for the quantification and imaging of coenzyme A. *ACS Sustainable Chemistry & Engineering.* 2017, vol. 5(6), 4992–5000.

61. Sun, D., Ban, R., Zhang, P. H., Wu, G. H., Zhang, J. R., Zhu, J. J. Hair fiber as a precursor for synthesizing of sulfur- and nitrogen-co-doped carbon dots with tunable luminescence properties. *Carbon.* 2013, vol. 64, 424–434.

62. He, Z. L., Yang, X. E., Stoffella, P. J. Trace elements in agroecosystems and impacts on the environment. *Journal of Trace Elements in Medicine and Biology.* 2005, vol. 19(2–3), 125–140.

63. Shallari, S., Schwartz, C., Hasko, A., Morel, J. L. Heavy metals in soils and plants of serpentine and industrial sites of Albania. *Science of the Total Environment.* 1998, vol. 209(2–3), 133–142.

64. Sträter, E., Westbeld, A., Klemm, O. Pollution in coastal fog at Alto Patache, Northern Chile. *Environmental Science and Pollution Research.* 2010, vol. 17(9).1563–1573.

65. Nriagu, J. O. A global assessment of natural sources of atmospheric trace metals. *Nature.* 1989, vol. 338(6210), 47–49.

66. Tchounwou, P. B., Centeno, J. A., Patlolla, A. K. Arsenic toxicity, mutagenesis, and carcinogenesis a health risk assessment and management approach. *Molecular and Cellular Biochemistry.* 2004, vol. 2595(1), 47–55.

67. Tchounwou, P. B., Ishaque, A. B., Schneider, J. Cytotoxicity and transcriptional activation of stress genes in human liver carcinoma cells (HepG2) exposed to cadmium chloride. *Molecular Mechanisms of Metal Toxicity and Carcinogenesis.* 2001, 21–28.

68. Patlolla, A. K., Barnes, C., Yedjou, C., Velma, V. R., Tchounwou, P. B. Oxidative stress, DNA damage, and antioxidant enzyme activity induced by hexavalent chromium in Sprague-Dawley rats. *Environmental Toxicology.* 2009, vol. 24(1), 66–73.

69. Yedjou, C., Tchounwou, P. N-acetyl-L-cysteine affords protection against lead-induced cytotoxicity and oxidative stress in human liver carcinoma (HepG2) cells. *International Journal of Environmental Research and Public Health.* 2007, vol. 4(2), 132–137.

70. Sutton, D., Tchounwou, P., Ninashvili, N., Shen, E. Mercury induces cytotoxicity and transcriptionally activates stress genes in human liver carcinoma (HepG2) cells. *International Journal of Molecular Sciences*, 2002, vol. 3(9), 965–984.

71. Martinhon, P. T., Carreño, J., Sousa, C. R., Barcia, O. E., Mattos, O. R. Electrochemical impedance spectroscopy of lead(II) ion-selective solid-state membranes. *Electrochimica Acta*. 2006, vol. 51(15), 3022–3028.

72. Pereira, F. M., Brum, D. M., Lepri, F. G., Cassella, R. J. Extraction induced by emulsion breaking as a tool for Ca and Mg determination in biodiesel by fast sequential flame atomic absorption spectrometry (FS-FAAS) using Co as internal standard. *Microchemical Journal*. 2014, vol. 117, 172–177.

73. Abbasi, S., Khodarahmiyan, K., Abbasi, F. Simultaneous determination of ultra trace amounts of lead and cadmium in food samples by adsorptive stripping voltammetry. *Food Chemistry*. 2011, vol. 128(1), 254–257.

74. Ju, J., Chen, W. Synthesis of highly fluorescent nitrogen-doped graphene quantum dots for sensitive, label-free detection of Fe(III) in aqueous media. *Biosensors and Bioelectronics*. 2014, vol. 58, 219–225.

75. Ryan, E., Meaney, M. Determination of trace levels of copper(II), aluminium(III) and iron(III) by reversed-phase high-performance liquid chromatography using a novel on-line sample preconcentration technique. *The Analyst*. 1992, vol. 117(9), 1435.

76. Aragay, G., Pons, J., Merkoçi, A. Recent trends in macro-, micro-, and nanomaterial-based tools and strategies for heavy-metal detection. *Chemical Reviews*. 2011, vol. 111(5), 3433–3458.

77. Zheng P. and Wu N. Fluorescence and sensing applications of graphene oxide and graphene quantum dots: a review. Chemistry—An Asian Journal. 2017, vol. 12 (18) 2343–2353.

78. Liu, S., Tian, J., Wang, L., Zhang, Y., Qin, X., Luo, Y., Asiri, A. M., Al-Youbi, A. O., Sun, X. Hydrothermal treatment of grass: a low-cost, green route to nitrogen-doped, carbon-rich, photoluminescent polymer nanodots as an effective fluorescent sensing platform for label-free detection of Cu(II) ions. *Advanced Materials*. 2012, vol. 24(15), 2037–2041.

79. Chen, S., Yu, Y. L.,Wang, J. H. Inner filter effect-based fluorescent sensing systems: a review. *Analytica Chimica Acta*. 2018, vol. 999, 13–26.

80. Sun, X., He, J., Yang, S., Zheng, M., Wang, Y., Ma, S., Zheng, H. Green synthesis of carbon dots originated from *Lycii fructus* for effective fluorescent sensing of ferric ion and multicolor cell imaging. *Journal of Photochemistry and Photobiology B: Biology*. 2017, vol. 175, 219–225.

81. Liu, W., Diao, H., Chang, H., Wang, H., Li, T., Wei, W. Green synthesis of carbon dots from rose-heart radish and application for Fe³⁺ detection and cell imaging. *Sensors and Actuators B: Chemical*. 2017, vol. 241, 190–198.

82. Carvalho, J., Santos, L. R., Germino, J. C., Terezo, A. J., Moreto, J. A., Quites, F. J., Freitas, R. G. Hydrothermal synthesis to water-stable luminescent carbon dots from Acerola fruit for photoluminescent composites preparation and its application as sensors. *Materials Research*. 2019, vol. 22(3).

83. OECD Reviews of Health Systems: Switzerland 2011. *OECD Reviews of Health Systems*.

84. Yang, X. Zhuo, Y., Zhu, S., Luo, Y., Feng, Y., Dou, Y. Novel and green synthesis of high-fluorescent carbon dots originated from honey for sensing and imaging. *Biosensors and Bioelectronics*. 2014, vol. 60, 292–298.

85. Bandi, R., Gangapuram, B. R., Dadigala, R., Eslavath, R., Singh, S. S., Guttena, V. Facile and green synthesis of fluorescent carbon dots from onion waste and their potential applications as sensor and multicolour imaging agents. *RSC Advances*. 2016, vol. 6(34), 28633–28639.

86. Chen, K., Qing, W., Hu, W., Lu, M., Wang, Y., Liu, X. On-off-on fluorescent carbon dots from waste tea: their properties, antioxidant and selective detection of CrO_4^{2-}, Fe^{3+}, ascorbic acid and L-cysteine in real samples. *Spectrochimica Acta Part A: Molecular and Biomolecular Spectroscopy*. 2019, vol. 213, 228–234.

87. Zhu, L., Yin, Y., Wang, C. F., Chen, S. Plant leaf-derived fluorescent carbon dots for sensing, patterning and coding. *Journal of Materials Chemistry C*. 2013, vol. 1(32), 4925.

88. Atchudan, R., Edison, T. N. J. I., Chakradhar, D., Perumal, S., Shim, J. J., Lee, Y. R. Facile green synthesis of nitrogen-doped carbon dots using *Chionanthus retusus* fruit extract and investigation of their suitability for metal ion sensing and biological applications. *Sensors and Actuators B: Chemical*. 2017, vol. 246, 497–509.

89. Atchudan, R., Edison, T. N. J. I., Aseer, K. R., Perumal, S., Lee, Y. R. Hydrothermal conversion of *Magnolia liliiflora* into nitrogen-doped carbon dots as an effective turn-off fluorescence sensing, multi-colour cell imaging and fluorescent ink. *Colloids and Surfaces B: Biointerfaces*. 2018, vol. 169, 321–328.

90. Vandarkuzhali, S. A. A., Jeyalakshmi, V., Sivaraman, G., Singaravadivel, S., Krishnamurthy, K. R., Viswanathan, B. Highly fluorescent carbon dots from pseudo-stem of banana plant: applications as nanosensor and bio-imaging agents. *Sensors and Actuators B: Chemical*. 2017, vol. 252, 894–900.

91. Shen, J. Shang, S., Chen, X., Wang, D., Cai, Y. Facile synthesis of fluorescence carbon dots from sweet potato for Fe^{3+} sensing and cell imaging. *Materials Science and Engineering: C*. 2017, vol. 76, 856–864.

92. Guo, Y., Zhang, L., Cao, F., Leng, Y. Thermal treatment of hair for the synthesis of sustainable carbon quantum dots and the applications for sensing Hg^{2+}. *Scientific Reports*. 2016, vol. 6(1).

93. Li, L. S., Jiao, X. Y., Zhang, Y., Cheng, C., Huang, K., Xu, L. Green synthesis of fluorescent carbon dots from Hongcaitai for selective detection of hypochlorite and mercuric ions and cell imaging. *Sensors and Actuators B: Chemical*. 2018, vol. 263, 426–435.

94. Bano, D., Kumar, V., Singh, V. K., Hasan, S. H. Green synthesis of fluorescent carbon quantum dots for the detection of mercury(ii) and glutathione. *New Journal of Chemistry*. 2018, vol. 42(8), 5814–5821.

95. Gu, D., Shang, S., Yu, Q., Shen, J. Green synthesis of nitrogen-doped carbon dots from lotus root for Hg(II) ions detection and cell imaging. *Applied Surface Science*. 2016, vol. 390, 38–42.

96. Zhang, R., Chen, W. Nitrogen-doped carbon quantum dots: facile synthesis and application as a 'turn-off' fluorescent probe for detection of Hg^{2+} ions. *Biosensors and Bioelectronics*. 2014, vol. 55, 83–90.

97. Huang, H., Lv, J. J., Zhou, D. L., Bao, N., Xu, Y., Wang, A. J., Feng, J. J. One-pot green synthesis of nitrogen-doped carbon nanoparticles as fluorescent probes for mercury ions. *RSC Advances*. 2013, vol. 3(44), 21691.

98. Yu, J., Song, N., Zhang, Y. K., Zhong, S. X., Wang, A. J., Chen, J. Green preparation of carbon dots by Jinhua bergamot for sensitive and selective fluorescent detection of Hg^{2+} and Fe^{3+}. *Sensors and Actuators B: Chemical*. 2015, vol. 214, 29–35.

99. Murugan, N., Prakash, M., Jayakumar, M., Sundaramurthy, A., Sundramoorthy, A. K. Green synthesis of fluorescent carbon quantum dots from *Eleusine coracana* and their application as a fluorescence 'turn-off' sensor probe for selective detection of Cu^{2+}. *Applied Surface Science*. 2019, vol. 476, 468–480.

100. Gedda, G., Lee, C. Y., Lin, Y. C., Wu, H. Green synthesis of carbon dots from prawn shells for highly selective and sensitive detection of copper ions. *Sensors and Actuators B: Chemical*. 2016, vol. 224, 396–403.

101. Bhamore, J. R., Jha, S., Park, T. J., Kailasa, S. K. Fluorescence sensing of Cu^{2+} ion and imaging of fungal cell by ultra-small fluorescent carbon dots derived from *Acacia concinna* seeds. *Sensors and Actuators B: Chemical*. 2018, vol. 277, 47–54.

102. Liu, Y., Zhao, Y., Zhang, Y. One-step green synthesized fluorescent carbon nanodots from bamboo leaves for copper(II) ion detection. *Sensors and Actuators B: Chemical*. 2014, vol. 196, 647–652.

103. Liu, L., Gong, H., Li, D., Zhao, L. Synthesis of carbon dots from pear juice for fluorescence detection of Cu^{2+} ion in water. *Journal of Nanoscience and Nanotechnology*. 2018, vol. 18(8), 5327–5332.

104. Kumari, A. Kumar, A., Sahu, S. K., Kumar, S. Synthesis of green fluorescent carbon quantum dots using waste polyolefins residue for Cu^{2+} ion sensing and live cell imaging. *Sensors and Actuators B: Chemical*. 2018, vol. 254, 197–205.

105. Niu, X., Liu, G., Li, L., Fu, Z., Xu, H., Cui, F. Green and economical synthesis of nitrogen-doped carbon dots from vegetables for sensing and imaging applications. *RSC Advances*. 2015, vol. 5(115), 95223–95229.

106. Das, P., Ganguly, S., Bose, M., Mondal, S., Das, A. K., Banerjee, S., Das, N. C. A simplistic approach to green future with eco-friendly luminescent carbon dots and their application to fluorescent nano-sensor 'turn-off' probe for selective sensing of copper ions. *Materials Science and Engineering: C*. 2017, vol. 75, 1456–1464.

107. Kumar, A., Chowdhuri, A. R., Laha, D., Mahto, T. K., Karmakar, P., Sahu, S. K. Corrigendum to Green synthesis of carbon dots from Ocimum sanctum for effective fluorescent sensing of Pb2+ ions and live cell imaging. Sensors and Actuators B: Chemical. 2018, vol. 263, 677.

108. Tan, X. W., Romainor, A. N. B., Chin, S. F., Ng, S. M. Carbon dots production via pyrolysis of sago waste as potential probe for metal ions sensing. *Journal of Analytical and Applied Pyrolysis*. 2014, vol. 105, 157–165.

109. Bhatt, S., Bhatt, M., Kumar, A. Vyas, G., Gajaria, T., Paul, P. Green route for synthesis of multifunctional fluorescent carbon dots from Tulsi leaves and its application as Cr(VI) sensors, bio-imaging and patterning agents. *Colloids and Surfaces B: Biointerfaces*. 2018, vol. 167, 126–133.

110. Arumugam, N., Kim, J. Synthesis of carbon quantum dots from Broccoli and their ability to detect silver ions. *Materials Letters*. 2018, vol. 219, 37–40.

111. Bhamore, J. R., Jha, S., Singhal, R. K., Park, T. J., Kailasa, S. K. Facile green synthesis of carbon dots from *Pyrus pyrifolia* fruit for assaying of Al^{3+} ion via chelation enhanced fluorescence mechanism. *Journal of Molecular Liquids*. 2018, vol. 264, 9–16.

112. Barceló, J., Poschenrieder, C. Fast root growth responses, root exudates, and internal detoxification as clues to the mechanisms of aluminium toxicity and resistance: a review. *Environmental and Experimental Botany*. 2002, vol. 48(1), 75–92.

113. Wen, X., Shi, L., Wen, G., Li, Y., Dong, C., Yang, J., Shuang, S. Green and facile synthesis of nitrogen-doped carbon nanodots for multicolor cellular imaging and Co^{2+} sensing in living cells. *Sensors and Actuators B: Chemical*. 2016, vol. 235, 179–187.

CHAPTER 5

Fluorescent Carbon Quantum Dots: Properties and Applications

K. A. ANN MARY[1,*], JOHNS NADUVATH[1], and A. P. SUNITHA[2]

[1]*Department of Physics, St. Thomas College, Thrissur 680001, Kerala, India*

[2]*Department of Physics, Government Victoria College, Palakkad 678001, Kerala, India*

Corresponding author. E-mail: ancyk06@gmail.com

ABSTRACT

Among fluorescent materials in nanoscale regime, nontoxic carbon quantum dots (CQDs) are explored extensively and widely studied for the past few years. CQDs prepared from board selection of inexpensive green carbonaceous precursors exhibits emission wavelengths ranging blue to red with high quantum yield. Moreover, excitation dependent emission, biocompatibility, aqueous solubility, and high photostability of CQDs activated more and more research to finely tune their properties. Due to these unique and fascinating qualities, CQDs could replace toxic organic dyes and quantum dots becoming promising materials for photovoltaics, photocatalysis, bioimaging, and chemical sensing applications. This chapter includes a brief description on certain remarkable properties of CQDs and their potential applications.

5.1 INTRODUCTION

Quantum dots (QDs) are semiconductor nanoparticles having a size less than ~10 nm and these quasi-zero dimensional materials have a difference in density of energy states compared to their 1D, 2D, or bulk structures.

Consequently, their size-dependent electronic and optical properties made them inevitable in diverse applications. In addition to size-tunable optical absorption thresholds, QDs have narrow and intense photoluminescence (PL) spectral features. The width of the spectral bandgap can be altered by varying the size of QDs [1]. Highly hydrophobic QDs can be hydrophilized by the postfabrication surface passivating techniques [1]. Hydrophilized QDs with excellent biocompatibility are efficient candidates for bioanalytical applications. Nanodots are utilized for promising biological applications such as immunofluorescence technology, cellular tracking, and bioimaging [2]. While nanodots of CdS, ZnS, ZnSe exhibited UV emissions, PbSe, CdTe, PbS nanocrystals span the infrared spectral regions [3]. QD-based biological tools have shown excellent potential for efficient drug delivery systems and cancer cell imaging techniques [4]. But recently, these toxic semiconducting nanocrystals are now replaced with carbon nanoallotropes such as graphene, fullerenes, diamond, carbon nanotube (CNT), and graphite due to the urge for more environmental friendliness. Carbon family members such as graphite and diamond consist of only carbon atoms, but with extremely strange properties. These differences can be explained on the basis of the way of connecting carbon atoms in each structure [5]. An opaque, soft graphitic structure having remarkable electrical conductivity is entirely different from sp^3 hybridized tetrahedral structure of diamond with exceptional brilliance. Graphite is a dense 3D pack of graphene monolayers combined via weak van der Waals forces. In contrast, graphene comprises of sp^2 carbon cores having a 2D hexagonal lattice [5]. Fullerene, also known as C60 possess a truncated icosahedral structure with each C60 molecules contains 60 sp^2 carbon atoms [5–7]. Carbon nanotubes are another promising carbon family member with prominent properties due to their tubular shape with interconnected carbon atoms [5, 8, 9]. Among nanosized luminescent carbon-based materials with at least one dimension less than 10 nm, carbon quantum dots (CQDs), graphene quantum dots (GQDs), and polymer dots (PDs) possess the same kind of fluorescence [10–12]. Though they consist of sp^2/sp^3 carbon, they have different internal structure and surface groups. CQDs are mostly spherical and either amorphous or crystalline (multiple-layer graphite structures) with particle size less than 10 nm. Evidently, GQDs has graphene lattices inside the dots and larger lateral dimension with chemical groups on the edges. While crystal lattice of GQDs exhibits a d-spacing value of 0.24 nm that can be assigned to (100) plane of single graphene, crystalline CQDs have an interplanar spacing of 0.34 nm corresponding to (002) plane of graphite [13]. Carbonized PDs are cross-linked or aggregated polymer with a hydrophobic core and hydrophilic polymer chains on the external surface. Among them, CQDs have a unique position due to its facile and green fabrication route with widely available carbonaceous precursors.

CQDs were first found by Xu et al. while cleaning arc synthesized CNTs via preparative electrophoresis method [14]. The CQDs have gained greater attention in the last decade because of its high fluorescence efficiency. Many composites can be formed using these CQDs to obtain improved optical, electronic, and magnetic properties. CQDs exhibit tunable fluorescence, high quantum yield, high water solubility, excellent chemical stability, nontoxicity, and good biocompatibility. CQDs are gaining more interest due to their potential applications in various diverse fields such as chemical sensing, bioimaging, catalysis, photovoltaics, and drug and gene delivery [15]. Moreover, excellent photon upconversion by two-photon absorption is observed in the near-infrared (NIR) region. The CQDs show bright and colorful fluorescence emissions over the visible and extending into the NIR (Figure 5.1).

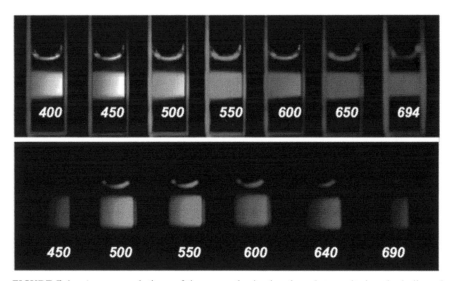

FIGURE 5.1 Aqueous solutions of the as-synthesized carbon dots excited at the indicated wavelengths in nm and photographed directly.

Source: Reprinted (adapted) with permission from Ref. [16]. © 2006 American Chemical Society.

CQDs can be prepared easily by green technology using various naturally available precursors [15]. CQDs are synthesized from a wide variety of precursors at different reaction temperatures such as from phenol by carbonization at 900 °C, from ascorbic acid by heat treatment at 90 °C, from citrate by carbonization at 300 °C, from polyethene glycol and saccharide by microwave treatment at 50 °C, and from watermelon peels by carbonization at 220 °C [17]. The low-cost precursors for CQDs also include organic edible carbon sources like orange juice, cabbage, honey, sugarcane juice, strawberry juice, and grass by hydrothermal treatment [18–23].

Various production strategies have been adopted for using commercial carbon nanopowders as the precursor material to prepare the smaller nanoparticles. According to Hu et al., direct laser irradiation of graphite powder was performed in an organic solvent (diamine hydrate, diethanol-amine, or PEG200N-amine-terminated polyethylene glycol oligomers of 200 average molecular weight) under ultrasonication for growing CQDs. The CQDs with high quantum yield as high as 7.8% were synthesized using the solvent diethanolamine [24]. Bourlinos et al. used thermal carbonization of ammonium citrate salts, for the preparation of carbon dots with the average size of 7 nm that exhibited fluorescence quantum yield of 3% at 495 nm excitation [25]. Similarly, Wang et al. synthesized CQDs via thermal oxidation of citric acid, in a molten lithium nitrate salt in an argon atmosphere, followed by the surface passivation reaction with PEG1500 (polyethylene glycol oligomers of 1500 average molecular weight). Effective surface passivation or functionalization of CQDs can widely change the optical properties of CQDs. The functionalized CQDs exhibit better fluorescence performance and spectroscopic properties with high quantum yield [26]. Liu et al. reported a different way to fabricate carbon nanoparticles from smoldering candles. The soot was treated with a nitric acid solution (5 M), followed by purification with polyacrylamide gel electrophoresis, and the resulting CQDs in an aqueous suspension exhibited PL over a broad spectral range [15].

Chattopadhyay and coworkers worked on amorphous carbon nanoparticles of size 4–30 nm, by heating bread, jaggery, corn flakes, or biscuits. CQDs are generally smaller carbon nanoparticles with various surface passivation produced via modification or functionalization [25]. Sun et al. doped the core carbon nanoparticle surface with inorganic salts (ZnO, ZnS, or TiO_2) that leads to the formations of CZnO-dots, CZnS-dots, or $CTiO_2$-dots. These hybrid dots with the organic functionalization exhibited enhanced fluorescence and remarkable photocatalytic properties [27].

5.2   OPTICAL PROPERTIES OF CARBON QUANTUM DOTS

For the last few years, nanocarbon cores gained cumulative interest and are emerging as a less toxic alternative to traditional transition metal-based QDs. Their unique fluorescence properties, good photostability, biocompatibility, and facile synthesis from any carbon-based sources made them potential materials in bioimaging, sensing, and other optoelectronic devices. Fluorescent properties of CQDs are influenced by several factors like synthesis route, precursors used, capping agents, and doping.

5.2.1   ABSORBANCE

Due to the presence of oxygen groups on the surface, CQDs can be easily functionalized with organic molecules that can alter their optical properties. For carbon dots, the strong optical absorption in UV region is attributed to π–π* transition in carbon core and n–π* transition of carbonyl groups on the surface. For doped CQDs hybridization with a heteroatom such as N, S, and P can also cause strong UV wavelength absorption. Absorption spectrum usually has a tail extending to the visible range and depends on surface passivation or doping.

5.2.2   PHOTOLUMINESCENCE

The most interesting and unique property of CQDs is their PL making them potential candidates in the fields of photocatalysis, chemical sensing, bioimaging, and photovoltaics. Generally, CQDs are blue luminescent nanoparticles whose fluorescence energy and intensity can be tuned with surface passivation or doping. A clear perceptive of PL origins is currently unresolved that many interpretations derived from quantum effect, surface traps, surface passivation, intramolecular H-bondings, and others are already reported [28, 29]. Even though excitation-dependent emission is widely investigated in CQDs, at some circumstances, CQDs also exhibit excitation independent. Recent reports showed excitation dependent PL in CQDs with multiple emissive sites and excitation independent PL with singlet surface state [30]. Presence of surface traps with different energy levels originated from surface groups like C–O, C = O, and –COOH cause CQDs to luminescence at different energies with different excitation wavelengths. For doped CQDs, the added atoms may be responsible to decrease original surface states and thereby trap supporting excitation-independent emission [28, 31]. However, more investigations have emerged from CQDs or GQDs where both excitation-dependent and excitation-independent PL can be obtained by surface engineering or tuning carbonization degree [13, 28]. Thus compared to semiconductor QDs, fluorescence from CQDs cannot be well-tuned with particle size and have much wider emission bandwidth.

The average time spent by CQDs in its excited state before returning to the ground state via photon emission is calculated as its PL decay lifetime. Longer decay lifetimes with high quantum yields (QY) are the necessary requirements for an efficient biosensor. QY is the ratio of the number of emitted photons to the number of absorbed photons and is an intrinsic property of a fluorophore. The QY of CQDs is improved after surface modification or passivation. By eliminating toxic QDs, organic dyes, the level of various biopigments, food dyes, and metal ions can be monitored through a sensing platform based on fluorescent CQDs. Moreover, nitrogen- and sulfur-doped CQDs exhibited temperature-dependent PL lifetimes decay despite of variations in pH, the concentration of CDs, and solution ionic strengths [28, 32]. Owing to good biocompatibility these CQDs can be potential candidates to probe intracellular temperature. To date, maximum quantum yield reported for CQDs is 93.3% with a long lifetime of 19.5 ns which has been applied for selective detection of mercury ions [33].. Solvent effects also played a crucial role in PL decay lifetime. Aromatic solvents have a moderate lifetime of 4–5 ns and it decreased for less polar solvents like alcohols. For high polar aprotic solvents such as acetonitrile decay lifetime values increased with polarity [28, 34]. Moreover, doping with nitrogen, sulfur, boron, or phosphorous can significantly alter the QY and PL lifetimes of CQDs. Nitrogen-doped CQDs have relatively longer lifetime of 10.6 ns with appreciable quantum yield due to electron-rich amino groups on the c dot surfaces [35].

5.2.3   CHEMILUMINESCENCE

Chemiluminescence (CL) is defined as a phenomenon by which photons are emitted by a substance during a chemical reaction by absorbing chemical energy. Zhen Lin and his team first reported CL property of CQDs in the presence of potassium permanganate ($KMnO_4$) and cerium(IV) [29, 36]. As evident from electron paramagnetic resonance, due to the injection of holes into the CQDs from oxidants, the rate of electron–hole annihilation is increased and emits photons as CL. CL properties of CQDs depend on surface groups and alkalinity of the solution. In basic medium, the effect of electron releasing property of CQDs and recombination of these injected electrons by emitting photons are responsible for observed CL behavior [29, 37, 38].

5.2.4 UPCONVERSION LUMINESCENCE

Upconversion fluorescence is anti-Stokes behavior involving multiple photon absorption at low energy excitations. So far, upconversion fluorescence materials reported are mostly rare-earth elements/organic dyes based materials. Compared to those costly and toxic materials, CQDs efficiently convert NIR light into high energy photons with extremely large two-photon cross-sections on laser excitations. Upconversion fluorescent materials have advantages of low optical penetration depth, spectral overlapping, photobleaching, and others over materials exhibiting down-conversion fluorescence. Signal-to-noise ratio can be increased efficiently due to the absence of autofluorescence from surroundings. To investigate upconversion fluorescence, lasers with sophisticated optical attachments are required as upconversion is a nonlinear optical process that requires high energy excitations. Some misinterpretations are already reported by observing fluorescence from spectrofluorimeter due to the excitation of second-order diffraction light from the noncoherent monochromators [39]. Under NIR excitation, a unique emission peak in visible wavelength with zero spectral shift was observed for CQDs [31]. Dependence of upconversion fluorescence on size effects or/and surface defects still remains controversial.

5.2.5 PHOSPHORESCENCE

Phosphorescence of a luminescent material has attracted tremendous attention due to the property of strong long-lived linked emissions. Their longer lifetimes typically of milliseconds to seconds can be attributed to the forbidden electronic transitions, that is, triplet to singlet phosphorescence. Phosphorescence of CQDs prepared from ethanolamine and EDA in the presence of phosphoric acid gave a lifetime to be 1.46 and 1.39 s, respectively [40, 41]. Observed phosphorescence can be attributed to phosphorous doping in CQDs and are most applicable in security feature printing. Additionally, phosphorescence decay lifetime of CQDs can also be significantly altered without doping phosphorous [28, 42]. At room temperature, when excited with UV light, CQDs in polyvinyl alcohol matrix having aromatic carbonyls groups on the surface exhibited phosphorescence due to the transitions from triplet excited states [43]. Phosphorescence of CQDs in aqueous solution showed higher lifetime than in the dry form as hydrogen bonding in water rigidifies the system [29, 43]. Phosphorescence depends on the pH of the medium too. Basic mediums showed ultralong lifetimes than acidic or neutral mediums due to the deprotonation of carboxylic groups with increased conjugation of electronic systems [28, 44].

5.3   APPLICATIONS OF CARBON QUANTUM DOTS

5.3.1   PHOTOVOLTAICS

The large consumption of electrical energy around the globe has created much demand in the field of energy production from renewable resources. The conversion of solar energy to electrical energy using the semiconducting materials that exhibit the photovoltaic property has become a cost-effective method of electrical energy production. A good absorber material in the case of terrestrial photovoltaic devices should have bandgap in between 1 and 1.5 eV so as to absorb sunlight of wavelength range 350–1000 nm [45]. The optical absorption must be in the range of 10^4–10^5 cm^{-1}. Those materials with long diffusion length and low rate of recombination with high quantum yield can be chosen as a good absorber material. Thin-film solar cells based on crystalline, amorphous, and microcrystalline silicon, chalcogenide materials, cadmium- and lead-based compounds, organic materials, dye-sensitized solar cells, cells based on titanium dioxide, copper- and tin-based materials, quantum dots, and others have been highly researched. Dye-based cells are cost-effective but have not been produced on large scale as the reliability on outdoors becomes a question. Quantum dot-based and plastic-based thin films that are known as third-generation thin films are still less attractive for manufacturing because of lower efficiency. Now to the efficiency remains high for solar cells based on silicon technology [45]. But the indirect bandgap of silicon of 1.1 eV cannot absorb more than 50% of visible spectra and needs for large absorber layer thickness are problems with silicon [46]. On considering the scarcity of the material, improvement of efficiency of the devices based on silicon and time taken for energy-pay-back, the research on new photovoltaic technologies and the search for cost-effective materials are significant [45]. The challenge is to develop an economically effective technology by which useful high-grade energy can be produced by earth-abundant, cost-effective materials to meet the terawatt demand and still it is an open problem [47]. CQDs of size less than 10 nm is an area of hot research nowadays in the field of photovoltaics [48]. High power conversion efficiency (PCE) of 9.10% is reported for silicon nanowire (Si NW) array/CQD core–shell heterojunction photovoltaic device [49]. The "green" nitrogen-doped carbon dots solar cell achieves the best PCE of 0.79% [50]. CQDs with excellent upconversion efficiency can be embedded into an active layer of polymer solar cells (PSCs) to improve their efficiency [51]. With optimized CQDs, the planar heterojunction perovskite solar cell devices showed improved electron mobility, electron extraction ability, and well-matched energy levels. This leads to remarkable increases on the short circuit current, open-circuit voltage, and PCE as high as ~ 19% [52].

The first step of conversion of light energy to electrical energy by photovoltaic method basically is the absorption of light energy by an appropriate semiconductor material, in a suitable form. Normally an internal electric field developed in the p–n junction will separate the charge carriers and these will flow to the external load through electrodes [46, 47]. The increase in carrier concentration will make a change in the Fermi level position of the p–n junction. The current (J_f) and voltage (V_f) are related in the case of a thin-film solar cell by the relation

$$J_f = j\left[\exp\frac{e\left(V_f - J_f R\right)}{K} - 1\right] + \left(\frac{V_f - RJ_f}{R_{shunt}}\right) - J_{load} \tag{5.1}$$

where J_{load} is current when electrons flow after illumination of the junction, R is the series resistance, j is the reverse saturation current, K is the product of diode factor, Boltzmann constant and absolute temperature that depends on the activation energy of the defect states ($E_{activation}$) also.

If $J_f = 0$ and $J_{load} = J_{short-circuit,}$ then from Equation (5.1), the open-circuit voltage V_{oc} can be calculated as

$$V_{open\ circuit} = \frac{E_{activation}}{e} + \frac{K}{e}\ln\left(\frac{J_{short-circuit}}{j_0}\right) \tag{5.2}$$

where j_0 is the reverse saturation current factor.

The efficiency of a solar cell is calculated from $J_{\text{short-circuit}}$, $V_{\text{ope circuit}}$, maximum voltage (V_{max}) and maximum current (I_{max}) as per the relation

$$\eta_{\text{solar cell}} = \frac{V_{\text{max}} I_{\text{max}}}{V_{\text{open circuit}} J_{\text{short-circuit}}} \times 100\% \tag{5.3}$$

$\dfrac{V_{\text{max}} I_{\text{max}}}{V_{\text{open circuit}} J_{\text{short-circuit}}}$ is known as the fill factor, which is a measure of the amount of power that can be realized from a solar cell. Figure 5.2 shows the model of current versus voltage graph of a solar cell under illumination. Heterojunction solar cells exhibit better characteristics than homojunction solar cells [43]. Heterojunctions and interfaces with suitable bandgap alignment will help in easy separation of charge carriers before recombining [48]. As per Schockley–Queisser method, the efficiency (E) of a solar cell can be calculated from the total input power (P_{input}) and the energy of photons generated as

$$E = \frac{h \upsilon_i N_i}{P_{\text{input}}} \tag{5.4}$$

where N_i is the number of photons with a frequency greater than υ_i [46].

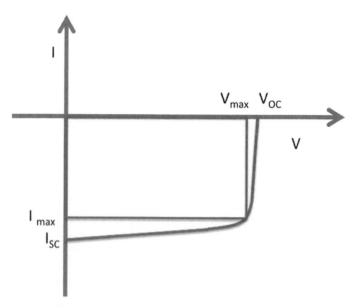

FIGURE 5.2   I–V characteristics of solar cell on illumination.

By incorporation of quantum dots and plasmons to conventional solar cells, the efficiency can be increased. Reports show that by increasing the weight percentage of CQDs in CQD/ZnO heterostructures the conversion efficiency of bare ZnO structure was increased by 7% [60]. In N-doped CQDs, when used in TiO_2 based dye-sensitized solar cells (DSSCs), an improved conversion efficiency is reported. N doping modifies the work function of CQDs. With increased work-function, it becomes easy to inject the charge carriers generated by light to the conduction band of TiO_2 (4.45 eV) [59]. The lowest unoccupied molecular orbital (LUMO) energy level of graphene QDs is reported to be in range of 3.5–3.7 eV [60]. Use of N-doped CQDs in DSSCs as electron donor layer has reported to convert the UV light to green light and this increased the efficiency of the solar cell by 10% [61]. CQD layers are experimented in bi-facial DSSCs and bi-tandem solar cells to enhance the conversion efficiency [62]. CQDs are used to enhance extraction of electrons, by incorporating with polyethyleneimine both in single junction and tandem solar cells and a maximum efficiency of 9.49% could be achieved [63].

QDs solar cells can be assumed to be having a p/QD/n structure as shown in Figure 5.3. It can be perceived as heterostructures containing donor/QD/acceptor interfaces [52]. The p material can act as an absorber and it absorbs photons of energy higher than the bandgap of the absorber. QDs of intermediate band energy absorbs photons of energy less than that of the bandgap of the absorber and transmits electrons from the valence band to intermediate bands. So by multiple transitions efficiency of the cell can be increased as the use of QD layer cause simultaneous production of more electron–hole pairs by absorption of photons of lesser energy [50]. So Equation (5.1) transforms by the introduction of an intermediate layer as

$$J_f = J_{load} - j_r - j\left[\exp\frac{eV_f}{nkT} - 1\right] \tag{5.5}$$

where j_r is a function of load current [50].

There are different structures like QDs immersed in electrolytes or QD sensitized structures adopted for solar cell engineering [52]. CQDs can be used in various varieties of QD solar cell structures and other solar energy converting technologies due to its wide range of properties like size tuned bandgap, good downconversion and upconversion PL, as electron acceptor, donor, and blocking region to avoid recombination [44].

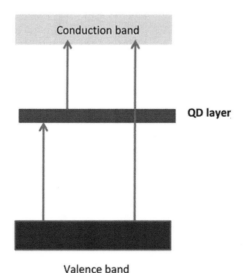

Valence band

FIGURE 5.3   Schematic diagram of intermediate band QD solar cell.

Si nanowire/carbon dot solar cell structure is showing an efficiency of 9.10% on AM1.5 global illumination. The heterojunction barrier height reported in this case was 0.75 eV. The rectification ratio calculated was 1000 at 0.8 V in the dark [60]. Xie et al. compare the various Si-CQD solar cell structures-SiNW-array (Schottky), SiNW-array/CQD heterojunctions, planar silicon/CQD heterojunctions. Devices were characterized by the thickness of the CQD layer constant for comparison. They could draw the best performance from the SiNW-array /CQD heterojunctions with a short circuit current density 30.09 mA/cm^2, open-circuit voltage 0.51 V and a fill factor of 0.593 [60]. The conversion efficiency in this case and that of silicon nanopillar/P3HT hybrid cell (9.2%) is comparable [60]. This structure reports a high zero bias photosensitivity of 3000 at 600 nm irradiation, large responsivity of 353 mA/W, a fast rise time of 20 μS, and a fall time of 40 μS [60].

The first step for characterizing CQDs for photovoltaic application is to check the PL [61]. Information about optically active states and mid bandgap states of high density can be drawn by this technique. By understanding the PL mechanism of the material, information about recombination rate can be understood. The recombination behavior influences the photovoltaic property of a material. By studying the Stoke's shift exhibited in PL the possible open-circuit voltage can be determined [61].

CQDs are incorporated in conventional solar cells to widen the spectrum of absorption as it is showing both downconversion and upconversion [48]. CQDs can also be used as photosensitizer due to the up and down conversion properties [62]. In downconversion CQDs are absorbing photons of higher energy, compared to the photons of visible spectra, which can be absorbed by a conventional solar cell. The absorption spectrum of CQDs is broad and it exhibits large absorption coefficient. This feature makes CQDs a good candidate for photovoltaic applications [63].

CQDs can be used as a spectral converter, photosensitizer, and electron mediator as the electron distribution in it enables it to act both as an electron acceptor and donor. CQDs have good electron storage capacity. Upconversion PL exhibited by CQDs makes them appreciable to be used as spectral converters [64]. So this structure can be used to block recombination of electron–hole pairs in solar cells [65]. The presence of carbonyl functional group decreases the rate of electron–hole recombination.

Another most interesting feature of CQDs that is applicable to photovoltaics is the size-tunable bandgap [48]. Due to the presence of infinite π network graphene sheets exhibit no bandgap and hence it is less appreciable to be used for semiconducting applications. But by breaking these infinite chains to finite ones or isolated ones imposes bandgap [66]. This isolation of sp^2 domains enhances the fluorescence emission by reducing quenching. These structures with the surface groups act as localizing centers to trap the charge carriers and hence CQDs become more attractive in solar cells. The bandgap depends on the size of the structure. As the size decreases, a blueshift in bandgap is observed [62].

Delayed exciton relaxation behavior makes CQDs applicable in photoelectrochemical cells (PECs), which convert solar energy to electrical energy for chemical reactions [48]. Through delayed fluorescence the recombination rate can be decreased [48]. CQDs can act as electron sinks and can be applied as photoanodes in solar cells [48]. CQD/TiO_2, $CQD/CdSe/TiO_2$, CQD/hematite structures are researched for application to PECs [48]. Considering the appreciable highest occupied molecular orbital (HOMO) and LUMO energy levels of CQDs, they are also researched as hole-transport medium and electron blocking layers in perovskite solar cells [67]. Solar cell efficiency depends on the geographical location also. In some places it will be too sunny and in some other it will not be so. Solar energy cannot be drawn during night also. But Meng et al. [68] report that CQDs synthesized from biomass can produce electricity in dark and light.

As CQD layers reduce series resistance in solar cells, they are experimented as electron transport layers in polymer-based solar cells [69]. CQDs synthesized by chemical vapor deposition are researched as electron acceptors as well in PSCs [70]. Depleted bulk heterojunction structure using CQDs has been proposed for solar cells to increase absorber thickness, to keep the CQD layer depleted, thereby managing both the charge collection and IR absorption [71]. The external quantum efficiency of such devices is reported to be 40% with an excitonic peak located at 950 nm. The device efficiency at AM 1.5 illumination was 5.5% [71]. Solar cells made with graphene QDs with high solubility/TiO_2 structure exhibited a large short circuit current as high as 200 $\mu A/cm^2$. The open-circuit voltage reported in this case is 0.48 V and the fill factor 0.58 under illumination by AM1.5 global light [72].

A comparative study of different CQD-based solar cell structure—glass/ITO/CQD/Al or Mg (Schottky), glass/FTO/TiO_2 or ZnO/CQD/MoO_3 or Au (depleted heterojunction), glass/FTO/TiO_2-CQD/Au or MoO_3 (depleted bulk heterojunction), glass/FTO/TiO_2/CQD-TiO_2-electrolyte/Pt or Au (CQD sensitized), glass/FTO/TiO_2/CQD/graded recombination layer/CQD/Au or MoO_3 (tandem), glass/FTO/TiO_2/CQD/quantum funnel/Au or MoO_3 (quantum funnel structure)—is discussed with spatial band diagram in each case by Kramer et al. [61]. The introduction of recombination layer into glass/FTO/TiO_2/CQD/graded recombination layer/CQD/Au or MoO_3 (tandem) help the electrons produced in the top cell to combine with the holes produced in the bottom cell. The depleted heterojunction structures always compromise the absorbed photon and extracted electron rate. This ratio can be improved by enhancing the quasi-neutral charge carrier collection [61]. Quantum funnel structure is such an architecture that utilizes the size effect of CQDs for improving this ratio. The SiNW/CQD heterojunction solar cell with as prepared efficiency 9.10%, which was kept in the air without encapsulation, when underwent for stability study in the air after 2 weeks exhibited a decrement in the photoconversion efficiency by 14%–7.82% [60]. The light to dark current ratio was observed to decrease from 3000 to 2700 [60]. CQDs are lightweight materials, which are easy to synthesize. The processing temperature is not high and solution-based processing technologies can be followed for growth of CQDs. But the controllability on doping remains a challenge. CQDs exhibit low carrier mobility and the possibilities of carrier trapping by surface defect states are high. The incomplete understanding of the transport and recombination mechanisms remains as a challenge to play with the material [61].

5.3.2 PHOTOCATALYSIS

Environmental pollution associated with the air, water, soil, and others and shortage of fresh water due to continual increase in world population have emerged as the most serious crisis facing in the world. In recent years, many research groups have focused on intense research to develop a potentially sustainable approach to overcome these issues. Nowadays, more and more attention has been paid to address water pollution by introducing economically viable, environmental friendly photosensitive catalysts. The photocatalysts have a capability to breakdown/convert organic impurities/contaminants in wastewater to harmless substances using photons from UV–visible light [73]. Additionally, worldwide efforts have been initiated for finding innovative methods for hydrogen production, because hydrogen is considered as ideal energy material for the future. One of the most effective and low-cost methods for the production of hydrogen is photocatalytic water splitting (refer Figure 5.4) [74]. Fujishima and Honda first reported the phenomenon of water splitting in 1972 [75]. Photocatalytic activity of semiconducting materials depends on the separation of charge carriers (hole and electron), charge carrier transport, and lifetime of carriers [76]. The separated holes react with adsorbed water molecules and generate hydroxyl radicals, which oxidize the organic molecules present in wastewater. The reaction of photoexcited electrons with oxygen leads to the creation of superoxide radical anions. But the recombination of carriers due to various defects of materials limits the photocatalytic efficiency [76–78]. Furthermore, it is difficult to develop photocatalytic materials with\ greater light absorption especially in the wavelength region of visible light. Titanium dioxide (TiO_2) is promising in this regard due to its low cost, nontoxicity, photoactivity, chemical stability, strong oxidizing abilities, and others [79]. Researchers attracted widespread interest in modification of wide bandgap semiconducting oxides such as TiO_2, ZnO to take advantage in absorption range in ultraviolet light and visible light [80].

CQDs, the novel zero-dimensional (0D) materials with size less than 10 nm in carbon family, have received enormous attention because of its fascinating properties such as excellent biocompatibility, low cost and toxicity, robust chemical inertness, and water solubility [82, 83]. Various breakthrough works related to carbon-based nanomaterials reported that electron-accepting and transport properties of these materials showed enhancement in the transport of charge carriers [84]. Most recently, in order to enhance the photocatalytic activity, semiconductor-based nanocatalysts are used as new candidates by tailoring morphology and size of these catalysts. The fabrication of carbon dots based efficient photocatalytic composites (TiO$_2$/CQDs, Fe$_2$O$_3$/CQDs) provides a drastic change in photocatalytic performance. These composites found to have versatile and advantageous properties such as extended light absorption range and reduction in the recombination sites. The large electron storage reservoir capacity of carbon dots leads to the formation of direct conduction paths for electrons, which may promote the production of a large amount of active oxygen radicals [85, 86].

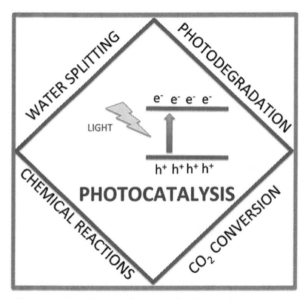

FIGURE 5.4 Diagram depicting various applications of photocatalysts under light irradiation [81].

During the last decade, an increasing number of research articles on the area of carbon dot-based TiO$_2$ photocatalysts have been published. Chen et al. prepared carbon dots (C-dots) decorated TiO$_2$ synthesized by hydrothermal-calcination that possess excellent upconversion PL. These C-dots decorated titanium dioxide showed 2.3 times faster photodegradation of gemfibrozil than pristine TiO$_2$ under visible light irradiation [87]. Furthermore, TiO$_2$ nanotube arrays anchored with carbon dots showed improved rhodamine B (RhB) degradation and water splitting. It was observed that the higher photo-catalytic activity of these catalysts is due to the upconversion properties of the carbon dots, extended spectral response, bonding interaction between the carbon dots, and the titanium dioxide nanotubes [88]. CQDs modified P25 TiO_2 composites with a "dyade"-like structure showed enhanced photo-catalytic H_2 evolution is because of dual role played by CQDs. Here CQDs act as an electron reservoir under UV–Vis light irradiation, thereby helping effective separation of charge carriers of P25. CQDs herein perform as a photosensitizer, leading to efficient absorption of visible light [89]. Li et al. synthesized nanohybrid composites of carbon dots and titanium dioxide using solvothermal method. Compare with commercial Degussa P25, the carbon dots-TiO_2 composites showed excellent photo-degradation on methylene blue (MB), under ultraviolet light irradiation methyl orange (MO) and RhB [90]. Formation of absorption bands in the 370–450 nm region due to doping of carbon dots in TiO_2 nanoparticles and the surface of g-C_3N_4 nanosheets through hydrothermal process promotes the photocatalytic degradation of enrofloxacin [91].

Researchers also designed and fabricated various complex photocatalysts by the integration of CQDs with various materials other than TiO_2. Novel Fe_2O_3/CQDs nanocomposites synthesized via a hydrothermal method exhibited enhanced photocatalytic capability for toxic gas degradation under the irradiation of visible light. Due to the large electron-storage capacity of carbon dots, electrons generated from Fe_2O_3 particles through photoexcitation transported freely through the CQD network. This leads to the production of a large amount of active oxygen radicals, with very strong oxidation capability, which provides photocatalytic degradation of gas-phase benzene and gas-phase methanol [92]. "Red-emitting magnesium-nitrogen-embedded carbon dots" (r-Mg-N-CD) prepared with the help of Bougainvillea plant leaves extract exhibited emissions around 678 nm, which makes them as a potential candidate for visible-light photocatalytic active material for the degradation of pollutant dye (MB) [93].

Wu et al. prepared Cu–N-doped carbon dots by one-step pyrolysis from the precursor $Na_2[Cu(EDTA)]$. These dots displayed photooxidation of 1,4-DHP in aqueous solution. Cu–N-doped carbon dots showed improved photocatalytic ability than pure carbon dots due to its better electron-accepting and donating capabilities. Xie and coworkers observed higher visible light-mediated photocatalytic degradation of tetracycline using carbon dots/g-C_3N_4/MoO_3 in contrast to pristine g-C_3N_4 and MoO_3/g-C_3N_4 composite [94]. In order to reduce nitro-benzene derivatives in the aqueous phase, heterojunction films prepared using carbon dots and CdS via electrophoretic and sequential chemical bath deposition method were employed. The electron donor–acceptor heterojunction formed in hybrid films can accelerate the separation of carriers and effectively reduce the chance for recombination, eventually leads to better photocatalytic efficiency [95]. De et al. synthesized thermostable and reusable carbon dot reduced Cu_2O nanohybrid/hyperbranched epoxy nanocomposite for the degradation of photodegradation of ethyl paraoxon organophosphate pesticide under solar light [96]. Novel ternary gCN-NS/CDs/AgCl nanocomposites fabricated by the decoration of carbon dots and AgCl particles over the nanosheets of gCN showed excellent photocatalytic degradations of four pollutants including two cationic dyes (RhB and MB), one anionic dye (MO), and one colorless pollutant (phenol) under visible light [97].

CQDs embedded in mesoporous α-Fe_2O_3 act as efficient photocatalysts, which showed up to 97% capacity retention even after 3 cycles [98]. CQDs/ hydrogenated TiO_2 nanobelt heterostructures show an active photocatalytic property in UV–Vis–NIR broad spectrum [99]. Composite of TiO_2 (P25) and N-doped CQDs (P25/NCQD)—act as a catalyst of the photo-oxidation of NO under UV and visible light irradiation [100]. The heterostructure of ZnO and CQDs can effectively increase the photodegradation efficiency compare to bare ZnO. It was observed that the enhancement in photocatalytic activity of ZnO/CQDs heterostructure strongly depends on the number of layers of CQDs. The improved performance of ZnO/CQDs heterostructure as photo-catalysts is attributed to enhanced carrier formation and reduction in the recombination of these charge carriers because of the electronic interaction between ZnO and CQDs [101]. Improved light absorption and the interfacial transfer of excited electrons result in significantly enhanced photocatalytic activities of carbon dots/BiOBr nanocomposites under both UV and visible light irradiation [102].

5.3.3   CHEMICAL SENSING

CQDs can be effectively used in chemical sensing like in the sensing of mercury ions—Hg^{2+}, cadmium ions—Cd^{2+}, zinc ions—Zn^{2+}, copper ions—Cu^{2+}, silver ions—Ag^+, gold ions—Au^{3+}, iron ions—Fe^{3+}, Fe^{2+}, cobalt ions—Co^{2+}, nickel ions, palladium ions, platinum ions, chromium—Cr^{3+}, Cr^{6+}, manganese ions—Mn(II), lead ions—Pb^{2+}, tin ions—Sn(II), bismuth ions—Bi^{3+}, aluminium ions—Al^{3+}, potassium ions and beryllium ions—K^+, Be^{2+} [103]. Mercury ions (Hg^{2+}), due to its strong toxicity and bioaccumu-lation, can lead to various diseases even at low concentrations, including headaches, hyperspasmia, renal failure, and so on. Based on the quenching effect of mercuric ions on the fluorescence of CQDs the presence of Hg^{2+} ions can be sensed. Zhang et al. reported a "turn-on" fluorescence nanosensor for selective determination of Hg^{2+} [104]. The wide nature of precursors for CQDs and their chemical sensing is shown in Table 5.1.

TABLE 5.1 Literature Review of CQDs Regarding the Precursor of Synthesis and Sensing Element

Sr. No.	Detection	CQDs Source	DOI
1	Acetone and dopamine (neurotransmitter)	NB-CQDs from citric acid, borax and p-phenylenediamine	10.1016/j.snb.2018.10.075
2	Ag+ ions	S/N-CDs synthesized from citric acid (CA) and guanidine thiocyanate (GITC)	10.1016/j.saa.2018.10.004
		1,3,6-Trinitropyrene and 3-mercaptopropionic acid (MPA)	10.1016/j.snb.2016.11.044
3	Alcohol vapors and volatile organic compounds	Rice husk	10.1016/j.colsurfa.2018.09.077
4	*Bacillus anthracis* marker (dipicolinic acid, DPA)	Citric acid and urea	10.1007/s10853–018-2955–3
5	Cr(VI) ions	Hexamethylenetetramine (HMTA) Lemon peel waste	10.1016/j.snb.2018.09.101 10.1039/C6RA10488F
6	Cr(VI) ions and dopamine (neurotransmitter)	phosphorus/nitrogen co-doped c-dot (PNCD) by using H_3PO_4 and ethylenediamine	10.1039/C8RA06120C
7	Cu^{2+} ions	g-C_3N_4 in H_2O_2 solution Mg-N-embedded C-dots Sugarcane juice Purified lemon extract and l-arginine Bamboo leaves Grass	10.1007/s10854–0180193–8 10.1039/C8NJ04754E 10.1039/C8AY00928G 10.1016/j.msec.2017.03.045 10.1016/j.snb.2014.02.053 10.1002/adma.201200164
8	Cu^{2+} ions and amine vapor	Black sesame seeds	10.1088/1757–899X/378/1/012003
9	Cu^{2+} ions, glutathione	N-CQDs from hexamethylenetetramine	10.1007/s00216–018-1387-x

TABLE 5.1   *(Continued)*

Sr. No.	Detection	CQDs Source	DOI
10	Cysteine	Cobalt-doped CDs (Co-CDs) by using folic acid (FA) as precursor and $CoCl_2$ as dopant	10.1016/j.snb.2018.10.029
11	Fe^{2+}/Fe^{3+} ions	Trisodium citrate dehydrate and urea	10.1016/j.jlumin.2018.10.001
		lime oil extract	10.1039/C7RA13432K
12	Fe^{3+} ions	N-CDs by using biomass waste (i.e., expired milk)	10.1021/acsomega.8b01919
		N-CDs from rose-heart radish	10.1016/j.snb.2016.10.068
		Aminosalicylic acid	10.1021/acsami.6b13954
		B,N,S-CDs from 2,5-diaminobenzenesulfonic acid and 4-aminophenyl boronic acid hydrochloride.	10.1021/acsami.6b15746
		N,S-CDs from citric acid & thiourea	10.1021/acsami.7b04514
		(N-CDs) from pyrolysis of konjac flour	10.1039/C4TB00368C
		Foxtail millet	10.1039/C8NJ01072B
		Honey	10.1016/j.bios.2014.04.046
		Coriander leaves	10.1039/C5AN00454C
		Onion waste	10.1039/C6RA01669C
		Garlic	10.1186/s11671–016-1326–8
13	Fe^{3+} and ascorbic acid (AA)	N,S-codoped c-dots from toxic cigarette butts (CBs)	10.1021/acsomega.8b01743
14	Glucose	Lignite (coal)	10.1038/s41598–018-32371–9
		Phenylboronic acid	10.1021/ac5001338
		AuNPs–rGO nanocomposites from willow bark	10.1039/C2CY20635H
15	Hg^{2+} and Ag^+ ions	Uric acid	10.1016/j.snb.2016.11.149

TABLE 5.1   *(Continued)*

Sr. No.	Detection	CQDs Source	DOI
16	Hg^{2+} and Cu^{2+} ions	Urine	10.1039/C5GC02032H
17	Hg^{2+} ions	Table sugar and honey Pineapple peel Pomelo peel Cucumber juice Flour Lotus root Strawberry juice Coconut milk Jackfruit	10.1016/j.jlumin.2018.09.061 10.1021/acsomega.8b01146 10.1021/ac3007939 10.1039/C4RA10885J 10.1016/j.snb.2013.04.079 10.1016/j.apsusc.2016.08.012 10.1039/C3RA43452D 10.1016/j.jlumin.2014.12.048 orcid.org/0000–0002-8903–5132
18	l-lyseine	Citric acid and ethylenediamine	10.1016/j.snb.2018.10.026
19	Methotrexate	Citric acid monohydrate and l-cysteine	10.1016/j.aca.2018.10.005
20	Nitrophenol (used to manufacture drugs, fungicides, insecticides, and dyes)	N and S co-doped C-dots (CQDs) prepared using palm shell powder (PSP) and triflic acid	10.1016/j.materresbull.2018.08.033
21	Phenobarbital (epilepsy drug)	Cedrus	10.1016/j.talanta.2018.09.069
22	Sudan I	Cigarette filters	10.1016/j.aca.2018.03.024
23	Different metal ions	Camphor	10.1039/C4RA10471D

5.3.4 BIOIMAGING

Visible excitation and emission wavelengths, fluorescence brightness at the individual dot level, and high photostability made CQDs potential candidates as drug carriers and for bioimaging system. Surface-passivated ZnS-doped CQDs are first injected mice for in vivo imaging by 470 or 545 nm wavelength excitation [105]. Due to low toxicity, surface-functionalized and heteroatom-doped CQDs are applied for drug/gene delivery and cancer diagnostics [106].

CQDs can be extensively used for biophotonic applications including cellular uptakes. The fluorescence brightness in the cellular environment and the output of in vivo imaging in mice models are excellent for CQDs when compared to toxic CdSe–ZnS QDs [107]. Based on investigations from cytotoxicity and in vivo toxicity analysis, CQDs exhibit competitive performance in cellular or in vivo imaging [107].

5.4 CONCLUSION

This chapter highlighted the optical properties of CQDs and new progress in their analytical applications. Even though various facile chemical routes have been exploited for synthesizing their amorphous and crystalline form, the dependence of their crystalline structure with optical properties is not well explored. Moreover, various factors regarding the origin of their tunable fluorescence are still a debating problem. Compared to traditional semiconducting quantum dots key qualities like low toxicity, biofriendliness are favoring wide applicability of CQDs but their low quantum yields are still an issue to be resolved.

KEYWORDS

- **carbon quantum dots**
- **photoluminescence**
- **chemiluminescence**
- **phosphorescence**
- **photovoltaics**
- **photocatalysis**

REFERENCES

1. Pawar, R. S., Upadhaya, P. G., and Patravale, V. B. Chapter 34—Quantum dots: novel realm in biomedical and pharmaceutical industry. Handbook of Nanomaterials for Industrial Applications, C. Mustansar Hussain, Ed. Elsevier. 2018, 621–637.
2. Grigore, M. E., Holban, A. M., and Grumezescu, A. M. 9—Nanotherapeutics in the management of infections and cancer. Nanobiomaterials Science, Development and Evaluation, M. Razavi and Thakor, A., Eds. Woodhead Publishing. 2017, 163–189.

3. Suri, S., Ruan, G., Winter, J., and Schmidt, C. E. Chapter I.2.19—Microparticles and nanoparticles. Biomaterials Science (Third Edition), Ratner, B. D., Hoffman, A. S., Schoen, F. J., and Lemons, J. E., Eds. Academic Press. 2013, 360–388.

4. Bose, A. and Wui Wong, T. Chapter 11—Nanotechnology-enabled drug delivery for cancer therapy. Nanotechnology Applications for Tissue Engineering, Thomas, S., Grohens, Y., and Ninan, N., Eds. Oxford: William Andrew Publishing. 2015, 173–193.

5. Georgakilas, V., Perman, J. A., Tucek, J., and Zboril, R. Broad family of carbon nanoallotropes: classification, chemistry, and applications of fullerenes, carbon dots, nanotubes, graphene, nanodiamonds, and combined superstructures. Chem. Rev. 2015, Vol. 115, 4744–4822.

6. Kroto, H. Space, stars, C60, and soot. Science, 1988, Vol. 242, no.4882, 1139–1145.

7. Krätschmer, W., Lamb, L. D., Fostiropoulos, K., and Huffman, D. R. Solid C 60: a new form of carbon. Nature, 1990, Vol. 347, no. 6291, 354–358.

8. Thostenson, E. T., Ren, Z., and Chou, T.-W. Advances in the science and technology of carbon nanotubes and their composites: a review. Compos. Sci. Technol., 2001, Vol. 61, no. 13, 1899–1912.

9. Dresselhaus, M. S., Dresselhaus, G., Eklund, P. C., and Rao, A. M. Carbon nanotubes. The Physics of Fullerene-Based and Fullerene-Related Materials, Andreoni, W., Ed. Dordrecht: Springer Netherlands. 2000, 331–379.

10. Dong, Y., Lin, J., Chen, Y., Fu, F., Chi, Y., and Chen, G., Graphene quantum dots, graphene oxide, carbon quantum dots and graphite nanocrystals in coals. Nanoscale, 2014, Vol. 6, no. 13, 7410–7415.

11. Tao, S., Feng, T., Zheng, C., Zhu, S., and Yang, B. Carbonized polymer dots: a brand new perspective to recognize luminescent carbon-based nanomaterials. J. Phys. Chem. Lett. 2019, Vol. 10, no. 17, 5182–5188.

12. Fan, Z., Li, S., Yuan, F., and Fan, L. Fluorescent graphene quantum dots for biosensing and bioimaging. RSC Adv. 2015, Vol. 5, no. 25, 19773–19789.

13. Zhu, S., Song, Y., Zhao, X., Shao, J., Zhang, J., and Yang, B. The photoluminescence mechanism in carbon dots (graphene quantum dots, carbon nanodots, and polymer dots): current state and future perspective. Nano Res. 2015, Vol. 8, no. 2, 355–381.

14. Xu, X. et al. Electrophoretic analysis and purification of fluorescent single-walled carbon nanotube fragments. J. Am. Chem. Soc. 2004, Vol. 126, no. 40, 12736–12737.

15. Gao, X., Du, C., Zhuang, Z., and Chen, W. Carbon quantum dot-based nanoprobes for metal ion detection. J. Mater. Chem. C 2016, Vol. 4, no. 29, 6927–6945.

16. Sun, Y.-P., et al. Quantum-Sized carbon dots for bright and colorful photoluminescence. J. Am. Chem. Soc. 2006, Vol. 128, no. 24, 7756–7757.

17. Ying Lim, S., Shen, W., and Gao, Z. Carbon quantum dots and their applications. Chem. Soc. Rev. 2015, Vol. 44, no. 1, 362–381.

18. Sahu, S., Behera, B., Maiti, T. K., and Mohapatra, S. Simple one-step synthesis of highly luminescent carbon dots from orange juice: application as excellent bio-imaging agents. Chem. Commun. 2012, Vol. 48, no. 70, 8835–8837.

19. A Alam, M., Park, B.-Y., Ghouri, Z. K., Park, M., and H.-Y. Kim. Synthesis of carbon quantum dots from cabbage with down- and up-conversion photoluminescence properties: excellent imaging agent for biomedical applications. Green Chem. 2015, Vol. 17, no. 7, 3791–3797.

20. Mandani, S., Dey, D., Sharma, B., and Sarma, T. K. Natural occurrence of fluorescent carbon dots in honey. Carbon 2017, Vol. 119, 569–572.

21. T. S. and R. S. D. Green synthesis of highly fluorescent carbon quantum dots from sugarcane bagasse pulp. Appl. Surf. Sci. 2016, Vol. 390, 435–443.
22. Huang, H. et al. One-pot green synthesis of nitrogen-doped carbon nanoparticles as fluorescent probes for mercury ions. RSC Adv. 2013, Vol. 3, no. 44, 21691–21696.
23. Liu, S., et al. Hydrothermal treatment of grass: a low-cost, green route to nitrogen doped, carbon-rich, photoluminescent polymer nanodots as an effective fluorescent sensing platform for label-free detection of Cu(II) ions. Adv. Mater. 2012, Vol. 24, no. 15, 2037–2041.
24. Hu, S.-L., Niu, K.-Y., Sun, J., Yang, J., Zhao, N.-Q., and Du, X.-W. One-step synthesis of fluorescent carbon nanoparticles by laser irradiation. J. Mater. Chem. 2009, Vol. 19, no. 4, 484–488.
25. Bourlinos, A. B., Stassinopoulos, A., Anglos, D., Zboril, R., Karakassides, M., and E. P. Giannelis. Surface functionalized carbogenic quantum dots. 2008, Vol. 4, no. 4, 455 458.
26. Liu, J., Liu, X., Luo, H., and Gao, Y. One-step preparation of nitrogen-doped and surface passivated carbon quantum dots with high quantum yield and excellent optical properties. RSC Adv. 2014, Vol. 4, no. 15, 7648–7654.
27. Sun, Y.-P., et al. Doped carbon nanoparticles as a new platform for highly photoluminescent dots. J. Phys. Chem. C. 2008, Vol. 112, no. 47, 18295–18298.
28. Mintz, K. J., Zhou, Y., and Leblanc, R. M. Recent development of carbon quantum dots regarding their optical properties, photoluminescence mechanism, and core structure. Nanoscale 2019, Vol. 11, no. 11, 4634–4652.
29. Wang, Y. and Hu, A. Carbon quantum dots: synthesis, properties and applications. Mater, J. Chem. C 2014, Vol. 2, no. 34, 6921–6939.
30. Sharma, A., Gadly, T., Gupta, A., Ballal, A., Ghosh, S. K., and Kumbhakar, M. Origin of excitation dependent fluorescence in carbon nanodots. J. Phys. Chem. Lett. 2016, Vol. 7, no. 18, 3695–3702.
31. Gogoi, S. and Khan, R. NIR upconversion characteristics of carbon dots for selective detection of glutathione. New J. Chem. 2018, Vol. 42, no. 8, 6399–6407.
32. Kalytchuk, S., et al. Carbon Dot Nanothermometry: intracellular photoluminescence lifetime thermal sensing. ACS Nano 2017, Vol. 11, no. 2, 1432–1442.
33. Zheng, C., An, X., and Gong, J. Novel pH sensitive N-doped carbon dots with both long fluorescence lifetime and high quantum yield. RSC Adv. 2015, Vol. 5, no. 41, 32319–32322.
34. Kumar, P. and Bohidar, H. B. Observation of fluorescence from non-functionalized carbon nanoparticles and its solvent dependent spectroscopy. J. Lumin. 2013, Vol. 141, 155–161.
35. Yang, Z., et al. Nitrogen-doped, carbon-rich, highly photoluminescent carbon dots from ammonium citrate. Nanoscale. 2014, Vol. 6, no. 3, 1890–1895.
36. Lin, Z., Xue, W., Chen, H., and Lin, J.-M. Classical oxidant induced chemiluminescence of fluorescent carbon dots. Chem. Commun. 2012, Vol. 48, no. 7, 1051–1053.
37. Dou, X., Lin, Z., Chen, H., Zheng, Y., Lu, C., and Lin, J.-M. Production of superoxide anion radicals as evidence for carbon nanodots acting as electron donors by the chemiluminescence method. Chem. Commun. 2013, Vol. 49, no. 52, 5871–5873.
38. Zhao, L., et al. Chemiluminescence of carbon dots under strong alkaline solutions: a novel insight into carbon dot optical properties. Nanoscale. 2013, Vol. 5, no. 7, 2655–2658.

39. Barati, A., Shamsipur, M., and Abdollahi, H. A misunderstanding about upconversion luminescence of carbon quantum dots. J. Iran. Chem. Soc. 2015, Vol. 12, no. 3, 441–446,

40. Jiang, K., Wang, Y., Gao, X., Cai, C., and Lin, H. Facile, quick, and Gram-scale synthesis of ultralong-lifetime room-temperature-phosphorescent carbon dots by microwave irradiation. Angew. Chem. Int. Ed. 2018, Vol. 57, no. 21, 6216–6220.

41. Bian, L., et al. Simultaneously enhancing efficiency and lifetime of ultralong organic phosphorescence materials by molecular self-assembly. J. Am. Chem. Soc. 2018, Vol. 140, no. 34, 10734–10739.

42. Han, Y., et al. Rational design of oxygen-enriched carbon dots with efficient room temperature phosphorescent properties and high-tech security protection application. ACS Sustain. Chem. Eng. 2019, Vol. 7, no. 24, 19918–19924.

43. Deng, Y., Zhao, D., Chen, X., Wang, F., Song, H., and Shen, D. Long lifetime pure organic phosphorescence based on water soluble carbon dots. Chem. Commun. 2013, Vol. 49, no. 51, 5751–5753.

44. Tan, J., Ye, Y., Ren, X., Zhao, W., and Yue, D. High pH-induced efficient room temperature phosphorescence from carbon dots in hydrogen-bonded matrices. J. Mater. Chem. C 2018, Vol. 6, no. 29, 7890–7895.

45. Poortmans, J. and Archipov, V. Preface: Thin film solar cells: fabrication, characterization and applications. Poortmans, J. and Archipov, V., Ed. John Wiley & Sons Ltd. 2006, xvii–xxix.

46. Kodigala, S. R., Chapter-1: Thin Film Solar Cells from Earth Abundant Materials, Elsevier. 2014, 1–13.

47. Chopra, K. L. and Das, S. R., Chapter-1: Thin Film Solar Cells, Springer Science & Business Media, LLC, 1983, 1–18

48. Molaei, M. J. The optical properties and solar energy conversion applications of carbon quantum dots: a review. Sol. Energy. 2020, Vol. 196, 549–566. doi:10.1016/j.solener.2019.12.036.

49. Xie, C., Nie, B., Zeng, L., Liang, F. X., Wang, M. Z., Luo, L., Feng, M., Yu, Y., Wu, C. Y., Wu, Y., and Yu, S. H., Core-shell heterojunction of silicon nanowire arrays and carbon quantum dots for photovoltaic devices and self-driven photodetectors, ACS Nano 2014, Vol. 8, 4015–4022.

50. Wang, H., Sun, P., Cong, S., Wu, J., Gao, L., Wang, Y., Dai, X., Yi, Q. and Zou, G., Nitrogen-doped carbon dots for "green" quantum dot solar cells. Nanoscale Res. Lett. 2016, 11–27.

51. Liu, C., Chang, K., W.Guo, Li, H., Shen, L., Chen, W., and Yan, D. Improving charge transport property and energy transfer with carbon quantum dots in inverted polymer solar cells. Appl. Phys. Lett. 2014, 105, 073306.

52. Li, H., Shi, W., Huang, W., Yao, E. P., Han, J., Chen, Z., Liu, S., Shen, Y., Wang, M., and Yang, Y. Carbon quantum dots/TiO_x electron transport layer boosts efficiency of planar heterojunction perovskite solar cells to 19%. Nano Lett. 2017, 17, no. 4, 2328–2335.

53. Neamen, D. A. and Biswas, D. Chapter-10: Semiconductor Physics and Devices, Tata McGraw Hill Education Private Limited, 2012, 417–421.

54. Rezaei, B., et al., The impressive effect of eco-friendly carbon dots on improving the performance of dye-sensitized solar cells. Sol. Energy. 2019, Vol. 182, 412–419.

55. Zhang, Y.-Q., et al. N-doped carbon quantum dots for TiO_2-based photocatalysts and dye-sensitized solar cells. Nano Energy. 2013, Vol. 2, 545–552.

56. Yan, X., et al. Independent tuning of the band gap and redox potential of grapheme quantum dots. J. Phys. Chem. Lett. 2011, Vol. 2, 1119–1124

57. Riaz, R., et al. Dye-sensitized solar cell (DSSC) coated with energy down shift layer of nitrogen-doped carbon quantum dots (N-CQDs) for enhanced current density and stability. Appl. Surf. Sci. 2019, Vol. 483, 425–431.

58. Duan, J., et al. Efficiency enhancement of bifacial dye-sensitized solar cells through bi-tandem carbon quantum dots tailored transparent counter electrodes. Electrochim. Acta 2018, Vol. 278, 204–209.

59. Kang, R., et al. High-efficiency polymer homo-tandem solar cells with carbon quantum-dot-doped tunnel junction intermediate layer. Adv. Energy Mater. 2018, Vol. 8, 1702165.

60. Xie, C., Nie, B., Zeng, L., Liang, F.-X., Wang, M.-Z., Luo, L., Feng, M., Yu, Y., Wu, C.-Y., Wu, Y. and Yu, S.-H. Core Å shell heterojunction of silicon nanowire arrays and carbon quantum dots for photovoltaic devices and, ACS Nano 2014, Vol. 8, 4015–4022. doi:10.1021/nn501001j.

61. Kramer, I. J, and Sargent, E. H. Colloidal quantum dot photovoltaics: a path forward, ACS Nano 2011, Vol. 5, 8506–8514. doi:10.1021/nn203438u.

62. Cao, L., et al., Photoluminescence properties of graphene versus other carbon nanomaterials. Acc. Chem. Res. 2012, Vol. 46, 171–180.

63. Mirtchev, P., et al. Solution phase synthesis of carbon quantum dots as sensitizers for nanocrystalline TiO_2 solar cells. J. Mater. Chem. 2012, Vol. 22, 1265–1269.

64. Li, H., et al. Near-infrared light controlled photocatalytic activity of carbon quantum dots for highly selective oxidation reaction. Nanoscale. 2013, Vol. 5, 3289–3297.

65. Wang, R., et al. Recent progress in carbon quantum dots: synthesis, properties and applications in photocatalysis. J. Mater. Chem. A. 2017, Vol. 5, 3717–3734.

66. Molaei, M. J. Carbon quantum dots and their biomedical and therapeutic applications: a review. RSC Adv. 2019, Vol. 9, 6460–6481.

67. Paulo, S., et al. Carbon quantum dots as new hole transport material for perovskite solar cells. Synth. Metals. 2016, Vol. 222. 10.1016/j.synthmet.2016.04.025.

68. Meng, Y., et al. Biomass converted carbon quantum dots for all-weather solar cells. Electrochim. Acta 2017, Vol. 257, 259–266.

69. Yan, L., et al. Synthesis of carbon quantum dots by chemical vapor deposition approach for use in polymer solar cell as the electrode buffer layer. Carbon 2016, Vol. 109, 598–607.

70. Cui, B., et al. Fluorescent carbon quantum dots synthesized by chemical vapor deposition: an alternative candidate for electron acceptor in polymer solar cells. Opt. Mater. 2018, Vol. 75, 166–173.

71. Barkhouse, D.A.R., Debnath, R., Kramer, I.J., Zhitomirsky, D., Pattantyus-abraham, A.G., Levina, L., Etgar, L., Grätzel, M., and Sargent, E. H.. Depleted bulk heterojunction colloidal quantum dot photovoltaics, Adv. Mater. 2011, Vol. 23, 3134–3138. doi:10.1002/adma.201101065.

72. Yan, X., Cui, X., Li, B. and Li, L., Large, Solution-process graphene quantum dots as light absorbers for photovoltaics, Nano Lett. 2010, Vol. 10, 1869–1873. doi:10.1021/nl101060h.

73. Zhu, S., Wang, D. Photocatalysis: basic principles, diverse forms of implementations and emerging scientific opportunities, Adv. Energy Mater. 2017, Vol. 7, 1700841.

74. Ni, M., Leung, M. K. H., Leung, D. Y. C., Sumathy, K. A review and recent developments in photocatalytic water-splitting using TiO_2 for hydrogen production. Renew. Sustain. Energy Rev. 2007, Vol. 11, 401–425.
75. Fujishima, A., Honda, K. Electrochemical photolysis of water at a semiconductor electrode, Nature 1972, Vol. 238, 37–38.
76. Byrne, C., Subramanian, G., Pillai, S. C. Recent advances in photocatalysis for environmental applications. J. Environ. Chem. Eng. 2018, Vol. 6, 3531–3555.
77. Ibhadon, A. O., Fitzpatrick, P. Heterogeneous photocatalysis: recent advances and applications. Catalysts, 2013, Vol. 3, 189–218.
78. Jafari, T., Moharreri, E., Amin, A. S., Miao, R., Song, W., Suib, S. L. Photocatalytic water splitting—the untamed dream: a review of recent advances. Molecules 2016, Vol. 21, 900.
79. Humayun, M., Raziq, F., Khan, A., Luo, W. Modification strategies of TiO_2 for potential applications in photocatalysis: a critical review. Green Chem. Lett. Rev. 2018, Vol. 11, 86–102.
80. Nakata, K., Fujishima, A. TiO_2 photocatalysis: design and applications. J. Photochem. Photobiol. C Photochem. Rev. 2012, Vol. 13, 169– 189.
81. Han, M., Zhu, S., Lu, S., Song, Y., Feng, T., Tao, S., Liu, J., Yang, B. Recent progress on the photocatalysis of carbon dots: classification, mechanism and applications. Nano Today 2018, Vol.19, 201–218
82. Zhu, S., Song,Y., Zhao, Zhu, X., S., Song, Y., Zhao, X., Shao, J., Zhang, J., Yang, B. The photoluminescence mechanism in carbon dots (graphene quantum dots, carbon nanodots, and polymer dots): current state and future perspective. Nano Res. 2015, Vol. 8, 355–381.
83. Prasannan, A., Imae, T. One-pot synthesis of fluorescent carbon dots from orange waste peels. Ind. Eng. Chem. Res. 2013, Vol. 52, 15673−15678.
84. Cao, S., Yu, J. Carbon-based H_2-production photocatalytic materials. J. Photochem. Photobiol. C Photochem. Rev. 2016, Vol. 27, 72–99.
85. Zhang, H., Ming, H., Lian, S., Huang, H., Li, H., Zhang, L., Liu, Y., Kang, Z., Lee, S. T. Fe_2O_3/carbon quantum dots complex photocatalysts and their enhanced photocatalytic activity under visible light. Dalton Trans. 2011, Vol. 40, 10822–10825.
86. Zhang, J., Kuang, M., Wang, J., Liu, R., Xie, S., Ji, Z. Fabrication of carbon quantum dots/TiO_2/Fe_2O_3 composites and enhancement of photocatalytic activity under visible light. Chem. Phys. Lett. Vol. 730, 2019, 391–398.
87. Chen, P., Wang, F., Chen, Z. F., Zhang, Q., Su, Y., Shen, L., Yao, K., Liu Y., Cai, Z., Lv, W., Liu, G. Study on the photocatalytic mechanism and detoxicity of gemfibrozil by a sunlight-driven TiO_2/carbon dots photocatalyst: the significant roles of reactive oxygen species. Appl. Catal. B Environ. 2017, Vol. 204, 250–259.
88. Wang, Q., Huang, J., Sun, H., Zhang, K. Q., Lai, Y. Uniform carbon dots@TiO_2 nanotube arrays with full spectrum wavelength light activation for efficient dye degradation and overall water splitting, Nanoscale, 2017, Vol. 9, 16046–16058.
89. Yu, H., Zhao, Y., Zhou, C., Shang, L., Peng, Y., Cao, Y., Wu, L. Z., Tung, C. H., Zhang, T. Carbon quantum dots/TiO_2 composites for efficient photocatalytic hydrogen evolution. J. Mater. Chem. A, 2014, Vol. 2, 3344–3351.
90. Li, F., Tian, F., Liu, C., Wang, Z., Du, Z., Li, R., Zhang, L. One-step synthesis of nanohybrid carbon dots and TiO_2 composites with enhanced ultraviolet light active photocatalysis. RSC Adv. 2015, Vol. 5, 8389–8396.

91. Su, Y., Chen, P., Wang, F., Zhang, Q., Chen, T., Wang, Y., Yao, K., Lv, W., Liu, G. Decoration of TiO_2/g-C_3N_4 Z-scheme by carbon dots as a novel photocatalyst with improved visible-light photocatalytic performance for the degradation of enrofloxacin, RSC Adv. 2017, Vol. 7, 34096–34103.

92. Yu, B. Y., Kwak, S. Y. Carbon quantum dots embedded with mesoporous hematite nanospheres as efficient visible light-active photocatalysts. J. Mater. Chem. 2012, Vol. 22, 8345–8353.

93. Bhati, A., Anand, S. R., Gunture, Garg, A. K., Khare, P., Sonkar, S. K. Sunlight-Induced photocatalytic degradation of pollutant dye by highly fluorescent red-emitting Mg-N-embedded carbon dots, ACS Sustain. Chem. Eng. 2018, Vol. 6, 9246–9256.

94. Xie, Z., Feng, Y., Wang, F., Chen, D., Zhang, Q., Zeng, Y., Lv, W. , Liu, G. Construction of carbon dots modified MoO_3 /g-C_3N_4 Z-scheme photocatalyst with enhanced visible-light photocatalytic activity for the degradation of tetracycline. Appl. Catal. B Environ. 2018, Vol. 229, 96–104.

95. Chai, N. N., Wang, H. X., Hu, C. X., Zhang, Q. W. H. L. Well-controlled layer-by-layer assembly of carbon dot/CdS heterojunctions for efficient visible-light-driven photocatalysis. J. Mater. Chem. A 2015, Vol. 3, 16613–16620.

96. De, B., Voit, B., Karak, N. Carbon dot reduced Cu_2O nanohybrid/hyperbranched epoxy nanocomposite: mechanical, thermal and photocatalytic activity. RSC Adv. 2014, Vol. 4, 58453–58459.

97. Khaneghah, S. A., Yangjeh, A. H., Abedi, M. Decoration of carbon dots and AgCl over g-C_3N_4 nanosheets: novel photocatalysts with substantially improved activity under visible light. Sep. Purif. Technol. 2018, Vol. 199, 64–77.

98. Yu, B. Y., Kwak, S. Y. Carbon quantum dots embedded with mesoporous hematite nanospheres as efficient visible light-active photocatalysts. J. Mater. Chem. 2012, Vol. 22, 8345.

99. Tian, J., Leng, Y., Zhaob, Z., Xiac, Y., Sang, Y., Hao, P., Zhan, J., Li, M., Liu, H. Carbon quantum dots/hydrogenated TiO_2 nanobelt heterostructures and their broad spectrum photocatalytic properties under UV, visible, and near-infrared irradiation. Nano Energy, 2015, Vol. 11, 419–427

100. Martinsa, N. C. T., Ângelo, J., Girão, A. V., Trindade, T., Andradea, L., Mendes, A. N-doped carbon quantum dots/TiO_2 composite with improved photocatalytic activity. Appl. Catal. B Environ. 2016, Vol. 193, 67–74.

101. Li, Y., Zhang, B. P., Zhao, J. X., Ge, Z. H., Zhao, X. K., Zou, L. ZnO/carbon quantum dots heterostructure with enhanced photocatalytic properties. Appl. Surf. Sci. 2013, Vol. 279, 367–373.

102. Du, Q., Wang, W., Wu, Y., Zhao, G., Ma, F., Hao, X. Novel carbon dots/BiOBr nanocomposites with enhanced UV and visible light driven photocatalytic activity. RSC Adv. 2015, Vol. 5, 31057–31063.

103. Gao, X., Du, C., Zhuang, Z., Chen, W., Carbon quantum dot-based nanoprobes for metal ion detection. J. Mater. Chem. C, 2016, Vol 4, 6927–45.

104. Yuan, C., Liu, B., Liu, F., M.-Y. Han, and Zhang, Z. Fluorescence 'turn on' detection of mercuric ion based on bis(dithiocarbamato)copper(II) complex functionalized carbon nanodots. Anal. Chem. 2014, Vol. 86, 1123–1130.

105. Yang, S. T., Cao, L., Luo, P. G., Lu, F., Wang, X., Wang, H., Meziani, M. J., Liu, Y., Qi, G., Sun, Y.-P. Carbon dots for optical imaging in vivo. J. Am. Chem. Soc. 2009, Vol. 131, 11308–11309.

106. Li, Q., Ohulchanskyy, T. Y., Liu, R., Koynov, K., Wu, D., Best, A., Kumar, R., Bonoiu, A., Prasad, P. N. Photoluminescent carbon dots as biocompatible nanoprobes for targeting cancer cells in vitro. J. Phys. Chem. C, 2010, Vol. 114, 12062–12068.
107. Luo, P. G., Yang, F., Yang, S. T., Sonkar, S. K., Yang, L., Broglie, J. J., Liu, Y., Sun, Y. P. Carbon-based quantum dots for fluorescence imaging of cells and tissues. RSC Adv. 2014, Vol. 4, 10791–10807.

PART II

Optical Films, Fibers, and Materials

CHAPTER 6

Latest Deposition Techniques for Optical Thin Films and Coatings

SHRIKAANT KULKARNI

Department of Chemical Engineering, Vishwakarma Institute of Technology, Pune 411037, India

*Corresponding author. E-mail: shrikaant.kulkarni@vit.edu

ABSTRACT

A thin-film optical coating is a technique aimed at modifying in a well-established procedure, the surface properties that support it, and the application of the coating. Although the applied coating is expected to impart a host of numerous properties, environmental, thermal, chemical, and acoustical, the major purpose is often a plethora of expected specular optical properties that affect the quality of the light modified by the surface without changing its direction. To obtain such type of coatings, the surface has to be smooth in terms of its optical behavior that means that the deviations in size can be of order lesser than the wavelength of the incident light. The coatings may consist of one or more thin layers of desired materials that collectively present the properties that are desired ones, primarily due to a cumulative effect of interference and the innate properties of the chosen materials. At times the materials may themselves be engineered in terms of their microstructure. The forces that are responsible for keeping together thin films and adhering them to their surfaces are of all short range order. However, these forces are very strong although of short range order and therefore a layer of contaminant one molecule thick can hinder the forces altogether. Thus both smooth and clean surface is the prerequisite for obtaining excellent surface coatings with desired specular optical properties.

This chapter takes a review of the enormously broad range of types of coatings and the evolution in optical coatings technology as well as merits and limitations of different kinds of coatings.

6.1   INTRODUCTION

Numerous processes have been in use for the application of optical coatings and to present an exhaustive list is difficult. The earliest one was the use of mirrors particularly during the later Roman Empire and was obtained by pouring lead in molten form over glass. The adequate lead was bound to the glass to improve significantly its reflectance but in the process the glass shattered and therefore the mirrors were small and irregular in shape. An improved technique came up in the 15th century that used a tin amalgam coating combined with the tremendous improvements in the quality of flat glass simultaneously that allowed for production of quality mirrors of modern times. Advent of chemical silvering technique took place and was preferably used in particular for front-surface mirrors until the 1930s. Simultaneously, a lot much of progress took place in photographic processes resulting into the Lipmann emulsion for the development of a thin-film interference filter. Synthetic tarnishing of glass surfaces to decrease their reflectance was also brought about to certain degree in the 19th century. However, other than the front-surface mirror coatings, the optical coatings were not much needed. However, with the rise in complexity of optical instruments and systems, which ask for better surface properties such as decrease in reflectance. It significantly enhances transmittance while getting rid off glare and ghost images. Vacuum deposition was adopted as the most sought after process. The field of optical coatings evolved rapidly in the later part of the 20th century.

Lot many developments took place in the field of optics, but the advent of the laser was considered as the most significant happening and lasers require coatings. The unprecedented expansion on the frontier of optics took place in the latter part of the 20th century in terms of evolution in optical coatings in the 21st century. Today, vacuum deposition is the dominant process for obtaining optical coatings, The process can be qualified based on the vapor source; physical vapor deposition uses a liquid or solid source and chemical vapor deposition uses a chemical vapor. While evaporation by thermal means has been the primary process for years together, we are now finding lot much use of alternative vacuum processes and processes that do not rely on vacuum.

6.2   EVOLUTION IN THE OPTICAL COATINGS DEPOSITION

Pohl and Pringsheim (1912) worked on the deposition of metallic mirrors by using a process referred to as vacuum distillation [1]. This had brought to the fore the principles of what is termed as thermal evaporation. Thermal evaporation is quite simple. Material is subjected to heating under vacuum to boiling, and the vapor generated then condenses on the desired substrates to be deposited, which are kept at temperatures reasonably below the freezing point of the evaporant. The evaporated material forms a thin solid film with immediate effect. Although Pohl and Pringsheim's work primarily was confined to metals, it was found that dielectric and semiconducting materials too could be well deposited by using such a technique. In general, this technique has been in use for the deposition of optical coatings since 1930s and has still enormous value. The thickness of the coated film is in harmony with the same rules as illumination, inverse square law combined with the cosines of incident and emission angles. Further to maintain the uniformity required for any single-layer antireflection coatings the substrates have to be moved during deposition. The movement involves either single or double rotation of the substrates to be deposited.

Dielectric coatings produced by the thermal evaporation technique presented two major problems such as lack of resistance to abrasion and marginally unstable spectral characteristics, usually settle the said problems however after a shift to higher wavelengths and such behavior was referred to as settling. In the 1940s, it was observed to enhance significantly the abrasion resistance by subjecting the substrates to heating during deposition [2] and that was the first major development as a part of this technique. Heating by irradiation is quite normal. The early evaporation sources involved the use of crucibles made up refractory metallic materials, like molybdenum, tantalum, or tungsten, heated by passing an electric current. The sources used to be stretched to connect to the ends of electrodes, called as boats given their semblance with boat. Boat sources are still in use, but often are vulnerable to corrosion on interacting with the heated charge. They are used at temperatures near their melting points thereby distort and alter the vapor distribution. However, there has been a gradual shift toward electron-beam heating over the time. The electron-beam source contains a metallic crucible, normally copper, with a lining of suitable material, which is kept cool with water. The charge in the crucible is heated with the help of a collimated beam of electrons sufficiently energetic. Beam current can have a magnitude of several amperes and the voltage several kilovolts and so the heat supplied to the charge is massive worth melting almost anything. However, interaction with the crucible is stifled by cooling with water. The electron source consists of a hot filament that too is amenable to corrosion on exposing it to the evaporant. It is therefore embodied in the structure of the source while the electron beam is bent through at least 180° using the magnetic field with pole pieces of the permanent magnet used to lie along the sides of the source. This makes the beam to streamline well enough. Further, additionally small, electromagnets too are quite often used so as to change the beam position and moved in a certain pattern over the charged surface.

Although the electron-beam source is costlier than the boat source, however, its stability and reliability can lower the running costs and high yield of acceptable products. Results of experimentation on both metals and dielectrics subsequently showed that the key parameter governing heating was the ratio of the substrate temperature to the melting point of the evaporant. Thin films so deposited by evaporation acquire a columnar microstructure while the way in which columns are packed is of utmost importance in determining the film property profile.

Movchan and Demshishin noticed three growth zones:

1. Zone one with a ratio for dielectrics <0.22–0.26 produces loose dendrimeric columns with poor mechanical properties, in particular abrasion resistance.
2. Zone two with a ratio of 0.45–0.5 produces the columnar structure very densely packed with exacerbated abrasion resistance and stability.
3. Zone three with ratios > 0.5 at which the columns are replaced with small recrystallized grains.

From the optical coatings perspectives, zone two is preferred much. Films developed in Zone three usually exhibit more scattering losses, and the temperature required for the various refractory materials is too high for normal deposition equipments. This structural zone model is significantly important in getting an understanding of film properties and it has evolved over the years [3] and is still a topic of further research [4]. For example, the energy of the evaporant has been given due recognition. A transition zone between zone one and two has been identified too [5].

Capillary condensation of moisture from the atmosphere in the pores in between the columns had been shown responsible for the drifting or settling of the optical coatings [6–9]. Energetic processes [10, 11] hinted that they might bring about some improvement.

6.3 THE ENERGETIC PROCESSES

The energetic processes are so-named as they are accompanied by the addition of kinetic energy to the condensation process. This energy addition can be done in various ways. It may be added upon by the evaporant itself or may be extracted from purposefully generated beams of energized ions made to the incident upon the substrate surface. The energetic processes help in increasing the properties like solidity, density, and adhesion of the decorated films. The genesis of the universal columnar structure of thin films that are formed by thermal evaporation is attributed to the constrained movement of the condensing moieties, which has been confirmed by computer modeling [12, 13].

The possible underlying mechanism for the inherent densification in the energetic processes was the increase in the mobility of the condensing species due to bombardment of the energetic beam, maybe due to a sort of annealing brought about by the thermal spikes from the collisions [14]. However, computer simulation studies show that the densification was ascribed to the transfer of momentum instead of thermal spikes [15] although thermal spikes might be instrumental in sealing of few pores in the films. Further experimentation confirmed that the mechanism for densification is the transfer of momentum primarily [16, 17]. The classification of thin-film coating techniques is shown in Figure 6.1.

6.4 ION-ASSISTED DEPOSITION (IAD)

Ion-assisted deposition (IAD) is one of the simplest energetic processes. It involves thermal evaporation accompanied by ion bombardment for growing film. Broad-beam ion sources had been developed basically by NASA for exploring them to be used in propulsion engines for outer space. These processes were adopted by industry later, mainly for ion-beam etching in the semiconductor industry, and therefore were available for use in optical coating machines too [18, 19]. The Kauffmann type broad-beam sources could be made a part of thermal evaporation machines without much modification. The ions derived from the source having a few hundred electron volts energy are positively charged that would charge any dielectric material rapidly, requiring no further bombardment. Electrons sufficient in number for the neutralization of charge are thus made available to the extracted beams thereby producing a neutral ion beam. Most of the times argon was used as the bombarding ion in the past but over the time a broad range of bombarding species are in use. Therefore it was observed soon that the bombardment in IAD could exacerbate the solidity of films with no moisture sensitivity [20]. The IAD was therefore rapidly adopted as it was characterized by absence of moisture-induced shifts, high packing density, and more stable optical constants.

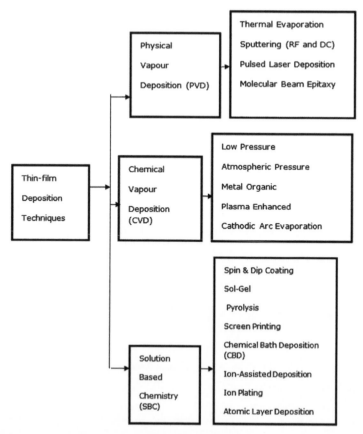

FIGURE 6.1   Classification of thin-film coating techniques.

The comparative advantages of IAD over other energetic processes are as follows:

- Significantly inexpensive;
- Almost a universal technique for optical coatings involving conversion from thermal evaporation in the early times;
- Addition of an ion source with its power supply does not ask for the scrapping of an incumbent, expensive machine;
- A technical advantage is that the bombardment parameters can be controlled keeping other deposition parameters intact;
- The Kaufman source having extraction grid produces with almost a continuum in the distribution of ion energies, thereby making it an exciting tool for research.

The refractory oxides are a very vital family of materials. These materials are characterized by a tendency to lose some oxygen thereby creating vacancies causing absorption in the evaporation process. The loss of oxygen can be compensated by supplying oxygen from outside of the machine in a process called reactive evaporation. Reactive evaporation can yield stoichiometric oxides too by the evaporation of metals in an oxidizing atmosphere that demands however a certain degree of judgment. The packing of the thin film is dependent upon regulated oxygen supply. Silicon and aluminum nitrides too are quite useful optical materials, although direct reactive evaporation does not work as the nitrogen is not much reactive. The reactivity of oxygen and nitrogen can be largely enhanced by allowing it to pass through the ion source. Pure oxygen can be highly reactive particularly with the Kaufman type hot filament; therefore when passed through the ion source, it is normally blended with the argon and thus diluted. Nitrogen, however, has an advantage on passing through the ion source as it becomes adequately reactive to yield highly transparent and quality nitride films [21–23].

There are numerous variants of IAD varying in terms of the design of the source. An ion source usually generates positive ions, using a grid. A plasma source produces the entire plasma. The electrons in the plasma are largely free and have an escaping tendency leaving the plasma with a positive charge that repels the positive ions such that they further can bombard the growing film, which is usually called plasma IAD distinct from IAD.

The recent developments involve either using radio frequency (13.6 MHz) [24, 25] or microwave (2.46 GHz) frequency [26, 27] for the excitation of the plasma in the source. It decreases the concentration of impurities in the ion beam and also allows operation of the source to be carried out for a longer time before any maintenance arises. The fluorides by virtue of their enormously broad range of transparency ranging from the far IR to the far UV are much-used thin-film materials. However, the bombardment as a prerequisite for the energetic processes is responsible for damaging the material by removal of fluorine in preference, loss of fluorine to a smaller extent is tolerable because the vacancies created are often healed by oxygen and subsequent minimization of optical absorption. Fluorine loss to a greater extent, however, leads to the formation of an oxyfluoride that possesses normally poor durability, narrow transparency regions, and high refractive index. It means that bombardment of the fluorides in energetic processes has to be checked. The gentle bombardment is advantageous and is instrumental in improving adhesion, increasing packing density, and inhibiting oxidation but not to the same extent the improvements shown by the oxides and nitrides [28–30]

6.5   ION PLATING

Ion plating is a term used for a host of processes rather in an imprecise manner. In the beginning, it was aimed at replacing electroplating [31]. The technique involves a source of thermal evaporation, a metallic substrate holder, and a glow discharge is provided for in between them that can sustain potential of many thousand volts, the anode is the source while the cathode is the substrate holder. Such an arrangement is referred to as high-voltage ion plating. The underlying mechanism was that the positive ions would be migrated through the plasma resulting into bombardment followed by a coating of the substrates. Deposits that are hard and tough with outstanding throw are found to be obtained. However, it is now known beyond doubt that the process is far more complex in nature [32] and the way it is implemented may differ. The degree of ionization in high-voltage ion plating is very small and the bombardment has a major role in the growing film by species (particles) sufficiently energetic, usually neutral and that acquire energy from collisions in the discharge. Such particles, primarily derived from the evaporant, are instrumental in forming the growing film, transferring momentum, and compacting it similar to IAD. The enhanced throwing power is attributed largely to the simultaneous gas scattering. This high-voltage variant of ion plating has so far been to a lesser extent for optical coatings, although Reid and colleagues did obtain quite tough zinc sulfide depositions for use in the infrared region [33].

Reactive low-voltage ion plating [34–36] is frequently used for the deposition of optical coatings. It involves a thermal evaporation machine and an electron-beam source with a reasonably conducting metal or suboxide as a substrate. The source is insulated and is connected to the positive end of a DC power supply that offers a large amount of current at a potential of a few hundred volts. The negative end of the power supply is attached to a plasma chamber, which is insulated from the machine as like source, which gives rise to an intense beam of electrons at low voltage which is migrated to the electron-beam source, at which intense plasma develops above the evaporating surface. The substrate holder too is electrically insulated from the machine and the highly mobile electrons, impart a negative charge to it, dragging positive ions from the plasma. These positive ions from the source material impinge on, and become part of, the growing film that acquires density due to the transfer of momentum. For getting transparency in films, surplus oxygen is fed to the plasma so as to become the process reactive. Nitrogen too can be used either to form an oxynitride or a nitride, possessing outstanding optical properties [37].

6.6   SPUTTERING

Sputtering is the oldest technique of all the energetic processes that originated in the middle of the 19th century [38] but was not used much in the deposition of optical coatings until the latter part of the 20th century. It consists of an evacuated chamber with an anode and a cathode, both of them metallic. The arrangement made is such that a glow discharge can be maintained in the gas in an evacuated chamber. A voltage of a few kilovolts, and a pressure of about ≥ 1 Pa (0.01 mbar), is what required roughly. Positive ions generated from the discharge are dragged toward the cathode and bombard it with high energy. Momentum transfer takes place due to the cascading collisions between bombarding ions and the molecules of the cathode that cause the ejection of molecules from the surface of cathode with significant kinetic energy. To obtain the best possible results, the atomic weight of the bombarding material has to resemble with that of the target material thereby optimizing the transfer of momentum. Ionization of the sputtered material takes place to a small degree and, ideally, such ionized molecules follow straight-line paths to hit on the anode, or substrate, and grow a thin solid film there. The extent of momentum transfer from the bombarding material decides the density of microstructure in the film grown. Figure 6.2 shows a simple diagram for sputtering technique.

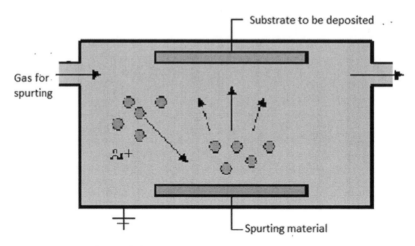

FIGURE 6.2   Sputtering technique.

There are, however, a few problems that cause deviation from the ideal conditions:

- The process works only for conducting metals. If we use dielectric materials in place of the metallic anode and cathode the surfaces quickly charge and resist any further sputtering.
- The surfaces must be discharged to improve sputtering.
- The capacitance of the surface of a dielectric material is abysmally low and the rate of charging very high. However, to maintain the rate of discharging equally high, the process wherein plasma is created with the help of a radio frequency field has to be used.
- The process smoothly takes place and has been in use for depositing optical coating but in particular and not in general because of many problems associated with radiofrequency enabled processes than with zero frequency.
- Another limitation is that the plasma becomes a barrier to the sputtered material that results in the scattering, which further reduces the energy of the condensing material and subsequently counteracts the desired densification.

Magnetron sputtering [39] can achieve better results at lower pressure, of the order of 0.1 Pa (0.001 mbar) as against 1 Pa or greater in conventional sputtering as well as the reduction in the required voltage to few hundreds of volts rather than some kilovolts. Magnetron sputtering involves forcing of the electrons into cycloidal paths under the influence of a magnetic field. Permanent magnets with alternating poles are mounted behind the cathodes such that electric and the magnetic fields are crossed over a short range, say 10–20 mm above the cathode surface. That is the region where the plasma forms due to the exacerbated ionization attributed to the longer cycloidal path followed by the electrons. The plasma gets localized in its degree and in the so-called erosion region. On plane rectangular-shaped cathodes, the erosion region traces a map of a racetrack and therefore it is so-called.

Reactive sputtering is a way out for overcoming the problems of dielectrics. Many dielectric materials are either metal oxides or nitrides. In such cases, the cathode can be the desired metal and reaction can be brought about by feeding oxygen or nitrogen to the background gas. The discharge activates the reacting gas so as to take its reaction with the growing film to almost completion resulting into high-quality dielectrics. Unfortunately, the reactive sputtering process is not much productive and reliable on using fluorides as a target as the fluorine sputters in preference and lose, either from the growing film or from a target containing fluoride, similar to the ion-assisted processes. In reactive sputtering, the reacting gas reacts with both the growing film as well as the sputtering target. The result of magnetron sputtering is a build-up of insulating material called as target poisoning while in the racetrack, any insulating material is removed by the bombardment that causes some hysteresis in the electrical characteristics that have to be taken care of in the control system, although serious problems occur outside the racetrack. The thin dielectric film behaves as a capacitor that charges up on the arrival of the positive ions. As the coating is thin, the capacitance is high and the charge shows a significant amount of energy. This energy is made available as and when required when a capacitor fails and the tremendous amount of stored energy results into an arc that is usually powerful enough to fuse the localized areas of the cathode that subsequently cause ejection of metallic globules, most of which arrive at the surface of the substrate. The high capacitance means a long time constant such that discharging the capacitors can be brought about at comparatively lower frequencies, of the order of a few tens of kHz. Typical power supplies that periodically alter their polarity to cause discharging of the capacitors can be a remedial solution.

Mid-frequency sputtering or dual magnetron sputtering technique is gaining much of ground in optical coating. It involves two magnetrons of similar kind attached to opposite sides of a power supply working at about 40 kHz. It means that after every half cycle the magnetrons switch from anode to cathode and vice versa that effectively discharges the capacitors and helps in getting over a problem associated with reactive sputtering called as the disappearing anode. The anode in the usual sputtering process is mostly the machine body or a rod. The anode slowly gets covered with an insulating film similar to the normal substrates and its activity ultimately reduces to abysmally low. In this type of sputtering technique every alternate target is the anode. This technique was introduced primarily for coating large surfaces, normally glass substrates, but over the time it has been in use for optical coating with precision.

Target poisoning can be rinsed off by scrubbing the entire cathode surface using the ions in the plasma. This demands moving either the plasma, or the substrate surface, or both. In the cylindrical rotating magnetron source, the target is cylindrical and is subjected to rotate about the permanent magnets. This offers an added advantage of removing the eroded area and eventually, the racetrack disappears and thereby the use of cathode is far more efficient. Smaller cylindrical magnetrons have made inroads into the optical field characterized by precision, typically in the mid-frequency configuration. Although they do not show the poisoning effects as that shown by reactive deposition, they can have the disappearing anode that can be removed by the application of the mid-frequency technique.

In 1975, a process termed as alternating ion plating was originated [40]. Although, it was not tended to optical coatings and did not drag much attention at that time, however, its effect has been far-reaching on the field of optics. The plasma treatment of the growing film that is observed in ion plating, IAD, sputtering, and host of other reactive energetic processes, interferes with the process of deposition of optical coatings. The bombardment of ion reaches out to a certain depth in the film and therefore to have an incremental growth in the film, bombardment has to be continued till we get a desired thickness of the coating. Such a technique can be very easily implemented by placing the substrates on a cylindrical drum and rotating it such that the substrates move through the source and are treated in stages for the deposition to take place. This typical cylindrical geometry with its various treatment stages has been accepted widely across the energetic processes. Radical-assisted sputtering is a specifically an interesting and innovative process devised by Shincron Co. Ltd. that makes use of this geometry [41–43]. In this case, atomic oxygen is used in the treatment phase of the growing film. The oxygen is extremely reactive that brings about oxidation quite efficiently. Films having exceptionally high optical quality are developed.

Closed-field magnetron sputtering developed by Applied Multilayers LLC employs a similar kind of cylindrical intermittent deposition system; however, the plasma treatment is done to the whole cylindrical surface. It is accomplished with the help of placing magnetron targets in the ring around the cylinder such that the permanent magnets present alternating poles like North-South–North, South–North_South, and so on. Such an arrangement generates a magnetic confinement system that spreads the plasma around the substrate carrier cylindrical in geometry and therefore the plasma treatment in this case is quite efficient. Exceedingly high-quality optical coatings are obtainable by this technique [44]. Deposition using intermittent treatment can also be attained with a substrate carrier having a flat circular configuration. The advantage of this technique is that by having a horizontal carrier the substrates are held in place under gravitational force, plasma treatment takes place in the usual way. It is then a very smooth operation as far as loading the carrier robotically is concerned without its exposure to the atmosphere. This is the underlying principle of the Helios sputtering system devised by Leybold Optics GmbH.

The latest development in magnetron sputtering is high power impulse magnetron sputtering [45]. Here, the process is subjected to pulses with a high pulse power that consists of a highly plasma ionized metal flux similar to the outcome of a cathodic arc. The optical films obtained by this technique have better adhesion strength, density, and smoothness and yield desired structures at lower temperatures, which otherwise are obtainable at much higher temperatures. The pulsing is done at a very low frequency (<1 kHz), and the pulse duration of a few hundred microseconds does not cause overheating of the targets such that even thermally labile substrates can be coated. Although the process has not been much applied in the deposition of optical coatings, the evidence shows that it can offer advantages in numerous optical applications, for example, TiO_2, sputtered from a ceramic-based target has been found to offer higher refractive index than with other incumbent processes. The indices achieved were 2.45 and 2.48 at 550 nm as against 2.38 with DC magnetron sputtering [46].

Ion-beam sputtering addresses a host of the problems various sputtering processes by employing a neutral ion beam obtained using a broad-beam source to cause sputtering the target material. Pure dielectric materials can be sputtered using this technique as it does not have the target-charging problem. Further, the plasma of which the ions are obtained is in an isolated chamber means that the plasma intervention noticed in the usual sputtering processes is completely ruled out. Movement of sputtered material from the target to the substrate is unobstructed completely. The films obtained are therefore of very high quality. At times, a second ion source is employed to bombard the growing film in dual ion-beam sputtering. This kind of treatment is aimed at further controlling the film properties, in particular intrinsic stress. The process is comparatively slower than its counterparts and the coated area, too, are small; however, due to the high quality combined with intrinsic accuracy of the films, it has come up as an important process for the deposition of high-performance coatings.

6.7   CATHODIC ARC EVAPORATION

Cathodic arc evaporation [47] consists of an arc (high-current and low-voltage) that is struck between an electrode and a cathode necessarily conducting. Since the cross-sectional area of the arc is smaller and eventually its impingement area on the cathode leading to a very high local temperature (hot spots) of the order of 15,000 °C. Evaporant increases the energy of the cathode to a large extent and gets ionized. The local hot spots move over the surface of the cathode, in a random motion. At times, a magnetic field is used to bring about the movement of the spots in a raster scan with a regular pattern.

The major advantage of the process has been very successfully used in the deposition of coatings on cutting tools. It has not much penetrated the optical thin-film field as yet except a few encouraging results (e.g. R.I. of 2.73 for TiO_2 at 630 nm) [48]. A problem of major concern with the simple cathodic arc technique is the release of droplets in the molten form of material that, if allowed to reach the growing film, will work as scattering defects that are detrimental in optical applications. For better optical applications, therefore, the evaporant can be filtered that, given the ionizing nature of the evaporant, undergoes change in its direction magnetically so as to separate it from the undesirable material particles [49]. If ion bombardment is added to this process similar to IAD then the resulting process is referred to as ion-assisted filtered cathodic arc evaporation [50].

6.8   PULSED LASER DEPOSITION (PLD)

High temperatures of the target materials over a local area are also attained by irradiating them with pulsed high-power lasers, such a technique is called pulsed laser deposition (PLD). Smith and Turner (1965) used this technique probably for the first time for the deposition of optical coatings [51], wherein they used a pulsed ruby laser. They thought that the stoichiometry of semi-conducting materials and alloys might be retained in the ultimate films formed although only some of their materials exhibited a similar composition as like the sources. This to a certain extent might have been due to their use of targets in the form of powder in crucibles and the lower pulse energy made available during the process in those days. However, problems were confronted for materials transparent to the electromagnetic radiation in the visible region of the spectrum. They observed that there had to be absorption to a certain extent at the ruby laser wavelength, and a few of the powders had to be blended along with a little amount of carbon to accomplish this.

The laser has to be placed outside the chamber and its output is passed through an optical system and is exposed to the evaporating material. The coating may not be obtained as expected at times the element is made to rotate from time to time better exposure. Smith and Turner found that the pulses brought about the reevaporation of the deposited material.

All kinds of high-power lasers have been put into use right from UV to far IR. The high fields due to the very high power and short pulses help in getting over the problem of transparency by effecting breakdowns in the target material. PLD therefore is considered as one of the viable techniques in the deposition of a host of optical coatings [52]. However, it could not join the mainstream and is used for some selective applications where it has comparative advantages over other techniques. The process can be adjusted to accomplish the best stoichiometry that is otherwise difficult to achieve say for alloys, and therefore the resulting films can offer excellent optical performance. Latest reports show outstanding transparency by zirconium dioxide and hafnium dioxide [53, 54] and vanadium dioxide [55], an agile material that can be change over its thermal behavior, for example, semiconductor properties below the transition point (68 °C) and metallic properties above it.

6.9   CHEMICAL VAPOR DEPOSITION (CVD)

In chemical vapor deposition (CVD) technique the deposition of thin-film is done by the chemical interaction of precursors [56]. The precursors are carried into a reaction vessel through a carrier gas stream. In this reactor chamber the precursors are made to react, the reaction product in itself is added to the growing film. In conventional CVD, the reaction was brought about by a hot substrate and is used sometimes. A limitation of this process is that an efficient reaction will take place too rapidly at a fast rate and therefore its accommodation to the growing film is a major issue that results in a relatively loose deposit having poor mechanical strength and optical properties. Thermal CVD is preferable for those reactions that do not take place too fast or are not so favorable. As impurities in such processes often induce reactions at a rapid pace, therefore conventional thermal CVD ask for the reactants that are ultrapure.

A better approach adopted nowadays is to use reactants that react more efficiently and to trigger the reaction by using short pulses [57]. The time lag between the pulses allows the accommodation of the reaction product in the film that results in getting films with high optical quality and density. Although the arrival of the reactants is pulsed into a thermal CVD process [58, 59], the better option is to initiate the reaction using either a radio frequency pulses or microwave frequency plasma. The process is commonly known as plasma impulse chemical vapor deposition although many other acronyms too are used for its recognition.

An exciting coupling of microwave with radiofrequency to excite the plasma for producing a narrowband rejection filter of silicon nitride with variation in packing density is done [60]. Homogenization in CVD reflects upon the flow pattern in the reactor vessel and usually each substrate geometry demands a distinct reactor design. This constrains the productivity of CVD in smaller batch processes. It is primarily used, therefore for production processes run for longer times and of similar products where the reactor design is stable, for example, the production of cold mirrors. The parabolic reflectors (50 mm) around quartz halogen projection lamps hold a cold mirror and although originally the combination of reflector-source was meant for projectors it has been in use for a host of lighting forms. The market potential may run into many million parts/year, and therefore the cold mirror signifies a sort of long production run that is an ideal candidate for CVD [61].

6.10   ATOMIC LAYER DEPOSITION (ALD)

Atomic layer deposition (ALD) [62, 63] involves the introduction of the precursors in alternate pulses into the reaction chamber. Each pulse forms a thin layer over the substrate surface. The vapor pressure of the material allows the evaporation of material to leave behind a monomolecular layer followed by the reaction of the second precursor with the material on the surface of the substrate. The process takes time as the reactor needs to be purged after each precursor pulse but it has the tremendous advantage of coating quite complex shapes with a significant degree of uniformity. In this case, the incremental additions to thickness are small but quite regular. The control of film thickness can be automated completely and in a normal and straightforward manner TiO_2 is easily deposited using precursors like $TiCl_4$ and water, and Al_2O_3 in the form of trimethylaluminum ($Al_2(CH_3)_6$) and water [64] and an excellent optical quality is obtained. SiO_2 is relatively difficult to deposit. Sound reliable processes similar to the TiO_2 and Al_2O_3 would help go a long way in the greater use in the field of optics. SiO_2 films doped with precursors such as Al_2O_3 using TMA and tris(tert-butoxy)silanol (($ButO)_3SiOH$) are successfully obtained [65].

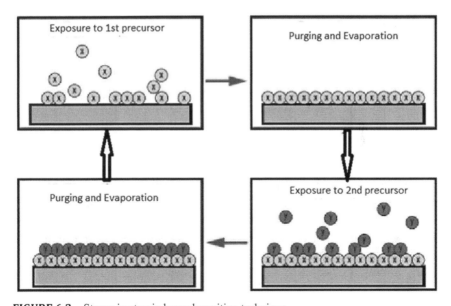

FIGURE 6.3   Stages in atomic layer deposition technique.

A latest report is in particular quite encouraging [66] wherein multilayer antireflection coatings were obtained by using HfO_2 and SiO_2 as materials for thin-films deposition. Tetrakis-[dimethylamido]-hafnium (TDMAH) and tris-[dimethylamino]-silane (3DMAS) as precursors and oxygen activated by plasma produced the HfO_2 and the SiO_2 deposition respectively. However, single-layer SiO_2 coating exercises yield further better results in respect of film properties using a commercially available precursor, AP-LTO®330, and

O_3 as the source of oxygen at a temperature of the substrate of about 200 °C. Figure 6.3 shows the stages in ALD technique.

6.11   SOL–GEL METHOD

Sol–gel processes use a solution that is converted into a gel followed by its deposition on a substrate surface [67]. The precursor is a metal alkoxide soluble in an appropriate solvent, at times ethanol. The small quantity of water is added to hydrolyze the material to form a polymeric jelly like mass and make the solution acidic slightly. Such a material is a loose gel having pores filled with liquid that can be deposited on a substrate surface by dipping. Evaporation of the liquid and compaction of the material is brought about subjecting it to a heat treatment. Higher the temperature, higher is the density of the film formed. Temperature from 600 to 1000 °C can be applied if maximum compaction is asked for. However, at times lower temperatures are employed to acquire some prestructure so as to reduce the index of the film making it amenable to use it as an antireflection coating. TEOS (tetra-ethylorthosilicate, $Si(OC_2H_5)_4$) for SiO_2 films and titanium tetraethoxide $[Ti(OC_2H_5)_4]$ for TiO_2 are typical precursors. Multilayer coatings can be derived although the blending of materials is a major cause of concern. It means drying in between each deposition and therefore frequent coating as a heterogeneous single layer of porous SiO_2 is obtained while the heterogeneity help improves the antireflecting performance of the coating.

There are two major reasons about why this process is of much interest:

- Discrete layers of broadband antireflection coatings for SiO_2 create problems due to the nonavailability of too low index materials.
- The laser damage threshold of sol–gel materials is significantly high [68]. A major cause that limits their use as antireflection coatings is that in environments that are not controlled, the moisture and other contaminants occupy pores such that their index changes with time. They are used to a greater extent when very large lasers are operated in controlled environments.

6.12   ETCHING

The etching is the oldest process of removal of material as against deposition. In the 18th century, for the first time reported a decrease in the reflectance of glass after treating it with an acid. Then in 1904, Taylor filed a patent on an etching process for lowering down in surface reflection. The etching process was evolved and achieved a broad range of spectacular coatings on borosilicate glasses by application of a high temperature to extract the glass constituents and then an acid treatment to remove all leaving behind the silica skeleton that that is attributed to formation of a coating with low reflectance over the wavelength range from 350 nm to 2.5 µm. Moreover, in 1980, a published work gave an account of how bombardment of plastic surfaces by energetic (2MeV) ions of He or C and then etching with NaOH solution yielded an array of needle-like structures that worked as high-performance antireflection coatings. In the recent past, etching techniques have been revived in particular simple but quite efficient plasma surface treatment for plastic substrates, for example, poly(methyl methacrylate) plastic is used in particular for coating and efficiently with this treatment process. The surfaces are subjected to treatment with an argon-oxygen plasma with the same machine used for plasma IAD. The state-of-the-artwork has produced hybrid coatings where inorganic films are part of a multilayer antireflection coating with an organic layer as ultimate one, and use of melamine, plasma treatment yield the low index, of 1.2, desired for low reflectance spanning over a broad-angle range [69].

6.14 MODERN TECHNIQUES

The principal techniques that are in use presently used for the deposition of optical coatings have been discussed so far. However, there are still so many more techniques that are used based on the circumstances, and are very selectively used and it is difficult to cover all of them. Glancing angle deposition is a form of thermal evaporation wherein the angle of incidence of the vapor with the surface of the substrate is intentionally made too high. This yields a marked columnar structure with poor packing density. The films are poor in mechanical strength but show a very low refractive index that is quite important in antireflection coatings [70]. However, it is essentially confined to internal surfaces unaffected by environmental effects. The films also show a marked birefringence that holds much of potential in polarization-sensitive applications. Other techniques such as spinning, spraying, plasma polymerization, electroplating, electrophoresis, anodization, Langmuir–Blodgett tend to form a floating monomolecular layer, molecular beam epitaxy, and processes depending upon the Lipmann emulsion [71]. These are a few representatives ones.

6.15 CONCLUSION

The beginning of the field of optical coatings took place with metallic coatings for mirrors but it was the 20th century and after the advent of vacuum processes, in particular thermal evaporation, there has massive and necessary expansion in the deposition of optical coatings into almost all aspects in the field of optics. Thermal evaporation is a very vital process but an array of other techniques have joined with it, such that a broad range of processes are presently used for the deposition of optical coatings. This chapter has emphasized the frequently used techniques along with those which are relatively much less frequently used in practice, although they hold a lot much of promise and potential. The major trend that is developed over the time is the proliferation of techniques, all of which bring with them their comparative advantages that facilitate in order to meet specific optical coating needs. One thing is certain that there is no universal technique that can be used across the board for all kinds of optical applications. The choice of method is driven by the typical optical application. Although the future is an uncertain activity, but we can keep on working for the continuous development and expansion of the deposition process protocols to meet the ever-increasingly demands of the field of optics at large.

KEYWORDS

- coating
- specular
- optical properties
- interference
- microstructure
- contaminant

REFERENCES

1. Alasaarela, T., Maula, J., Sneck, S., 2009. Optical coating of glass tubes by atomic layer deposition. In: 52nd Annual Technical Conference Proceedings, Santa Clara. Society of Vacuum Coaters, pp. 478–482.
2. Anders, A., 2005. Plasma and ion sources in large area coatings: a review. Surf. Coat. Technol. 200, 1893–1906.

3. Anders, A., 2010. A structure zone diagram including plasma-based deposition and ion etching. Thin Solid Films 518, 4087–4090.

4. Bach, H., Krause, D. (Eds.), 1997. Thin Films on Glass. Springer-Verlag, Berlin, Heidelberg. Bovard, B.B., Ramm, J., Hora, R., Hanselmann, F., 1989. Silicon nitride thin films by low voltage reactive ion plating: optical properties and composition. Appl. Opt. 28, 4436–4441.

5. Briggs, R.M., Pryce, I.M., Atwater, H.A., 2010. Compact silicon photonic waveguide modulator based on the vanadium dioxide metal-insulator phase transition. Opt. Express 18, 11192–11201.

6. Cheung, J.T., Sankur, H., Chang, T., 1992. Applications of pulsed laser deposition to optics. Opt. Photon. News 3, 24–27.

7. Choy, K.L., 2003. Chemical vapour deposition of coatings. Prog. Mater. Sci. 48, 57–170.

8. Dirks, A.G., Leamy, H.J., 1977. Columnar microstructure in vapor-deposited thin films. Thin Solid Films 47, 219–233.

9. Fraunhofer, J.V., 1888. Versuche €uber die Ursachen des Anlaufens und Mattwerdens des Glases und die Mittel, denselben zuvorzukommen. In: Joseph von Fraunhofer's Gesammelte Schriften. Verlag der Koniglich Bayerischen Akademie der Wissenschaften, München.

10. Fulton, M.L., 1999. New ion-assisted filtered cathodic arc deposition (IFCAD) technology for producing advanced thin-films on temperature-sensitive substrates. Proc. SPIE 3789, 29–37.

11. Gibson, D.R., Brinkley, I., Waddell, E.M., Walls, J.M., 2008a. Closed field magnetron sputtering: new generation sputtering process for optical coatings. Proc. SPIE 7101 (710108), 1–12.

12. Gibson, D.R., Brinkley, I.T., Waddell, E.M., Walls, J.M., 2008b. Closed field magnetron sputter deposition of carbides and nitrides for optical applications. In: 51st Annual Technical Conference Proceedings. Society of Vacuum Coaters, Chicago.

13. Grove, W.R., 1852. On the electro-chemical polarity of gases. Philos. Trans. R. Soc. B142, 87–101.

14. Hagedorn, H., Klosch, M., Reus, H., Zoeller, A., 2008. Plasma ion-assisted deposition with radio frequency powered plasma sources. Proc. Soc. Photo-Opt. Instrumen. Eng. 7101, 710109–11 to 710109–6.

15. Harper, J.M.E., Cuomo, J.J., Kaufman, H.R., 1982. Technology and applications of broad-beam ion sources used in sputtering. Part II. Applications. J. Vac. Sci. Technol. 21, 737–756.

16. Hausmann, D., Becker, J., Wang, S., Gordon, R.G., 2002. Rapid vapor deposition of highly conformal silica nanolaminates. Science 298, 402–406.

17. Hirsch, E.H., Varga, I.K., 1978. The effect of ion irradiation on the adherence of germanium films. Thin Solid Films 52, 445–452.

18. Hirsch, E.H., Varga, I.K., 1980. Thin film annealing by ion bombardment. Thin Solid Films 69, 99–105.

19. Hwangbo, C.K., Lingg, L.J., Lehan, J.P., Macleod, H.A., Suits, F., 1989. Reactive ion-assisted deposition of aluminum oxynitride thin films. Appl. Opt. 28, 2779–2784.

20. Ives, H.E., 1917. Lippmann color photographs as sources of monochromatic illumination in photometry and optical pyrometry. J. Opt. Soc. Am. 1, 49–63.

21. Jacobson, M.R. (Ed.), 1988. Modeling of optical thin films. Proc. SPIE 821, 233.

22. Kaufman, H.R., Cuomo, J.J., Harper, J.M.E., 1982a. Technology and applications of broad-beam ion sources used in sputtering. Part I. Ion source technology (note: Part II listed under harper). J. Vac. Sci. Technol. 21, 725–736.
23. Kaufman, H.R., Harper, J.M.E., Cuomo, J.J., 1982b. Developments in broad-beam, ion-source technology and applications. J. Vac. Sci. Technol. 21, 764–767.
24. Kelly, P.J., Arnell, R.D., 2000. Magnetron sputtering: a review of recent developments and applications. Vacuum 56, 159–172.
25. Kelly, P.J., Bradley, J.W., 2009. Pulsed magnetron sputtering—process overview and applications. J. Optoelectron. Adv. Mater. 11, 1101–1107.
26. Koch, H., 1965. Optische Untersuchungen zur Wasserdampfsorption in Aufdampfschichten (inbesondere in MgF2 Schichten). Phys. Status Solidi 12, 533–543.
27. Lee, C.C., 1983. Moisture Adsorption and Optical Instability in Thin Film Coatings. PhD Dissertation, University of Arizona.
28. Lyon, D. A., 1946. Method for coating optical elements. US patent 2,398,382. April 16, 1946.
29. Macleod, H.A., Richmond, D., 1976. Moisture penetration patterns in thin films. Thin Solid Films 37, 163–169.
30. Martin, P.J., Macleod, H.A., Netterfield, R.P., Pacey, C.G., Sainty, W.G., 1983. Ion-beam assisted deposition of thin films. Appl. Opt. 22, 178–184.
31. Martin, P.J., Netterfield, R.P., Kinder, T.J., Desc^otes, L., 1991. Deposition of TiN, TiC, and TiO$_2$ films by filtered arc evaporation. Surf. Coat. Technol. 49, 239–243.
32. Mattox, D.M., 1967. Apparatus for coating a cathodically biassed substrate from plasma of ionized coating material. US patent 3,329,601.
33. Mattox, D.M., 1979. Mechanisms of ion plating. In: Proceedings of International Conference on Ion Plating and Allied Techniques (IPAT 79), London. CEP Consultants Ltd., Edinburgh.
34. Maula, J., Alasaarela, T., Sneck, S., 2009. Atomic layer deposition (ALD) for optical coatings. 52nd Annual Technical Conference Proceedings, Santa Clara. Society of Vacuum Coaters, pp. 486–491.
35. Messerly, M.J., 1987. Ion-Beam Analysis of Optical Coatings. PhD Dissertation, University of Arizona.
36. Minot, M.J., 1976. Single-layer, gradient refractive index antireflection films effective from 0.35 to 2.5 micrometers. J. Opt. Soc. Am. 66, 515–519.
37. Moll, E., Pulker, H.K., Haag, W., 1986. Method and apparatus for the reactive vapor deposition of layers of oxides, nitrides, oxynitrides and carbides on a substrate. US patent 4,619,748. October 28.
38. Motovilov, O.A., Lavrischev, A.P., Smirnov, A.N., 1974. Stable narrow-band interference filters for the visible region. Sov. J. Opt. Technol. 41, 278–279.
39. Movchan, B.A., Demchishin, A.V., 1969. Study of the structure and properties of thick vacuum condensates of nickel, titanium, tungsten, aluminium oxide and zirconium dioxide. Fiz. Met. Metalloved. 28, 653–660.
40. Müller, K.-H., 1986a. Model for ion-assisted thin-film densification. J. Appl. Phys. 59, 2803–2807.
41. Müller, K.-H., 1986b. Monte Carlo calculation for structural modifications in ion-assisted thin film deposition due to thermal spikes. J. Vac. Sci. Technol. A 4, 184–188.

42. Nagae, E., Sakurai, T. and Matsumoto, S. (2001). RAS (radical assisted sputtering) system and its application on depositing optical thin films. Topical Meeting on Optical Interference Coatings, Banff, Canada. Optical Society of America, MB5.

43. Netterfield, R.P., Martin, P.J., Sainty, W.G., 1986. Synthesis of silicon nitride and silicon oxide films by ion-assisted deposition. Appl. Opt. 25, 3808–3809.

44. Ogura, S., Macleod, H.A., 1976. Water sorption phenomena in optical thin films. Thin Solid Films 34, 371–375.

45. Pfeiffer, K., Shestaeva, S., Bingel, A., Munzert, P., Ghazaryan, L., Helvoirt, C., et al., 2016. Comparative study of ALD SiO_2 thin films for optical applications. Opt. Mater. Exp. 6, 660–670.

46. Pohl, R., Pringsheim, P., 1912. €Uber der Herstellung von Metallspiegeln durch Destillation im Vakuum. Verh. Dtsch. Phys. Ges. 14, 506–507.

47. Pulker, H.K., 1992. Ion plating as an industrial manufacturing method. J. Vac. Sci. Technol. 10, 1669–1674.

48. Pulker, H.K., Guenther, K.H., 1995. Reactive physical vapor deposition processes. In: Flory, F.R. (Ed.), Thin Films for Optical Systems. Marcel Dekker, Inc., New York, Basel, Hong Kong.

49. Reid, I.M., Macleod, H.A., Henderson, E., Carter, M.J., 1979. The Ion Plating of Optical Thin Films for the Infrared. Proceedings of International Conference on Ion Plating and Allied Techniques (IPAT 79). London, July 1979. CEP Consultants Ltd., Edinburgh.

50. Sakudo, N., 1998. Microwave ion sources for material processing. Rev. Sci. Instrum. 69, 825–830.

51. Sarakinos, K., Alami, J., Wuttig, M., 2007. Process characteristics and film properties upon growth of TiOx films by high power pulsed magnetron sputtering. J. Phys. D 40, 2108–2114.

52. Saxe, S.G., 1985. Ion-Induced Processes in Optical Coatings. PhD Dissertation, University of Arizona. Schiller, S., Heisig, U., Goedicke, G., 1975. Alternating ion plating—a method of high-rate ion vapor deposition. J. Vac. Sci. Technol. 12, 858–864.

53. Schulz, U., Munzert, P., Leitel, R., Wendling, I., Kaiser, N., T€unnermann, A., 2007. Antireflection of transparent polymers by advanced plasma etching procedures. Opt. Expr. 15, 13108–13113.

54. Schulz, U., Präfke, C., Gödeker, C., Kaiser, N., Tünnermann, A., 2011. Plasma etched organic layers for antireflection purposes. Appl. Opt. 50, C31–C35.

55. Segner, J., 1995. Plasma impulse chemical vapor deposition. In: Flory, F.R. (Ed.), Thin Films for Optical Systems. Marcel Dekker, Inc., New York, Basel, Hong Kong, pp. 203–229.

56. Smith, H.M., Turner, A.F., 1965. Vacuum deposited thin films using a ruby laser. Appl. Opt. 4, 147–148.

57. Song, Y., Sakurai, T., 2004. High-rate, low-temperature radical-assisted sputtering coater and its applications for depositing high-performance optical filters. Vacuum 74, 409–415.

58. Song, Y., Sakurai, T., Maruta, K., Matusita, A., Matsumoto, S., Saisho, S., Kikuchi, K., 2000. Optical and structural properties of dense SiO_2, Ta_2O_5 and Nb_2O_5 thin-films deposited by indirectly reactive sputtering technique. Vacuum 59, 755–763.

59. Spiller, E., Haller, I., Feder, R., Baglin, J.E.E., Hammer, W.N., 1980. Graded-index AR surfaces produced by ion implantation on plastic materials. Appl. Opt. 19, 3022–3026.

60. Tang, W.T., Ying, Z.F., Hu, Z.G., Li, W.W., Sun, J., Xu, N., Wu, J.D., 2010. Synthesis and characterization of HfO_2 and ZrO_2 thin films deposited by plasma assisted reactive pulsed laser deposition at low temperature. Thin Solid Films 518, 5442–5446.

61. Targove, J.D., 1987. The Ion-Assisted Deposition of Optical Thin Films. PhD Dissertation, University of Arizona.

62. Targove, J.D., Macleod, H.A., 1988. Verification of momentum transfer as the dominant densifying mechanism in ion-assisted deposition. Appl. Opt. 27, 3779–3781.

63. Targove, J.D., Lingg, L.J., Lehan, J.P., Hwangbo, C.K., Macleod, H.A., Leavitt, J.A., Mcintyre Jr., L.C., 1987. Preparation of aluminum nitride and oxynitride thin films by ion-assisted deposition. In: Materials Modification and Growth using Ion Beams Symposium, Anaheim, CA, USA. Materials Research Society, Pittsburgh, PA, pp. 311–316.

64. Targove, J.D., Lingg, L.J., Macleod, H.A., 1988. Verification of momentum transfer as the dominant densifying mechanism in ion-assisted deposition. In: Optical Interference Coatings. Optical Society of America, Tucson, AZ, pp. 268–271.

65. Taylor, H.D., 1904. Lenses. United Kingdom patent 29561.

66. Thomas, I.M., 1993. Sol-gel coatings for high power laser optics: past present and future. Proc. SPIE 2114, 232–243.

67. Thornton, J.A., 1974. Influence of apparatus geometry and deposition conditions on the structure and topography of thick sputtered coatings. J. Vac. Sci. Technol. 11, 666–670.

68. Vernhes, R., Zabeida, O., Klemberg-Sapieha, J.E., Martinu, L., 2004. Single-material inhomogeneous optical filters based on microstructural gradients in plasma-deposited silicon nitride. Appl. Opt. 43, 97–103.

69. Westmoreland, D. (1993). CVD method for semiconductor manufacture using rapid thermal pulses. US patent 5,227,331. July 13, 1993.

70. Woo, S.-H., Park, Y.J., Chang, D.-H., Sobahan, K.M.A., Hwangbo, C.K., 2007. Wideband antireflection coatings of porous MgF2 films by using glancing angle deposition. J. Korean Phys. Soc. 51, 1501–1506.

71. Zhang, W., Gan, J., Hu, Z., Yu, W., Li, Q., Sun, J., Xu, N., Wu, J., Ying, Z., 2011. Infrared and Raman spectroscopic studies of optically transparent zirconia (ZrO_2) films deposited by plasma-assisted reactive pulsed laser deposition. Appl. Spectrosc. 65, 522–527.

CHAPTER 7

Applications of Polymer Optical Fibers

SHRIKAANT KULKARNI

Department of Chemical Engineering, Vishwakarma Institute of Technology, Pune 411037, India

*Corresponding author. E-mail: shrikaant.kulkarni@vit.edu

ABSTRACT

Polymer-based fibers are becoming a frontier technology and because of the richness of phenomena observed make them a perpetually hot area of research. It all begun with the history of how polymer fibers originated, evolved, and the advantage of theory of wave propagation of light in fibers. When it comes to use of polymers for a host of applications the basic physics and materials science-related issues have to be understood such as how to make polymer fibers, how do they work, and how can they be used to advantage for practical applications. In order to develop new polymers for making fibers, there are usually challenges in terms of materials processing aspects. The physical properties of these numerous types of fibers differ significantly in terms of the materials used, and the possible waveguide geometries, decide the basis for the difference in their applications. Sometime rudimentary methods are preferred over even those trying to produce a compendium of all the innovative materials and techniques which employ drawing techniques, governed by the viscoelastic properties of polymers. Further, synthesis of fibers should be followed by characterization since they will always play an important role in ensuring whether the fibers made can be used for specific applications or basic research. In several devices, we can make use of different types of fibers like electrooptic fibers and fiber lasers. There is a range of smart materials developed as an application of polymer optical fibers. In particular, a large part of this chapter reviews applications of polymer fiber optics in various forms and further discusses the possible science-fiction-like applications by virtue of various phenomena involved in it.

7.1 INTRODUCTION

7.1.1 HISTORY OF POLYMER FIBERS

Glass fibers have been used extensively in long-haul telecommunications, but they are having their own problems [1]. In this field, what is expected is that the fiber optic cables have to be flexible such that they can be laid to cover the terrain. Since glass is rigid from within, a diameter of about 125 μm is required to bring in flexibility in the fiber. However, glass possesses a very high tensile strength, although it is highly brittle. Glass fibers are not easy to handle and are vulnerable to damage by impact and abrasion particularly at smaller diameters. Moreover, they have a tendency that their surface creates Griffith [2] cracks, which causes fracture, demands that their surface be covered with polymer elastic in nature. To protect the fragile single-mode glass fiber, a multilayer cable is preferably used

Polymers are characterized by better flexibility and lower brittleness as compared to glass. It is easy and straightforward to derive a large-diameter polymer optical fiber (POF) with minimum number of protective layers. It is also very simple to allow light to propagate into such a fiber on using a core with a larger diameter. In contrast, joining single-mode fibers in unison demand their positioning in a precise manner and thereby escalate the product cost. The problem with coupling light into a small fiber core, is not only associated with its size, but also the angle at which the light is required to be launched. Figure 7.1 shows a ray in a large diameter core multimode fiber that brings about wave-guiding at the critical angle, θ_C. If the angle of incidence is greater than θ_A, the light will not be guided. The term numerical aperture can be defined as the sine of the largest angle of incidence that will facilitate total internal reflection at the cladding core interface, or

$$N_A = \sin \theta_A. \tag{7.1}$$

Applying Snell's law for the ray that enter the core ($\sin \theta_A = n_1 \sin (90 - \theta_C)$), and the condition for the critical angle ($n_1 \sin \theta_C = n_2$), Equation (7.1) can be solved for the refractive index in regard to the refractive indices of the core (n_1) and cladding (n_2)

$$N_A = (n_1^2 - n_2^2)^{1/2} \qquad (7.2)$$

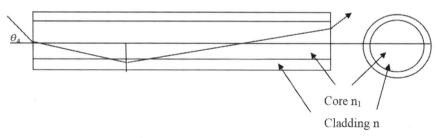

FIGURE 7.1   The numerical aperture of a fiber.

The small-core single-mode fibers have a small difference in refractive index so is the numerical aperture, asking for quite precise launch conditions. Applications that do not require high bandwidth or need fiber to connect devices that are in close vicinity for them multimode POF is the ideal candidate. Polymer fibers are hence employed for interconnecting optically consumer electronics and cater to the needs of fiber-to-the-home applications. Due to polymers' lower mass density and higher robustness, polymer fibers have a comparative advantage over glass in applications that demand lightweight parts or constrained by the space limitations. Of course, polymers find use as light plates in illuminated displays of commercial aircraft. Polymer fibers specifically used for placard illumination on aircrafts, and for optically interconnecting electronic parts in automobiles [3].

7.2   MULTIMODE FIBERS

Multimode polymer fibers have been developed commercially for the first time in the 1970s by DuPont and Mitsubishi Rayon in the United States and Japan, respectively. DuPont in fact both patented as well as commercialized a host of fibers [4]. Mitsubishi, however, developed a fiber traded it under the name Eska (TM), which consists of a poly(methyl methacrylate) (PMMA) core while poly(fluoroalkylmethacrylate) cladding. This cladding has low refractive index and is rugged and moisture resistant. Eska Extra is presently a product having 125 dB km^{-1} loss at 567 nm.

In the 1980s [5], pioneering studies were undertaken by Oikawa and Kaino and were instrumental in determining underlying the loss in the multimode fibers and this knowledge was built upon the earlier work carried out by Oikawa [6]. Both employed an extrusion process to develop the core. Kaino's process involves the polymerization of monomer in situ, followed by extrusion and coating with the cladding in a single pot. While Oikawa, in contrast, synthesized the polymer separately in an ampoule, extruded it, and then coated with a silicon resin. Oikawa carried out studies on polystyrene as core materials while Kaino focused on PMMA. Indeed the optical loss minimizes in most polymers the visible region of the spectrum. The optical loss can be reduced by derivatizing with a heavy atom to replace hydrogen, thus shifting the modes toward the infrared region. Of course, deuterated and fluorinated polymers show far less optical loss in the visible region that can be measured. Other mechanisms according to Kaino responsible for optical loss may be attributed to imperfections in waveguide. Specifically, the minimum loss is observed near 650 nm in visible region. This is referred to as the "transparency window" that is frequently employed in polymer fiber devices that ask for lower loss.[7]

7.3   GRADIENT-INDEX FIBERS

Gradient-index rods and spheres have been in use in the form of lenses over the years. Since any plastic rod can be subjected to heating and drawn as a fiber, a gradient-index polymer rod can be converted into a fiber. The graded-index (GI) fiber is advantageous in terms of both a larger numerical aperture (readily light can be coupled) and a higher bandwidth. Further, a parabolic refractive index distribution profile gives rise to the maximum bandwidth. A better understanding of the bandwidth associated with a parabolic refractive index profile can be achieved by considering that the rays of light prorogate. As the refractive index is greater on axis its axial propagation will take place more slowly in contrast to a ray in the lesser refractive index region. Further the rays that are not launched along the axis will oscillate sinusoidal wave form as they propagate through the fiber. Although the rays travel sinusoidally further, excursions are observed in lower refractive index regions, and so is the velocity component [8–11].

7.4   SINGLE-MODE FIBERS

These polymeric optical fibers were exhibited for the first time by Kuzyk and coworkers in the early 1990s [12]. These fibers possess cores that are dye-doped that lead to the higher refractive index and a potential high intensity dependent refractive index. Dyes used were like Disperse Red 1 Azo dye, Squarylium dyes, and Pthalocyanine dyes. The core and cladding have a diameter of 8 and 125 μm, respectively. It presents a fiber both with the refractive index and single-mode intensity profile.

These fibers were characterized for the linear properties like loss and waveguide mode, extensively characterized [13]. Moreover, the intensity related refractive index was measured using a Sagnac interferometer based technique and found to be high [14]. Single-mode fibers with copolymers as both core and cladding materials were used by Bosc and Toinen [15]. In experimental work, the contribution of the copolymer in both the core and cladding were maintained so as to obtain a refractive index distribution profile that produced a single-mode guide. This work contradicted the study of Kuzyk [16] and Garvey, [17] wherein the dye itself is employed to adjust the refractive index of the core. Both dye and copolymer composition can be maintained to regulate the nonlinear optical properties as well as refractive index of single-mode fibers [18]. Single-mode polymer fibers hold a lot much of promise in making in-line devices compatible with those made out of single-mode silica fiber.

7.5   FIBERS WITH ADVANCED STRUCTURE

A structured fiber consists of various cores, holes, electrodes, and others, for example the electrooptic fiber.

7.6   IMAGING FIBERS

An imaging fiber consists of numerous noninteracting cores such that the intensity profile making way at one end exits, unaltered, from the other end. Such a fiber is particularly made by putting together multiple small fibers. An imaging fiber, therefore, stretches the focal plane from one site to another. These fibers are used in endoscopes that are employed in medical applications for imaging organs in the interior. To make it productive, an imaging fiber has to be comprised of smaller cores to optimize the resolution, larger numerical aperture so as to collect the light in an effective manner, and the material should have enough flexibility to allow the imaging fiber to be readily stuffed in tight spaces. Medical applications demand bundle of fibers with the lowest possible diameter to keep the patient comfortable apart from allowing an access to tight places. In fact, the individual fibers that constitute an imaging bundle have to be built from cores and cladding with high- and low-refractive index, respectively. Moreover, the cladding should be wafer thin to provide for the maximum permissible packing density of the waveguides [19].

Redrawing is a commonly used method for obtaining imaging fiber. First, a fiber is drawn having a desired aspect ratio and difference in the refractive index between core and cladding. The fiber is then chopped off into smaller ones, which are put together in a bundle in a honeycomb arrangement to develop a preform, which is drawn as the imaging fiber. The fiber so obtained is called a second-generation fiber as the fibers in the bundle are pulled for the second time. In turn, the imaging fiber that is obtained can be chopped off to form a preform that is pulled again. Each generation fiber results into smaller waveguides. However, if any errors incorporated in the process of forming preforms introduce more defects in further generations of fiber. Usually, the advantages of the smaller core size are less as compared to the more number of misregistration after some three generations. Given the better flexibility in a polymer, deformation is possible of an imaging fiber so as to modify the image. For example, an adiabatic taper wherein the diameter of the fiber gradually decreases over its length that leads to a smaller image at the exit of the fiber. Similarly, a bundle of the fibers can be made to twist at higher temperature to develop an image inverter as produced by paradigm optics. As the image inverters are light in weight and quite thin [20], they find use in imaging applications wherein the size and weight both are of critical importance.

7.7   CAPILLARY TUBES

A capillary tube consists of a hole in a small diameter cylinder that drags a liquid because of forces that act between the liquid and walls of the hole. The column weight of liquid increases with the volume and the interactions between the liquid and walls that in turn increases with the surface area, a small-diameter cylindrical hole presents the highest surface-to-volume ratio, thus generating the greater drawing force. This effect is termed as a capillary action [21].

Numerous applications make use of the effect of capillary action. In the latest research developments, the mapping of the human genome has been done and the future challenge is to unravel how the millions of proteins are encoded by using about 30,000 genes. Such investigations are referred to as proteomics. A key tool in the science of proteomics laboratory is capillary action. As a solitary capillary tube is employed to carry out a test, using an array of small-diameter capillary tubes enhances the throughput of such testing exercises. Polymer capillary tubes have been better placed over glass particularly in proteomics applications, in the form of polymer capillary arrays as the ideal candidates [22]. It explains how is the process of making preforms that can be used into arrays of holes and presents an array of polymer capillary tubes from the following perspectives:

- Polymer fibers can be manipulated for customized applications. For example, one can make an endoscope using a bundle of fibers for imaging, electrodes (embedded within the fiber) to supply current for cauterization, and a capillary tube, to administer medicines or collect body fluid samples [23].
- Electrodes can wrap around capillary tubes for the application of an electric field to the liquid placed in the capillary tube. For example, a small-diameter capillary tube between two indium electrodes with spherical geometry [24].
- In the drawing process, the cross-sectional areas of both the electrodes and capillary tubes are similar to those in the preform prior to drawing. Any enhancements in the quality of POF structures demand study of processes that are brought about meticulously and in a tedious, time-consuming exercise [25].

7.8   HOLEY FIBER

Holey fiber consists of a 2D periodic structure of holes along the axis of a fiber without a hole at the center of the fiber that works as an imperfection and the light will propagate along it. The holey fiber was first prepared by using silica glass in 1996 and reported by a group of researchers from the University of Southampton [125]. Further, such kind of structures made out of polymer fibers were first reported by researchers at the Sydney University [126] A preform was made out of Disperse Red 1 doped PMMA or DR1-doped PMMA polymer consisting of six concentric hexagonal layers of holes by drilling a DR1-doped PMMA cylinder [26].

Holey fibers have numerous advantages over conventional waveguides that are made within a higher refractive index region as follows:

- Knight and coworkers [27] found that holey fibers will confirm to the single-mode condition over a wider region of the electromagnetic spectrum.
- They are suitable for applications that ask for waveguiding of many colors like parametric frequency generators and spectrophotometers. For example, the light guiding in the fiber appears white due to the presence of wideband of colors.
- Holey fibers consisting of hexagonal array of holes are found to work as polarization-maintaining fibers for two orthogonal linear polarizations [28]. Even on bending or twisting the fiber, the polarization retains its linearity. However, elliptical light is found to be observed for any other polarization,
- The hexagonal pattern of holes as well as uniformity in hole diameter vary from perfection. Such imperfections and given that this fiber has three layers, the waveguiding mode is sound. However, the field leaks do take place away from the center and therefore from hexagonality like at the upper left part of the fiber.
- Such fibers can be used as stress sensors. The polymer fiber becomes birefringent on squeezing from the side, and a part of the light then scatters from the core.

7.9   APPLICATIONS OF POLYMER OPTICAL FIBERS (POFS)

7.9.1   SENSING

Various sensors have been designed and developed using POF, for example, strain and mechanical deformation [29, 30], temperature [31], humidity [32], and the detection of ozone [33] sensors to name a few. POF sensors are commercialized into devices from their laboratory prototypes, and the automobile industry presently is the key customer. For example, developing systems for protecting pedestrians and recognizing seat-occupancy [34]. Reports on POF sensors [35–37] emphasize upon the role of fiber modifications in enhancing their performance efficiency. Key changes such as Long Period Gratings, writing Fiber Bragg Gratings or doping either the core or cladding with dyes that are fluorescent and developing fiber tapers to improve the field of evanescence. Table 7.1 gives a comparison between electrical and optical devices in terms of their function.

TABLE 7.1   Distinction Between Optical and Electrical Devices

Function	Electrical Devices	Optical Devices
Sensing	Thermistor	Interferometer
	Stress sensor	Interferometer
Transmission capacity	Cu metal wires	Optical fibers
	Patterned conductors	
		Optical waveguides
Logic	Transistors	Sagnac loop
	Switches	
		Electrooptic switch
		All-optical switches
Actuation	Piezoelectric materials	Photomechanical fiber-based

7.9.2   DISTRIBUTED SENSORS

Distributed sensor systems are aimed at measuring an environmental parameter at different locations. The method involves developing a distributed temperature/strain detector by placing pairs of Bragg gratings along the optical fiber longitudinal axis directed towards the source, at which a splitter sends the signal to a spectrometer. The spectrum profile changes in the form of characteristic peaks and their intensity even when a small amount of a chemical agent is present [38].

7.9.3   CHEMICAL SENSORS

Fibers are used to combine waveguides of silica glass and photosensitive polymer coatings so as to develop hybrid detectors. Currently, the commonest method used for preparing a fiber that can sense a chemical agent, is to coat a glass fiber with a polymer of whose fluorescence spectrum gets altered on detection of a chemical agent like a nerve agent, pesticide, hazardous molecule, or a banned substance. The polymer coating is subjected to presensitization to allow only one target molecule to come in the close proximity of the fluorescing site. Within the body of the sensing coating, there are voids, as imprinting within the polymer in the form molecule of the target chemical agent to be detected. A fluorescing molecule site is available within every void. When a chemical agent molecule gets into the polymer, it occupies the void, altering the fluorescence spectrum. Only a molecule with a specific geometry of a chemical agent-specific fits into the voids. Jenkins has noted that the selectivity is of such a high order that a molecule other than the target but with similar kind of geometry does not find much place to fit in and therefore produce a very weak signal in the fluorescence spectrum [39].

Light when allowed to pass through either the glass fiber or the polymer-coated fibers both exhibit fluorescence phenomenon. Some of the fluorescence signals combine in the fiber and is guided back toward the source, where a splitter sends the signal to a spectrophotometer. The imprinted polymer is made by coordinating the molecules of the target with ligands. The encapsulated molecules combined with a suitable monomer concentration and are partially polymerized (cross-linked) composition. The glass fiber is coated with this polymerized liquid composition and allowed the polymerization process to complete on the glass fiber surface. The glass fiber thus coated is soaked in alcohol: water blend such that swelling takes place, the target molecules then escape leaving behind free voids. The sensor on drying then is ready to use [40].

7.9.4   ARTIFICIAL NOSE

The artificial nose is a device that can sense a single chemical agent. However, it is difficult to sense multiple agents using this kind of design. Holey POFs possessing an array of cores can be of use in developing sensors for multi-agents sensing as follows:

- Multimode holey fibers guide a wide range of wavelengths, and the structure can be so designed so as to make the light to propagate in the proximity of the holes. Thus a holey fiber along its full length acts as a sensor.
- The longer length and higher sensor volume enhance the sensitivity of the device
- The holes which define the waveguides carry the material to be detected to the detector [41]. The exacerbated surface area made available for exposure to the chemical agent too boosts the sensitivity and reduces the time lag in obtaining the signal.

The major advantage of the holey POF is its integration potential; which means a host of devices can be coupled with the help of a single POF. Such Thus the artificial nose is a device of such a kind that is designed to sense and quantify many chemical agents and can estimate the chemical composition of an environment to which it is exposed. Applications of it cover wide range right from personal safety, security, to scientific instruments. Further by functionalization of holey fibers with multiple functionalities, complex integrated devices can be made out of it that can couple various functions such as sensing, actuation, and logic in designing an architecture of high order [42].

7.9.5   SMART MATERIALS

A smart material has the ability to respond to the stimuli like heat, electric current, light, etc in terms of either change in shape or other properties. This kind of behavior can be used to the advantage for developing devices without moving any individual components. For example, a motor possesses nonmoving components such as wire windings, magnets, and gears etc. Materials exhibiting such properties have the ability to translate the stimulus into the commensurate amount of mechanical energy and thereby can be made in much smaller form and are also more reliable. Bell used materials to obtain musical tones by converting the stimulus in the form of light pulses which is a best example of a so defined smart material. Bell could generate the acoustical waves in the material used in response to the energy absorbed from a beam of light radiation [43].

7.9.6   SMART STRUCTURES

A smart structure consists of highly interconnected smart materials so as to bring about better interaction among them on exposing to external stimuli. A smart material can be a smart structure when the components are micro-scopic in scale or single molecules. The prerequisite for a smart material or structure to function efficiently and to carry out complex operations is high interconnectivity among components. Light is the ideal candidate for carrying information because of its high bandwidth and capacity of parallel processing. Development of a smart structure is one that is interconnected with light; however, there are many issues such as actuation, logic, and sensing that are dependent upon light-sensitive materials [44].

Many applications of such materials are in the foresight, some of them are related to science fiction. For example, a smart skin on wing of an aircraft. The wing on an aircraft faces turbulence and the distribution profile of nonuniform stress. A smart structure comprised of a 2D array of sensors/

actuators could sense the stress profile and make up for the ideal wing profile to experience a smooth ride. Smart wallpaper can be designed using the same concept that prohibits sound from entering and thereby preventing acoustical disturbance from taking place. Smart structures find use as vibration checkers on satellite platforms to prevent the mechanical vibrations from building up which otherwise are difficult to dissipate in space and also on sensitive optics [45].

7.9.7   SMART PHOTONIC MATERIALS

Smart materials or structures have the capability to respond to external stimuli having a defined functionality. Both the stimulus and response can have chemical, acoustical, optical, thermal, or mechanical, in nature. A host of components can be made use of to design such a material or structure such as piezoelectric materials that respond to a stimulus as an electrical field by undergoing change in shape; magnetostrictive materials respond to a magnetic field as a stimulus while the fiber sensors wherein the fibers transmit or reflect light depending upon stress, temperature, or the chemical environment; liquid crystals whose optical reflectivity may be manipulated on the application of a potential or light; shape memory alloys that undergo change in elasticity and shape based on temperature or light exposure. An interesting characteristic feature of a smart material is that their detecting and response happen at one physical point. The response shown by the material, either local or collective while each point works independently as an element of the interconnected structure [46]. Availability of a host of smart materials of different types would certainly foresee the design of smart materials or structures with numerous innovative and quite exciting applications.

The fundamental components of a traditional smart mechanical device are sensors, logic units, and actuators. The sensor transforms input information like stress or strain into an electrical signal that is analyzed by a logic circuit unit like an integrated circuit, and an electrical output as a response from it is instrumental in driving a motor or transducer. Light-encoded information is processed at a faster rate and with better flexibility than that by electronics. Thus the power of optical computing can be used to advantage for the standard smart devices, however, it would require two more elements: a converter that converts electricity into light and another one that vice versa. All-optical devices are unique as they encode mechanical information directly into light and vice versa [47]. An added advantage of MPU-based devices is that the functions like sensing, logic, and actuation are coupled into only mesoscopic monolithic unit. On combining all such units into a smart structure, a highly interconnected network results wherein interactions between a host of units lays the foundation for the functioning of complex computing units. The intelligence imparted to such materials or structures would be of high orders in terms of speed and sophistication than electronics-based devices. Photo-mechanical devices have proved themselves worthwhile as they bridge the technological gap required to devise comprehensive optical technology base [48].

7.10   PHOTONIC CRYSTAL FIBERS

A photonic crystal is a periodic structural arrangement with a period of the order of a given wavelength of light. Light interacts with a photonic crystal similar to the way electrons interact with an atomic crystal. In order to understand the principle of light interaction with a periodic structure, two different possibilities are taken into account such as either light guiding in parallel waveguides or light moving perpendicular to a periodic refractive index distribution profile [49].

7.11   SMART THREADS AND FABRICS

A set of Manitowoc Public Utilities (MPUs) defined in a fiber would show a complex smart response depending upon the states of the all of the MPUs. To understand the complexity of the response intensive modeling and the numerical analysis techniques are required. This is called a smart thread. Such smart threads can be coupled to develop further complex systems. For example, the on waving threads a smart fabric develops, which then can then be reinforced in some other material to develop a composite [50]. The fabric can be reinforced in a composite and can be used as a free-standing material. Moreover, the individual threads can be powered with light or connected together, based on the desired functionality. Even a sandwich structure can be formed from two composite woven fabrics that can cover the surface of an object.

The sandwich structure can be made to bend by allowing light to propagate through the fabric. In addition, the geometry of the structure can be retained even after vibrations. While every unit in the smart structure responds to local stress, the information is transferred to adjoining MPUs, which then can respond prior to reaching the stress or pressure to the latter units. For example, a smart skin that is used to cover an aircraft wing. When the leading edge of the aircraft wing experiences a change in either pressure or turbulence, the smart fabric in the smart skin is so designed that it adjusts the trailing edge of the wing prior to reaching the pressure wave to it [51].

7.12   TRANSVERSE ACTUATION

There are numerous geometries that can be explored that caters to the needs of a desired application or can better mechanical leverage. Transverse motion of a fiber can be more advantageous in specific applications than a change in length. Such a device can be developed by using an asymmetric fiber waveguide, wherein the axis of a high refractive index guiding region (defined as a doped dye region) is opposed by the fiber axis. In such a fiber, light generates a waveguide that is consisted of a dye-doped region, which plays the dual role of increasing the refractive index to permit waveguiding and to pass on the photomechanical effect to the desired region. A light beam brings about uniaxial deformation of the guide. As there is no deformation in the adjoining area, the difference in stress will bring about bending of the fiber in the transverse or perpendicular direction (similar to the way a bimetallic strip responds to temperature variations). A smart positioner can be developed using this concept having two degrees of freedom [52].

7.13   CRACK FORMATION SENSING AND PREVENTION

Smart threads but smaller in lengths can be designed for highly selective and specialized applications. For example, a smart thread can be designed into a sensor for detecting local stresses that occur when materials fracture with the help of an array of distributed actuators that respond to the development of internal local stresses. A transparent MPU that although is positioned away from the potential cracking site is designed with a suitable geometry that generates maximum actuating stress in order to oppose crack formation [53]. The distributed array of actuators is prepared by using a photomechanical material. In addition to responding to local and distributed stresses from the other actuators, the design of the system can be made in such a way that it can respond intensely to minor changes in strain because of the proximity of sensors to the potential crack. The performance efficiency of the smart thread can be diagnosed and quantified by using a photodetector that can help in evaluating material fatigue. The output of the smart thread can be communicated as an input to another array of devices farther from the crack location [54–56].

7.14 NOISE REDUCTION

Suppression of vibration gives rise to a decrease in acoustical noise. Such noise reduction systems are commercialized that consists of one or more microphones to diagnose noise and a control system that can activate a host of speakers to generate a field of noise in contrast phase to the original one. The MPU can act as a microphone (by the pressure detection), the logical control system, and speaker (by the sound generation of the right phase to nullify the noise), simultaneously [57–59]. Thus it is very much possible to design structures that generate an automated noise-free environment. Such a possibility can be explored for harnessing the potential dual-use applications. For example, smart wallpaper developed by using a distribution of MPUs can cut down noise indoors. Such wallpapers can be of great used in innumerable industrial noisy environments as well as to check the quantum of sound generating from a mechanical area and is responsible for creating noise pollution. The programing of a smart fabric can also be done for highly sophisticated applications. For example, a smart wallpaper can be programed to minimize noise from the sources if they are known beforehand and distribution of noise is fixed and thereby will not disturb a speech. That would help facilitate individuals to communicate verbally among themselves in those areas that are prone to greater levels of background noise. Smart skins and wallpaper certainly hold a lot much of promise and potential in various other applications like echo suppression (such as designing better concert halls) and surveillance of intruders [60–62] .

7.15 ILLUMINATION

Illumination optics has come to the fore as one of the biggest applications of POF and is attributed to several advantages of POF. POF illumination fibers possess larger cores, more light tapping capacity, and high flexibility. In the past POF has been used in illumination applications as point sources, for example, as localized spotlights in either buildings or those areas having limited space. A potential growing market is foreseen in future by the exploration of side scattering POF technology for strip lighting. These are created by the embedding the scattering particles in the polymer and used at distances of the order of 5 m, normally by coupling with a LED source [63–65].

7.16   HIGH-SPEED DATA TRANSMISSION

Most prominent growth area for POF is a short distance and high speed data transmission. The demand is on the rise for broader bandwidth to provide for the versatility in applications as ranging from the automotive, consumer electronics industry to fiber to the home and local area networks [66]. In all such application, the advantage of easy connectivity and installation of the smart POF fibers is taken. Step-index POF has been so far in use for most of the data transmission applications. They offer data transmission at a speed of 400 Mbits s^{-1} over a distance of 50 m and are very much put into use in consumer electronics and industrial electronics, apart from automotive industry. The substantial contribution made by the latter and is therefore called as "killer application" for POF, with the huge potential for the market in the foreseeable future to grow [66]. With the increase in bit rates, however, the step-index fibers will not deliver, and therefore the GI fibers are used as the better substitute [67].

The key applications for GI POF have been derived because of their versatility in use from streets to homes or offices. Further, it has the capacity to give an impetus to rise in usable bandwidth. The demand for more bandwidth is on the rise continuously, at the instance of strong governmental support. It has become a possibility now that the telephone, Internet, and television (both video on demand and broadcasts) can be run using the same connection in near future (Triple Play) [68]. This will demand data transmission at higher rates of the order of 10 Gbits s^{-1} over a longer distance say 100 m or more. The advent of high definition television and huge video screens offers a further better opportunity, asking for the data rates at 3–5 Gbits s^{-1} for linking displays with video players. Internet server "farms" too demand huge data transmission, and replacing the present copper-based facilities with optical fiber-based ones will go a long way in substantial cost cuttings. Japan presently is at the forefront in the world as far as installation of FTTH networks are concerned, with more than 40 lacks households linked by 2005 itself, and the whole country has been connected almost a decade ago. The government of Japan has adopted digital broadcasting television system using FTTH since 2006. Korea Telecom had an ambitious program of connecting 10 million households with FTTH by 2010 itself [69–72].

In Europe, POF-ALL: Paving the Optical Future with Affordable Lightning Fast Links, is a new consortium exploring POF for furthering FTTH applications. A host of strategies are presently being unraveled. Silica fiber as a backbone and connection to the consumer with wireless or copper cables is provided within an apartment building. Smart POFs that are fluorinated and possessing higher bandwidth and operating in the IR region have also been in use. The POF-ALL consortium, however, is aimed at using PMMA-based POF working in the visible region, with the rate of data transmission 100 Mbit s^{-1} over a distance of 300 m at 520 nm wavelength, and 1 Gbit s^{-1} over 100 m distance at 650 nm [73].

The ultimate choice of fibers depend upon life of the fiber, resistance to bending of tight, and installation cost, ease with which connection can be made, how far sources are economical and detectors other than optical necessities such as availability of bandwidth and transmission over adequate length. Researchers in this field are of the opinion that larger core fibers working in the visible region are in great demand in particular for household applications [74]. Indeed, the adoption of FTTH and other shorter distance, but high bandwidth applications offer tremendous opportunities for POF, but are constrained by more technical considerations as against those needed in the past. In order to specify such requirements and acquire better consistency to the industry, additionally a host of international standard (IEC SC86A) too was introduced, in the 2006 [75–77].

7.18 CONCLUSION

Glass fibers have been used in long-haul telecommunications for years together but they have their own limitations could not stand the test of time. As the demand of the time is the higher speed of data transmission over a longer distance and hassle-free as well as a common connection for television (video on demand and broadcasting), Internet, and telephone as a triple play. POFs are characterized by flexibility and lower brittleness as compared to conventional glass fibers. Moreover, POFs are easy and straightforward to derive from polymeric materials with a large diameter core and minimum number of protective layers in fibers. It facilitates better propagation of light and at faster pace too. POFs in various forms such as single-mode, multimode, gradient-index, imaging, holey, smart threads, capillary tubes, and photonic crystal fibers having varieties of architectures that find widespread applications in various walks of life such as automotive industry, electronics, sensing, and imaging. Further, there is an ample scope and opportunities as well as challenges to explore POFs for designing new configurations and architectures for meeting the requirements of different sections of the society at an affordable cost. POF technology has a bright foreseeable future and will scale up new heights and will go a long way in serving the humanity.

KEYWORDS

- **polymer fibers**
- **visco-elastic**
- **electrooptic**
- **science fiction**
- **phenomena**

REFERENCES

1. Bell, A. G., "On the Production and Reproduction of Sound by Light," Proceedings of the American Association for the Advancement of Science **29**, 115–136 (1881).
2. Rayleigh, L., "On the Passage of Electric Waves Through Tubes or the Vibrations of Cylinders," Phil. Mag. **43**, 125 (1897).
3. Hondros, D. and Debye, P., "Elektromagnetishe Wallen an Dielektrishen Dr¨ahten," Ann. Phys. **32**, 465 (1910).
4. Schriever, O., "Elektromagnetishe Wallen an Dielektrishen Dr¨ahten," Ann. Phys. **63**, 645 (1920).
5. Hopkins, H. H. and Kapany, N. S., "A Flexible Fiberscope Using Static Scanning," Nature **173**, 39–41 (1954).
6. Baird, J. L., Brit. Pat. Spec. No. 20, 969/27 (1927).
7. Snitzer, E., "Cylindrical Dielectric Waveguide Modes," J. Opt. Soc. Am. **51**, 491–498 (1961).
8. Snitzer, E. and Osterberg, H., "Observation of Dielectric Waveguide Modes in the Visible Spectrum," J. Opt. Soc. Am. **51**, 499–505 (1961).
9. Emslie, C., "Review Polymer Optical Fibres," J. Mater. Sci. **23**, 2281–2293 (1988).
10. Griffith, A. A., "The Phenomena of Rupture and Flow in Solids," Phil. Trans. R. Soc. **221a**, 163 (1921).
11. Hager, T. C., Brown, R. G., and Derick, B. N., "Automotive Fibers" Trans. Soc. Automotive Eng. **76**, 581 (1976).
12. Kaino, T., Fujiki, M., Oikawa, S., and Nara, S. "Low-Loss Plastics Optical Fibers," Appl. Opt. **20**, 2886–2888 (1981).
13. Oikawa, S., Fujiki, M., and Katayama, Y., "Polymer Optical Fiber with Improved Transmittance," Electron. Lett. **15**, 830–831 (1979).

14. Ohtsuka, Y., Senga, T., and Yasuda, H., "Light-Focusing Plastic Rod with Low Chromatic Aberration," Appl. Phys. Lett. **25**, 659 (1974).
15. Ohtsuka, Y., Sugano, T., and Terao, Y., "Studies on the Light-Focusing Rod 8: Copolymers of Diethylene Glycol *bis* (Allkyl Carbonate) with Methacrylic Ester of Flourine Containing Alcohol," Appl. Opt. **20**, 2319 (1981).
16. Ohtsuka, Y. and Koike, Y., "Determination of the Refractive-Index Profile of Light-Focusing Rods: Accuracy of Method Using Interphako Interference Microscopy," Appl. Opt. **19**, 2866–2867 (1980).
17. Nihei, E., Ishigure, T., and Koike, Y., "High-Bandwidth, Graded-Index Polymer Optical Fiber for Near-Infrared Use," Appl. Opt. **35**, 7085–7090 (1996).
18. Kuzyk, M. G., Paek, U. C., and Dirk, C.W., "Guest-Host Fibers for Nonlinear Optics," Appl. Phys. Lett. **59**, 902 (1991).
19. Garvey, D. W., Zimmerman, K., Young, P., Tostenrude, J., Townsend, J. S., Zhou, Z., Lobel, M., Dayton, M., Wittorf, R., and Kuzyk, M. G., "Single-Mode Nonlinear-Optical Polymer Fibers," J. Opt. Soc. Am. B **13**, 2017–2023 (1996).
20. Garvey, D. W., Li, Q., Kuzyk, M. G., Dirk, C. W., and Martinez, S., "Sagnac Interferometric Intensity-Dependent Refractive-Index Measurements of Polymer Optical Fiber," Opt. Lett. **21**, 104–106 (1996).
21. Bosc, D. and Toinen, C., "Full Polymer Single-Mode Optical Fiber," IEEE Photonics Technol. Lett. **4**, 749–750 (1992).
22. Welker, D., Personal Communication (2000).
23. Whitaker, N. A. J., Gabriel, M. C., Avramopolous, H., and Huang, A., "All-Optical, All-Fiber Circulating Shift Register with an Inverter," Opt. Lett. **24**, 1999 (1991).
24. Xiong, Z., Peng, G. D., and Chu, P. L. "Nonlinear Coupling and Optical Switching in a β-Carotene-Doped Twin-Core Polymer Optical Fiber," Opt. Eng. **39**, 624–627 (2000).
25. Snyder, A. W., "Couple-Mode Theory for Optical Fibers," J. Opt. Soc. Am. **62**, 1267–1277 (1972).
26. Snyder, A. W. and Y. Chen, "Nonlinear Fiber Couplers: Switches and Polarization Beam Splitters," Opt. Lett. **14**, 517–519 (1989).
27. P. L. Chu, "Polymer Optical Fiber Bragg Gratings," Opt. Photon. News **16**, 53–56 (2005).
28. Knight, P. L., Birks, T. A., Russel, P., S. J., and Atkin, D. M., "All-Silica Single-Mode Optical Fiber with Photonic Crystal Cladding," Opt. Lett. **21**, 1547–1549 (1996).
29. van Eijkelenborg, M. A., Large, C. J., Argyros, A., Zagari, J., Manos, S., Issa, N. A., Bassett, I., Fleming, S., McPhedran, R. C., de Sterke, C. M., and Nicorovici, N. A. P. "Microstructured Polymer Optical Fibre," Opt. Express **9**, 319–327 (2001).
30. Fink, Y., Ripin, D. J., Fan, S., Chen, C., Joannopoulos, J. D., and Thomas, E. L., "Guiding Optical Light in Air Using an All-Dielectric Structure," J. Light. Technol. **17**, 2039–2041 (1999).
31. Temelkuran, B., Hart, S. D., Benoit, G., Joannopoulos, J. D., and Fink, Y. "Wavelength-Scalable Hollow Optical Fibres with Large Photonic Bandgaps for CO_2 Laser Transmission," Nature **420**, 650–653 (2002).
32. Dumont, M., Sekkat, Z., Loucif-Saibi, R., Nakatani, K., and Delaire, J. A., "Photoisomerization, Photoinduced Orientation and Orientational Relaxation of Azo Dyes in Polymeric Films," Nonlinear Opt. **5**, 395–406 (1993).
33. Sekkat, Z. and Dumont, M., "Polarization Effects in Photoisomerization of Azo Dyes in Polymeric Films," Appl. Phys. B **53**, 121–123 (1991).

34. Sekkat, Z. and Dumont, M., "Poling of Polymer Films by Photoisomerization of Azo Dye Chromophores," **2**, 359–362 (1992).
35. Sekkat, Z. and Dumont, M., "Photoassisted Poling of Azo Dye Doped Polymeric Films at Room Temperature," Appl. Phys. B **54**, 486–489 (1992).
36. Sekkat, Z. and Knoll, W., "Creation of Second-Order Nonlinear Optical Effects by Photoisomerization of Polar Azo Dyes in Polymeric Films: Theoretical Study of Steady-State and Transient Properties," J. Opt. Soc. Am. B. **12**, 1855–1867 (1995).
37. Bian, S., Zhang, W., and Kuzyk, M. G., "Erasable Holographic Recording in Photosensitive Polymer Optical Fibers," Opt. Lett. **28**, 929–931 (2003).
38. Bian, S. and Kuzyk, M. G., "Phase Conjugation by Low-Power Continuous-Wave Degenerate Four-Wave Mixing in Nonlinear Polymer Optical Fibers," App. Phys. Lett. **84**, 858–860 (2004).
39. Froggatt, M., "Distributed Measurement of the Complex Modulation of a Photoinduced Bragg Grating in an Optical Fiber," Appl. Opt. **35**, 5162–5164 (1996).
40. Jenkins, A. L., Uy, O. M., and Murra, G. M. Y, "Polymer-Based Lanthanide Luminescent Sensor for Detection of the Hydrolysis Product of the Nerve Agent Soman in Water," Anal. Chem. **71**, 373–378 (1999).
41. Arnold, B. R., Euler, A. C., Jenkins, A. L., O. M. Uy, and Murray, G. M., "Progress in the Development of Molecularly Imprinted Polymer Sensors," Johns Hopkins APL Tech. Digest **20**, 190–198 (1999).
42. Jenkins, A. L., R. Yin, and Jensen, J. L., "Molecularly Imprinted Polymer Sensor for Pesticide and Insecticide Detection in Water," Analyst **126**, 798–802 (2001).
43. Uchino, K. and Cross, L. E., "Electrostriction and Its Interrelation with Other Anharmonic Properties of Materials," Jpn. J. Appl. Phys. **19**, 171–173 (1980).
44. Uchino, K., "Photostrictive Actuator," Ultrason. Symp., 721–723 (1990).
45. Uchino, K., "Ceramic Actuators: Principles and Applications," MRS Bull., April **29**, 42 (1993).
46. M. Camacho-Lopez, Finkelmann, H., P. Palffy-Muhoray, and Shelley, M., "Fast Liquid-Crystal Elastomer Swims into the Dark," Nat. Mater. **3**, 307–310 (2004).
47. Buckland, E. L. and R. W. Boyd, "Electrostrictive Contribution to the Intensity-Dependent Refractive Index of Optical Fibers," Opt. Lett. **21**, 1117–1119 (1996).
48. Poga, C. and Kuzyk, M. G., "Quadratic Electroabsorption Studies of Third-Order Susceptibility Mechanisms in Dye-Doped Polymers," J. Opt. Soc. Am. B **11**, 80–91 (1994).
49. Kuzyk, M., Moore, R. C., and L. A. King, "Second-Harmonic-Generation Measurements of the Elastic Constant of a Molecule in Polymer Matrix," J. Opt. Soc. Am. B **7**, 64 (1990).
50. D. J.Welker and Kuzyk, M. G., "Photomechanical Stabilization in a Polymer Fiber-Based All-Optical Circuit," Appl. Phys. Lett. **64**, 809–811 (1994).
51. Kuzyk, M. G., Welker, D. J., and Zhou, S., "Photomechanical Effects in Polymer Optical Fibers," Nonlinear Opt. **10**, 409–419 (1995).
52. Welker, D. J. and Kuzyk, M. G., "Optical and Mechanical Multistability in a Dye-Doped Polymer Fiber Fabry-Perot Waveguide," Appl. Phys. Lett. **66**, 2792–2794 (1995).
53. Welker, D. J. and Kuzyk, M. G., "All-Optical Switching in a Dye-Doped Polymer Fiber Fabry-Perot Waveguide," Appl. Phys. Lett. **69**, 1835–1836 (1996).

54. Dumont, M. and Sekkat, Z., "Dynamic Study of Photoinduced Anisotropy and Orientation Relaxation of Azo Dyes in Polymeric Films. Poling at Room Temperature." SPIE Proc. **1774**, 1–12 (1992).

55. Dumont, M., G. Froc, and Hosotte, S., "Alignment and Orientation of Chromophores by Optical Pumping," Nonlinear Opt. **9**, 327–338 (1995).

56. Dumont, M., Hosotte, S., Froc, G., and Sekkat, Z., "Orientational Manipulation of Chromophores Through Photoisomerization," SPIE Proc. **2042**, 1–12 (1993).

57. Welker, D. J. and Kuzyk, M. G., "Suppressing Vibrations in a Sheet with a Fabry-Perot Photomechanical Device," Opt. Lett. **22**, 417–418 (1997).

58. Bian, S., Robinson, D., and Kuzyk, M. G., "Optically Activated Cantilever Using Photomechanical Effects in Dye-Doped Polymer Fibers," J. Opt. Soc. Am. B **23**, 697–708 (2006).

59. Bohren, C.F. and Huffman, D.R., "Absorption and Scattering of Light by Small Particles," John Wiley & Sons (New York, 1983).

60. Kaino, T., Fujiki, M., Oikawa, S., and Nara, S., "Low Loss Plastic Fibers," Appl. Opt. **20**, 2886–2888 (1981).

61. Kuang, K. S. C and Cantwell, "W J. The Use of Plastic Optical Fibre Sensors for Monitoring the Dynamic Response of Fibre Composite Beams," Meas. Sci. Technol., **14**, 736–745 (2003).

62. Kuang, K. S. C., Cantwell, W. J., and Scully, P. J. "An Evaluation of a Novel Plastic Optical Fiber Sensor for Axial Strain and Bend Measurements" Meas. Sci. Technol, **13**, 1523–1534 (2002).

63. Kumar, V. V. Ravi Kanth, George, A. K., Knight, J. C., and Russell, P St J. "Tellurite photonic crystal fiber" Opt. Express, **11**(20), 2641–2645 (2003).

64. Kumar, V. V. Ravi Kanth, George, A. K., Reeves, W. H., Knight, J. C., Russell, P St J, Omenetto, F. G., and Taylor, A. J. "Extruded soft glass photonic crystal fiber for ultrabroad super-continuum generation" Opt. Express, **10**(25), 1520–1525 (2002).

65. Kuriki, K., Shapira, O., Hart, S., Benoit, G., Kuriki, Y., Viens, J., Bayindir, M., Joannopoulos, J., and Fink, Y. "Hollow Multilayer Photonic Bandgap Fibers for NIR Applications." Opt. Express, **12**(8), 1510–1507 (2004).

66. Large, M. C. J, Ponrathnam, S, Argyros, A, Bassett, I, Punjari, N. S., Cox, F, Barton, G. W., and van Eijkelenborg, M. A. (2006). Microstructured polymer optical fibres: New opportunities and challenges. In Burillo, G, Ogawa, T, Rau, I, and Kajzar, F, editors, Molecular Crystals and Liquid Crystals Journal, Special issue, Proceedings of the 8th international conference on frontiers of polymers and advanced materials, volume 446, pages 219–31.

67. Limpert, J., Schreiber, T., Nolte, S., Zellmer, H., T¨unnermann, A., Iliew, R., Lederer, F., Broeng, J., Vienne, G., Petersson, A., and Jakobsen, C. "High-power Air-clad Large-moderate Photonic Crystal Fiber Laser" Opt. Express, **11**(7), 818–823 (2003).

68. Liu, H. Y., Liu, H. B., and Peng, G. D. "Polymer Optical Fibre Bragg Gratings based Fibre Laser." Opt. Commun., **266**(1), 132–135 (2006).

69. Liu, H. Y., Liu, H. B., Peng, G. D., and Chu, P. L. "Observation of Type I and Type II Gratings Behavior in Polymer Optical Fiber" Optics Commun., **220**(4–6), 337–343 (2003).

70. MacChesney, J. B., O'Connor, P. B., and Presby, H. M. (1974). "A New Technique for Preparation of Low-loss and Graded Index Optical Fibers" Proc. IEEE, **62**(9), 1280–1281.

71. Mach, P, Dolinski, M, Baldwin, K. W., Rogers, J. A., Kerbage, C, Windeler, R. S., and Eggleton, B. J. "Tunable Microfluidic Optical Fiber" Appl. Phys. Lett., **80**(23), 4294–4296 (2002).

72. Marcatili, E. A. J (1973). Air clad optical fiber waveguide. US Patent 3712705.

73. Monro, T. M., West, Y. D., Hewak, D. W., Broderick, N. G. R, and Richardson, D. J. "Chalcogenide Holey Fibres" Electron. Lett., **36**(24), 1998–2000 (2000).

74. Murofushi, M (1996). Low loss perfluorinated POF. In Proceedings of the International Plastic Optical Fibres Conference, pages 17–23, Paris, France.

75. Muto, S., Sato, H., and Hosaka, T. "Optical Humidity Sensor Using Fluorescent Plastic Fiber and Its Application to Breathing Condition Monitor" Jpn. J. Appl. Phys., **33**(10), 6060–6064 (1994).

76. Myaing, M. T., Ye, J. Y., Norris, T. B., Thomas, T, Jr, J. R. Baker, Wadsworth, WJ, Bouwmans, G, Knight, J. C., and Russell, P St J. "Enhanced Two Photon Biosensing with Double-clad Photonic Crystal Fiber" Opt. Lett., 28(14), 1224–1226 (2003).

77. Nocivelli, A. (2006). "Plastic Fibre Promises Ubiquitous Optical Access" *Fibre Systems Europe in association with* LIGHTWAVE Europe, page 14.

CHAPTER 8

Rare Earth Oxalates: A Promising Optical Material Without Luminescence Quenching

DINU ALEXANDER[1,*] and CYRIAC JOSEPH[2]

[1]*Department of Physics, Newman College, Thodupuzha, Kerala, India*

[2]*School of Pure and Applied Physics, Mahatma Gandhi University, Kottayam, Kerala, India*

Corresponding author. E-mail: dinualexanderd@gmail.com

ABSTRACT

The quenching free photoluminescence in fully concentrated rare-earth-based optical materials makes them excellent candidates for various optoelectronic applications. The factors leading to quenching free luminescence is the focal theme of this chapter. It covers the basic description of different quenching processes like concentration quenching, cross-relaxation quenching, phonon-assisted quenching, and hydroxyl quenching in luminescent materials. This chapter also describes the photoluminescence properties of fully concentrated rare-earth oxalate with terbium oxalate as a representative case. The crystal structure of rare-earth oxalate facilitates well-separated luminescence centers and there is least possibility for energy migration between identical luminescent centers thus weakening concentration quenching in this matrix. The quenching free nature of terbium oxalate is analyzed further by studying the variation of photoluminescence emission with the concentration of Tb^{3+} incorporated in lanthanum oxalate matrix up to the full concentration of terbium.

8.1   INTRODUCTION

Rare-earth-based inorganic luminescent materials found widespread applications in optoelectronic devices such as display devices, lighting, luminescence-based sensors, and bioimaging. In addition to this, they play a significant role in drug delivery, optical imaging, biolabeling, and in fluorescence resonance energy transfer assays [1–3]. Luminescent materials are available in a variety of forms like single crystals, nanocrystals, microcrystals, polymers, metal complexes, and coordination compounds. Among these nanostructured materials have attracted much attention due to its controllable size and morphology that have a great impact on its optical, magnetic, and electronic properties.

The nanophosphor materials exhibit remarkably superior optical properties than bulk counterpart due to quantum confinement effects. As the particle size is reduced to nanoregime the nonradiative losses decreases considerably than in bulk phosphors, which enhances the radiative efficiency of the luminescent material. In this aspect, the synthesis and optical characterization of rare-earth-based inorganic nanophosphor deserve special attention [4–6]. One of the main prerequisite for an efficient phosphor material is its intense emission-without luminescence quenching. Even though, fully concentrated matrices are a promising choice for designing nano phosphor with intense emission—this field is not much explored. The reports on fully concentrated luminescent materials are less despite of their guaranteed intense emission [7]. In fully concentrated systems or stoichiometric phosphors, the activator concentration can be 100%, thereby providing maximum luminescent centers.

In this chapter, different types of luminescence quenching mechanisms and the importance of quenching free luminescent materials are discussed. The advantages of the oxalate host and the spectroscopic properties of efficient green-emitting terbium oxalate in the platform of single-crystal structure are detailed. Moreover, the structural specificity leading to the quenching free nature of fully concentrated terbium oxalate is described.

8.2   RARE-EARTH-BASED LUMINESCENT MATERIALS: MERITS AND DEMERITS

The characteristic feature of rare-earth ions is their partially filled 4f shells; with unique 4f shell for each rare-earth ion. Thus the rare-earth ions are characterized by a neutral Xenon core [Xe], unfilled f shell, and some outer shells that screen the 4f shell from outside. Since, these partially filled electronic level (f shells) of rare-earth ions are well shielded from the external perturbing field by the completely filled outer levels $5s^2 5p^6$ levels, the rare-earth ions exhibit their sharp and well-defined narrow peaks irrespective of the host material. Thus the rare-earth-doped inorganic luminescent materials are of interest due to their excellent luminescence properties like sharp and intense emission, high luminous efficiency, high color purity, and due to the different emission colors originating from the f–f or f–d transitions in rare-earth ions. Moreover, dopant ions can be selected based on the requirement of applications as different activators give different emission of light from ultraviolet to near-infrared range. For example, the doping of rare-earth, Eu^{3+}, Pr^{3+}, Sm^{3+}, in host matrices gives red emission, the rare-earths Er^{3+} and Tb^{3+} can be selected for green emission, Dy^{3+} ion is used predominantly for producing yellow emission and the rare-earths Nd^{3+} and Er^{3+} emit light in the infrared region [8, 9].

One of the most important limitations of rare-earth-based luminescent material is luminescence quenching at lower concentration of activator ions. List of some important Tb/Dy/Eu activated rare-earth-based luminescent materials with their optimum concentration, excitation wavelength, and emission wavelength is given in Table 8.1. From Table 8.1, it is clear that in most of the materials luminescence quenching occurs at lower concentration of activator ion, thereby limiting further addition of activator ions in the matrix. In this aspect, fully concentrated quenching free luminescent materials deserve special attention to produce intense emission in luminescent materials [10–23].

TABLE 8.1 List of Some Important Tb/Eu/Dy Luminescent Materials

Material	Optimum Concentration (%)	Excitation Wavelength (nm)	Emission Wavelength (nm)
$Ca_3Si_2O_7$:Tb	$x = 0.12$	254	542
$LiMgPO_4$:Tb^{3+}	$x = 1.5$	352	543
$SrLa_{1-x}AlO_4$:xTb^{3+}	$x = 0.5$	228	548
$Y_4Si_2O_7N_2$:Tb^{3+}	$x = 15$	275	544
$Ba_2Ca\,WO_6$:xEu^{3+}	$x = 0.2$	314	614
Sr_2CeO_4:Eu^{3+}	$x = 5$	280	616
Gd_2O_3: Eu^{3+},Li+	$x = 6$	265	612
$InBO_3$:xEu^{3+}	$x = 5$	263	616
$Y_4Al_2O_9$: Dy^{3+}	$x = 1$	447	572
Ca_2PO_4Cl:Dy^{3+}	$x = 5$	363	576
$La_6Ba_4(SiO_4)6F_2$:Dy^{3+}	$x = 0.18$	353	579
$Ca_3Al_2O_6$:Dy^{3+}	$x = 0.5$	351	573

8.2.1   QUENCHING OF LUMINESCENCE

The luminescence emission can be quenched in different ways such as concentration quenching, cross-relaxation quenching, hydroxyl quenching, and multiphonon relaxation.

The phenomenon by which the luminescence emission starts to decrease after a critical concentration of activator ion is called concentration quenching and occurs due to the efficient resonant energy migration between identical luminescent ions as represented in Figure 8.1. When the concentration of activator ion is high, aggregation of luminescent centers occurs, which leads to a small average distance between identical luminescent ions and thereby causing enhanced probability for energy migration between ions until it reaches a site where the nonradiative transition occurs. These sites will act as an energy sink and these processes lead to quenching of emission while transferring the excitation energy. Host matrices that provide small average distance between the luminescent centers promote energy transfer and quenching of luminescent emission at low activator concentration.

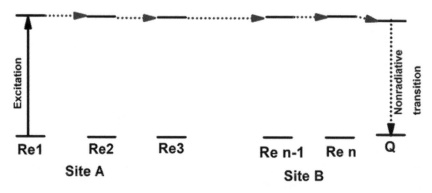

FIGURE 8.1   Schematic of energy transfer by energy migration.

The energy transfer by the cross-relaxation process usually occurs between identical luminescent ions in a material when there exists a pair of energy levels that have the same energy separation. The cross-relaxation process between rare-earth ions is graphically represented in Figure 8.2. In this mechanism, same ion act both as sensitizer and activator. The energy difference between the levels $(E_4–E_3)$ in one system is get transferred to another system by which the second system acquires this excitation energy and moving to a higher state (E_2) from the ground state (E_1) as depicted in Figure 8.2. Thus in the first system, the excited ions return to the lower level E_3 from the excited level E_4 and in the second system the ground state atoms in E_1 moves to the excited state E_2 during the cross-relaxation mechanism. This type of nonradiative energy transfer becomes prominent at higher doping concentrations in rare-earth ions. The energy transfer through cross-relaxation becomes prominent when there exist resonant energy level in the two systems (donor and acceptor) or the excitation energy needed for the second system (acceptor) is slightly lower than that of energy given by the first system (donor). Thus the cross-relaxation type of quenching does not involve the actual migration of excitation energy among identical luminescent centers in a host material [24].

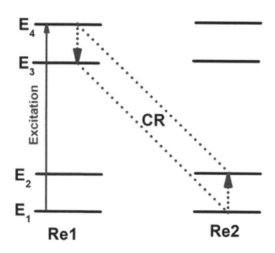

FIGURE 8.2   Schematic representation of cross-relaxation mechanism.

Water molecules have a very significant effect on the luminescent properties due to the possibility of vibrational quenching. The OH oscillators in the water molecules act as efficient quenchers of the f–f luminescence. But among the luminescent lanthanides, Tb and Eu are comparatively less susceptible to quenching by OH oscillators since the energy gaps between the emissive levels and the ground levels for Tb and Eu are of the order 14,700 and 12300 cm^{-1}. Due to the larger energy gap between the lowest excited level and highest ground-state manifold, terbium and europium are not as sensitive to OH vibrations as the other lanthanides like dysprosium and samarium with energy gaps 7840 and 7400 cm^{-1} respectively. In multiphonon relaxation, the excited ion decays to the ground state by the interaction with lattice phonons.

If the energy difference between an excited state and the next lower lying state is greater than the phonon energy of the lattice, two or more phonons bridge the energy gap by using the most energetic vibrations of the host lattice [25, 26].

8.2.2 QUENCHING FREE LUMINESCENCE

The prevailing mechanisms for luminescence quenching are concentration quenching, cross-relaxation, hydroxyl quenching, and multiphonon relaxation. The cross-relaxation process depends only on the energy levels of rare-earth ion while the multiphonon relaxation depends on both the phonon energy of the host matrix and the energy difference between the emitting level and ground state of the particular rare-earth ion. The number of overtones of OH vibrations required to bridge the emitting level and ground level of the rare-earth ion is the determining factor in hydroxyl quenching. Concentration quenching is the prime factor limiting quenching of luminescence and depends only on the structure of host matrix irrespective of the energy level of rare-earth ion. Thus structural features of host matrix play a significant role in reducing the quenching effect of active luminescent centers. The luminescent material with high quenching concentration of rare-earth ion is necessary for achieving high luminescence output. In a certain type of host materials, the emission intensities of RE^{3+} ion is found to be increasing with the doping concentration and the intensity reaches the maximum when the host lattice occupies full concentration of rare-earth ion.

Some of the reports on fully concentrated materials include: A red emitting phosphor of fully concentrated Eu^{3+}-based molybdenum borate $Eu_2MoB_2O_9$ [27]; the red luminescence characteristics of fully concentrated Eu-based phosphor $EuBaB_9O_{16}$ [28]; breakthrough in concentration quenching threshold of up conversion luminescence via spatial separation of the emitter doping area for bioapplications [29]; layered structure produced nonconcentration quenching in a novel Eu^{3+}-doped phosphor [30]. All these materials opened up new class of luminescent materials for producing intense emission without any limitation of adding activators to the host matrix.

Also, there exist reports in the literature describing the inherent structural features of host matrix that provide well separation between identical luminescent centers due to the layered structure and thereby preventing energy migration between identical luminescent centers and the subsequent concentration quenching. Jiao et al. reported absence of concentration quenching for Tb^{3+}, Eu^{3+}, Tm^{3+} in $Ca_3Bi(PO_4)_3$ host material [31]. According to them, Tb^{3+} /Eu^{3+}/ Tm^{3+} ions in $Ca_3Bi(PO_4)_3$ material are well isolated by the PO_4 group and because of this isolation energy transfer between these ions that occurs very hardly thereby eliminating quenching of emission. Qin et al. reported that in $Na_5Ln(WO_4)_4$ host when the Ln^{3+} sites are fully replaced by Tb^{3+}/Eu^{3+} ions the luminescence intensity is increasing with the Tb^{3+}/Eu^{3+} concentration without concentration quenching. In $Na_5Tb(WO_4)_4$ and Na_5Eu $(WO_4)_4$ the layered scheelite structure of the host material provides large separation between Tb^{3+}–Tb^{3+} and Eu^{3+}–Eu^{3+} pairs, providing quenching free luminescence even at full concentration of Tb^{3+} and Eu^{3+} [32]. D. Qin and W. Tang in their report on $Na_3La(PO_4)_2$:Tb^{3+}/Eu^{3+} phosphor verified that the $Na_3RE(PO_4)_2$ structure is built up by helical ribbons of $[REO_y]$ linked through $[PO_4]$ tetrahedron leading to a layered structure. This layered structure plays a significant role in weakening the quenching effect by providing a separation between RE–RE such that energy migration is hard to occur [33]. From the above reports, it is found that the structure of host materials plays a decisive role in determining the optical properties. Thus the host materials that permit heavy doping of rare-earth ions are significant in developing promising inorganic phosphors for intense emission.

Lanthanide metal-organic frameworks with a layered structure providing large energy separation between luminescent centers are another alternative for achieving intense luminescence output. Design and synthesis of lanthanide coordination polymers with intense fluorescence emissions are of great significance for fluorescence materials. The synthesis and luminescent properties of plenty of complexes containing oxalic acid as one of the flexible ligands have been continuously reported [34, 35]. Compared with the complicated synthesis procedure and structural uncertainties of the rare-earth-carboxylate complexes, rare-earth oxalate crystals can emerge out as an alternative. The synthesis process of rare-earth oxalate is very simple and the composition and crystal structure are well defined, but it retains all the advantages of complex metal-oxalate frameworks. Thus rare-earth oxalates ensuring the structural features of MOF's will be a possible candidate for obtaining higher quenching concentration of rare-earth ions.

8.3   RARE-EARTH OXALATE

Among the various rare-earth compounds, rare-earth oxalates are of significance due to its luminescent, magnetic behavior, and its use as a precursor for preparing superconducting materials. Recently, the rare-earth oxalates have evoked much attention due to their remarkable optical properties, which find applications in solid-state laser due to its high active ion concentration in the order of 10^{21} ions cm^{-3}. This attractive feature of the fully concentrated rare-earth oxalate crystals can be exploited to derive intense luminescence emission when compared with other doped systems. Moreover, as an important bridging ligand, oxalate anion makes intriguing structures among the rare-earth compounds and they exhibit superior optical properties [36]. Single crystals of cerium oxalate, neodymium praseodymium oxalate, dysprosium praseodymium oxalate, and samarium oxalate were studied in recent years but the detailed spectroscopic investigations in the platform of single-crystal structure were not attempted [37–40]. Therefore it is of scientific interest to investigate the luminescence properties of rare-earth ions in oxalate matrix in the platform of crystal structure of host material. The following discussion on this chapter is devoted to the photoluminescence characteristics of green-emitting terbium oxalate crystals.

8.3.1   TERBIUM OXALATE

Generally, terbium-activated phosphors are used for producing green emission because of its sharp emission around 543 nm corresponding to the $^5D_4$ $^7F_5$ transition of Tb^{3+} ion. Efficient green-emitting phosphors are highly desired for various optoelectronic applications like solid-state lighting, pc-converted LEDs and in-display devices. Moreover, the green color plays a major role among the tricolor centers constituting red, green, and blue with a pixel ratio of 3:6:1 for display applications [7, 41]. Thus photoluminescence studies on green-emitting terbium oxalate crystals deserve special attention.

The intense green luminescence exhibited by fully concentrated terbium oxalate nanocrystals were reported by our research group [7]. The nanocrystals were synthesized by facile microwave-assisted coprecipitation method and the optimized synthesis parameters were given in the report. The microwave-assisted coprecipitation method is the prominent one because the size and shape of the crystals can be easily controlled by changing reaction parameters and it ensures the formation of nanocrystals with the almost same size and shape. Further, single crystals of terbium oxalate were grown by single diffusion gel technique and crystal structure was solved and reported [42, 43]. This report focuses on the structure-dependent photoluminescence behavior of terbium oxalate by analyzing structural features.

8.3.2 SYNERGISTIC EFFECT OF CRYSTAL STRUCTURE ON PHOTOLUMINESCENCE

The crystal structure analysis of terbium oxalate revealed that the crystal system is monoclinic with space group $P2_1/c$. In the structure, the Tb metal center is linked to three carboxylate oxygen group and is coordinated to three water molecules. The layer structure of terbium oxalate is built up by the hydrogen-bonded stacking of terbium–oxalate layers along crystallographic b direction. The polyhedral view of layer structure is given in Figure 8.3. The coordination polyhedral of terbium oxalate can be described as distorted monocapped square antiprismatic arrangement with each central terbium atom is coordinated to nine oxygen atoms in the structure [43].

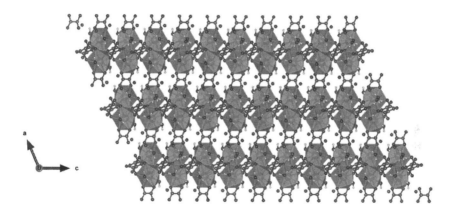

FIGURE 8.3 The layer structure of terbium oxalate.

The peculiar feature of the structure is the well isolation of the Tb centers by the oxalate ligand and coordinated water molecules. This can be more clearly analyzed by examining the crystal structure of terbium oxalate in the ac plane.

The Tb-oxalate layers in the ac plane are the building blocks and are extended along the b direction. The spatial arrangement of terbium atoms is depicted in Figures 8.4 and 8.5. For each Tb atoms, there are eight nearest neighbors (marked as Tb_1-Tb_8) in the same layer and one each from the immediate top and bottom layers. The distance between the Tb atoms is tabulated in Table 8.2.

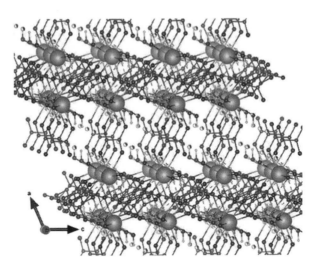

FIGURE 8.4   Spatial arrangement of Tb atoms in terbium oxalate.

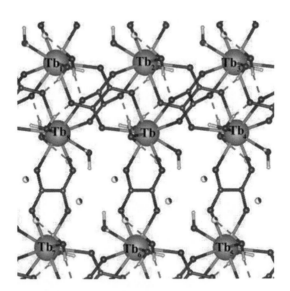

FIGURE 8.5   View of the structure in the ac plane.

The most probable luminescence limiting factor in matrices with high doping of luminescence centers is the concentration quenching, which arises due to the energy migration from one activator to another until an energy sink in the lattice is reached, which is related to the interaction between the activators [44]. According to Blasse, when the activator concentration exceeds the limiting value, energy transfer takes place resulting in concentration quenching [45]. Blasse assumed that for the critical concentration, the average shortest distance between nearest activator ions is equal to the critical distance that can be evaluated from the concentration quenching data using the following equation

$$R_{c} = 2\left[\frac{3V}{4\pi\chi_{c}N}\right]^{\frac{1}{3}}$$

where V is the volume of unit cell, N is the number of available dopant sites, and cc is the critical concentration. Ye et al. described the luminescence properties of terbium-doped yttrium oxalate ($Y2-x(C2O4)3:xTb^{3+}$) and found that concentration quenching occurs in this system when $x = 1.4$ that corresponds to 70% of Tb in the Y sites [46]. In the present case of fully concentrated terbium oxalate the critical concentration cc assume its maximum value 1, and thus the minimum separation between the Tb–Tb centers to induce concentration quenching can be estimated as 7.70 Å using the parameters $V = 956$ Å^3 and $Z=4$ obtained from structure analysis.

TABLE 8.2   Distance Between Terbium Atoms

Tb–Tb Pair	Separation (Å)	Mean Separation (Å)
Tb–Tb1	10.8140	
Tb–Tb2	6.1205	
Tb–Tb3	10.7312	
Tb–Tb4	6.3710	
Tb–Tb5	8.9775	
Tb–Tb6	10.7312	8.8884
Tb–Tb7	9.6105	
Tb–Tb8	6.3742	
Tb–Tb9	9.5774	
Tb–Tb10	9.5774	

The average separation between the Tb centers in fully concentrated terbium oxalate crystals is 8.888 Å and is much longer than the critical distance of 7.70 Å, the minimum separation between the Tb–Tb centers to induce concentration quenching calculated through concentration equation. This clearly confirms that energy transfer cannot happen between Tb ions in oxalate matrix and hence concentration quenching is not effective in oxalate matrix. Similar observations were reported in the case of $Tb_{1-x}Eu_x(DPA)$-(HDPA) and Eu_2WO_6 [47, 48].

In order to confirm this peculiar feature—absence of the concentration quenching—the further investigation was carried out on Tb^{3+}-doped lanthanum oxalate nanocrystals by studying the variation of luminescence intensity with the doping concentration of Tb^{3+}. Lanthanum oxalate is the ideal neutral matrix to study the effect of variation of photoluminescence emission of Tb^{3+} on account of the identical crystal structure as that of terbium oxalate. Moreover Ln^{3+} is an optically inactive rare-earth ion, hence ensures the inherent photoluminescence excitation and emission characteristics of Tb^{3+}. Terbium-doped lanthanum oxalate decahydrate $(La_{2-x}Tb_x(C_2O_4)_3.10H_2O)$ nanocrystals with the percentage of incorporation of Tb^{3+} in the range of 5%–100% ($x = 0.1$–2) was prepared by the microwave-assisted coprecipitation method by taking stoichiometric amounts of lanthanum nitrate and terbium nitrate.

The photoluminescence excitation spectra of the representative samples with 20% and 80% terbium content (x =0.4 and 1.6) monitoring the 543 nm $(^5D_4-^7F_5)$ emission of terbium are shown in Figure 8.6. There is markable enhancement in the intensity of the charge transfer band and f–f transitions as the terbium content is increased in the nanocrystals.

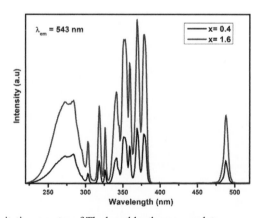

FIGURE 8.6   Excitation spectra of Tb-doped lanthanum oxalate.

The photoluminescence emission spectra of lanthanum oxalate nanocrystals with terbium incorporated in the range of 5%–100% ($x = 0.1$–2) excited with the strongest excitation peak at 369 nm ($^7F_6 \rightarrow ^5L_{10}$) is shown in Figure 8.7 and Figure 8.8 shows the variation of the maximum intensity peak at 543 nm with terbium content in the sample.

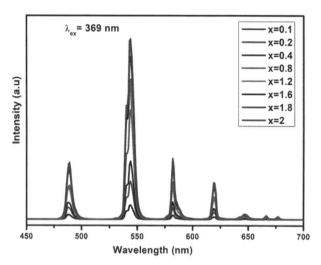

FIGURE 8.7 Emission spectra of $La_{2-x}Tb_x(C_2O_4)_3 \cdot 10H_2O$ nanocrystals ($\lambda_{ex} = 369$ nm).

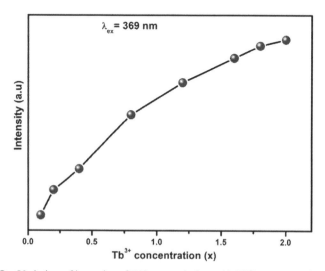

FIGURE 8.8 Variation of intensity of 543 nm emission with Tb^{3+} concentration on exciting at 369 nm.

From Figures 8.7 and 8.8 it is well evident that as the concentration of Tb^{3+} is increasing, the luminescence intensity is increasing, and is maximum for the fully concentrated terbium oxalate nanocrystals. Similar trend in the luminescence emission is observed for the charge transfer band excitation (275 nm) also, as evidenced from Figures 8.9 and 8.10. The regular increase in the emission intensity of Tb^{3+} emission with concentration confirms the absence of concentration quenching in oxalate matrix.

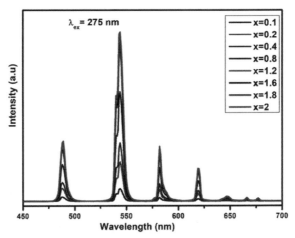

FIGURE 8.9   Emission spectra of $La_{2-x}Tb_x(C_2O_4)_3.10H_2O$ nanocrystals ($\lambda_{ex} = 275$ nm)

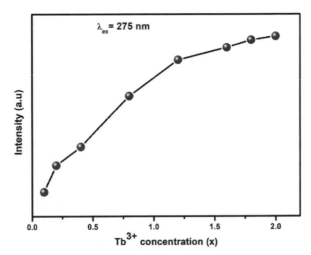

FIGURE 8.10   Variation of intensity of 543 nm emission with Tb^{3+} concentration on exciting at 275 nm.

The above discussions point to the fact that photoluminescence emission of terbium oxalate nanocrystals is favored by the structural specificity of the crystal. The various features such as facile synthesis method, absence of concentration quenching, and lesser response to OH vibrational quenching of terbium oxalate nanocrystals along with the intense and sharp green emission with high color purity makes it a promising phosphor material for specific applications.

8.3.3   COLORIMETRIC ANALYSIS

The quality of a color can be assessed by using the procedures given by the Commission Internationale de I'Eclairage (CIE) scheme and is the most common international approach to describe the human eye perception. The CIE coordinate of the emission from the sample is marked in Figure 8.11 and the digital photograph of emission from the sample is shown in the inset.

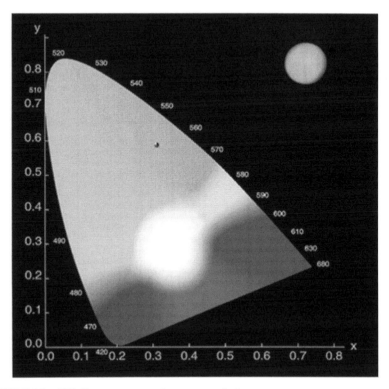

FIGURE 8.11   CIE diagram representing green emission.

8.4 CONCLUSION

The quenching free luminescence exhibited by fully concentrated terbium oxalate is experimented by analyzing the photoluminescence emission of Tb^{3+} in terbium-doped lanthanum oxalate synthesized with different terbium concentrations. The studies revealed that emission intensity increases regularly with Tb concentration and is maximum when lanthanum is fully replaced by terbium ions. This indicates that the commonly observed concentration quenching phenomenon is not significant in fully concentrated terbium oxalate. This is corroborated with the help of the crystal structure of terbium oxalate. Well separated Tb centers reduce energy migration and are further proved by calculating the critical distance using Blasse's theory. This finding is of much importance since rare-earth oxalates are isostructural in nature with the layer structure as the unique characteristic. The absence of concentration quenching is of particular importance in designing favorable luminescent materials.

KEYWORDS

- **quenching**
- **luminescence**
- **crystal structure**

REFERENCES

1. M. Rai, G. Kaur, S. K. Singh, S. B. Rai, Probing a new approach for warm white light generation in lanthanide doped nanophosphors, Dalton Trans. 44 (2015) 6184.
2. F. Meiser, C. Cortez, F. Caruso, Biofunctionalization of fluorescent rare-earth–doped lanthanum phosphate colloidal nanoparticles, Angew. Chem. Int. Ed. 43 (2004) 5954.
3. Y. Guo, B.K. Moon, B. C. Choi, J.H. Jeong, J.H. Kimb, Multi-wavelength excited white-emitting $K2Gd(1-x)(PO4)(WO4):xDy^{3+}$ phosphors with satisfactory thermal properties for UV-LEDs, RSC Adv. 7 (2017) 23083.
4. R. Kubrin, Nano phosphor coatings: Technology and applications, Opportunities and Challenges, Kona Powder Part J. 31 (2014) 22.
5. H. A. Hopee, Recent developments in the field of inorganic phosphors, Angew. Chem. Int. Ed. 48 (2009), 3572.
6. E.T. Goldburt, B. Kulkarni, R.N. Bhargava, J. Taylor, M. Libera, Size dependent efficiency in Tb doped Y2O3 nanocrystalline phosphor, J. Lumin. 74 (1997) 190.

7. D. Alexander, K. Thomas, S. Sisira, P.R. Biju, N. V. Unnikrishnan, M. A .Ittyachen, Synthesis and optical characterization of sub-5 nm Terbium oxalate nanocrystals: A novel intense green emitting phosphor, Dyes Pigm, 148 (2018) 386–393.

8. G. Blasse, Chemistry and Physics of R-activated Phosphors, Handbook on the Physics and Chemistry of Rare Earths, Chapter 34, North-Holland Publishing Company, 1979

9. B. G. Wybourne, The fascination of rare earths—then, now and in the future. J. Alloys Compd. 380 (2004) 96

10. G.Q. Wang, X.H. Gong, Y.J. Chen, J.H. Huang, Y.F. Lin, Z.D. Luo, Y.D. Huang, Novel red phosphors $KBaEu(XO_4)_3$ (X= Mo, W) show high color purity and high thermostability from a disordered chained structure, Dalton Trans. 46 (2017) 6776.

11. C. Qin, Y.G. Huang, G. Chen, L. Shi, X. Qiao, J. Gan, H.J. Seo, Luminescence properties of a red phosphor europium tungsten oxide Eu_2WO_6, Mater. Lett. 63 (2009) 1162.

12. Z. Y. Mao, Y. Zhu, Y. Zeng, L. Gan, Y. Wang, Concentration quenching and resultant photoluminescence adjustment for $Ca_3Si_2O_7:Tb^{3+}$, J. Lumin, 143 (2013), 587–591.

13. M. Shi, D. Zhang, C. Chang, Tunable emission and concentration quenching of Tb^{3+} in magnesium lithium phosphate, J. Alloys Compd., 627 (2015) 25–30.

14. X. Huang, C. He, X. Wen, Z. Huang, Y. Liu, M. Fang, X. Wu, X. Min, Preparation, Structure, luminescence properties of terbium doped pervoskite–like structure green emitting phosphors $SrLa AlO_4:Tb^{3+}$, Opt. Mater., 95 (2019) 109191.

15. F.C. Lu, L. J. Bai, Y. Lu, W. Dang, Z. P. Yang, P. Lin, Photoluminescence mechanism and thermal stability of Tb^{3+} doped $Y_4Si_2O_7N_2$ green emitting phosphor, J. Am. Ceram. Soc, 1–6 (2014).

16. E. Sreeja, S. Gopi, V. Vidyadharan, P. R. Mohan, C .Joseph, N. V. Unnikrishnan, P. R. Biju, Luminescence properties and charge transfer mechanism of host sensitized $Ba_2CaWO_6:Eu^{3+}$ phosphor, Powder Technol., 323 (2018) 445–453.

17. S. K. Gupta, M. Sahu, K. Krishnan, M. K. Saxena, V. Natarajan, S. V. God bole, Bluish white emitting Sr_2CeO_4 and red emitting $Sr2CeO4:Eu^{3+}$ nanoparticles: optimization of synthesis parameters, characterization, energy transfer and photoluminescence , J. Mater. Chem. C, (2013) 1,7054

18. Z. W. Chiu, Y.J. Hsaio, T.H. Fang, L. W. Ji, Photoluminescence and optoelectronic characteristic of Eu doped $InBO_3$ Nanocrystals, Int. J. Electrochem. Sci., 10 (2015), 2391–2399.

19. R. G. Abhilash Kumar, S. Hata, K. Ikeda, K. G. Gopchandran, Organic mediated synthesis of highly luminescent Li^+ ion compensated $Gd_2O_3:Eu^{3+}$ nanophosphors and its Judd-Ofelt analysis, RSC Adv., 6 (2016) 67295–67307.

20. Y. S. Lian, Y. Wang, J. Li, Z. Zhu, Z. Y. You, C. Yang, Y. D. Xu, Structural and fluorescence features of $Dy^{3+}:Y_4Al_2O_9$ phosphors for yellow colour emitting displays, Vaccum, 173 (2020), 109165.

21. W. Zhijun, L. I. Panlai, Y. Zhiping, G. Qinglin, L. Xu, T. Feng, A novel white light emitting phosphor $Ca_2PO_4Cl:Dy^{3+}$, Luminescence , concentration quenching and thermal stability, J Rare Earth 11 (2015) 1137.

22. J. Zhang, Q. Guo, L. Liao, Y. Wang, M. He, H. Ye, L. Mei, H. Liu, T.Zhou, B.Ma, Structure and luminescence properties of $La_6Ba_4(SiO_4)6F2:Dy^{3+}$ phosphor with apatite structure, RSC Adv., 3 (2018) 38883.

23. A.N. Yerpude, S.J. Dhoble, Luminescence properties of micro $Ca_3Al_2O_6:Dy^{3+}$, Micro Nano Lett. 7 (2012) 268–270.

24. G. Morgan, D. Huber, W. Yen, quenching of fluorescence by cross relaxation in LaF$_3$:Pr^{3+}, J. Phys. Colloq. 46 (1985) 29.

25. J.G. Bunzlia, On the design of highly luminescent lanthanide complexes, Coordin. Chem. Rev. 19 (2015) 293.

26. M.J. Weber, Multiphonon relaxation of rare-earth ions in yttrium ortho aluminate, Phys. Rev. B, 8 (1973) 54.

27. N. Xie, Y. Huang, X. Qiao, L. Shi, H.J.Seo, A red emitting phosphor of fully concentrated Eu^{3+} based molybdenum borate Eu$_2$MoB$_2$O$_9$, Mater. Lett. 64 (2010) 1000–1002.

28. H. Lin, Y. Huang, H.J. Seo, The red luminescence characteristics of fully concentrated Eu-based phosphor EuBaB$_9$O$_{16}$, Phys. Status Solid A 207 (2010), 1210–1215.

29. X.Liu, X. Kang, Y. Zhang, L. Tu, Y. Wang, Q. Zeng, C. Li, Z. S.hi, H. Zhang, Breakthrough in concentration quenching threshold of up conversion luminesence via spatial separation of the emitter doping area for bio applications, Chem. Commun. 47 (2011) 11957–11959.

30. J. Li,Q. Liang, Y. Cao, J. Yan, J. Yan, J. Zhou,Y. Xu, L. Dolgov, Y. Meng, J.Shi, M. Wu, Layered structure produced Nnon concentration quenching in a novel Eu^{3+} doped phosphor, Appl. Mater. Interfaces 10 (2018) 41479–41486.

31. M. Jiao, N. Guo, W. Lu, Y. Jia, W. Lv, Q. Zhao, B. Shao, H. You, Synthesis, structure and photoluminescence of europium, terbium and thulium doped Ca$_3$Bi(PO$_4$)$_3$ phosphors, Dalton Trans. 42 (2013) 12395.

32. D. Qin, W. Tang, Energy transfer and multicolour emission in the single phased Na$_5$Ln(WO$_4$)$_{4-z}$(MoO$_4$)z:Tb^{3+}, Eu^{3+} (Ln=La, Y, Gd) phosphors, RSC Adv. 6 (2016) 45376.

33. D. Qin, W. Tang, Crystal structure, tunable luminescence and energy transfer properties of Na$_3$La(PO$_4$)2:Tb^{3+}, Eu^{3+} phosphors, RSC Adv. 7 (2017) 2494.

34. R. Decadt, K. V. Hecke, D. Depla, K. Leus, D. Weinberger, I. V. Driessche, P. V. D. Voort, R.V. Deun, Synthesis, crystal structures, and luminescence properties of carboxylate based rare-earth coordination polymers, Inorg. Chem. 51 (2012) 11623.

35. X. Jing, X. Zhou, T. Zhao, Q. Huo, Y. Liu, Construction of Lanthanide–Organic Frameworks from 2-(Pyridine-3-yl)-1H-4,5 imidazoledicarboxylate and Oxalate, Cryst. Growth Des. 12 (2012) 4225.

36. A.J. Calahorro, D.F. Jimenez, A.S. Castillo, M. Viseras, A.R. Dieguez, Novel 3D lanthanum oxalate metal-organic-framework: synthetic, structural, luminescence and adsorption properties, Polyhedron, 52 (2013) 315.

37. M.C. Mary, G. Vimal, K. P. Mani, G. Jose, P. R. Biju, C. Joseph, Growth and characterization of Sm^{3+} doped cerium oxalate single crystals. J. Mater. Res. Tech. 5 (2016) 268.

38. C. Joseph, G. Varghese, M.A. Ittyachen, Growth and characterization of mixed neodymium praseodymium oxalate decahydrate crystals in silica gel, Cryst. Res Technol. 30 (1995)159.

39. V. Thomas, A. Elizebeth, G. Jose, H. Thomas, C. Joseph, N.V. Unnikrishnan, M.A. Ittyachen, Studies on the growth and optical characterization of dysprosium praseodymium oxalate single crystals, J. Optoelectron. Adv. Mater. 7 (2005) 2687.

40. G. Vimal, K.P. Mani, G. Jose, P.R. Biju, C. Joseph, N.V. Unnikrishnan, M. A. Ittyachen, Growth and spectroscopic properties of samarium oxalate single crystals, J. Cryst. Growth 404 (2014) 20.

41. A. Potdevin, G. Chadeyron, R. Mahiou, Tb^{3+}-doped yttrium garnets: promising tunable green phosphors for solid-state lighting, Chem. Phys. Lett. 490 (2015) 50.
42. D. Alexander, K. Thomas, S. Sisira, G. Vimal, K. P. Mani, P.R. Biju, N. V. Unnikrishnan, M. A. Ittyachen, C. Joseph, Photoluminescence properties of fully concentrated terbium oxalate: a novel efficient green emitting phosphor, Mater. Lett., 189 (2017) 160–163.
43. D. Alexander, K. Thomas, S. Sisira, P.R. Biju, N. V. Unnikrishnan, M. A .Ittyachen, C. Joseph, Efficient green luminescence of terbium oxalate crystals: A case study with Judd-Ofelt theory and single crystal structure analysis and the effect of dehydration on luminescence, J. Solid State Chem, 262, 68–78.
44. S. Omagari, T. Nakanishi, Y. Hirai, Y. Kitagawa, T. Seki, K. Fushimi, H.Ito, Y. Hasegawa, Origin of Concentration quenching in Ytterbium Coordination polymers: phonon-assisted energy transfer, Eur. J. Inorg. Chem. (2018) 561.
45. G. Blasse, Energy transfer between inequivalent Eu^{2+} ions, J. Solid State. Chem. 62 (1986) 207.
46. M. Ye, L. Zhou, F. Hong, L. Li, Q. Xia, K. Yang, X. Xiong, Synthesis and photo luminescent properties of cuboid–like Y2 (C_2O_4)3:Tb^{3+} green-emitting phosphors, Opt. Mater. 47 (2015) 161.
47. M.O. Rodrigues, J. Dutra, L. Nunes, G. Sá, W. Azevedo, P. Silva, F. Paz, R. Freire, S.A. Júnior, $Tb^{3+} \rightarrow Eu^{3+}$ Energy transfer in mixed-lanthanide-organic frameworks, J. Phys. Chem. C 116 (2012) 19951.
48. C. Qin, Y. Huang, G. Chen, L. Shi, X. Qjao, J. Gan, H.J. Seo, Luminescence properties of a red phosphor europium tungsten oxide Eu_2WO_6, Mater. Lett. 63 (2009) 1162.

PART III

Optical Properties of Advanced Materials

The Optical Properties and Versatile Biomedical Applications of Semiconductor Nanoparticles: Imaging, Sensing, and Photothermal Therapy

S. PRASANTH

Centre for Excellence in Advanced Materials, Cochin University of Science and Technology, Cochin, India

Corresponding author. E-mail: prasanthshanmu@gmail.com

ABSTRACT

Nanomaterials are increasingly important and are at the forefront of modern research. The unique physical and chemical properties of nanomaterials influence their different applications. The unique properties of the nanomaterials are due to the quantum confinement effects. Because of this quantum confinement effects, these nanoparticles possess novel electrical and optical properties. Optical properties of nanomaterials are the most important and encouraging properties for a variety of innovative applications. Of late, optical properties of nanomaterials are increasingly explored in biotechnology and medicine being widely used in therapeutics and diagnostics. Among the various inorganic nanoparticles semiconductor nanoparticles have profound interest because of their intriguing optical properties and can be more readily integrated into different applications. The most important and promising application of semiconductor nanomaterials is the design of versatile nanostructures such as super lattices and multi modal agents which offer new opportunities in the biomedical field. In this chapter, we discuss recent advances in the semiconductor nanoparticles and its optical properties. We also discuss the recent biological applications of semiconductor nanoparticles in biological imaging, fluorescence sensing and therapeutics such as drug delivery and photothermal therapy.

9.1   INTRODUCTION

Nanomaterials are increasingly important and are at the forefront of modern research. The unique physical and chemical properties of nano-materials like surface structure, chemical composition, increased control light spectrum, high chemical reactivity, and higher strength influence their different applications [1–6]. Optical properties of nanomaterials are the most important and encouraging properties for a variety of innovative applications. Of late, optical properties of nanomaterials are increasingly explored in biotechnology and medicine being widely used in therapeutics and diagnostics [8]. Nanoparticles (NPs) can be synthesized from inor-ganic or organic materials. The inorganic nanomaterials are semiconductor materials and nanomaterials of metals. Carbon-based nanomaterials like fullerene and carbon nanotubes, supramolecular assemblies, proteins, and nucleic acids are examples of organic- and bionanomaterials. Over the last few years different inorganic nanomaterials with superior properties have been created for multifunctional applications [7]. Among the various inorganic NPs semiconductor NPs have a profound interest because of their intriguing properties and can be more readily integrated into different applications. The most promising and exciting applications of these nano-materials are in the biomedical field due to its peculiar properties include narrow emission lines, high photostability, long fluorescence lifetime, high surface free energy, and others. The confined size and high density of corner and edge surface state oxide NPs exhibit substantial physical and chemical properties. The diverse applications of these materials highly depend on their size dependent optical, mechanical, transport, and the surface properties [9–14].

Semiconductor nanocrystals (NCs) are classified as II–VI, III–V, or IV–VI semiconductor NCs, according to the periodic table groups into which these elements are formed [15]. These semiconductor NPs have elicited substantial interest because they have promising applications in solar cells, electronic devices, light-emitting diodes (LEDs), laser technology, wave-guide, biosensor, molecular and cellular imaging, therapeutics, catalysis, and others [16–21].

In this chapter, we would like to discuss some optical properties of semi-conductor NPs and also their biological applications like optical sensing, imaging and some therapeutic applications.

9.2 SIGNIFICANCE OF SEMICONDUCTOR NANOPARTICLES

Semiconductors nanomaterials have generated popular research interest in many areas of physics, chemistry, life science, and material science. These nanostructures show dimension-dependent properties and are building blocks in devices and processes. The optical properties of semiconductor NPs highly depend on the size, for example, the absorption and emission spectra are tune-able with particle size [22]. By altering the shape and height of the potential the confinement of electrons and holes can be tuned in semiconductor NPs [23]. In semiconductor nanoparticles, indirect-gap group IV semiconductor NPs and II–VI compound semiconductor NPs have got considerable atten-tion because of their optical and electronic properties. The optical, electrical, and structural properties of the semiconductors NPs can be modified by doping and doped semiconductors have been widely investigated because of their different applications like cathode ray tubes, fluorescent lamps, X-ray detectors, LED, and laser materials [24]. The doping of impurity ions creates excess electrons or holes or sometimes a lattice distortion can take place and gives rise to localized levels in the bandgap of a semiconductor and these states are responsible for many properties like photoluminescence, electroluminescence (ECL), and transport properties [25].

The morphological features and their size-dependent properties play a crucial role for developing the next-generation optoelectronic devices. The optical properties of semiconductor NPs such as reflection, transmission, broad absorption, and narrow light emission, high photostability and high quantum yield are used increasingly for different applications like biological imaging, sensing, and labeling [26]. The band gap of the semiconductor nanoparticles can be turned to precise energy depending on their dimen-sionality.. Figure 9.1 shows the variations in the bandgap of CdSe quantum dots, wires, and wells [24]. Liang-shi Li et al. studied the variations in the band energy by changing the width and length for the CdSe rods and the photoluminescence spectra show redshift on increase in the length of the rod (Figure 9.2) [27]. This is due to quantum confinement and "band-mixing" in CdSe NCs [28]. In CdSe, the elongation causes the symmetry breaking and thus modifies the band mixing.

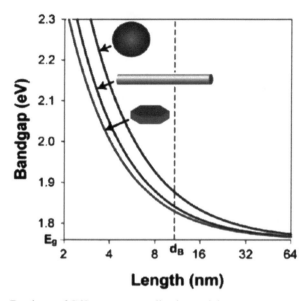

FIGURE 9.1   Band gap of CdSe quantum well, wire, and dots.

Source: Reprinted with permission from Ref. [24] © 2010 American Chemical Society.

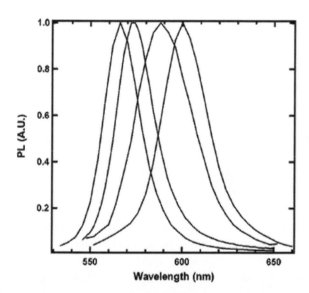

FIGURE 9.2   Photoluminescence spectra CdSe quantum rods with different lengths, 9.2, 11.5, 28.0, and 37.2 nm, respectively (from left to right).

Source: Reprinted with permission from Ref. [27]. © 2001 American Chemical Society.

By investigating the optical properties of nanomaterials one can get a deep understanding of their structural information because the optical properties depend on the internal electronic structure of nanomaterials. The atoms on the surface of the crystal and the dangling bond highly influence their optical properties [29]. The electronic states with the bandgap are due to the dangling bonds at a semiconductor surface or interface. These mid-gap states fill up to the Fermi level with electrons that originate in the bulk of the material. An electric field is created due to the accumulation of charge at the surface and a depletion region that leads to bending of the valence and conduction band edges. The generated electron–hole pairs prohibit the radiative recombination [30].

9.3 METHOD OF SYNTHESIS OF SEMICONDUCTOR NANOPARTICLES

There are different methods for synthesizing and processing various semiconductor NPs like binary, core shell, and alloyed, which have different biological applications. One of the most important synthesis methods is, semiconductor NPs from has been using organometallic compounds in coordinating solvents at high temperature in air-free environment [31]. Homogeneous nucleation is produced when the organometallic reagents were injected into the solvent and a nearly monodisperse NPs were obtained. The major disadvantage of this technique is temperature control and homogeneous mixing for large-scale production of quantum dots is particularly difficult because temperature control and homogeneous mixing are difficult in large volume [32, 33].

Aerosol-assisted methods have been widely used for the synthesis of different inorganic and organic materials. Didenko et al. synthesized CdSe NPs by continuous chemical aerosol flow synthesis [34]. They obtained size-controlled and highly fluorescent CdSe NPs by changing the combinations of Cd and Se precursors, solvents, and surface stabilizers (Figure 9.3). The microwave-assisted method is used to synthesize monodisperse nanoparticles. This technique has significant advantages over conventional heating techniques like faster reaction time, the overheating of the sample can be reduced and the heating efficiency is higher. Oleic acid-functionalized CdSe quantum dots were carried out with diesel as a solvent using microwave synthesis method [35]. Black phosphorous (BP) quantum dots were synthesized by microwave method for electrocatalytic oxygen evolution reactions [36].

Synthesis of a water-soluble nanoparticle is highly relevant for biological applications. Water-soluble L-cysteine capped CdTe quantum dots were synthesized by the one-pot method [37, 38]. Water-soluble ZnO NPs were synthesized by a one-pot polyol hydrolysis method for cell labeling applications [39]. Biosynthesis of semiconductor NPs like ZnO and TiO_2 were achieved. TiO_2 NPs were synthesized by Biotemplated method, ordered mesoporous TiO_2 NPs was achieved on the cell membrane of yeast cells [40].

The solvothermal method is one of the widely accepted methods for the synthesis of luminescent nanoparticles. This method is based on the reaction between solids in liquid

precursors at high pressure and mild temperature and is usually carried out in a Teflon-lined autoclave [41].

CuS NPs are increasingly considered for many biological applications and can be synthesized by different methods such as hydrothermal/solvothermal method, microwave irradiation, chemical precipitation, and sonochemical synthesis [44]. In order to obtain high yield and uniform NPs, the hydrothermal method is preferred. Chemical precipitation method is one of the most widely used synthesis technique for semiconductor nanoparticles. The main advantages of the chemical precipitation method are its easy processability at ambient conditions and the possibility of doping different kinds of impurities with high doping concentrations even at room temperature [42]. CuS NPs are synthesized by a simple chemical precipitation method for different biomedical applications (Figure 9.4) [43, 44].

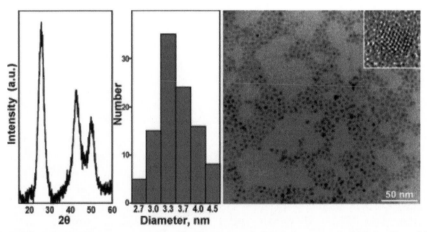

FIGURE 9.3   XRD spectrum, size distribution and TEM images of CdSe/steric acid nanoparticles by chemical aerosol flow method (copyright @ACS).

Source: Reprinted with permission from Ref. [34]. © 2005 American Chemical Society.

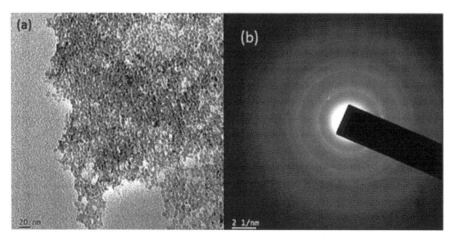

FIGURE 9.4   TEM image and SAED pattern of L-cysteine capped CuS nanoparticles synthesized by chemical precipitation method (copyright @RSC).

Source: Reprinted with permission from Ref. [42]. © 2016 The Royal Society of Chemistry.

9.4   OPTICAL PROPERTIES OF SEMICONDUCTOR NANOPARTICLES

The optical properties of nanomaterials are the major aspect of both fundamental and technological reason. Semiconductor NPs or quantum dots exhibit unique optical properties because of the quantum confinement [45, 46]. The random on-off blinking phenomena exhibited by the single-particle is one of the most striking properties, which highly influences the tracking of particles in biological environments. The blinking is a result of the repeated ionization of the nanocrystal due to an Auger process or from the trapping of one of the charge carriers at or near the particle surface. The off-time can be reduced in low dielectric media that poorly solvate an expelled charge and by neutralize the surface traps for charge carriers by adsorbing thiolate ligands and amine. In addition, create a deeper and wider potential well around the NCs by growing an insulating shell around the nanocrystal [47, 48].

Semiconductor NPs have become attractive in different applications due to their distinguished properties like high quantum yield and long fluorescence lifetime, simultaneous excitation of multiple NPs by a single light source, narrow, symmetrical. and broad-spectrum spanning from the ultraviolet to infrared region, high photostability, tunable emission, and resistance to photobleaching [49]. The fluorescent semiconductor NPs are widely used in many biological applications like biolabeling of DNA, proteins, and cells [50, 51]. Cadmium selenide (CdSe), zinc sulfide (ZnS), cadmium telluride (CdTe), silver sulfide (Ag$_2$S) zinc oxide (ZnO), zinc selenide (ZnSe), lead selenide (PbSe), and so forth are some of the important semiconductor NPs among others.

When NPs are excited with a particular wavelength, an electron–hole pair is produced and their recombination causes an emission of light. An exciton electrostatically bounded electron–hole pair and is generated in a semiconductor when an electron in the valence band absorbs a photon with energy greater than or equal to the bandgap to conduction band [52, 53]. When the size of the NPs becomes approximately equal to the Bohr exciton radius, the band structure changes to discrete energy level and is due to the quantum confinement effect. As the particle size decreases the energy difference between the highest occupied level and lowest unoccupied level widens. The discrete energy level results in the discrete absorption of quantum dots, that is, when the particle size is smaller more is the blueshift in their luminescence (Figure 9.5) [54].

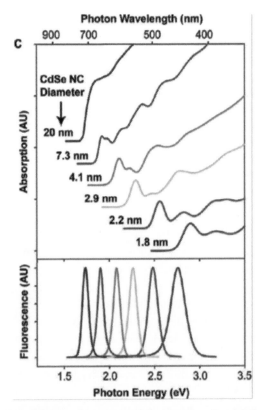

FIGURE 9.5   Tuning of photoluminescence and absorbance on the variations in the size of NPs.
Source: Reprinted with permission from Ref. [24] © 2010 American Chemical Society.

9.5 BIOLOGICAL APPLICATIONS OF SEMICONDUCTOR NANOPARTICLES

The novel properties of semiconductor NPs make them a suitable candidate for diverse applications, ranging from energy conversion and storage to biomedical imaging. The NPs (<10 nm) easily taken up and excreted, and show longer blood circulation times in comparison with the bulk materials. The different biological applications of semiconductor NPs are described below.

9.5.1 SEMICONDUCTOR NANOPARTICLES FOR BIOIMAGING

The bioimaging process uses specific biochemical strategies to label the biomolecule with a contrast agent in order to be detected with the read-out system. The label and biorecognition element are the major part of bioimaging probe. The major requirements for a suitable label are having a good analytical signal, soluble in suitable buffer or bio matrixes, have site-specific functional group, have reported data about its photophysics, and reproducible quality [55].

The quantum dots can replace the organic fluorophore for biological applications because of its exceptional photostability, bright emission, size-tunable colon, broad absorption and narrow emission bands, and large two-photon absorption cross-section. The most of the NPs designed for in vivo imaging and therapy absorb in the near-infrared (NIR) wavelength because the light absorption from biologic tissue component is minimized at NIR region [44]. The two kinds of photoluminescence mechanisms are down conversion and upconversion. In downconversion, NPs absorb high energy photon and emit a low energy photon. In contrast, upconversion via multiple absorptions or energy transfer processes converts the absorbed low energy light into higher energy emission. The fluorescence from both processes has been used in molecular imaging. NPs are also used as contrast agents, for magnetic resonance imaging (MRI), optical coherence tomography, photoacoustic imaging, and two-photon luminescence spectroscopy [56–59].

9.5.1.1 MAGNETIC RESONANCE IMAGING (MRI)

Nuclear magnetic resonance is a powerful noninvasive approach for the accurate detection and diagnosis of variety of conditions from torn ligaments to tumors. It is based on the detection of nuclear spin reorientations in a magnetic field. Here the incident radiation consists of low energy radio frequencies at the right angle to a static magnetic field, therefore MRI causes no adverse effect on the biological tissues and also it provides exceptional penetration depth. MRI is usually used for detection of water protons which are present, in higher molar concentration, in biological specimens. The contrast enhancement agents are used to enhance and improve the contrast between the normal and diseased tissue. In clinical applications paramagnetic Gd complexes are used as MRI contrast agents. However, the free Gd complexes have very short blood circulation time and have toxic side effects like nephrogenic systemic fibrosis and cerebral deposition. Therefore magnetic NPs can be considered as an alternative for the conventional Gd-based contrast agents [60]. The NPs provide better biocompatibility and have greater sensitivity in the micromolar or nanomolar range. Among the various magnetic NPs Mn-based and Fe-based NPs are considered ideal substitutes for Gd-based contrast agents [56].

MRI is based on the nuclear magnetic resonance of protons abundant in living organisms, notably in the form of water. The proton nuclear magnetic moments are aligned in the presence of a strong magnetic field. By the application of a radio frequency pulse they are selectively deflected in the transversal plane. Now we can define a term known as relaxation time, it is the time taken for magnetic moments to come back to the original longitudinal direction of the static magnetic field. There are two different relaxation times namely T1 and T2. T1 is the longitudinal recovery and T2 is the transversal decay, gadolinium and manganese are usually associated with T1 imaging while iron is T2 weighted images [61].

Gadolinium can be replaced by manganese for the development of MRI T1 contrast agents. However, the manganese complex suffers less stability in order to improve the stability of the Mn complex manganese oxide NPs can be used. The toxic effect of induced by Mn^{2+} can be overcome attaching bisphosphonate dendrons (PDns) to the surface of MnO NPs. These NPs will rapidly eliminate through feces and urine and thus PDns functionalized NPs are a suitable candidate for positive MRI contrast agent [62]. The biocompatibility of MnO NPs can be improved by coating polyethylene glycol (PEG) on the surface of the NPs and can also the targeting capacity can be improved by conjugated with specific polypeptides and other aptamers. After connected with specific targeting molecule, AS1411 aptamer, AS1411-PEG-MnO nanoprobe realized promising potential as T1 MRI contrast agent (Figure 9.6) [63].

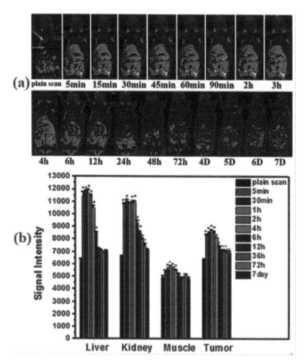

FIGURE 9.6 Pseudocolor MR T1 images of mice having renal carcinoma tumors pre- and postinjection of AS1411-PEG-MnO NPs. The red, yellow, and green arrows indicate tumor, heart, and liver, respectively. (b) The signal intensities of different organs (liver, kidney, muscle and tumor) before (plain scan) and after (5 min, 30 min, 60 min, 2 h, 4 h, 6 h, 12 h, 24 h, 36 h, 72 h, 7D) intravenous injection of AS1411-PEG-MnO nanoprobe.

Source: Reprinted with permission from Ref. [63]. © 2018 Elsevier.

Magnetic iron oxide nanoparticles (γ-Fe$_2$O$_3$) are extensively studied for use in nanomedicine. For MRI applications super-paramagnetic iron oxide NPs (SPIONs) are widely used. It is reported that using SPIONs the presence of amyloid deposits in the mouse and rat brain has been successfully identified ex vivo [64].

9.5.1.2 *SEMICONDUCTOR NANOPARTICLES FOR FLUORESCENCE IMAGING*

Fluorescence bioimaging is one of the promising imaging techniques on comparing with other methods like computerized tomography or MRI. The fluorescence imaging technique has different advantages such as the excitation source is of nonionizing radiation, real-time imaging and data processing is simple, and cost-effective [65]. Quantum dots are widely used for cell imaging applications, there are nonspecific or cell-specific cell biological applications of quantum dots and is dependent on the property of the molecules attached to the surface of the quantum dots. The small organic molecules capped quantum dots can directly interact with the cells and can enter into cytosol [66]. The peptide conjugated quantum dots facilitate the targeted intracellular labeling and imaging of nucleus or mitochondria.

Lipids, liposomes, and surfactants transport quantum dots into the cytosol efficiently. The molecules like cationic or amine-rich copolymers, and carbohydrates such as chitosan, glucose, and cationic polysaccharides can be used for nonspecific intracellular transport of quantum dots and this is useful for the detection and imaging of any cell type. For the practical applications like detection and imaging of cancer cells the biomarker specific targeted cell labeling is often used [67]. For the imaging of biomolecular functioning in cells the two valuable processes are fluorescence resonance energy transfer (FRET) and bioluminescence resonance energy transfer (BRET).

Miyeon Lee et al. synthesized stable blue-emitting BP quantum dots by a simple sonication-assisted liquid-phase exfoliation method. They show very low cytotoxicity and also excellent fluorescence cellular imaging capability (KB tumor cells) (Figure 9.7) [68]. H. U. Lee and coworkers synthesized biocompatible BP quantum dots for fluorescent imaging. By the uptake of BP quantum dots into HeLa cells, blue and green fluorescence in the cells were observed under UV and visible light excitation.

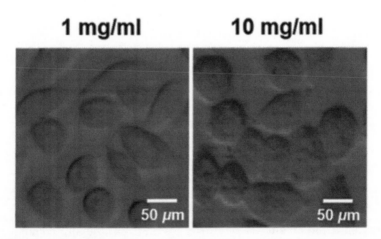

FIGURE 9.7   Confocal fluorescence image of KB tumor cells treated with PEG-BPQDs.
Source: Reprinted with permission from Ref. [68]. © 2017 American Chemical Society.

The NIR emitting Ag_2S NPs bioimaging applications due to their reduced autofluorescence and negligible tissue scattering in this region [69]. Surface functionalized PbS NPs are also used for in vivo imaging [70].

9.5.2 SEMICONDUCTOR NANOPARTICLES FOR BIOSENSING

A biosensor consists of three parts (1) a component that recognizes the analyte and produces signals, (2) signal transducer, and (3) a device that record signal. The sensor element used in the sensor should be highly specific and sensitive to the analyte that is to be detected. Semiconductor NPs become a key component for biological sensing and it offers an alternative to molecular fluorescent labels because of their unique optical properties such as broad absorption spectra, narrow and size-tunable emission, and photostability [71]. These spectral characteristics of quantum dots allow simultaneous excitation using a single wavelength source, resulting in emission at multiple wavelengths [72]. The surface functionalization of quantum dots (QDs) has been performed for the integration of QDs into biorecognition strategies. For the sensitive and fast sensing QDs incorporated microbeads (fluorescence-encoded microbeads) can be used. Depending upon the functionalization strategy the analyte can quench or enhance the fluorescence of the QDs.

9.5.2.1 FUNCTIONALIZATION OF NANOPARTICLES

For the commercialization of the nanoproducts and their effective use in different applications in the medical, engineering, and electronics fields, it is necessary to functionalize NPs with suitable functionalizing agents. The surface modification of the NPs can improve/alter the several properties, such as the hydrophilicity, hydrophobicity, conductivity, and photoluminescence. The biocompatibility of the NPs can be improved by surface functionaliza-tion. Different functionalizing agents can be used for the functionalization of NPs such as small molecules, surfactants, polymers, and biomolecules. By the conjugation of biomolecules with NPs will improve the biocompatibility and specific recognition. The small molecules and polymers incorporate drugs, contrast agents, and biomolecules by covalent or noncovalent conju-gation [73–76].

By modifying the surface of NPs with organic molecules we can improve the water-solubility, photo-stability, and quantum yield of the NPs and this is highly desirable for many applications like and sensing, imaging, and drug delivery. Most of the NPs surface are hydrophobic in nature and are not suitable for biological applications. Therefore it is essential to hydrophilize the surface of the NPs for their effective use in such applications. Different types of organic coatings provide water-soluble layer on the surface of NPs. The various functional groups like –COOH, –NH$_2$, and –OH on the surface coating is important for sensor applications because these functional groups allow the interactions with the analyte molecules, resulting the variations in the optical properties of NPs. The organic capping controls the sensitivity and selectivity of the sensors. Moreover, it stabilizes the NPs in aqueous media. The surface functionalization also improves the luminescence efficiency by decreasing the surface defects [77].

In sensing applications, the interactions of analyte molecules with the NPs lead to the changes in the luminescence properties. The affinity toward the different analyte can be determined by using different capping agent. Oleic acid-functionalized QDs were used for the detection of nitroaromatic explosive molecules like 2,4,6-trinitrotoluene (TNT) or nitrobenzene (NB). TNT or NB can quench the fluorescence of QDs. S. Huang et al. reported N-acetyl-L-cysteine (NAC)-capped CdTe/CdS/ZnS core/shell/shell QDs for the selective detection of L-ascorbic acid (AA) [78]. The AA quenches the fluorescence of QDs via static quenching mechanism. Helena Goncalves et al. reported the detection of Hg(II) ions in aqueous media using PEG 2000 and NAC-coated CDs [79].

For the efficient delivery of biomacromolecules with minimal cytotoxicity into the specific sites, biomolecule coated NPs were used. Proteins and antibodies functionalized NPs can bind to cell surface receptors, providing targeted delivery. Herceptin-coated multivalent gold and silver NPs were used for targeting particles for imaging and photothermal therapy (PTT) [80].

9.5.2.2   THE MECHANISM OF SENSING

QD is used as optical or electrical transduction based on the variations in the photoluminescence or photocurrent intensity of the QDs in the presence of the analyte. In fluorescent bioassays the variations in the photoluminescence can be achieved by different mechanisms like FRET, charge transfer quenching, and ECL.

FRET is a nonradiative energy transfer process between the donor and an acceptor molecule, the donor molecule is a fluorophore and the acceptor is either fluorophore or quenchers. QDs are used as donors in some FRET-based biosensors, since the QDs exhibit size-dependent emission, therefore we can improve spectral overlap with the absorption of acceptor molecule [77].

Zhang and coworkers designed a QD-FRET nanosensor for DNA detection. They used streptavidin-coated CdSe–ZnS QD as a donor and organic fluorescence dye Cy5 as an acceptor [81]. This QD-FRET biosensor showed excellent performance with a detection limit of 4.8 fM. Kattke and group have developed an amine-derivatized (PEG)-coated CdSe/ZnS QDs-FRET-based immunoassays for spores in the solution of *Aspergillus amstelodami*. The QDs were linked to the antibody (ant-Aspergillus) through SMCC cross-linker and is interacted with BHQ-3-labeled mold analytes with lower affinity than *A. amstelodami* spores [82].

Alireza Kalarestaghi et al. report a highly sensitive competitive immunoassay for the detection of aflatoxin B1 using FRET. They used a magnetic/silica core shell as a signal intensifier, and the detection is based on FRET from antibody immobilized Cd/Te quantum dots to Rhodamine 123. The Rhodamine 123 fluorophore and the QDs come close spatial proximity due to the specific immune reaction between the antibody and aflatoxin B1 and upon excitation of the QDs FRET occurs. Here Rhodamine 123 acting as the acceptor and the QDs as the donor. The obtained signal is intensified by the magnetic/silica core shell to NPs[83].

9.5.2.3 DIRECT FLUORESCENCE

Fluorescence is one of the emerging optical molecular probes for sensing and imaging in vitro and in vivo. The fluorescence emission is highly depending on its surrounding condition and therefore highly sensitive toward the surface state of the NPs. If there are any changes in the surface of the NPs due to the chemical or physical interactions with the given chemical species (analytes), it will alter the fluorescence emission. The alteration in the emission spectrum can provide quantitative information about the identity or quantity of the analyte of interest. The direct monitoring of changes in fluorescence of NPs for sensing applications is one of the easiest and most reported approaches because it is very easy to perform. The change in the fluorescence spectrum is the quenching or enhancement in the emission spectrum. The enhancement in the emission spectrum is due to the formation of a shell by the analyte on the surface of NPs, which eliminate the nonradiative recombination of charge carriers. The reduction in the fluorescence intensity is known as quenching. There are two types of quenching: static and dynamic. In static quenching, a nonfluorescent ground-state complex is formed between fluorophore (NPs) and the quencher (analyte) and this complex absorbs light and immediately returns to the ground state without emission of photons. In dynamic or collisional quenching, the collisional encounter between the fluorophore and the analyte leads to the reduction in the fluorescence intensity. In some cases, the quenching may be due to both by collisional and a complex formation with the same quencher [84–87].

Liu et al have reported a facile "switch-on" fluorescent probe for AA detection in food samples using carbon quantum dots–MnO_2. Here the MnO_2 nanosheet quenches the fluorescence intensity of carbon dots through an inner filter effect (IFE) to form a CQDs-MnO_2 probe. When AA was added into the quenched CQDs solution, due to the redox reaction the added MnO_2 was destroyed and the fluorescence of the system was recovered (Figure 9.8). The limit of detection for AA was found to be 42 nM, with a linear range of 0.18–90 μM [88]. W. Gue and coworkers reported the detection of mercury ion using BP quantum dots using IFE of tetraphenyl porphyrin tetra sulfonic acid toward BP quantum dots [89] (Figure 9.9).

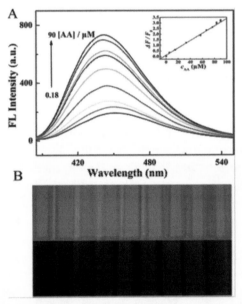

FIGURE 9.8   (A) Fluorescence emission spectra of CQDs-MnO_2 solution upon addition of various concentrations of AA in the presence of 98 μM MnO_2 nanosheets. Inset: Stern Volmer plot (B) Photographs of the colored products with different concentrations of AA under visible light (top) and UV light (bottom), with an excitation wavelength of 365 nm.

Source: Reprinted with permission from Ref. [88]. © 2016 American Chemical Society.

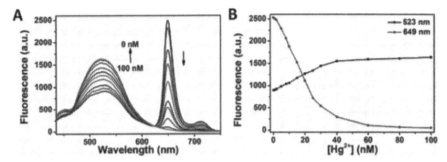

FIGURE 9.9 (a) Fluorescence responses of BP QDs-based assay toward different concentration of Hg^{2+}. (b) Fluorescence intensity changes for BP QDs at 523 nm (black line) and tetraphenyl porphyrin tetra sulfonic acid at 649 nm (red line).

Source: Reprinted with permission from Ref. [89]. © 2017 American Chemical Society.

9.5.2.4 BIOLUMINESCENCE RESONANCE ENERGY TRANSFER (BRET)

BRET is a cell-based assay for studying protein–protein interactions. BRET deals with the transfer of energy from a bioluminescent donor to an acceptor species in the presence of a substrate but requires no external source of illumination to excite the donor molecule. Herein, a bioluminescence protein as a donor can emit light using biochemical energy. FRET requires an external excitation source, whereas there is no external excitation source needed for BRET and is one major advantage of BRET over FRET. The most commonly used donor system in BRET is Renillaluciferase (Rluc). Rluc oxidizes its substrate coelenterazine, which generate emission maximum at 480 nm. The analogs of coelenterazine can generate other emission wavelengths, for example, coelenterazine 400a (CLZ400A substrate) that has an emission at around 395 nm [77].

Fluorescence NPs are used in the BRET for sensing applications, the NPs act as acceptors and are chosen in such a way whose absorption spectrum overlapping with the protein emission wavelength. M. Kumar and group designed a BRET-based sensing system for the nucleic acid detection. The sensing system consists of two small antidense oligonucleotide probes that were covalently attached with Rluc and QD. The emission spectrum of Rluc overlaps with the absorption spectrum of QDs absorption that causes enhancement in the fluorescence intensity of the QDs. The increase in intensity due to BRET acts as a sensing signal and provides quantitative information about the analyte [90]. S. Tsuboi et al. developed a NIR optical sensor for the detection of apoptosis cells using the BRET method.

They used HisRLuc·Annexin V as bioluminescence label for the imaging probe. The QDs (glutathione-coated CdSeTe/CdS) were attached to the HisRLuc·Annexin V via their histidine tags because of the high affinity between histidine and Cd^{2+} ions. The Annexin-conjugated QDs have strong emission wavelength at 830 nm width. This can bind and recognize the phosphatidylserine in the plasma membrane of apoptotic cells, enabling the specific detection of apoptotic cell [91].

9.5.2.5   PHOTON-INDUCED ELECTRON TRANSFER

Photon-induced electron transfer (PET) is an excited state electron transfer process by which an excited electron is transferred from donor to acceptor. The absorption of a photon activates the molecule to undergo redox reaction. In order to initiate the electron transfer from the donor to an acceptor molecule, it should absorb visible or UV light. The molecule that absorbs light get excited and is referred as sensitizer and the other molecule is known as quencher.

Recently PET mechanisms are widely employed in sensing applications. The fluorescence NPs forming a NP–spacer–receptor assembly format is shown in Figure 9.10. The NPs grafted to a receptor via a spacer forming the above format. The excitation of the NPs causes the transfer of an electron from receptor to the NPs under this configuration. The transfer of electrons from the NPs back to the ground state causes the photon emission will be eliminated, which in turn reduces the fluorescence intensity and some times no fluorescence occurs. When the analyte is added, it binds the receptor and electrostatically attract the electrons and thus the thermodynamics for PET become no longer favorable. This resumes the transfer of electrons back to the ground state from excited NPs and generates fluorescence [77].

Sara Raichlin and group applied electron transfer quenching mechanism for the detection of DNA, aptamer–substrate complexes, and telomerase activity. Doxorubicin (DB) intercalated in double-stranded nucleic acids as an electron-transfer quencher of CdSe/ZnS QDs. The electron-transfer will take place from different sized QDs (CdSe/ZnS or CdTe, CdS) to DB.

Upon interaction with the DNA the luminescence quenching (Figure 9.11) and the sensing through the electron-transfer quenching of the QDs by the intercalated DB will occur [92].

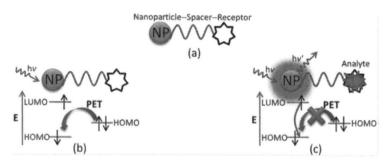

FIGURE 9.10 Schematic representation of (a) format of typical PET using fluorescent nanoparticle, (b) PET occurring by the transfer of electron from the receptor to the fluorescent nanoparticle, and (c) the enhancement of fluorescence emission of the fluorescent nanoparticle the as PET is blocked in the presence of analyte that binds to the receptor.

Source: Reprinted with permission from Ref. [77]. © 2015 The Royal Society of Chemistry.

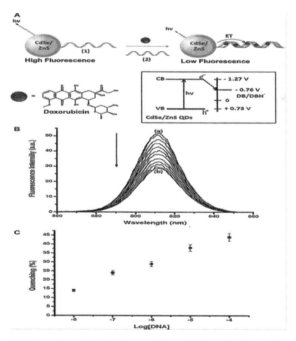

FIGURE 9.11 (a) Detection of DNA by nucleic acid-functionalized QDs and DB, as a quencher. Inset: Energy level diagram corresponding to the electron transfer quenching of the CdSe/ZnS QDs by DB. (b) Time-dependent quenching of the luminescence of the QDs upon interaction with the analyte DNA (c) Calibration curve corresponding to the analysis of different concentrations of (2) by following the degree of quenching of the luminescence of the QDs by DB, after a fixed time-interval of 60 min.

Source: Reprinted with permission from Ref. [92] © 2011 Elsevier.

The quenching of the QDs by the DB intercalated into duplex DNA domains in the vicinity of the QDs was further implemented for the development of new optical aptasensor (cocaine) configurations (Figure 9.12). Finally, for the detection of telomerase in cancer cells, they applied the electron transfer quenching mechanism of QDs by DB (Figure 9.13).

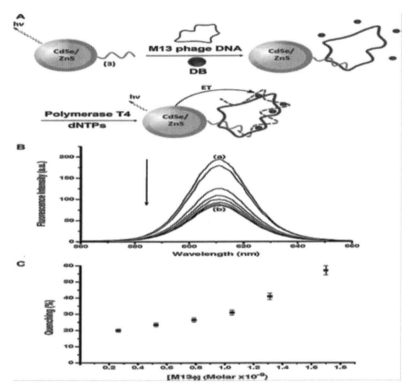

FIGURE 9.12 (a) Amplified detection of M13φ DNA by the (3)-modified QDs through the replication of the analyte DNA and the incorporation of DB into the resulting duplex. (b) Time-dependent luminescence quenching of the QDs upon analyzing M13 φDNA. (c) Calibration curve corresponding to the analysis of variable concentrations of M13φDNA through the replication of the analyte, incorporation of DB into the resulting duplex, and following the electron transfer quenching of the QDs (copyright @ Elsevier).

Source: Reprinted with permission from Ref. [92]© 2011 Elsevier.

9.5.3 SEMICONDUCTOR NPS FOR PHOTOTHERMAL THERAPY (PTT)

PTT is a promising and extensive research area in recent years and is minimally invasive, harmless, and highly effective therapeutic treatment. The main advantages of PTT over conventional radiotherapy or chemotherapy are the capability for deep tissue penetration and minimal effect of nonselective cell death on the surrounding healthy tissue. PTT relies on an optical absorbing agent or photosensitizer, the activation of photosensitizer is by a pulsed laser irradiation at NIR and convert it into heat for thermal ablation of tumor cells. A photosensitizer should have large absorption cross-section, low toxicity, solubility, and ease in functionalization [93].

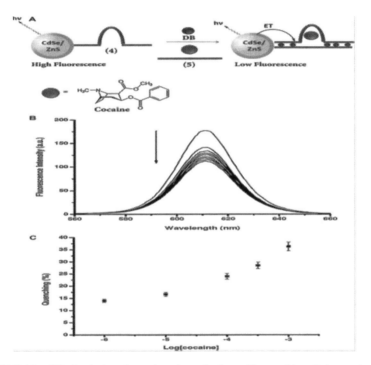

FIGURE 9.13 (A) Analysis of cocaine through the self-assembly of the anti-cocaine aptamer subunits complex on the QDs, and the incorporation of DB into the resulting duplex regions. (B) Time-dependent quenching of the luminescence of the QDs upon interaction of the (4)-modified QDs in the presence of cocaine. (C) Calibration curve corresponding to the analysis of different concentrations of cocaine by following the degree of quenching of the luminescence of the QDs by DB, after a fixed time-interval of 60 min.

Source: Reprinted with permission from Ref. [92]© 2011 Elsevier.

Different types of nanomaterials fulfil most of the requirements of PTT agents. Recently, different types of nanomaterials are used as PTT agents including noble metal nanostructures, carbon-based materials, and semiconductor nanomaterials. Among the various metal NPs gold (Au) nanostructure like Au nanorods, nanoshells, and nanostars are widely used as PTT agents; however, Au nanostructure have low photostability and these are very expensive and the synthesis procedure is highly complicated. Carbon-based nanomaterials are also used in PTT agents but photothermal conversion efficiency is very poor. The semiconductor NPs are considered as the alternative for the metal and carbon-based nanostructures for PTT agents. Among the various semiconductor NPs W18O49 nanowires, copper sulfide (CuS), and copper selenide (CuSe) NPs are promising candidates for PTT [93, 94].

CuS, a p-type semiconductor with unique properties like high stability, biocompatibility, low cost, and high photothermal efficiency under excitation at 808 nm laser, has attracted elicited great interest from biomedical research. The maximum absorbance of CuS NPs is around 900 nm due to the d–d energy band transition of Cu^{2+} ions and is not influenced by the surrounding environment. Under the excitation of at 808 nm leads to an increase in temperature of CuS NPs, which could make use for the foundation of PTT.

Huiting Bi et al. designed a chemo-photothermal therapeutic system which is based on DB-conjugated CuS NPs. The DB release depends on the hydrazine bond as the cleavable link under different pH values. The designed system has a high drug release rate (88.0%) (Figure 9.14) with the photothermal effect under NIR irradiation in mildly acidic environments, which is highly desirable for chemotherapy and photothermal ablation. The in vitro cytotoxicity of DB-CuS NPs was studied in HeLa cancer cells under 808 nm laser irradiation (Figure 9.15). The in vivo treatment was performed on 22 tumor-bearing mice and high tumor inhibition efficacy has been achieved after 14 days of treatment [95].

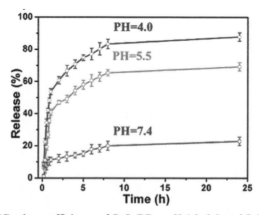

FIGURE 9.14   DB release efficiency of CuS–DB at pH 4.0, 5.5, and 7.4 PBS buffers.

Source: Reprinted with permission from Ref. [95]. © 2016 The Royal Society of Chemistry.

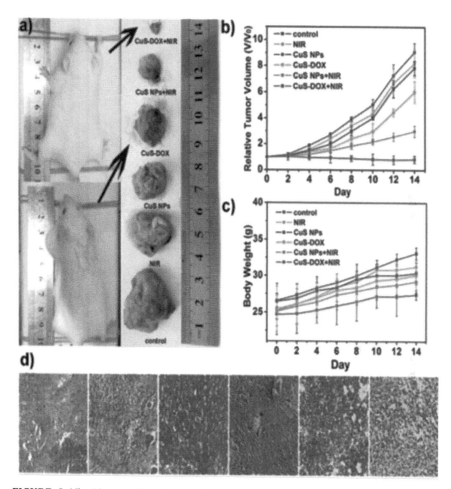

FIGURE 9.15 Photographs of excised tumors from representative mice after 14-day treatment and the digital photos of the mice after injection of the CuS–DB NPs with and without NIR irradiation (a). The relative tumor volumes of H22 tumor-bearing mice (b) and the body weights in different groups after treatment (c). H&E stained tumor sections after 14-day treatment from different groups (d).

Source: Reprinted with permission from Ref. [42]. © 2016 The Royal Society of Chemistry.

Liu and coworkers prepared cysteine-coated CuS NPs for PTT applications. The prepared cysteine-coated CuS NPs have good photostability and biocompatibility and they exhibit an intense absorbance in the NIR region. The NPs have good photothermal conversion efficiency (38.0%). Because of the excellent photothermal effect, the cysteine-coated NPs (low concentration 50 ppm) can effectively inhibit the tumor growth (Figure 9.16) [96].

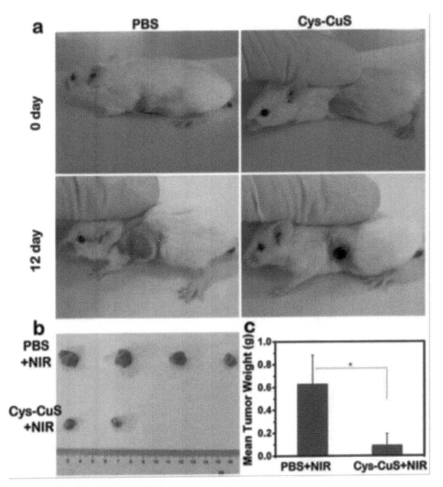

FIGURE 9.16   (a) Representative photos of mice bearing osteosarcoma tumors before and after treatments of PBS, Cys-CuS. (b) Photos of the tumors collected from PBS and Cys-CuS groups of mice at the end of treatments (day 12). (c) Mean weights of tumors collected from mice at the end of PBS and Cys-CuS treatments. ($p < 0.05$ was considered to be statistically significant difference.)

Source: Reprinted with permission from Ref. [96]. © 2014 The Royal Society of Chemistry.

Colin M. Hessel and group investigated the PTT applications of CuSe NPs. Amphiphilic polymer-coated $Cu_{2-x}Se$ Nps were synthesized by a colloidal hot injection method. The prepared NPs are highly water soluble and exhibit strong NIR optical absorption at 980 nm. They show a significant photothermal heating with a photothermal transduction efficiency of 22%.

The destruction of human colorectal cancer cell (HCT-116) after 5 min of laser irradiation at 33 W/cm² (Figure 9.17) [97].

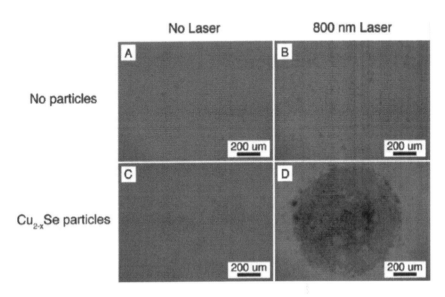

FIGURE 9.17 Comparison of photothermal destruction of human colorectal cancer cells (HCT-116) without (top row, a and b) and with (bottom row, c and d) the addition of 2.8×10¹⁵ Cu_{2-x} Se nanocrystals/L. Cells irradiated at 30 W/cm² with an 800 nm diode laser for 5 min (circular spot size of 1 mm) were stained with Trypan blue to visualize cell death and imaged with an inverted microscope in bright field mode.

Source: Reprinted with permission from Ref. [97]. © 2011 American Chemical Society.

9.6 CONCLUSION

The physical and chemical properties of the NPs are highly depending on their size and shape due to their large surface area or quantum size effect. The unique physical and chemical properties of nanomaterials influence their different application in many fields including health care and medicine. Optical properties of nanomaterials are the most important and encouraging properties for a variety of innovative applications. Of late, optical properties of nanomaterials are increasingly explored in biotechnology and medicine being widely used in therapeutics and diagnostics. Among the various inorganic NPs semiconductor, NPs have profound interest because of their intriguing optical properties and can be more readily integrated into different applications. The most important and promising application of semiconductor nanomaterials is the design of versatile nanostructures such as superlattices and multimodal agents that offer new opportunities in the biomedical field. In this account, we have discussed the optical properties and the biological applications of semiconductor nanoparticles.

KEYWORDS

- **semiconductor nanoparticles**
- **photoluminescence**
- **fluorescence sensing**
- **PTT**
- **MRI**
- **bioimaging**
- **quenching**

REFERENCES

1. Roco, M.C., Bainbridge, W.S., J. Nanoparticle Res., 7(2005), 1–13.
2. Schummer, J., Scientometrics, 59(2004), 425–465.
3. Valyansky, S.I., Naimi, E.K., Kozhitov, L.V., Mod. Electron. Mate., 2(2017), 79–84.
4. Cao, Y.C., Nanomedicine, 3(2008), 467–469.
5. Das, S., Mitra, S., S.M.P. Khurana, Debnath, N., Front. Life Sci., 7(2013), 90–98.
6. Nair, A.K., Mayeen, A., Shaji, L.K., 2018. Optical Characterization of Nanomaterials. Elsevier Ltd.
7. Biju, V., Itoh, T., Anas, A., Anal. Bioanl. Chem. 391 (2008), 2469–2495.
8. Salata, O. V, J. Nanobiotechnology 6 (2004), 1–6.
9. Rehman, F.U., Zhao, C., Jiang, H., Wang, X., Biomater. Sci., 4 (2016), 40–54.
10. Alexson, D., Chen, H., M. Cho, Dutta, M., Y. Li, P. Shi, Raichura, A., D. Ramadurai, Parikh, S., Stroscio, M.A., Vasudev, M., J. Phys. Condens. Matter, 17 (2005), 637–656.
11. Soloviev, V.N., Eichho, A., J. Am. Chem. Soc., 122 (2000), 2673–2674.
12. Xiong, H., Adv. Mater., 25 (2013), 5329–5335.
13. Liu, B. Li, F. Fu, K. Xu, Zou, R., Wang, Q., Zhang, B., Chen, Z., J. Hu, Dalt. Trans., 43 (2014), 11709–11715.
14. Matsuyama, K., Ihsan, N., Irie, K., Mishima, K., Okuyama, T., Muto, H., J. Colloid Interface Sci., 399 (2013), 19–25.
15. Suresh, S., Nanosci. Nanotechnol. 3 (2013), 62–74.
16. Hassan, A., Irfan, M., Jiang, Y., Mater. Lett., 210 (2018), 358–362.
17. Dhahri, R., Hjiri, M., El Mir, L., Alamri, H., Bonavita, A., Iannazzo, D., J. Sci. Adv. Mater. Devices, 2 (2017), 34–40.

18. Ezhilarasi, A.A., Vijaya, J.J., Kaviyarasu, K., Maaza, M., Ayeshamariam, A., Kennedy, L.J., J. Photochem. Photobiol. B, 164 (2016), 352–360.

19. Padmanabhan, P., Kumar, A., Kumar, S., Kumar, R., Gulyás, B., Acta Biomater., 41 (2016), 1–16.

20. Nazir, S., Hussain, T., Ayub, A., Rashid, U., Macrobert, A.J., Nanomed. Nanotechnol. Biol. Med., 10 (2014), 19–34.

21. Ahmad, M.S., Pandey, A.K., Rahim, N.A., Mater. Lett., 195(2017), 62–65.

22. Kumar Sahu, M., Int. J. Appl. Eng. Res., 14 (2019), 491–494.

23. Pal, S., Sarkar, S., Saha, S., Sarkar, P., 2012. Size-Dependent Electronic Structure of Semiconductor Nanoparticles.

24. Andrew, S. Smith, M., S. Nie, Acc. Chem. Res., 43(2010), 190–200.

25. Zhang, B., 2018. Optical Properties of Nanomaterials, Physical Fundamentals of Nanomaterials.

26. Fu, A., Gu, W., Larabell, C., Alivisatos, A.P, Curr. Opin. Neurobiol., 15 (2005), 568–575.

27. Li, L.S., Hu, J., Yang, W., Alivisatos, A.P., Nano Lett., 1(2001), 349–351.

28. Planelles, J., Rajadell, F., Climente, J.I.,. J. Phys. Chem. C, 114 (2010), 8337–8342.

29. Puzder, A., Williamson, A. J., Reboredo, F. A., & Galli, G., Phys. Rev. Lett., 91(2003), 157405.

30. Bhagyaraj, S.M., Oluwafemi, O.S., 2018. Nanotechnology: The Science of the Invisible, Synthesis of Inorganic Nanomaterials. Elsevier Ltd.

31. Obonyo, O., Fisher, E., Edwards, M., Douroumis, D.,. Crit. Rev. Biotechnol., 30 (2010), 283–301.

32. Ponce, A., Mejia-rosales, S., Jose-Yacaman, M., NPsBiol. Med. Methods Protoc., 906(2012), 453–471.

33. Xinyan Wang, Qiang Ma, B.L., Su, Y.L. and X., Luminescence, 22 (2006), 320–386.

34. Didenko, Y.T., Suslick, K.S., J. Am. Chem. Soc., 127(2005), 12196–12197.

35. Moghaddam, M.M., Baghbanzadeh, M., Keilbach, A., Kappe, C.O., Nanoscale, 4 (2012), 7435–7442.

36. Batmunkh, M., Myekhlai, M., Bati, A.S.R., Sahlos, S., Slattery, A.D., Benedetti, T.M., Gonçales, V.R., Gibson, C.T., Gooding, J.J., Tilley, R.D., Shapter, J.G., J. Mater. Chem. A, 7(2019), 12974–12978.

37. Wang, J., Li, D., Liu, X., Qiu, Y., Peng, X., Huang, L., Wen, H., Hu, J., New J. Chem., 42(2018), 15743–15749.

38. Wei, C., Li, J., Gao, F., Guo, S., Zhou, Y., Zhao, D., J. Spectrosc., 2015 (2014), 1–7.

39. Tang, X., Shi, E., Choo, G., Li, L., Ding, J., Xue, J., Langmuir, 25 (2009), 5271–5275.

40. Cui, J., He, W., Liu, H., Liao, S., Yue, Y., Colloids Surf. B Biointerfaces, 74 (2009), 274–278.

41. Zhong, H., Mirkovic, T., Scholes, G.D., 2011. Nanocrystal Synthesis.

42. Anugop, B., Prasanth, S., Raj, D.R., Vineeshkumar, T. V, Pranitha, S., Pillai, V.P.M., Sudarsanakumar, C., Opt. Mater., 62(2016), 297–305.

43. Prasanth, S., Rithesh Raj, D., Vineeshkumar, T. V., Thomas, R.K., Sudarsanakumar, C., RSC Adv., 6(2016), 58288–58295.

44. Goel, S., Chen, F., Cai, W., Small, 10(2013), 1–15.

45. Malik, M.A., Brien, P.O., Revaprasadu, N., S. Afr. J. Sci., 96(2000), 55–60.

46. Massadeh, S., Alaamery, M., n.d., NPs outline, in: Nanomedicine. pp. 357–373.

47. Suresh, S., Nanosci. Nanotechnol., 3(2013), 62–74.

48. Alivisatos, A.P., Science, 271 (1996), 933–936.

49. Kairdolf, B.A., Smith, A.M., Stokes, T.H., Wang, M.D., Young, A.N., Nie, S.,. Annu. Rev. Anal.Chem., 6(2013), 143–162.
50. Banerjee, A., Pons, T., Lequeux, N., Dubertret, B., Marie, P., Dubertret, B.,. Interface Focus, 6(2016), 1–16.
51. Pereira, G., Monteiro, C.A.P., Albuquerque, G.M., Cabrera, M.P., Filho, P.E.C., Fontes, A., Santos, B.S., JJ. Braz. Chem. Soc., 30 (2019), 2536–2560.
52. Zhang, L.Z., Sun, W., Cheng, P., Molecules, 8 (2003), 207–222.
53. Singh, M., Goyal, M., Devlal, K.,. J. Taibah Univ. Sci. 3655(2018), 470–475.
54. Neikov, O.D., Yefimov, N.A., 2019. Nanopowders, 2nd ed, Handbook of Non-Ferrous Metal Powders. Elsevier Ltd.
55. Martynenko, I. V., Litvin, A.P., Purcell-Milton, F., Baranov, A. V., Fedorov, A. V., Gun'Ko, Y.K., 2017. J. Mater. Chem., B 5, 6701–6727.
56. Martynenko, I. V., Litvin, A.P., F. Purcell-Milton, Baranov, A.V., Fedorov, A.V., and Y. K. Gun'Ko. J.Mat. Chem. B 5, 33 (2017), 6701–6727.
57. Liu, Liwei, Rui Hu, Wing-Chueng Law, Indrajit Roy, Jing Zhu, Ling Ye, Siyi Hu, Xihe Zhang, and Ken-Tye Yong. Analyst, 138, 20 (2013), 6144–6153.
58. Vlasceanu, G., Grumezescu, A.M., Gheorghe, I., Chifiriuc, M.C., Holban, A.M., 2017. Chapter 18—Quantum dots for bioimaging and therapeutic applications, Nanostructures for Novel Therapy. Elsevier Inc.
59. Yokoyama, Hiroyuki, Hengchang Guo, Takuya Yoda, Keijiro Takashima, Ki-ichi Sato, Hirokazu Taniguchi, and Hiromasa Ito. Opt. Exp., 14, 8 (2006), 3467–3471.
60. Caravan, P., Ellison, J.J., Mcmurry, T.J., Lauffer, R.B., Chem. Rev., 99 (1999), 2293–2352.
61. Na, H. Bin, Song, I.C., Hyeon, T., Adv. Mater., 21(2009), 2133–2148.
62. Chevallier, P., Walter, A., Garofalo, A., Veksler, I., Lagueux, J., Bégin-Colin, S., Felder-Flesch, D., Fortin, M.A., J. Mater. Chem., B 2 (2014), 1779–1790.
63. Li, J., Wu, C., Hou, P., Zhang, M., Xu, K.,. Biosens. Bioelectron., 102 (2018), 1–8.
64. Batmunkh, M., Myekhlai, M., Bati, A.S.R., Sahlos, S., Slattery, A.D., Benedetti, T.M., Gonçales, V.R., Gibson, C.T., Gooding, J.J., Tilley, R.D., Shapter, J.G., J. Mater. Chem., A 7, (2019), 12974–12978.
65. Kim, Min Woo, Hwa Yeon Jeong, Seong Jae Kang, In Ho Jeong, Moon Jung Choi, Young Myoung You, Chan Su Im, Theranostics, 9, 3 (2019), 837.
66. Doane, Tennyson L., Clemens Burda. Chem. Soc. Rev., 41, 7 (2012), 2885–2911.
67. Kim, M.W., Jeong, H.Y., Kang, S.J., Choi, M.J., You, Y.M., Im, C.S., Lee, T.S., Song, I.H., Lee, C.G., Rhee, K.J., Lee, Y.K., Park, Y.S., Sci. Rep., 7 (2017), 1–11.
68. Lee, M., Park, Y.H., Kang, E.B., Chae, A., Choi, Y., Jo, S., Kim, Y.J., Park, S.J., Min, B., An, T.K., Lee, J., In, S. Il, Kim, S.Y., Park, S.Y., In, I., ACS Omega, 2 (2017), 7096–7105.
69. Shen, Y., Lifante, J., Ximendes, E., Santos, H.D.A., Ruiz, D., Juárez, B.H., Zabala Gutiérrez, I., Torres Vera, V., Rubio Retama, J., Martín Rodríguez, E., Ortgies, D.H., Jaque, D., Benayas, A., Del Rosal, B.,. Nanoscale, 11(2019), 19251–19264.
70. Hu, R., Law, W.C., Lin, G., Ye, L., Liu, Jianwei, Liu, Jing, Reynolds, J.L., Yong, K.T., Theranostics, 2(2012), 723–733.
71. Biju, V., Chem. Soc. Rev., 43(2014), 744–764.
72. Wegner, K.D., Hildebrandt, N., Wegner, K.D., Hildebrandt, N.,. Chem. Soc. Rev., 44(2015), 4792–4834.
73. Chandrasekaran, P., Viruthagiri, G., Srinivasan, N., J. Alloys Compd. 540 (2012), 89–93.

74. Hodlur, R.M., Rabinal, M.K., Mohamed Ikram, I., J. Lumin., 149 (2014), 317–324.
75. Zhang, Y. hai, Zhang, H. shan, Guo, X. feng, Wang, H., Microchem. J., 89 (2008), 142–147.
76. Fernández-Argüelles, M.T., Wei, J.J., Costa-Fernández, J.M., Pereiro, R., Sanz-Medel, A., 2005. Anal. Chim. Acta, 549, 20–25.
77. Wegner, K.D., Hildebrandt, N., Wegner, K.D., Hildebrandt, N., Chem. Soc. Rev., 44 (2015), 4792–4834.
78. Huang, S., Zhu, F., Xiao, Q., Su, W., Sheng, J., Huang, C., Hu, B., RSC Adv., 4(2014), 46751–46761.
79. Gonçalves, H., Jorge, P.A.S., Fernandes, J.R.A., Esteves da Silva, J.C.G., 2010. Sens. Actuators B Chem., 145, 702–707.
80. Rubul Mout, Daniel F. Moyano, Subinoy Rana, and V.M.R.,. Chem. Soc. Rev., 41 (2012), 2539–2544.
81. Zhang, C.Y., Yeh, H.C., Kuroki, M.T., Wang, T.H., Nat. Mater., 4 (2005), 826–831.
82. Kattke, M.D., Gao, E.J., Sapsford, K.E., Stephenson, L.D., Kumar, A., Sensors, 11(2011), 6396–6410.
83. Kalarestaghi, A., Bayat, M., Hashemi, S.J., Razavilar, V., Iran. J. Biotechnol., 13(2015), 25–31.
84. Sudarsanakumar, C., Thomas, S., Mathew, S., Arundhathi, S., Raj, D.R., Prasanth, S., Thomas, R.K., Mater. Res. Bull., 110 (2019), 32–38.
85. Prasanth, S., RitheshRaj, D., Vineeshkumar, T. V., Sudarsanakumar, C., Chem. Phys. Lett., 700 (2018), 15–21.
86. Prasanth, S., Raj, D.R., Thomas, R.K., Vineeshkumar, T. V., Sudarsanakumar, C., RSC Adv., 6 (2016), 105010–105020.
87. Prasanth, S., Sudarsanakumar, C., New J. Chem. 41 (2017), 9521–9530.
88. Juanjuan Liu, Yonglei Chen, Weifeng Wang, Jie Feng, Meijuan Liang, Sudai Ma, and X.C., J. Agric. Food Chem., 64 (2016), 371–380.
89. Gu, W., Pei, X., Cheng, Y., Zhang, C., Zhang, J., Yan, Y., Ding, C., Xian, Y., ACS Sens., 2 (2017), 576–582.
90. Gu, W., Pei, X., Cheng, Y., Zhang, C., Zhang, J., Yan, Y., Ding, C., Xian, Y., ACS Sens., 2(2017), 576–582.
91. Bioluminescence, T., Energy, R., Bret, T., Tsuboi, S., Jin, T., ChemBioChem, 18 (2017), 2231–2235.
92. Raichlin, S., Sharon, E., Freeman, R., Tzfati, Y., Willner, I., Biosens. Bioelectron., 26 (2011), 4681–4689.
93. Rosal, B., Haro-gonzalez, P., Benayas, A., Nanoscale 6 (2014), 9494–9530.
94. Fong, J.F.Y., Ng, Y.H., Ng, S.M., 2018. Chapter 7. Carbon dots as a new class of light emitters for biomedical diagnostics and therapeutic applications, Fullerens, Graphenes and Nanotubes: A Pharmaceutical Approach. Elsevier Inc.
95. Bi, H., Dai, Y., Lv, R., Zhong, C., He, F., Gai, S., Gulzar, A., Yang, G., Yang, P., Doxorubicin-conjugated CuS nanoparticles for efficient synergistic therapy triggered by near-infrared light. Dalt. Trans., 45 (2016), 5101–10.
96. Liu, X., Li, B., Fu, F., Xu, K., Zou, R., Wang, Q., Zhang, B., Chen, Z., Hu, J., Facile synthesis of biocompatible cysteine-coated CuS nanoparticles with high photothermal conversion efficiency for cancer therapy, Dalt. Trans., 43 (2014), 11709–11715.
97. Hessel, C.M., Pattani, V.P., Rasch, M., Panthani, M.G., Koo, B., Tunnell, J.W., Korgel, B.A., Nanoletters, 11 (2011), 2560–2566.

CHAPTER 10

Optical Properties of Pr^{3+}-Doped Polyvinylidene Fluoride–Titanium Dioxide (PVDF–TiO$_2$) Hybrid

RANI GEORGE[1,2], SUNIL THOMAS[3], P. R. BIJU[2], N. V. UNNIKRISHNAN[2], and C. SUDARSANAKUMAR[2,*]

[1]*Department of Physics, St. Aloysius College, Edathua, Alappuzha 689573, Kerala, India*

[2]*School of Pure and Applied Physics, Mahatma Gandhi University, Kottayam 686560, Kerala, India*

[3]*Department of Physics, United Arab Emirates University, P.O. Box 15551, Al Ain, United Arab Emirates*

Corresponding author. E-mail: c.sudarsan.mgu@gmail.com

ABSTRACT

Polyvinylidene fluoride–titanium dioxide (PVDF–TiO$_2$) hybrid material is a new class of materials in the field of optoelectronics. Rare earth doping in this hybrid makes it useful in optical devices. In this chapter, we describe the synthesis and optical properties of Pr^{3+}-doped PVDF–TiO$_2$ hybrid. A nonhydrolytic sol–gel method was used for the synthesis. The spectroscopic properties of PVDF–TiO$_2$:Pr^{3+} were characterized using absorption and photoluminescence spectra. Direct and indirect bandgaps of the material were determined from the absorption spectrum. The nephelauxetic ratio and bonding parameter were calculated to identify the Pr^{3+}–ligand bond in the material. The oscillator strengths of the observed absorption peaks, Judd–Ofelt parameters, and spectroscopic quality factor were estimated to analyze the potential of the material to be used in optical devices. The emission characteristics were studied at an excitation wavelength of 446 nm. The colorimetric analysis based on the emission spectrum revealed that PVDF–TiO$_2$:Pr^{3+} emit bluish-green light upon 446 nm excitation.

10.1   INTRODUCTION

Polymer–inorganic hybrids have attracted much attention due to its ability to combine the properties of polymers and inorganic components to a single entity. Polymers have their intrinsic properties such as low processing temperature, low density, flexibility, and better film-forming capacity, whereas inorganic compounds have high refractive index, transparency, and better crystallinity [1]. The properties of polymer-inorganic hybrid strongly depend on its synthesis process and the microstructure of the material. The sol–gel method is one of the successful synthesis processes for hybrid materials, where polymer chains are interpenetrated into the inorganic network [2–4]. The desired optical, mechanical, and electrical properties of polymer–inorganic hybrid can be achieved by tuning the sol–gel process [1]. The applications of polymer–inorganic hybrids in display devices [5, 6] motivate the investigation of the optical properties of polyvinylidene fluoride–titanium dioxide (PVDF–TiO$_2$).

TiO$_2$ is notable among the inorganic compounds due to its unique properties such as nontoxicity, high optical transparency, corrosion-resistance, and better mechanical properties [7–9]. TiO$_2$ finds its applications in opto-electronic devices, photocatalysts, sensors, solar cells, and photochromic devices [10], and shows intense absorption in the ultraviolet (UV) region and excellent transparency in the visible region. TiO$_2$ can exist in three different phases namely, anatase (tetragonal), rutile (tetragonal), and brookite (ortho-rhombic). Anatase TiO$_2$ shows better stability, high reactivity, and a wide bandgap [7].

PVDF is of interest owing to its peculiarities such as high hydrophobicity, mechanical strength, thermal stability, and chemical resistance compared to other commercialized polymeric materials [11]. PVDF membrane has been extensively applied in ultrafiltration and microfiltration [11]. Studies have also been done on the piezoelectric and pyroelectric properties of PVDF [12]. The optical properties of Eu^{3+}-doped PVDF have been explored in a recent study [13]. For another example, Pogreb et al. [14] studied the luminescence of PVDF doped with Eu(III) (NO$_3$)$_3$ (o-Phen)$_2$ complex.

When doped with rare earth (RE), a polymer–inorganic hybrid can be used in solid-state lasers, optical sensors, waveguides, and display devices

[15]. This is because of the RE ions in the host matrix act as active luminescent centers. RE ions exist in solid matrices either as RE^{3+} or RE^{2+}. The electronic configuration of RE^{3+} is 4f^n 5d 5s^2 5p^6 and RE^{2+} is 4f^{n-1} 5d 5s^2 5p^6. The shielding of 4f orbitals by the filled 5s^2 5p^6 orbitals is responsible for the unique optical properties of RE ions. The transitions of 4f electrons emit characteristic colors of RE ions in the visible region of the spectrum. Of different RE ions, praseodymium is notable due to its blue, green, and red emissions depending on the host matrix [15]. Jose et al. [15] and Vidyadharan et al. [16] studied the optical properties and Judd–Ofelt (JO) analysis of Pr^{3+}-doped zirconia/polyethylene glycol composite and Sr$_{0.5}$Ca$_{0.5}$TiO$_3$ phosphor, respectively.

In this chapter, we report the optical properties of sol–gel synthesized Pr^{3+}-doped PVDF–TiO$_2$ hybrid. The optical bandgap, bonding nature, absorption, emission, and colorimetric analysis of PVDF–TiO$_2$:Pr^{3+} are described.

10.2   EXPERIMENTAL

Pr^{3+}-doped PVDF–TiO$_2$ hybrid was prepared by a nonhydrolytic sol–gel method. The sol–gel process is an efficient method to synthesize polymer–inorganic hybrids. In this study, the weight ratio between the inorganic and organic components was fixed at 50:50. The synthesis process is illustrated in Figure 10.1. First, PVDF was dissolved in dimethylformamide (DMF). In another beaker, titanium isopropoxide (TIP) was dissolved in ethanol (1:10 molar ratio). This TIP+ethanol solution was mixed with the PVDF+DMF solution. For Pr^{3+} doping, praseodymium nitrate hexahydrate (5 wt.%) was dissolved in ethanol and added to the above mixture. Nitric acid was added to accelerate the gelation process and the mixture was stirred for another ~15 min to form a viscous transparent liquid. This was then sealed in a polypropylene container and kept on a shelf for aging ~24 h. The resultant material was annealed at 80 °C for ~10 h and powdered thereafter.

The absorption spectrum of PVDF–TiO$_2$:Pr^{3+} was measured using Cary 5000 UV–visible–NIR spectrophotometer in the range 200–2200 nm at a step size of 1 nm. The spectrum was measured in the reflectance mode, which was later converted to absorbance. The photoluminescence of PVDF–TiO$_2$:Pr^{3+} was recorded on Horiba Scientific Fluoromax-4 Spectrofluorometer in the range 460–650 nm at the excitation wavelength of 446 nm.

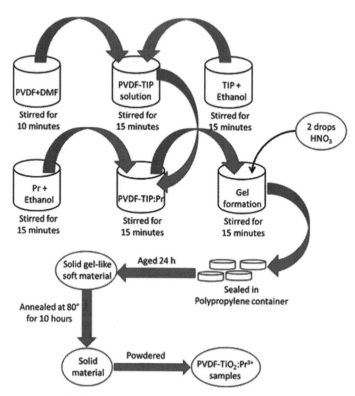

FIGURE 10.1   Synthesis procedure for Pr^{3+}-doped PVDF–TiO$_2$ hybrid.

10.3   RESULTS AND DISCUSSION

10.3.1   ABSORPTION SPECTRUM AND OPTICAL BANDGAP

The optical absorption of PVDF–TiO$_2$:Pr^{3+} as a function of wavelength is depicted in Figure 10.2A and B. Figure 10.2A corresponds to the UV–visible region and Figure 10.2B shows the NIR region. The broad absorption band in the UV region (240–370 nm) is due to the absorption of the host matrix. The other eight absorption peaks in the spectrum are due to the electronic transitions in Pr^{3+} from its ground level $^3H_4$ to different higher energy levels. These higher energy levels are identified with reference to the previous reports of other Pr^{3+}-doped materials [17, 18]. The peaks in the visible region of the spectrum at 446, 470, 485, and 593 nm are due to the transitions from the ground state $^3H_4$ to $^3P_2$, $^3P_1$, $^3P_0$, and $^1D_2$, respectively. The absorption peaks in the NIR region at 1013, 1455, 1562, and 1932 nm are due to the transitions from $^3H_4$ to $^1G_4$, $^3F_4$, $^3F_3$, and $^3F_2$, respectively. Similar absorption transitions are observed in Pr^{3+}-doped fluoroindate glass reported by Florez et al. [19].

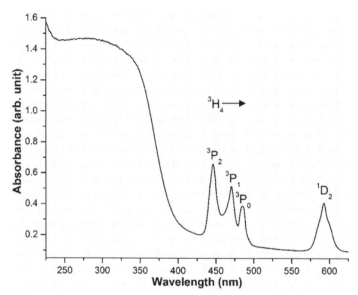

FIGURE 10.2A Absorption spectrum of PVDF–TiO$_2$:Pr^{3+} in the UV–Vis region.

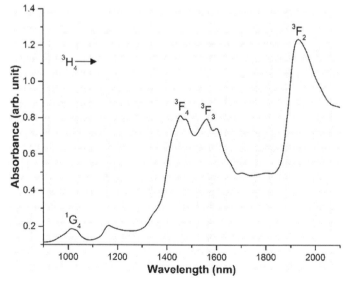

FIGURE 10.2B Absorption spectrum of PVDF–TiO$_2$:Pr^{3+} in the NIR region.

All the absorption transitions in the UV–visible region are purely electric dipole, which follow the selection rule of electric dipole transition ($|\Delta J| \leq 6$) but not the magnetic dipole transition ($|\Delta J| \leq 1$). In the NIR region, three peaks, $^3H_4 \rightarrow {}^1G_4$, $^3F_4$, and $^3F_3$ that follow $|\Delta J| \leq 1$ have magnetic dipole contribution too but $^3H_4 \rightarrow {}^3F_2$ are purely electric dipole. All the absorption transitions, except $^3H_4 \rightarrow {}^1D_2$ and $^3H_4 \rightarrow {}^1G_4$, are spin-allowed transitions ($\Delta S = 0$). The intensity of certain absorption bands in RE elements is very sensitive to its host matrix and these transitions are termed hypersensitive transitions, which obey the selection rule of quadrupole transitions ($|\Delta S| = 0$, $|\Delta L| \leq 2$, and $|\Delta J| \leq 2$). The absorption transitions from $^3H_4$ to $^3F_4$, $^3F_3$, and $^3F_2$, in the NIR region, are hypersensitive [20]. Judd and Ofelt [20, 21] reported that hypersensitive transitions are associated with high values of the reduced matrix elements $\|U^{(2)}\|$.

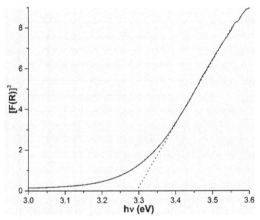

FIGURE 10.3A   A plot of $[F(R)]^2$ versus $h\nu$ to estimate the direct bandgap.

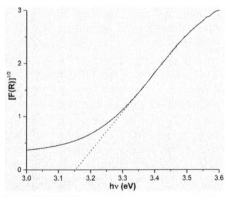

FIGURE 10.3B   A plot of $[F(R)]^{1/2}$ versus $h\nu$ to estimate indirect bandgap.

The large absorption in the higher energy region of the absorption spectrum (240–370 nm) is due to the absorption of electrons in the valence band and the resultant transitions to the conduction band. The type of the electronic transition (direct or indirect) gives the relation of the absorption coefficient to the photon energy. During the transition, if the momentum is conserved, the transition is direct; otherwise, it is indirect. The bandgap and the type of band-to-band transition are important to reveal the optical properties of PVDF–TiO$_2$:Pr^{3+}. The optical bandgap (E_g) of PVDF–TiO$_2$:Pr^{3+} was determined by Kubelka–Munk (K–M or $F(R)$) method. According to the K–M method [22]

$$\left(\frac{\alpha}{S}\right) = F(R) = \frac{(1-R)^2}{2R}$$ (10.1)

where R is the diffuse reflectance, $F(R)$ is the remission function, α is the absorption coefficient, and S is the dispersion factor. Absorption coefficient (α) is related to the incident photon energy by the equation

$$\alpha = A\left(h\nu - E_g\right)^\gamma$$ (10.2)

where A is a material-dependent constant, $h\nu$ is the photon energy, and γ is a constant. For an allowed direct transition $\gamma = 1/2$ and for an indirect transition $\gamma = 2$. Therefore, for a direct transition, Equation (10.1) becomes

$$\left[F(R)\right]^2 = \left(\frac{A}{S}\right)^2 \left(h\nu - E_g\right)$$ (10.3)

and for an indirect transition

$$\left[F(R)\right]^{1/2} = \left(\frac{A}{S}\right)^{1/2} \left(h\nu - E_g\right)$$ (10.4)

In the graphs of $[F(R)]^2$ versus $h\nu$ (Figure 10.3A) and $[F(R)]^{1/2}$ versus $h\nu$ (Figure 10.3B), the bandgaps of PVDF–TiO$_2$:Pr^{3+} are obtained by extrapolating the absorption edge on to the energy axis. In the direct transition, the bandgap obtained is 3.29 eV, whereas, in the indirect transition, the bandgap is 3.15 eV. These values are comparable to that of TiO$_2$–polyethylene composite reported by Romero-Saez et al. [23].

The refractive index (n) of the material is calculated from the band gap using the equation [24]

$$\frac{n^2-1}{n^2+2} = 1 - \sqrt{E_g/20} \qquad (10.5)$$

The dielectric constant [25] is expressed as

$$\varepsilon = n^2 \qquad (10.6)$$

The electronic susceptibility (χ) is determined using the equation

$$\chi = \frac{n^2-1}{4\pi} \qquad (10.7)$$

The reflective loss (R) is estimated using the refractive index [25] by

$$R = \left(\frac{n-1}{n+1}\right)^2 \times 100 \qquad (10.8)$$

The values of optical band gap (E_g), refractive index (n), dielectric constant (ε), electronic susceptibility (χ), and reflection loss (R) are listed in Table 10.1. The reflection loss of PVDF–TiO$_2$:Pr^{3+} is lower than that of some borate glasses [26, 27].

TABLE 10.1 The Values of the Optical Bandgap (E_g), Refractive Index (n), Dielectric Constant (ε), Electronic Susceptibility (χ), and Reflection Loss (R)

E_g (eV)	n	ε	χ	R (%)
3.29	2.323	5.396	0.350	15.9
3.15	2.358	5.560	0.363	16.4

TABLE 10.2 The Values of v_c, v_a, β, and δ

S. No.	S′L′J′ $^3H_4 \rightarrow$	Peak (nm)	v_c (cm^{-1})	v_a (cm^{-1})	β
1	$^3P_2$	446	22422	22520	0.9956
2	$^3P_1$	470	21277	21300	0.9989
3	$^3P_0$	485	20619	20750	0.9937
4	$^1D_2$	593	16863	16840	1.0014
5	$^1G_4$	1013	9872	9900	0.9972
6	$^3F_4$	1455	6873	6950	0.9889
7	$^3F_3$	1562	6402	6500	0.9849
8	$^3F_2$	1932	5176	5200	0.9954

$\beta' = 0.9945$

$\delta = 0.5528$

10.3.2 NEPHELAUXETIC RATIO AND BONDING PARAMETER

Nephlauxetic ratios [25] were estimated from the absorption spectrum to analyze the Pr^{3+}–ligand bond in the matrix. It is the ratio of energy (cm^{-1}) for a particular transition in the host matrix (v_c) to the energy (cm^{-1}) for the same transition in an aqueous medium (v_a)

$$\beta = \frac{v_c}{v_a} \tag{10.9}$$

The nephelauxetic ratios (β) for all the transitions in the absorption spectrum are given in Table 10.2. The average nephelauxetic ratio is denoted by β'. The bonding parameter (δ) [15] can be derived from the value of β' as

$$\delta = \frac{1 - \beta'}{\beta'} \times 100 \tag{10.10}$$

A positive value of δ means that Pr^{3+}–ligand bond in the host matrix is covalent type, whereas a negative value indicates ionic nature. The bonding parameter of PVDF–TiO$_2$:Pr^{3+} is 0.5528, which indicates covalent type Pr^{3+}–ligand bond in the material. Similar bonding nature is observed in Sm^{3+}-doped PVDF–ZrO$_2$ hybrid membrane [5].

10.3.3 JUDD–OFELT INTENSITY ANALYSIS

In 1962, Judd and Ofelt [20, 21] reported a theoretical model for the electronic transitions in REs. The model was based on the approximation that central ion interacts with the surrounding host ions via a static electric field, which is treated as a perturbation on the free-ion Hamiltonian. According to JO theory, the oscillator strength for a transition from ΨJ to $\Psi' J'$ [25] is given by

$$f_{cal} = \frac{8\pi^2 mcv}{3h(2J+1)} \frac{\left(n^2 + 2\right)^2}{9n} \sum_{\lambda=2,4,6} \Omega_\lambda \left| \left\langle \Psi J U^{(\lambda)} \Psi' J' \right\rangle \right|^2 \tag{10.11}$$

where Ω_λ are the JO intensity parameters, $U^{(\lambda)}$ are the doubly reduced matrix elements, m is the electron mass, c is the speed of light, h is the Planck's constant, $(2J+1)$ is the degeneracy of the ground state, and n is the refractive index. The experimental oscillator strength f_{exp} of a transition is expressed as

$$f_{exp} = 4.32 \times 10^{-9} \int \varepsilon(v) dv \qquad (10.12)$$

where $\varepsilon(v)$ is the molar absorptivity of a band at a wavenumber v (cm^{-1}) [28]. The rms deviation (σ) between f_{exp} and f_{cal} is

$$\sigma = \left[\frac{\sum (f_{exp} - f_{cal})^2}{N_i} \right]^{1/2} \qquad (10.13)$$

where N_i denotes the total number of transitions. The JO intensity parameters Ω_λ were calculated through a least-square fitting of the experimental oscillator strengths.

TABLE 10.3 The Experimental (f_{exp}) and Calculated (f_{cal}) Oscillator Strengths, Judd–Ofelt Intensity Parameters (Ω_λ), and Spectroscopic Quality Factor (Q)

S'L'J' $^3H_4 \rightarrow$	E_{exp} (cm^{-1})	E_{calc} (cm^{-1})	ΔE (cm^{-1})	f_{exp} ($\times 10^{-6}$)	f_{cal} ($\times 10^{-6}$)	Δf ($\times 10^{-6}$)
$^3P_2$	22422	22535	−113	0.8048	0.6696	0.1352
$^3P_1$	21277	21330	−53	0.3746	0.2627	0.1119
$^3P_0$	20619	20706	−87	0.2299	0.2577	-0.0278
$^1D_2$	16863	16840	23	0.6619	0.2070	0.4550
$^1G_4$	9872	9885	−13	0.2027	0.0606	0.1421
$^3F_4$	6873	6973	−100	0.7344	0.7240	0.0104
$^3F_3$	6402	6540	−138	1.1232	1.1542	-0.0310
$^3F_2$	5176	5149	27	0.8711	0.8691	0.0020

$\sigma = 0.18 \times 10^{-6}$
$\Omega_2 = 9.3781 \times 10^{-21}$ cm^2
$\Omega_4 = 3.0001 \times 10^{-21}$ cm^2
$\Omega_6 = 8.3392 \times 10^{-21}$ cm^2
$Q = 0.36$

Table 10.3 lists the experimental (f_{exp}) and theoretical (f_{cal}) oscillator strengths along with the JO parameters. The low value of σ (0.18×10^{-6}) indicates that f_{cal} values match well with f_{exp}. The JO intensity parameters of PVDF–TiO$_2$:Pr^{3+} follow the trend $\Omega_2 > \Omega_6 > \Omega_4$. A similar trend was found in PVDF–ZrO$_2$:Sm^{3+} hybrid membrane [5]. Ω_4 and Ω_6 are long-range parameters related to the bulk properties (like viscosity and basicity) of the host matrix

[25]. The spectroscopic quality factor ($Q = \Omega_4 / \Omega_6$) of PVDF–TiO$_2$:Pr^{3+} was evaluated as 0.36. This value is higher than that of Pr^{3+}-doped phosphors reported by Vidyadharan et al. [16] and PVDF–ZrO$_2$:Sm^{3+} hybrid membrane reported by George et al. [5]. This value is useful to predict the stimulated emission in a laser-active medium. The Ω_2 parameter is usually associated with the covalency of the RE–oxygen bond and asymmetry of the ligand field around the RE ion site [28]. The high value of Ω_2 indicates the covalent nature of Pr^{3+}–ligand bond. This result is consistent with the covalent nature determined from the bonding parameter in the previous section.

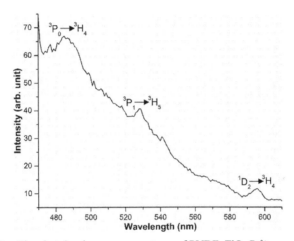

FIGURE 10.4A The photoluminescence spectrum of PVDF–TiO$_2$:Pr^{3+}.

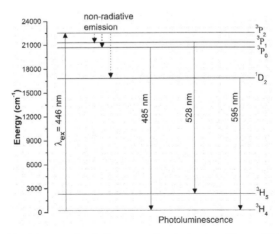

FIGURE 10.4B Partial energy level diagram and emission transitions of Pr^{3+} in PVDF–TiO$_2$ hybrid.

10.3.4   EMISSION SPECTRUM AND COLORIMETRIC ANALYSIS

The room temperature fluorescence spectrum for PVDF–TiO$_2$:Pr^{3+} is shown in Figure 10.4A. Since the absorption spectrum of the sample has an intense peak at 446 nm, the emission spectrum is recorded at this excitation wavelength. The spectrum consists of three characteristic peaks of Pr^{3+} namely, $^3P_0 \rightarrow {}^3H_4$, $^3P_1 \rightarrow {}^3H_5$, and $^1D_2 \rightarrow {}^3H_4$ at wavelengths 485, 528, and 595 nm, respectively. All the transitions are purely electric dipole transitions ($|\Delta J| \leq$ 6). The transitions $^3P_0 \rightarrow {}^3H_4$ and $^3P_1 \rightarrow {}^3H_5$ are spin-allowed ($\Delta S = 0$). The upward sloping of the spectrum might be due to the host emission. The branching ratios, the ratio of relative intensities, for the transitions $^3P_0 \rightarrow {}^3H_4$, $^3P_1 \rightarrow {}^3H_5$, and $^1D_2 \rightarrow {}^3H_4$ are 0.54, 0.27, and 0.19, respectively. A peak with a branching ratio greater than 0.5 can be suggested for lasing action [25]. It has to be noted that the branching ratio of $^3P_0 \rightarrow {}^3H_4$ is greater than 0.5.

The partial energy diagram shown in Figure 10.4B describes the emission mechanism of Pr^{3+} in PVDF–TiO$_2$ hybrid. When excitation energy of wavelength 446 nm is incident, Pr^{3+} ions absorb the energy and are transferred to the $^3P_2$ level. From this excited energy level, Pr^{3+} ions non-radiatively relax to $^3P_1$, $^3P_0$, and $^1D_2$ levels. The radiative transitions from these energy levels, as shown in the figure, appeared as emission peaks at 485, 528, and 595 nm in the spectrum.

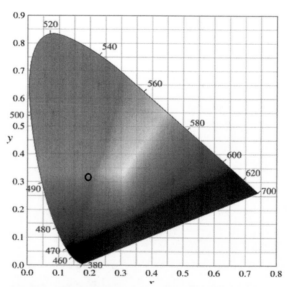

FIGURE 10.5   CIE chromaticity diagram for PVDF–TiO$_2$:Pr^{3+}.

The Commission Internationale de l'Eclairage (CIE) coordinate position of a sample express its emission color and is represented in (x, y) pair. This was evaluated using ColorCalculator 7.49 software. The obtained color coordinate (0.1951, 0.3151) corresponds to the bluish-green region in the CIE chromaticity diagram (Figure 10.5). Color purity (p) is the ratio of the distance between the given color coordinates (x, y) and the white point (0.33, 0.33) to the distance of the white point from the dominant wavelength point (x_d, y_d) [25]

$$p = \frac{\sqrt{(x-0.33)^2 + (y-0.33)^2}}{\sqrt{(x_d-0.33)^2 + (y_d-0.33)^2}}$$

(10.14)

The color purity obtained for PVDF–TiO$_2$:Pr^{3+} is 47%.

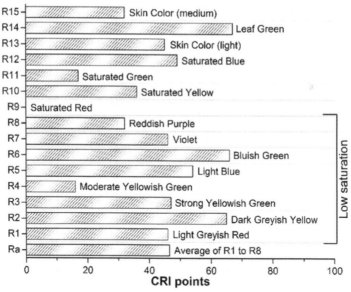

FIGURE 10.6 CRI points of PVDF–TiO$_2$:Pr^{3+}.

The color rendering index (CRI) is the measure of the ability of a source to render color compared to a standard light source [29]. Figure 10.6 shows the CRI rating map of PVDF–TiO$_2$:Pr^{3+}. Since the value of R9 (red index) is a negative one (–96), it is not shown in the figure. A general CRI, R_a is the average value of the emitting source from $R1$ to $R8$ in the CRI rating map. The value of R_a is found to be 47 for PVDF–TiO$_2$:Pr^{3+}.

10.4   CONCLUSION

Pr^{3+}-doped PVDF–TiO$_2$ hybrid was prepared by a nonhydrolytic sol–gel method. The absorption spectrum of the sample showed eight characteristic peaks of Pr^{3+}. The bonding parameter and higher Ω_2 value, calculated from the observed bands in the absorption spectrum, suggest that the nature of Pr^{3+}–ligand bond in PVDF–TiO$_2$ hybrid is covalent. The JO intensity parameters calculated follow the trend $\Omega_2 > \Omega_6 > \Omega_4$. The photoluminescence spectrum upon the excitation wavelength of 446 nm showed three emission peaks at wavelengths 485, 528, and 595 nm due to the transitions $^3P_0 \rightarrow {}^3H_4$, $^3P_1 \rightarrow {}^3H_5$, and $^1D_2 \rightarrow {}^3H_4$, respectively. The color coordinates $(0.1951, 0.3151)$ revealed the emission of PVDF–TiO$_2$:Pr^{3+} in the bluish-green region. The results suggest that PVDF–TiO$_2$:Pr^{3+} is amenable for use in optical devices.

ACKNOWLEDGMENT

The authors are thankful to the Department of Science and Technology (DST), Government of India for the financial support through its Promotion of University Research and Scientific Excellence (PURSE) program—Phase II.

KEYWORDS

- **polymer–inorganic hybrid**
- **sol–gel method**
- **PVDF**
- **TiO$_2$**
- **Judd–Ofelt**
- **analysis**
- **praseodymium**

REFERENCES

1. Schottner, G. Hybrid sol-gel-derived polymers: applications of multifunctional materials. Chemical Materials. 2001, Vol. 13, 3422–3435.

2. Yu, L., Shen, H., Xu, Z. PVDF–TiO$_2$ composite hollow fiber ultrafiltration membranes prepared by TiO$_2$ sol–gel method and blending method. Journal of Applied Political Science. 2009, Vol. 113, 1763–1772.

3. Bandyopadhyay, A., Sarkar M.D.E., Bhowmick A.K. Poly (vinyl alcohol)/silica hybrid nanocomposites by sol-gel technique: synthesis and properties. Journal of Material Sciences. 2005, Vol. 40, 5233–5241.

4. Zelazowska, E., Rysiakiewicz-pasek, E., Borczuch-laczka, M.,Cholewa-kowalska, K. Sol-gel derived hybrid materials doped with rare earth metal ions. Optical Materials. 2011, Vol. 33, 1931–1937.

5. George, R., Thomas, S., Joseph ,C., Biju, P.R., Unnikrishnan, N.V., Sudarsanakumar, C. Spectroscopic properties of Sm^{3+} -doped PVDF-ZrO2 hybrid membrane. Materials Today: Proceedings. 2020, In-press, https://doi.org/10.1016/j.matpr.2019.12.248

6. Dridi, C., Haouari, M., Ouada, H.B., Legrand, A.P., Davenas, J., Bernard, M., Andre, J.J., Said, A.H., Mattoussi, F. Spectroscopic investigations on hybrid nanocomposites : CdS:Mn nanocrystals in a conjugated polymer. Material Science and Engineering C. 2006, Vol. 26, 415–420.

7. Tanaka, K., Capule, M.F.V., Hisanaga, T. Effect of crystallinity of TiO$_2$ on its photocatalytic action. Chemical Physics Letters. 1991, Vol. 187, 2–5.

8. Cámara, R.M., Crespo, E., Portela, R., Suárez, S., Bautista, L., Gutiérrez-Martín, F., Sánchez, B. Enhanced photocatalytic activity of TiO$_2$ thin films on plasma-pretreated organic polymers. Catalysis Today. 2014, Vol. 230, 145–151.

9. Jing, F., Harako, S., Komuro, S., Zhao, X. Luminescence properties of Sm^{3+}-doped TiO$_2$ thin films prepared by laser ablation. Journal of Physics D: Applied Physics. 2009, Vol. 42, 085109–085116.

10. Augustynski, J.A.N. The role of the surface intermediates in the photoelectrochemical bahaviour of anatase and rutile TiO$_2$. Electrochemica Acta. 1993, Vol. 38, 43–46.

11. Liu, F., Hashim, N.A., Liu, Y., Abed, M.R.M., Li, K. Progress in the production and modification of PVDF membranes. Journal of Membrane Science. 2011, Vol. 375, 1–27.

12. Liu, J., Yang, B., Liu, J. Development of environmental-friendly BZT–BCT/P(VDF–TrFE) composite film for piezoelectric generator. Journal of Materials Science: Materials in Electronics. 2018, Vol. 29, 17764–17770.

13. Itankar, S.G., Dandekar, M.P., Kondawar, S.B., Bahirwar, B.M. Comparative photoluminescent study of PVDF/Eu^{3+} and PEO/Eu^{3+} electrospun nanofibers in photonic fabric. AIP Conference Proceedings. 2019, Vol. 2104, 020032-8.

14. Pogreb, R., Whyman, G., Musin, A., Stanevsky, O., Bormashenko, Y., Sternklar, S., Bormashenko, E. The effect of controlled stretch on luminescence of Eu(III) (NO$_3$)$_3$ (o-Phen)$_2$ complex doped into PVDF film. Material Letters. 2006, Vol. 60, 1911–1914.

15. Jose, S.K., Gopi, S., Thomas, V., Sreeja, E., Joseph, C., Unnikrishnan, N.V., Biju, P.R. Optical characterization and Judd-Ofelt analysis of Pr^{3+} ions in sol-gel derived zirconia/polyethylene glycol composite. Optical Materials. 2018, Vol. 76, 184–190.

16. Vidyadharan, V., Gopi, S., Mohan, P.R., Thomas, V., Joseph, C., Unnikrishnan, N.V., Biju, P.R. Judd-ofelt analysis of Pr^{3+} in Sr$_{1.5}$Ca$_{0.5}$SiO$_4$ and Sr$_{0.5}$Ca$_{0.5}$TiO$_3$ host matrices. Optical Materials. 2016, Vol. 51, 62–69.

17. Carnall, W.T., Fields, P.R., Waybourne, B.G. Spectral intensities of the trivalent lanthanides and actinides in solution. I. Pr^{3+}, Nd^{3+}, Er^{3+}, Tm^{3+}, and Yb^{3+}. Journal of Chemical Physics. 1965, Vol. 42, 3797–3806.

18. Pragash, R., Unnikrishnan, N.V., Sudarsanakumar, C. Spectroscopic properties of Pr^{3+} -doped erbium oxalate crystals. Pramana. 2011, Vol. 77, 1119–1126.

19. Florez, A. Judd-Ofelt analysis of Pr^{3+} ions in fluoroindate glasses : influence of odd third order intensity parameters. Journal of Non-Crystalline Solids. 1997, Vol. 213 & 214, 3–8.

20. Judd, B.R. Optical absorption intensities of rare-earth ions. Physics Review. 1962, Vol. 127, 750–761.

21. Ofelt, G. Intensities of crystal spectra of rare earth ions. Journal of Chemical Physics. 1962, Vol. 37, 511–520.

22. Valencia, S., Marín, J.M., Restrepo, G. Study of the bandgap of synthesized titanium dioxide nanoparticules using the sol-gel method and a hydrothermal treatment. Open Material Science Journal. 2010, Vol. 4, 9–14.

23. Romero-Sáez, M., Jaramillo, L.Y., Saravanan, R., Benito, N., Pabón, E., Mosquera, E., Gracia, F. Notable photocatalytic activity of TiO_2-polyethylene nanocomposites for visible light degradation of organic pollutants, Express Polymer Letters. 2017, Vol. 11, 899–909.

24. Abdelaziz, M. Investigations on optical and dielectric properties of PVDF/ PMMA blend doped with mixed samarium and nickel chlorides. Journal of Material Sciences. 2013, Vol. 24, 2727–2736.

25. Thomas, S., George, R., Rasool, S.N., Rathiah, M., Venkataramu, V., Joseph, C., Unnikrishnan, N.V. Optical properties of Sm^{3+} ions in zinc potassium fluorophosphate glasses. Optical Materials. 2013, Vol. 36, 242–250.

26. Sekhar, K.C., Hameed, A., Ramadevudu, G., Chary, M.N., Shareefuddin, M. Physical and spectroscopic studies on manganese ions in lead halo borate glasses, Modern Physics Letters B. 2017, Vol. 1750180, 1–13.

27. Pawar, P.P., Munishwar, S.R., Gedam, R.S. Physical and optical properties of Dy^{3+}. Pr^{3+} co-doped lithium borate glasses for W-LED. Journal of Alloys and Compounds. 2016, Vol. 660, 347–355.

28. Görler-Walrand, C., Binnemans, K. Handbook on the Physics and Chemistry of Rare Earths, Vol. 25. (Gschneidner, K.A., Eyring, L., eds.). New York: Elsevier–Publishing Inc; 1998.

29. Chen, L., Cheng, W., Tsai, C., Chang, J., Huang, Y., Huang, J., Cheng, W. Novel broadband glass phosphors for high CRI WLEDs. Optics Express. 2014, Vol. 22, 1040–1047.

CHAPTER 11

Synthesis and Optical Properties of Green Carbon Dots

MAMATHA SUSAN PUNNOOSE AND BEENA MATHEW*

School of Chemical Sciences, Mahatma Gandhi University, Kottayam, Kerala, India

Corresponding author. E-mail: beenamscs@gmail.com

ABSTRACT

Carbon dots (CDs) are a new class of the nanocarbon materials displaying a wide range of significant advantages such as low toxicity, chemical inertness, good water solubility, and physicochemical properties. They are very small particles often below 10 nm in size. CDs are of much interest in fields of nanomedicine, optoelectronics, biomedical, energy storage, and sensing. These are attributed to their excellent optical properties especially the intense and tunable luminescence and the easiness to modify their photoluminescence by means of functionalization and doping. Global energy issues have led the utilization of natural or renewable raw materials for the synthesis of CDs in order to minimize environmental pollution and toxicity. This chapter demonstrates the use of various natural sources as carbon precursor for the synthesis of CDs. The structural and optical properties of green CDs are also discussed.

11.1 INTRODUCTION

Carbon dots (CDs) are a new class of carbon nanomaterials having a size less than 10 nm, which is mainly composed of sp^2 hybridized carbon atoms [1]. They have promising applications than any other nanocarbon materials like carbon nanotubes, fullerenes, graphenes, and nanodiamonds. In recent years, the CDs are gaining attraction in the fields of sensors, optics, electronics, catalysis, energy, drug delivery, and cell imaging [2–5]. The outstanding water solubility, low cell toxicity, high biocompatibility, photostability, easy functionalization, and efficient optical properties of CDs have contributed to tremendous attracting applications [6, 7]. These properties also contributed well toward the high fluorescence of CDs compared to fluorescent organic dyes and semiconductor quantum dots.

The CDs can be synthesized by using top-down and bottom-up strategies. "Top-down" approach involves methods in which breaking down of larger mass carbon materials into individual nanoparticles occur. Several top-down methods include laser ablation, oxidation, arc discharge, electrochemical, and plasma treatment [8–10]. Graphite, multiwalled carbon nanotubes, candle soot, activated carbon, and lampblack are the main sources for the synthesis. Though these methods were successfully used for the preparation of CDs, they have certain drawbacks including high cost, time consumption, sophisticated instrumentation, high temperature, and reaction conditions like large amounts of toxic chemicals, and sometimes low product yield [11].

The "bottom-up" strategic methods like hydrothermal treatment, pyrolysis, and microwave synthesis are currently receiving growing attention. In view of the "green chemistry" approach, the CDs were successfully generated using renewable resources as raw materials, which avoids the use of toxic chemicals and complex post-treatment processes. The green CDs synthesized via green strategies is an inexpensive and ecofriendly approach for the large scale production. The raw materials and reaction conditions play an important role in carbonation and surface passivation during the formation of CDs. Subsequently, they contribute well toward the chemical structure and optical properties of CDs.

From the reported works of literature, it is found that hydrothermal treatment is more efficient while using biomasses such as fruits, vegetables, and plants as the carbon source. Hydrothermal treatment involves simple operation, easy control of the reaction, efficient and low energy consumption [12]. Hence, this method is widely adopted for the large scale production of reproducible and stable CDs of narrow size distribution with good quantum yields (QY). It can provide continuous heating energy even up to a high temperature that benefits the carbonation of bio-resource. In the microwave technique, electromagnetic energy is used for the production of heating effect. This process offers energy-efficient, rapid, homogenous, and uniform heating of the reaction medium, which shortens the reaction time and leads to the formation of uniform sized particles [13, 14]. This method has some limitations in terms of lack of control over irradiation power and absorption properties of the renewable precursors that may result in poor reproducibility. Pyrolysis method involves the carbonization of raw materials at higher temperature. This thermal treatment includes dehydration, polymerization, and carbonization that eventually leads to the formation of CDs. This method is highly advantageous due to the easy mode of operation and a short reaction time.

This chapter includes a few "bottom-up" methods leading to the synthesis of CDs through quite simple, facile, cost-effective, and eco-friendly approach by utilization of different green biomasses. It also deals with the structural and optical properties of the green synthesized CDs.

11.2   SYNTHESIS OF CARBON DOTS USING DIFFERENT GREEN PRECURSORS

11.2.1   FRUITS AS CARBON PRECURSOR

The resultant CDs from the hydrothermal carbonization of orange juice at 120 °C for 150 min can be easily separated into monodispersed particles of high fluorescence and coarse particles of less fluorescence by simply controlling the centrifugation speed. The monodispersed spherical CDs show an average size distribution between 1.5 and 4.5 nm with a maximum population of about 2.5 nm [15]. Apple juice [16] and pear juice [17] derived CDs by hydrothermal carbonization at 150 °C for 12 and 2 h, respectively, showed an average size of 4.5 and 10 nm. Simple heating of banana juice in an oven at 150 °C for 4 h resulted in the CDs formation. Plausible mechanism of synthesis involves hydrolysis, dehydration, and decomposition of different carbohydrates like glucose, fructose, and sucrose present in the carbon source resulted in the formation of soluble furfural aldehydes, ketones, and various organic acids. Polymerization, aromatization, and carbonization of these compounds at a supersaturation point led to a nuclear burst that yielded water-soluble CDs (Figure 11.1) [18]. The presence of ascorbic acid in banana makes the reaction to be complete within a short time. Transmission electron microscopy (TEM) studies show that the average size of CDs to be 3 nm. EDX analysis shows the minor presence of potassium from the banana source in addition to the carbon and oxygen signals.

Water dispersible CDs of QY 5.76% were synthesized via hydrothermal method using *Saccharum officinarum* juice as the precursor [19]. The bright blue fluorescent CDs show a mean diameter of 3 nm. Carvalho et al. reported the facile synthesis of CDs using acerola fruit [20]. Ultrasmall CDs are obtained from *Acacia concinna* seeds by microwave method [21]. The seeds are mixed with water and methanol in 1:1 ratio and subjected to microwave irradiation at 800 W for 2 min. After the reaction crude product was filtered through 0.45 μm syringe filter paper and dialyzed in milli-Q-purified water. TEM analysis prove the average size of synthesized ultrasmall particles to be 2.5 ± 0.7 nm. Spherical CDs of good dispersibility are obtained by the hydrothermal treatment of *Pyrus pyrifolia* and *Lycii fructus* fruits [22, 23]. They show an average size of 2.0 ± 1.0 and 3.3 nm respectively. AFM studies prove the particle size to be 5 nm for *Cydonia oblonga*-derived CDs, produced via microwave synthesis [24].

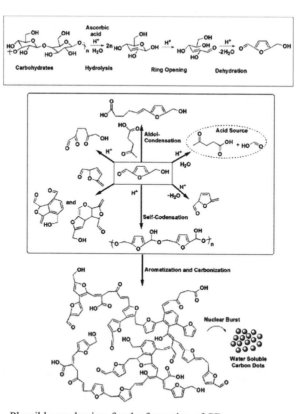

FIGURE 11.1   Plausible mechanism for the formation of CDs.

Source: Reprinted with permission from De et al. [18]. © 2011. The Royal Society of Chemistry.

The doping of nitrogen, phosphorus, boron, and sulfur on CDs have gained much attention due to their remarkable enhancement in QY and fluorescence ability. Aqueous ammonia is used as the nitrogen dopant in spherical N-CDs derived from *Actinidia deliciosa* commonly known as kiwi fruit [25]. Average size is measured to be 3.59 nm. The 74.59% carbon, 6.88% nitrogen, and 18.53% oxygen content were estimated from EDS analysis. Pomegranate juice [26] acted as the carbon source, while L-cysteine provided nitrogen and sulfur for the hydrothermal synthesis of S/N-CDs. The resultant solution was centrifuged, filtered, and freeze-dried to obtain co-doped CDs. They found application in the detection of cephalexin in human urine and raw milk samples. Ethylenediamine acts as the nitrogen precursor for the hydrothermal carbonization of natural peach gum polysaccharide [27]. The resultant N-CDs show an increase in QY from 5.31% to 28.46% by the surface passivation. The well dispersed spherical CDs show an average size in the 2–5 nm range.

For the hydrothermal carbonization synthesis of N-CDs using *Hylocereus undatus* and *Chionanthus retusus* fruits, aqueous ammonia is used as the nitrogen precursor [28, 29]. Average diameter size was 2.86 nm and 5 ± 2 nm, respectively. Both N-CDs show applications in biology. The *Prunus avium* and *Prunus mume*-derived N-CDs show an average size of 7 and 9 nm, respectively [30, 31]. The phytochemical analysis proves the presence of various organic acids, phenolic compounds, and sugars in the fruit extracts, which act as a good source of carbon. N-CDs synthesized with lemon juice using L-arginine as the nitrogen source show a narrow size distribution between 2.5 and 4.5 nm with a mean population of 3.5 nm [32]. N-CDs derived from *Phyllanthus emblica* [33] show catalytic ability toward the detoxification of a mixture of textile effluents using $NaBH_4$. Reduction of the dyes by the N-CDs can be explained using Langmuir–Hinshelwood mechanism [34]. From the HR-TEM the mean size of the particles is found to be 4.08 nm. EDX analysis shows the presence of 8.65 % nitrogen, 73.45% carbon, and 17.90% oxygen.

11.2.2 VEGETABLES AS CARBON PRECURSOR

Ascorbic acid, carotenoids, carbohydrate, and other carbonaceous organic matters present in sweet pepper [35] make it an important green source for CDs synthesis. A QY of 19.3% was obtained from the hydrothermal reaction of pepper and water at 180 °C in 4: 1 ratio for 5 h. A possible mechanism for

CDs formation involves the aromatization and carbonization of carbonaceous organic matter. Nanoparticle size histogram shows the average diameter of the synthesized CDs to be 4.6 nm, which is quite similar to the ones obtained from ascorbic acid [36]. The lattice parameter of 0.20 nm is attributed to the diffraction planes of sp^2 carbon.

Sweet-potato-derived CDs via hydrothermal method show a QY of 8.64% and were successfully used for ferric ion sensing in HeLa and HepG2 living cells [37]. *Hongcaitai*-derived-CDs also shows cytotoxicity toward HepG2 cells [38]. They are obtained by the hydrothermal heating in a reaction kettle at 240 °C for 20 h and were divided into parts. One part is soluble in both ethanol and water, whereas the other part is soluble in water only. Both parts showed successful cell imaging.

Well separated spherical CDs of monodispersed nature having an average diameter in the range of 2–6 nm was obtained by the hydrothermal treatment of Broccoli juice [39]. These CDs act as selective sensing tool for Ag^+ ions with a LOD of 0.5 μM. Chelate complex formation via the energy transfer between metal ions and CDs surface leads to strong fluorescence quenching. Vasimalai et al. [40] reported the average diameter of the CDs from the Cinnamon, red chili, turmeric, and black pepper sources to be 3.4 ± 0.5, 3.1 ± 0.2, 4.3 ± 0.5 and 3.5 ± 0.1 nm accordingly. The one-pot green hydrothermal synthesis involves pyrolysis, carbonization, and passivation [41]. TEM results give the average size of CDs to be 3.37, 3.14, 4.32, and 3.55 nm, respectively. All these CDs show strong potential for bioanalytical and clinical applications.

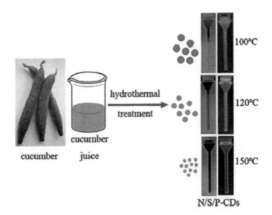

FIGURE 11.2   Formation of N/S/P-CDs by hydrothermal treatment of cucumber juice.

Source: Reprinted with permission from Wang et al. [49]. © 2011 The Royal Society of Chemistry.

Self-passivated N-CDs of 38.8% QY were synthesized using black soya beans [42] by means of pyrolysis method. It involves heating of soya beans in a crucible for 4 h at 200 °C in a muffle furnace. An average diameter of 5.16 ± 0.30 nm was calculated from TEM analysis. Moreover, they show dual responsive properties towards Fe^{3+} and free radicals. The N-CDs of natural green pak choi precursor [43] have an average size of 1.8 nm, show high stability toward various pH and various concentrations of NaCl. Li et al. reported the synthesis of N-CDs from different sources like Chinese yam [44], ginger [45] and rose-heart radish [46]. The QY was found to be 9.3%, 13.4%, and 13.6%, respectively.

Heteroatom doped S/N-CDs were first reported by the hydrothermal heating of water chestnut and onion together in an autoclave at 180 °C [47]. In order to obtain a QY of 12%, ratio of water chestnut and onion was adjusted to 2:3. Hydrothermal treatment of garlic at 200 °C for 3 h resulted in blue fluorescent S/N-CDs of 17.5% QY [48]. Garlic cloves act as the sole source for carbon, nitrogen, and sulfur. TEM images prove the average size to be 11 nm.

Wang et al. reported the synthesis of N/S/P-CDs by one-pot hydrothermal treatment of cucumber juice without any additives like salts, acids, or alkalies [49]. The synthesis was carried at three different temperatures 100, 120, and 150 °C that correspondingly emit with blue, greenish-blue, and green colors, respectively (Figure 11.2). This reveals the temperature dependence on the distribution of emission wavelength. The difference in emission color may be attributed to the different size, shape, and defects of N/S/P-CDs. TEM analysis supports the different morphologies of the synthesized CDs. A decrease in the fluorescence intensity of N/S/P-CDs at 120 °C was observed upon addition of Hg^{2+}ions. The strong affinity of mercuric ions toward the carboxylic group on N/S/P-CDs surface may result in the formation of complexes by means of coordination.

11.2.3   PLANT PARTS AS CARBON PRECURSOR

Hydrothermal treatment of coriander leaves [50] resulted in CDs formation showing bright green fluorescence under UV light. Elemental analysis of CDs show the presence of 50.8% carbon, 5.3% hydrogen, 4.07% nitrogen, and 39.83% oxygen, which proves the presence of higher content of carbon and oxygen-containing functional groups in the precursor. The average diameter was found to be 2.387 nm within the size distribution from 1.5 to 2.98 nm. *Ocimum sanctum* leaves derived CDs show a QY of 9.3% [51]. The hydrothermal treatment was optimized at 180 °C for 4 h, which resulted in particles having 3 nm in diameter. The amorphous nature of the CDs is indicated by the broad XRD peak at $2\theta = 25°$. Selective detection of tartrazine in food can be done using CDs derived from freshwater *aloe* by means of the hydrothermal method [52]. After heating, solution is filtered through a 0.22 μm membrane followed by washing with dichloromethane solvent in order to remove the unreacted organic moieties. These bright yellow emitting spherical CDs having an average size of 5 nm show a QY of 10.37%.

The elemental composition of Tulsi-leaves-derived CDs show 41.43% carbon, 4.81% hydrogen, 4.83% nitrogen, and 48.92% oxygen [53]. High percentage of carbon and oxygen indicates the presence of a large number of carbon and oxygen functional groups. They are of spherical morphology and have an average size of 5 nm. Hydrothermal treatment of *Jinhua bergamot* plant as the carbon source produced CDs of 10 nm size and 50.78% QY [54]. These act as a successful fluorescent probe for both Fe^{3+} and Hg^{2+} ions. Hydrothermal carbonization of willow bark [55] at 200 °C resulted in CDs with a diameter range of 1–4 nm and having a QY of 6%. These CDs act as effective photocatalyst for fabricating gold nanoparticles/reduced graphene oxide nanocomposites for electrochemical detection of glucose. The detection limit was found to be 45 μM.

Eutrophic algal blooms act as the carbon source for the microwave-assisted synthesis of CDs. [56]. The average diameter is estimated to be 8.5 ± 5.6 nm. The I_D/I_G value of 1.91 implies the amorphous nature of CDs with plenty of structural defects. Zhu et al. reported the CDs synthesis through pyrolysis of different leaves of oriental plane, lotus, and pine needles [57]. The respective QY obtained were 16.4%, 15.3%, and 11.8%. Out of the different temperatures used, pyrolysis at 350 °C was found to be the optimum. At low temperature incomplete carbonization occurred, whereas high temperature caused over oxidation of CDs surface.

Magnolia liliiflora [58] flower upon hydrothermal treatment resulted in the formation of spherical N-CDs of average size 4 ± 1 nm. The highly fluorescent N-CDs show high selectivity toward Fe^{3+} ions. CDs are derived from rose petals by means of microwave irradiation [59]. Dried flowers of *Osmanthus fragrans* Lour by means of the hydrothermal method produce CDs of QY 18.53% [60]. After thermal treatment at optimized conditions of 240 °C for 5 h solution was cooled to room temperature, centrifuged, and filtered through a 0.22 μm membrane filter. TEM image shows an average particle size of 2.23 nm. The crystallinity of the formed CDs was evident from the XRD analysis.

Polyethylenimine capped CDs are synthesized via hydrothermal carbonization of bamboo leaves [61] for 6 h at 200 °C. The average particle size of the capped CDs was 3.6 nm and had a QY of about 7.1%. Interlayer spacing of 0.32 nm proves the amorphous nature of *Tamarindus indica* leaves derived CDs [62]. It showed a QY of 46.6% and an average size of 3.4 ± 0.5 nm. L-cysteine, ethylene diamine and glycine functionalized *Coccinia indica* derived CDs show high selectivity toward Cu^{2+}, Pb^{2+}, and Fe^{3+} [63]. The corresponding detection limits were 0.045, 0.27, and 6.2 µM, respectively. TEM analysis proved the average size of CDs to be 8.9, 7.03, and 8.4 nm, respectively. Self-passivated CDs having an average size of 5.2 nm were used for the successful detection of Hg^{2+}. Microwave treatment of lotus root [64] without any surface passivating agents resulted in the formation of N-CDs. The average diameter is found to be 9.41 nm and has a nitrogen content of 5.23%.

11.2.4   WASTE MATERIALS AS CARBON PRECURSOR

Refluxing of *Trapa bispinosa* [65] peel extract at 90 °C for 2 h, followed by centrifugation and suspension in NaOH resulted in the formation of fluorescent CDs of average size ranging from 5 to 10 nm. A total of 7.1% QY was obtained for CDs derived from watermelon peel [66]. Resultant CDs from the hydrothermal treatment of waste tea extract show selective detection toward Fe^{3+}, CrO_4^{2-}, ascorbic acid, and L-cysteine in real samples [67]. CDs obtained from peanut shells [68] via pyrolysis approach showed a QY of 9.91%. The average size is found to be 1.62 nm that is much smaller than the CDs obtained from peanut skin [69]

The composite of ZnO with CDs derived from orange waste peels [70] is used as a photocatalyst for the degradation of naphthol blue-black dye (Figure 11.3). CDs synthesis involve the collection of peels and drying in a hot air oven at 150 °C for 10 h. It is then washed with aqueous H_2SO_4, filtered and dried in an oven for 150 °C for 2 h. Orange peels were then oxidized using sodium hypochlorite solution and washed with water until neutral pH is reached. It is then subjected to hydrothermal carbonization at 180 °C for 12 h. The obtained brown solution was filtered, washed with dichloromethane, centrifuged and dried at 100 °C for 2 h. The formed homogeneously dispersed stable water-soluble CDs show an average diameter in the range 2–7 nm.

FIGURE 11.3   Schematic representation for the formation of CDs from orange waste peels.
Source: Reprinted with permission from Prasannan et al. [70]. © 2013 American Chemical Society.

Water-soluble photoluminescent carbon dots are obtained by the hydrothermal carbonization of wheat straw [71]. DLS measurement showed an average hydrodynamic size of 1.7 ± 0.23 nm. Hydrothermal approach including ultrasonic-assisted chemical oxidation of waste polyolefins residue resulted in CDs of the average diameter of 2.5 nm [72]. The average size was in the range of 1.5–3.5 nm. The EDX analysis proves 88.79% and 11.21% of carbon and oxygen, respectively. And the amorphous nature of CDs is suggested by both XRD and SAED pattern. Room temperature ultrasound irradiation of food waste resulted in green CDs of uniform spherical shape with a diameter in the range of 1–7 nm [73]. AFM studies prove the size to be 4.6 nm.

Spherical CDs of lemon peel waste have an average diameter between 1 and 3 nm [74]. After the hydrothermal reaction, CDs sample is cooled, washed with dichloromethane, centrifuged, and dried at 100 °C. Conversion of carbonaceous organic materials in lemon peels to CDs includes dehydration, fragmentation, condensation, aromatization, and carbonization processes (Figure 11.4). CDs are mixed with TiO_2 nanofibers to form composites having photocatalytic efficiency in the degradation of methylene blue dye.

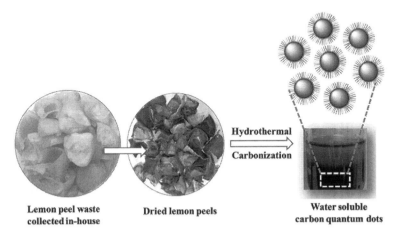

FIGURE 11.4 Illustration for the formation of CDs from the hydrothermal treatment of lemon peel waste.

Source: Reprinted with permission from Tyagi et al. [74]. © 2014 The Royal Society of Chemistry.

Thermal pyrolysis of sago waste [75] at different temperatures from 250 °C to 450 °C. It was noted that beyond 300 °C complete carbonization of the sample occurred. Heating beyond 400 °C may cause complete decomposition of the precursor into ashes instead of carbon-rich residues. The optimum temperature of carbonization was determined to be 400 °C. Uniform spherical CDs of particle size in the range 37–66 nm were obtained at the optimum temperature show the strongest fluorescence emission at 390 nm for an optimum excitation wavelength of 315 nm. The waste frying oil is used as the precursor for amorphous S-CDs [76]. These uniformly sized particles, of 2.6 nm, show a linear luminescence in the pH range between 3 and 9. The obtained QY was 3.66%.

11.2.5 ANIMAL DERIVATIVES AS CARBON PRECURSOR

Monodisperse CDs of average diameter 4 nm were obtained by the hydrothermal carbonization of prawn shells [77]. Dried shells after acid treatment were washed with water till neutral pH is attained. It is then refluxed with NaOH, then cooled and filtered through Whatman filter paper. The obtained product is dispersed in acetic acid and kept in hydrothermal treatment at 200 °C for 8 h. These CDs show selective sensitivity toward Cu^{2+} ions with a very low detection limit of 5 nM. A QY of 54% was obtained for the Indian-prawn-derived CDs [78]. The average size was found to be 6 nm.

Well dispersed egg white [79] derived CDs of diameter 2.1 nm show excellent pH stability. The hydrothermal treatment of egg white resulted in yellow CDs suspension of QY 61%. Proteins in the egg white were hydrolyzed to small low-molecular-weight peptides and amino acids. Hydrothermal treatment resulted in polymerization of amino acids and was then carbonized into CD cores. Honey-derived [80] CDs show sensing toward Fe^{3+} ions. They show blue fluorescence and are of 2 nm in size. Fluorescent intensities of these exhibited no variation upon the change in concentrations of NaCl, pH, and time.

Fragrant CDs of size 1–2 nm were synthesized through the hydrothermal method from three different bee pollens of rapeseed, lotus, and camellia flowers [81]. Depending on the species of bee pollen and reaction time, the QY varied between 6.1% and 12.8%. Well-dispersed spherical CDs were obtained by one-step microwave-assisted pyrolysis of wool [82]. TEM analysis shows that they are of 2.8 nm.

Microwave-assisted pyrolysis of chitosan, acetic acid, and 1,2-ethylenediamine resulted in N-CDs [83]. Both CDs and N-CDs are of spherical morphology with average diameters of 2.48 ± 0.40 nm and 4.27 ± 0.50 nm, respectively. Milk-derived N-CDs [84] show uniform dispersion of particles in the narrow range of 2–4 nm with a maximum population of 3 nm. Its cell cytotoxicity experiments were studied using human brain glioma tumor (U87) cell lines.

11.2.6   FOOD ITEMS AS CARBON PRECURSOR

Gd-CDs were derived from sucrose by making use of microwave-assisted polyol synthesis method [85]. A 30% sucrose solution is mixed with con. H_2SO_4 and $GdCl_3$ and was added to diethylene glycol (DEG). The mixture was sonicated and then heated in a domestic microwave oven. Formation of dark brownish color indicated the formation of CDs. It was then centrifuged and filtered. The doped Gd disorders the carbon rings and produces more emission energy traps is the plausible reason for the increase in QY of doped

CDs than bare CDs. TEM studies indicated that the spherical morphology of the well-separated CDs shows an average size of 5 nm.

Dried samples of *Eleusine coracana* upon pyrolysis produced spherical CDs of bright blue luminescence [86]. The size distribution is found to be in the range of 3–8 nm, with abundance maximum of 6 nm. Graphitic nature of produced CDs was evident from the XRD and SAED patterns. Corn flour-derived [87]. CDs show monodispersity with an average size of 3.5 nm.

The quasi-spherical N-CDs with an average diameter of 3.37 and 5.8 nm are produced using Konjac flour [88] and barley [89] as sources. N-CDs are prepared from konjac flour by means of pyrolysis at 470 °C for 1.5 h. The resultant carbonized product is mixed with ethanol and filtered. The solution was filtered, dried, dispersed in water and then filtered through a 0.22 μm filter membrane. Nitrogen content was estimated to be 7.03%. Ethylenediamine is used as the nitrogen source for N-CDs produced from Highland barley. An aqueous solution of N-CDs under UV light of 254 nm produces a blue emission at 480 nm. Table 11.1 provides an overview on the various methods for the green synthesis of carbon dots along with their respective applications. Table 11.2 provides the various precursors used for the hydrothermal preparation of green carbon dots.

TABLE 11.1   Overview of Different Green CDs: Method of Synthesis and Their Applications

Precursors	Method	QY (%)	Applications	References
Food waste	Ultrasound	2.85	Cytotoxicity, cell imaging, plant seed germination	73
Waste frying oil	Carbonization	23.2	pH sensing, bioimaging	76
Rose petals Microwave		13.5	Detection of tetracycline	59
Lotus root Microwave		19	Detection of Hg^{2+}, cell imaging	64
Cydonia oblonga	Microwave	8.55	Detection of As^{3+}, cell imaging of HT-29	24
Acacia concinna	Microwave	10.20	Detection of Cu^{2+}, bioimaging of fungal cells	21
Algal blooms	Microwave	13	In vitro imaging	56
Watermelon peel	Microwave	7.1	Bioimaging	66

TABLE 11.1   *(Continued)*

Precursors	Method	QY (%)	Applications	References
Sago waste Pyrolysis		–	Detection of Cu^{2+} and Pb^{2+}	75
Peanut shells Pyrolysis		9.91	Multi colour imaging of HepG2 cells	68
Eleusine coracana Pyrolysis		–	Detection of Cu^{2+}	86
Konjac flour	Pyrolysis	22	Detection of Fe^{3+}, cell imaging	88
Black soya beans Pyrolysis		38.8	Detection of Fe^{3+}, radical scavenging, cell imaging	42
Different leaves Pyrolysis		–	Detection of Fe^{3+}	57
Wool	Microwave-assisted pyrolysis	16.3	Detection of glyphosate	
Chitosan	Microwave-assisted pyrolysis	20.1	Detection of Fe^{3+}, cellular imaging	83
Sucrose	Microwave-assisted polyol	5.4	Cell labeling	85

TABLE 11.2   Hydrothermal Method for the Synthesis of Green CDs Using Various Green Precursors

Precursor	Procedure	QY (%)	Applications	References
Orange juice	120 °C, 1.5 h	26	Bioimaging	15
Apple juice	150 °C, 12 h	4.27	Cell imaging	16
Pear juice	150 °C, 2 h	–	Detection of copper	17
Banana juice	150 °C, 4 h	8.95	–	18
Lemon juice	200 °C, 3 h	7.7	Detection of Cu^{2+}	32
Cucumber juice	100–150 °C, 6 h	–	Detection of Hg^{2+}	49
Brocoli juice	190 °C, 6 h	–	Detection of Ag^+	39
Pomengranate	120°C, 5 h	4.8	Detection of cephalexin	26
Saccharum officinarum	120 °C, 3 h	5.76	Cell imaging	19

TABLE 11.2 *(Continued)*

Precursor	Procedure	QY (%)	Applications	References
Phyllanthus emblica	180 °C, 12 h	–	Catalytic detoxification of textile effluents	33
Peach gum	180 °C, 16 h	28.46	Detection of Au^{3+}	27
Acerola fruit	180 °C, 18 h	8.64	Detection of Fe^{3+}	20
Pyrus pyrifolia	180 °C, 6 h	10.8	Detection of Al^{3+}, cell imaging	22
Lycii Fructus	200 °C, 5 h	17.2	Detection of Fe^{3+}, multicolor cell imaging	23
Prunus avium	180 °C, 5 h	13	Detection of Fe^{3+}	30
Prunus mume	180 °C, 5 h	16	Cell imaging	31
Chionanthus retusus	180 °C, 6 h	9	Detection of Fe^{3+}, cytotoxicity	29
Green pakchoi	150 °C, 12 h	37.5	Detection of Cu^{2+}	43
Sweet potato	180 °C, 18 h	8.64	Detection of Fe^{3+}, cell imaging	37
Ocimum sanctum	180 °C, 4h	9.3	Detection of Pb^{2+}	51
Coriander	240 °C, 4 h	6.48	Detection of Fe^{3+}	50
Aloe	180 °C, 11 h	10.37	Detection of tartrazine in food samples	52
Tulsi	200 °C, 4 h	3.06	Detection of Cr^{6+}, bioimaging	53
Coccinia indica	180 °C, 7 h	0.54	Detection of Hg^{2+}	63
Tamarindus indica	210 °C, 5 h	46.6	Detection of glutathione and Hg^{2+}	62
Jinhua bergamot	200 °C, 5 h	50.78	Detection of Hg^{2+}, Fe^{3+}	54
Cinnamon	200 °C, 12 h	35.7	Bioimaging, cytotoxicity	40
Red chili	200 °C, 12 h	26.8	Bioimaging, cytotoxicity	40
Turmeric	200 °C, 12 h	38.3	Bioimaging, cytotoxicity	40
Black pepper	200 °C, 12 h	43.6	Bioimaging, cytotoxicity	40

TABLE 11.2   *(Continued)*

Precursor	Procedure	QY (%)	Applications	References
Sweet pepper	180 °C, 5 h	19.3	Detection of hypochlorite	35
Garlic	200 °C, 3 h	17.5	Cellular imaging, free radical scavenging	48
Rose heart radish	180 °C, 3 h	13.6	Detection of Fe^{3+}	46
Trapa bispinosa peel	90 °C, 2 h	1.2	Cytotoxicity	65
Waste polyolefins	120 °C, 12 h	4.84	Detection of Cu^{2+}, cell imaging	72
Milk	190 °C, 2 h	12	Cell imaging	84
Honey	100 °C, 2 h	19.8	Detection of Fe^{3+}, cell imaging, coding	80
Bamboo leaves	200 °C, 6 h	7.1	Detection of Cu^{2+}	61
Corn flour	180 °C, 5 h	7.7	Detection of Cu^{2+}, cell imaging	87
Barley	200 °C, 24 h	14.4	Detection of Hg^{2+}	89
Prawn shells	200 °C, 8 h	9	Detection of Cu^{2+} ions	77
Indian prawns	170 °C for 12 h	54	Bioimaging	78
Chinese yam	200 °C,	9.3	Detection of 6-mercaptopurine and Hg^{2+}	44
Ginger	300 °C, 2 h	13.4	Bioimaging	45
Bee pollen	180 °C, 24 h	6–12	Bioimaging	81
Hylocereus undatus	180 °C, 12 h	–	Cytotoxicity, catalytic degradation of methylene blue dye	28
Hongcaitai	240 °C, 20 h	21	Detection of hypochlorite and Hg^{2+}, cell imaging	38
Actinidia deliciosa	180 °C, 12 h	–	Catalytic reduction of rhodamine dye, cell cytoxicity	25
Willow bark	200 °C, 4 h	6	Forms Au/reduced grapheme oxide nanocomposites for the detection of glucose	55

TABLE 11.2 *(Continued)*

Precursor	Procedure	QY (%)	Applications	References
Magnolia liliiflora	240 °C, 12 h	11	Detection of Fe^{3+}, fluorescent ink, and multicolor imaging	58
Osmanthus fragrans	240 °C, 5 h	18.53	Detection of Fe^{3+}, ascorbic acid and cell imaging	60
Orange waste peel	180 °C, 12 h	36	Composite with ZnO used for the degradation of Naphthol blue-green azo dye	70
Lemon peel waste	200 °C, 12 h	14	Detection of Cr^{6+}, composites with TiO_2 show degradation of methylene blue	74
Wheat straw	250 °C, 10 h	20	Biolabeling, fluorescent ink, bioimaging, detection of Fe^{3+}	71
Waste tea extract	150 °C, 6 h	–	Detection of Fe^{3+}, CrO_4^{2-}, ascorbic acid and l-cysteine	67
Egg white	220 °C, 48 h	64	Living cell imaging, detection of Fe^{3+}, forms nanocomposites for thermosensitive luminescent properties	79

11.3 PROPERTIES OF GREEN CARBON DOTS

11.3.1 STRUCTURAL PROPERTIES

Depending upon the natural source used and synthetic method adopted the properties of CDs slightly varies accordingly. CDs usually consist of sp^2 hybridized crystalline carbon core. The different functional groups present on the carbon surface in addition to core carbon framework, due to the surface functionalization and heteroatom doping cause enhancement in the fluorescence. The amorphous nature of CDs are indicated by a broad peak in X-ray diffraction (XRD) pattern having a 2θ value between $20°$ and $25°$ showing an interlayer spacing (d) between 0.31 and 0.38 nm [51, 62]. The average particle diameter is found to be below 10 nm with spherical morphology. The particle size distribution, dispersibility, and morphology of synthesized green CDs are examined using TEM technique. Lemon juice-derived CDs show a broad diffraction peak at 25° in XRD with a d-spacing value of 0.32 nm and are of an average particle size of 4.6 nm (Figure 11.5) [90]

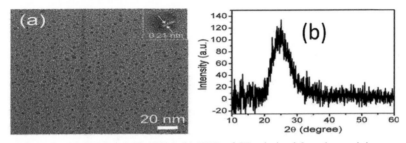

FIGURE 11.5   (a) TEM and HR-TEM, (b) XRD of CDs derived from lemon juice.

Source: Reprinted with permission from Ding et al. [90]. © 2013 The Royal Society of Chemistry.

Raman spectrum of *Prunus avium* derived N-CDs show two vibrational bands at 1360 and 1560 cm⁻¹, which corresponds to the D-band and G-band respectively (Figure 11.6) [30]. G and D bands correspond to sp² carbon and graphitic carbon atoms, respectively. The I_D/I_G ratio implies the crystallization and graphitization of the core [91]. The ratio of 0.78 implies the moderate graphitic structure of the synthesized N-CDs.

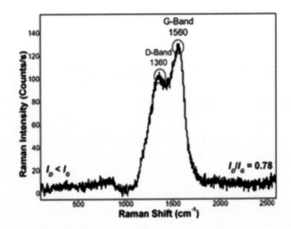

FIGURE 11.6   Raman spectra of *Prunus avium* derived N-CDs.

Source: Reprinted with permission from Edison et al. [30]. © 2016 Elsevier.

The surface functional groups and chemical composition of the green CDs are determined using X-ray photoelectron spectroscopy (XPS) and Fourier transform infrared (FT-IR) spectroscopy. C–O, C=O, and C–OH moieties may arise from the precursor molecules during the process of CDs synthesis. Some specific functional groups arise due to the heteroatom doping or surface functionalization. XPS spectrum of garlic-based CDs shows the presence of carbon, oxygen, nitrogen, and sulfur without the addition of any reagent [48]. The C/O/N/S atom ratio was found to be 72.2/19.9/6.9/1.0, respectively. The FT-IR peaks at 3427, 2940, and 1400 cm⁻¹ are attributed to the stretching vibrations of –OH, C–H, and C–N, respectively. And the absorptions at 1720, 1620, and 680 cm⁻¹ is ascribed to the existence of C=O, C=C, and C–S groups, respectively (Figure 11.7).

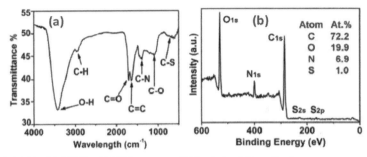

FIGURE 11.7   (a) FT-IR, (b) XPS spectra of CDs derived from Garlic.

Source: Reproduced with permission from Zhao et al. [48]. © 2015 American Chemical Society.

11.3.2   OPTICAL PROPERTIES

11.3.2.1   ABSORBANCE

CDs are generally identified by their absorbance peak in the UV–visible spectrum in 230–280 nm range with a tail extending in the visible region [33, 51]. A second peak appears around 300–340 nm range due to n–Л* transitions of C=O functional groups. The first peak arises due to the Л–Л* transitions of C=C groups [51]

11.3.2.2   TUNABLE PHOTOLUMINESCENCE (PL)

It is the most amazing optical property exhibited by CDs. They usually exhibit blue or green luminescence [43, 46, 65]. Commonly CDs display an excitation dependent emission character. With the increase of excitation wavelength from 280 to 500 nm a redshift in emission peak was observed. And the strongest emission peak was obtained at an excitation wavelength of 330 nm [46]. A red-shifted fluorescence from 463 to 534 nm with increase in excitation wavelength from 300 to 480 nm was observed for natural green pakchoi derived N-CDs [43]. An excitation maximum is obtained at 380 nm. *P. emblica* mediated N-CDs show a gradual increase in emission peak intensity upon variation of excitation wavelength between 270 and 320 nm due to the Л–Л* transitions of the carbon core [33]. A gradual decrease in intensity of photoluminescence (PL) intensity accompanied by a bathochromic shift was observed upon varying the excitation wavelength from 320 to 400 nm. The excitation at 320 nm resulted in a maximum intense PL emission at 400 nm (Figure 11.8).

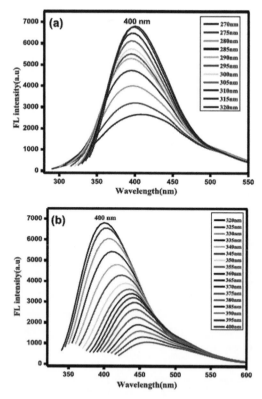

FIGURE 11.8   PL emission spectra of N-CDs at various excitation wavelengths from (a) 270 to 320 nm and (b) from 320 to 400 nm.

Source: Reprinted with permission from Arul et al. [33] © 2019 American Chemical Society.

The plausible explanation for the tunable PL emission of CDs involves nonuniformity in size, quantum confinement, surface defects, and surface states. By varying the CDs synthesis temperature from 160 to 240 °C, excitation independent spectra changes to excitation dependent photoluminescence [92]. The fluorescence of CDs may vary depending on the green precursor used, reaction conditions including temperature, surface passivation, and presence of heteroatoms [93]

11.3.2.3 UPCONVERSION PHOTOLUMINESCENCE (UCPL)

This phenomenon involves sequential absorption of two or more photons that leads to three emission of light at a wavelength lower than the excitation wavelength. This type of emission is quite similar to the anti-Stokes type. The process upconversion photoluminescence (UCPL) has various applications in various fields including bioimaging and cancer treatment. UCPL is exhibited by different green CDs [16, 35, 80].

11.4 CONCLUSION

Facile green synthesis of CDs is possible by using different renewable sources such as fruits, plants, vegetables, animal derivatives, food, and waste materials as the carbon precursors. Green chemistry approach toward the CDs synthesis via renewable and abundant natural sources reduces cost, avoids the usage of toxic chemicals, and reduces waste. The biogreen precursors could eliminate the requirement for chemical functionalization of CDs. Majority among the green CDs are self-passivated because of the availability of various functional groups in carbohydrates and proteins that are naturally present in the green sources itself. These green CDs offer aqueous dispersibility, high biocompatibility, unique luminescence properties, photostability, high QY, and upconversion fluorescence. CDs have benefits over conventional semiconducting quantum dots and organic fluorophores hence used for bioimaging and sensing applications due to their easy synthesis, cytotoxicity, and extensive optical properties. The excellent structural and optical properties of CDs make them a promising candidate for a wide variety of applications in the fields of nanosensing, photocatalysis, cell imaging, energy, and storage.

KEYWORDS

- **green strategy**
- **hydrothermal**
- **carbon precursors**
- **optical properties**
- **photoluminescence**

REFERENCES

1. Wang, Y., Hu, A. Carbon quantum dots: synthesis, properties and applications. *Journal of Materials Chemistry*. 2014, vol. 2(34), 6921–6939.
2. Zhang, J., Yu, S. H. Carbon dots: large-scale synthesis, sensing, and bioimaging. *Materials Today*. 2016, vol. 19(7), 382–393.
3. Lin, L., Rong, M., Luo, F., Chen, D., Wang, Y., Chen, X. Luminescent graphene quantum dots as new fluorescent materials for environmental and biological applications. *Trends in Analytical Chemistry*. 2014, vol. 54, 83–102.
4. Wang, Q., Huang, X., Long, Y., Wang, X., Zhang, H., Zhu, R., Liang, L., Teng, P., Zheng, H. Hollow luminescent carbon dots for drug delivery. *Carbon*. 2013, vol. 59, 192–199.
5. Lai, C. W., Hsiao, Y. H, Peng, Y. K., Chou, P. T. Facile synthesis of highly emissive carbon dots from pyrolysis of glycerol; Gram scale production of carbon dots/mSiO2 for cell imaging and drug release. *Journal of Materials Chemistry*. 2012, vol. 22(29), 14403.
6. Zhou, M., Zhou, Z., Gong, A., Zhang, Y., Li, Q. Synthesis of highly photoluminescent carbon dots via citric acid and Tris for iron (III) ions sensors and bioimaging. *Talanta*. 2015, vol. 143, 107–113.
7. Chen, J., Wei, J. S., Zhang, P., Niu, X. Q., Zhao, W., Zhu, Z. Y., Ding, h., Xiong, H. M. Red-emissive carbon dots for fingerprints detection by spray method: coffee ring effect and unquenched fluorescence in drying process. *ACS Applied Materials Interfaces*. 2017, vol. 9, 18429–18433.
8. Liu, H., Ye, T., Mao, C. Fluorescent carbon nanoparticles derived from candle soot. *Angewandte Chemie*. 2007, vol. 119(34), 6593–6595.
9. Sun, Y. P., Zhou, B., Lin, Y., Wang, W., Fernando, K. A. S., Pathak, P., Meziani, M. J., Harruff, B. A., Wang, X., Wang, H., Luo, P. G., Yang, H., Kose, M. E., Chen, B., Veca, L. M., Xie, S. Y. Quantum-sized carbon dots for bright and colorful photoluminescence. *Journal of the American Chemical Society*. 2006, vol. 128(24), 7756–7757.
10. Bottini, M., Balasubramanian, C., Dawson, M. I., Bergamaschi, A., Bellucci, S., Mustelin, T. Isolation and characterization of fluorescent nanoparticles from pristine and oxidized electric arc-produced single-walled carbon nanotubes. *The Journal of Physical Chemistry B*. 2006, vol. 110(2), 831–836.
11. Qu, K., Wang, J., Ren, J., Qu, X. Carbon dots prepared by hydrothermal treatment of dopamine as an effective fluorescent sensing platform for the label-free detection

of iron(III) ions and dopamine. *Chemistry—A European Journal*. 2013, vol. 19(22), 7243–7249.

12. Subramanian, V., Zhu, H., Vajtai, R., Ajayan, P. M., Wei, B. Hydrothermal synthesis and pseudocapacitance properties of MnO_2 nanostructures. *The Journal of Physical Chemistry B*. 2005, vol. 109(43), 20207–20214.

13. Sebastian, M., Aravind, A., Mathew, B. Green silver nanoparticles based multi-technique sensor for environmental hazardous Cu(II) ion. *BioNanoScience*. 2019, vol. 9(2), 373–385.

14. Joseph, S., Mathew, B. Microwave assisted facile green synthesis of silver and gold nanocatalysts using the leaf extract of *Aerva lanata*. *Spectrochimica Acta Part A: Molecular and Biomolecular Spectroscopy*. 2015, vol. 136, 1371–1379.

15. Sahu, S., Behera, B., Maiti, T. K. Mohapatra, S. Simple one-step synthesis of highly luminescent carbon dots from orange juice: application as excellent bio-imaging agents. *Chemical Communications*. 2012, vol. 48(70), 8835.

16. Mehta, V. N., Jha, S., Basu, H., Singhal, R. K., Kailasa, S. K. One-step hydrothermal approach to fabricate carbon dots from apple juice for imaging of mycobacterium and fungal cells. *Sensors and Actuators B: Chemical*. 2015, vol. 213, 434–443.

17. Liu, L., Gong, H., Li, D., Zhao, L. Synthesis of carbon dots from pear juice for fluorescence detection of Cu^{2+} ion in water. *Journal of Nanoscience and Nanotechnology*. 2018, vol. 18(8), 5327–5332.

18. De, B., Karak, N. A green and facile approach for the synthesis of water soluble fluorescent carbon dots from banana juice. *RSC Advances*. 2013, vol. 3(22), 8286.

19. Mehta, V. N., Jha, S., Singhal, R. K. One-pot green synthesis of carbon dots by using *Saccharum officinarum* juice for fluorescent imaging of bacteria (*Escherichia coli*) and yeast (*Saccharomyces cerevisiae*) cells. *Material Science and Engineering, C*. 2014, vol. 38, 20–27.

20. Carvalho, J., Santos, L. R., Germino, J. C., Terezo, A. J., Moreto, J. A., Quites, F. J., Freitas, R. G. Hydrothermal synthesis to water-stable luminescent carbon dots from Acerola fruit for photoluminescent composites preparation and its application as sensors. *Materials Research*. 2019, vol. 22(3).

21. Bhamore, J. R., Jha, S., Park, T. J., Kailasa, S. K. Fluorescence sensing of Cu^{2+} ion and imaging of fungal cell by ultra-small fluorescent carbon dots derived from *Acacia concinna* seeds. *Sensors and Actuators B: Chemical*. 2018, vol. 277, 47–54.

22. Bhamore, J. R., Jha, S., Singhal, R. K., Park, T. J., Kailasa, S. K. Facile green synthesis of carbon dots from *Pyrus pyrifolia* fruit for assaying of Al^{3+} ion via chelation enhanced fluorescence mechanism. *Journal of Molecular Liquids*. 2018, vol. 264, 9–16.

23. Sun, X., He, J., Yang, S., Zheng, M., Wang, Y., Ma, S., Zheng, H. Green synthesis of carbon dots originated from *Lycii fructus* for effective fluorescent sensing of ferric ion and multicolor cell imaging. *Journal of Photochemistry and Photobiology B: Biology*. 2017, vol. 175, 219–225.

24. Ramezani, Z., Qorbanpour, M., Rahbar, N. Green synthesis of carbon quantum dots using quince fruit (*Cydonia oblonga*) powder as carbon precursor: application in cell imaging and As^{3+} determination. *Colloids and Surfaces A: Physicochemical and Engineering Aspects*. 2018, vol. 549, 58–66.

25. Arul, V., Sethuraman, M. G. Facile green synthesis of fluorescent N-doped carbon dots from Actinidia deliciosa and their catalytic activity and cytotoxicity applications. *Optical Materials*. 2018, vol. 78, 181–190.

26. Akhgari, F., Samadi, N., Farhadi, K., Akhgari, M. A green one-pot synthesis of nitrogen and sulfur co-doped carbon quantum dots for sensitive and selective detection of cephalexin. *Canadian Journal of Chemistry*. 2017, vol. 95(6), 641–648.

27. Liao, J., Cheng, Z., Zhou, L. Nitrogen-doping enhanced fluorescent carbon dots: green synthesis and their applications for bioimaging and label-free detection of Au^{3+} ions. *ACS Sustainable Chemistry & Engineering*. 2016, vol. 4(6), 3053–3061.

28. Arul, V., Edison, T. N. J. I., Lee, Y. R., Sethuraman, M. G. Biological and catalytic applications of green synthesized fluorescent N-doped carbon dots using *Hylocereus undatus*. *Journal of Photochemistry and Photobiology B: Biology*. 2017, vol. 168, 142–148.

29. Atchudan, R., Edison, T. N. J. I., Chakradhar, D., Perumal, S., Shim, J. J., Lee, Y. R. Facile green synthesis of nitrogen-doped carbon dots using Chionanthus retusus fruit extract and investigation of their suitability for metal ion sensing and biological applications. *Sensors and Actuators B: Chemical*. 2017, vol. 246, 497–509.

30. Edison, T. N. J. I., Atchudan, R., Shim, J.-J., Kalimuthu, S., Ahn, B. C., Lee, Y. R. Turn-off fluorescence sensor for the detection of ferric ion in water using green synthesized N-doped carbon dots and its bio-imaging. *Journal of Photochemistry and Photobiology B: Biology*. 2016, vol. 158, 235–242.

31. Atchudan, R., Edison, T. N. J. I., Sethuraman, M. G., Lee, Y. R. Efficient synthesis of highly fluorescent nitrogen-doped carbon dots for cell imaging using unripe fruit extract of *Prunus mume*. *Applied Surface Science*. 2016, vol. 384, 432–441.

32. Das, P., Ganguly, S., Bose, M., Mondal, S., Das, A. K., Banerjee, S., Das, N. C. A simplistic approach to green future with eco-friendly luminescent carbon dots and their application to fluorescent nano-sensor 'turn-off' probe for selective sensing of copper ions. *Materials Science and Engineering: C*. 2017, vol. 75, 1456–1464.

33. Arul, V., Sethuraman, M. G. Hydrothermally green synthesized nitrogen-doped carbon dots from *Phyllanthus emblica* and their catalytic ability in the detoxification of textile effluents. *ACS Omega*. 2019, vol. 4(2), 3449–3457.

34. Joseph, S., Mathew, B. Facile synthesis of silver nanoparticles and their application in dye degradation. *Materials Science and Engineering: B*. 2015, vol. 195, 90–97.

35. Yin, B., Deng, J., Peng, X., Long, Q., Zhao, J., Lu, Q., Chen, Q., Li, H., Tang, H., Zhang, Y., Yao, S. Green synthesis of carbon dots with down- and up-conversion fluorescent properties for sensitive detection of hypochlorite with a dual-readout assay. *The Analyst*. 2013, vol. 138(21), 6551.

36. Zhang, B., Liu, C., Liu, Y. A Novel One-Step Approach to synthesize fluorescent carbon nanoparticles. *European Journal of Inorganic Chemistry*. 2010, vol. 2010(28), 4411–4414.

37. Shen, J. , Shang, S., Chen, X., Wang, D., Cai, Y. Facile synthesis of fluorescence carbon dots from sweet potato for Fe^{3+} sensing and cell imaging. *Materials Science and Engineering: C*. 2017, vol. 76, 856–864.

38. Li, L. S., Jiao, X. Y., Zhang, Y., Cheng, C., Huang, K., Xu, L. Green synthesis of fluorescent carbon dots from Hongcaitai for selective detection of hypochlorite and mercuric ions and cell imaging. *Sensors and Actuators B: Chemical*. 2018, vol. 263, 426–435.

39. Arumugam, N., Kim, J. Synthesis of carbon quantum dots from Broccoli and their ability to detect silver ions. *Materials Letters*. 2018, vol. 219, 37–40.

40. Vasimalai, N., Vilas-Boas, V., Gallo, J., Cerqueira, M. de F., Menéndez-Miranda, M., Costa-Fernández, J. M., Diéguez, L., Espiña, B., Fernández-Argüelles, M. T. Green synthesis of fluorescent carbon dots from spices for in vitro imaging and tumour cell growth inhibition. *Beilstein Journal of Nanotechnology.* 2018, vol. 9, 530–544.

41. Sun, X., Li, Y. Colloidal Carbon Spheres and Their Core/Shell Structures with Noble-Metal Nanoparticles. *Angewandte Chemie International Edition.* 2004, vol. 43(5), 597–601.

42. Jia, J., Lin, B., Gao, Y., Jiao, Y., Li, L., Dong, C., Shuang, S. Highly luminescent N-doped carbon dots from black soya beans for free radical scavenging, Fe3+ sensing and cellular imaging. *Spectrochimica Acta Part A: Molecular and Biomolecular Spectroscopy.* 2019, vol. 211, 363–372.

43. Niu, X., Liu, G., Li, L., Fu, Z., Xu, H., Cui, F. Green and economical synthesis of nitrogen-doped carbon dots from vegetables for sensing and imaging applications. *RSC Advances.* 2015, vol. 5(115), 95223–95229.

44. Li, Z., Ni, Y., Kokot, S. A new fluorescent nitrogen-doped carbon dot system modified by the fluorophore-labeled ssDNA for the analysis of 6-mercaptopurine and Hg (II). *Biosensors and Bioelectronics.* 2015, vol. 74, 91–97.

45. Li, C. L., Ou, C. M., Huang, C. C., Wu, W. C., Chen, Y. P., Lin, T. E., Ho, L. C., Wang, C. W., Shih, C. C., Zhou, H. C., Lee, Y. C., Tzeng, W. F., Chiou, T. J., Chu, S. T., Cang, J., Chang, H. T. Carbon dots prepared from ginger exhibiting efficient inhibition of human hepatocellular carcinoma cells. *Journal of Materials Chemistry B.* 2014, vol. 2(28), 4564–4571.

46. Liu, W., Diao, H., Chang, H., Wang, H., Li, T., Wei, W. Green synthesis of carbon dots from rose-heart radish and application for Fe³⁺ detection and cell imaging. *Sensors and Actuators B: Chemical.* 2017, vol. 241, 190–198.

47. Hu, Y., Zhang, L., Li, X., Liu, R., Lin, L., Zhao, S. Green preparation of S and N co-doped carbon dots from water chestnut and onion as well as their use as an off-on fluorescent probe for the quantification and imaging of coenzyme A. *ACS Sustainable Chemistry & Engineering.* 2017, vol. 5(6), 4992–5000.

48. Zhao, M. Lan, X. Zhu, H. Xue, T.-W. Ng, X. Meng, C.-S. Lee, P., Zhang, W. Green synthesis of bifunctional fluorescent carbon dots from garlic for cellular imaging and free radical scavenging. *ACS Applied Materials & Interfaces.* 2015, vol. 7(31), 17054–17060.

49. Wang, C., Sun, D., Zhuo, K., Zhang, H., Wang, J. Simple and green synthesis of nitrogen-, sulfur-, and phosphorus-co-doped carbon dots with tunable luminescence properties and sensing application. *RSC Advances.* 2014, vol. 4(96), 54060–54065.

50. Sachdev, A., Gopinath, P. Green synthesis of multifunctional carbon dots from coriander leaves and their potential application as antioxidants, sensors and bioimaging agents. *The Analyst.* 2015, vol. 140(12), 4260–4269.

51. Kumar, A., Chowdhuri, A. R., Laha, D., Mahto, T. K., Karmakar, P., Sahu, S. K. Corrigendum to Green synthesis of carbon dots from *Ocimum sanctum* for effective fluorescent sensing of Pb²⁺ ions and live cell imaging. *Sensors and Actuators B: Chemical.* 2018, vol. 263, 677.

52. Xu, H., Yang, X., Li, G., Zhao, C., Liao, X. Green synthesis of fluorescent carbon dots for selective detection of tartrazine in food samples. *Journal of Agricultural and Food Chemistry.* 2015, vol. 63(30), 6707–6714.

53. Bhatt, S., Bhatt, M., Kumar, A. , Vyas, G., Gajaria, T., Paul, P. Green route for synthesis of multifunctional fluorescent carbon dots from Tulsi leaves and its application as Cr(VI) sensors, bio-imaging and patterning agents. *Colloids and Surfaces B: Biointerfaces.* 2018, vol. 167, 126–133.

54. Yu, J., Song, N., Zhang, Y. K., Zhong, S. X., Wang, A. J., Chen, J. Green preparation of carbon dots by Jinhua bergamot for sensitive and selective fluorescent detection of Hg^{2+} and Fe^{3+}. *Sensors and Actuators B: Chemical.* 2015, vol. 214, 29–35.

55. Qin, X., Lu, W., Asiri, A. M., Al-Youbi, A. O., Sun, X. Green, low-cost synthesis of photoluminescent carbon dots by hydrothermal treatment of willow bark and their application as an effective photocatalyst for fabricating Au nanoparticles–reduced graphene oxide nanocomposites for glucose detection. *Catalysis Science & Technology.* 2013, vol. 3(4), 1027–1065.

56. Ramanan, V., Thiyagarajan, S. K., Raji, K., Suresh, R., Sekar, R., Ramamurthy, P. Outright Green synthesis of fluorescent carbon dots from eutrophic algal blooms for in vitro imaging. *ACS Sustainable Chemistry & Engineering.* 2016, vol. 4(9), 4724–4731.

57. Zhu, L., Yin, Y., Wang, C. F., Chen, S. Plant leaf-derived fluorescent carbon dots for sensing, patterning and coding. *Journal of Materials Chemistry C.* 2013, vol. 1(32), 4925.

58. Atchudan, R., Edison, T. N. J. I., Aseer, K. R., Perumal, S., Karthik, N., Lee, Y. R. Hydrothermal conversion of *Magnolia liliifora* into nitrogen-doped carbon dots as an effective turn-off fluorescence sensing, muliticolour cell imaging and fluorescent ink. *Colloids and Surface B: Biointerfaces.* 2018, vol. 169, 321–328.

59. Feng, Y., Zhong, D., Miao, H., Yang, X. Carbon dots derived from rose flowers for tetracycline sensing. *Talanta.* 2015, vol. 140, 128–133.

60. Wang, M., Wan, Y., Zhang, K., Fu, Q., Wang, L., Zeng, J., Xia, Z., Gao, D. Green synthesis of carbon dots using the flowers of *Osmanthus fragrans* (Thunb.) Lour. as precursors: application in Fe^{3+} and ascorbic acid determination and cell imaging. *Analytical and Bioanalytical Chemistry.* 2019, vol. 411(12), 2715–2727.

61. Liu, Y., Zhao, Y., Zhang, Y. One-step green synthesized fluorescent carbon nanodots from bamboo leaves for copper(II) ion detection. *Sensors and Actuators B: Chemical.* 2014, vol. 196, 647–652.

62. Bano, D., Kumar, V., Singh, V. K., Hasan, S. H. Green synthesis of fluorescent carbon quantum dots for the detection of mercury(ii) and glutathione. *New Journal of Chemistry.* 2018, vol. 42(8), 5814–5821.

63. Radhakrishnan, K., Panneerselvam, P., Marieeswaran, M. A green synthetic route for the surface-passivation of carbon dots as an effective multifunctional fluorescent sensor for the recognition and detection of toxic metal ions from aqueous solution. *Analytical Methods.* 2019, vol. 11(4), 490–506.

64. Gu, D., Shang, S., Yu, Q., Shen, J. Green synthesis of nitrogen-doped carbon dots from lotus root for Hg(II) ions detection and cell imaging. *Applied Surface Science.* 2016, vol. 390, 38–42.

65. Mewada, A., Pandey, S., Shinde, S., Mishra, N., Oza, G., Thakur, M., Sharon, M., Sharon, M. Green synthesis of biocompatible carbon dots using aqueous extract of *Trapa bispinosa* peel. *Materials Science and Engineering: C.* 2013, vol. 33(5), 2914–2917.

66. Zhou, J., Sheng, Z., Han, H., Zou, M., Li, C. Facile synthesis of fluorescent carbon dots using watermelon peel as a carbon source. *Materials Letters.* 2012, vol. 66(1), 222–224.

67. Chen, K., Qing, W., Hu, W., Lu, M., Wang, Y., Liu, X. On-off-on fluorescent carbon dots from waste tea: Their properties, antioxidant and selective detection of CrO_4^{2-}, Fe^{3+}, ascorbic acid and L-cysteine in real samples. *Spectrochimica Acta Part A: Molecular and Biomolecular Spectroscopy.* 2019, vol. 213, 228–234.
68. Ma, X., Dong, Y., Sun, H., Chen, N. Highly fluorescent carbon dots from peanut shells as potential probes for copper ion: the optimization and analysis of the synthetic process. *Materials Today Chemistry.* 2017, vol. 5, 1–10
69. Saxena, M., Sarkar, S. Synthesis of carbogenic nanosphere from peanut skin. *Diamond and Related Materials.* 2012, vol. 24, 11–14.
70. Prasannan, A., Imae, T. One-Pot Synthesis of fluorescent carbon dots from orange waste peels. *Industrial & Engineering Chemistry Research.* 2013, vol. 52(44), 15673–15678.
71. Yuan, M., Zhong, R., Gao, H., Li, W., Yun, X., Liu, J., Zhao, X., Zhao, G., Zhang, F. One-step, green, and economic synthesis of water-soluble photoluminescent carbon dots by hydrothermal treatment of wheat straw, and their bio-applications in labeling, imaging, and sensing. *Applied Surface Science.* 2015, vol. 355, 1136–1144.
72. Kumari, A., Kumar, A., Sahu, S. K., Kumar, S. Synthesis of green fluorescent carbon quantum dots using waste polyolefins residue for Cu^{2+} ion sensing and live cell imaging. *Sensors and Actuators B: Chemical.* 2018, vol. 254, 197–205.
73. Park, S. Y., Lee, H. U., Park, E. S., Lee, S. C., Lee, J. W., Jeong, S. W., Kim, C. H., Lee, Y. C., Huh, Y. S., Lee, J. Photoluminescent green carbon nanodots from food-waste-derived sources: large-scale synthesis, properties, and biomedical applications. *ACS Applied Materials & Interfaces.* 2014, vol. 6(5), 3365–3370.
74. Tyagi, A., Tripathi, K. M., Singh, N., Choudhary, S., Gupta, R. K. Green synthesis of carbon quantum dots from lemon peel waste: applications in sensing and photocatalysis. *RSC Advances.* 2016, vol. 6(76), 72423–72432.
75. Tan, X. W., Romainor, A. N. B., Chin, S. F., Ng, S. M. Carbon dots production via pyrolysis of sago waste as potential probe for metal ions sensing. *Journal of Analytical and Applied Pyrolysis.* 2014, vol. 105, 157–165.
76. Hu, Y., Yang, J., Tian, J., Jia, L., Yu, J. S. Waste frying oil as a precursor for one-step synthesis of sulfur-doped carbon dots with pH-sensitive photoluminescence. *Carbon.* 2014, vol. 77, 775–782.
77. Gedda, G., Lee, C. Y., Lin, Y. C., Wu, H. Green synthesis of carbon dots from prawn shells for highly selective and sensitive detection of copper ions. *Sensors and Actuators B: Chemical.* 2016, vol. 224, 396–403.
78. Sharma, V., Tiwari, P., Mobin, S. M. Sustainable carbon-dots: recent advances in green carbon dots for sensing and bioimaging. *Journal of Materials Chemistry B.* 2017, vol. 5, pp. 8904–8924.
79. Zhang, Z., Sun, W., Wu, P. Highly Photoluminescent carbon dots derived from egg white: facile and Green synthesis, photoluminescence properties, and multiple applications. *ACS Sustainable Chemistry & Engineering.* 2015, vol. 3(7), 1412–1418.
80. Yang, X., Zhuo, Y., Zhu, S., Luo, Y., Feng, Y., Dou, Y. Novel and green synthesis of high-fluorescent carbon dots originated from honey for sensing and imaging. *Biosensors and Bioelectronics.* 2014, vol. 60, 292–298.
81. Zhang, J., Yuan, Y., Liang, G., Yu, S. H. Scale-up synthesis of fragrant nitrogen-doped carbon dots from bee pollens for bioimaging and catalysis. *Advanced Science.* 2015, vol. 2(4), 1500002–1500008.

82. Wang, L., Bi, Y., Hou, J., Li, H., Xu, Y., Wang, B., Ding, H., Ding, L. Facile, green and clean one-step synthesis of carbon dots from wool: application as a sensor for glyphosate detection based on the inner filter effect. *Talanta*. 2016, vol. 160, 268–275.

83. Gong, X., Lu, W., Paau, M. C., Hu, Q., Wu, X., Shuang, S., Dong, C., Choi, M. M. F. Facile synthesis of nitrogen-doped carbon dots for Fe^{3+} sensing and cellular imaging. *Analytica Chimica Acta*. 2015, vol. 861, 74–84.

84. Wang, L., Zhou, H. S. Green synthesis of luminescent nitrogen-doped carbon dots from milk and its imaging application. *Analytical Chemistry*. 2014, vol. 86(18), 8902–8905.

85. Gong, N., Wang, H., Li, S., Deng, Y., Chen, X., Ye, L., Gu, W. Microwave-assisted polyol synthesis of gadolinium-doped green luminescent carbon dots as a bimodal nanoprobe. *Langmuir*. 2014, vol. 30(36), 10933–10939.

86. Murugan, N., Prakash, M., Jayakumar, M., Sundaramurthy, A., Sundramoorthy, A. K. Green synthesis of fluorescent carbon quantum dots from Eleusine coracana and their application as a fluorescence 'turn-off' sensor probe for selective detection of Cu^{2+}. *Applied Surface Science*. 2019, vol. 476, 468–480.

87. Wei, J., Zhang, X., Sheng, Y., Shen, J., Huang, P., Guo, S., Pan, J., Feng, B. Dual functional carbon dots derived from cornflour via a simple one-pot hydrothermal route. *Materials Letters*. 2014, vol. 123, 107–111.

88. Teng, X., Ma, C., Ge, C., Yan, M., Yang, J., Zhang, Y., Morais, P. C., Bi, H. Green synthesis of nitrogen-doped carbon dots from konjac flour with 'off–on' fluorescence by Fe^{3+} and l-lysine for bioimaging. *Journal of Materials Chemistry B*. 2014, vol. 2(29), 4631.

89. Xie, Y., Cheng, D., Liu, X., Han, A. Green hydrothermal synthesis of N-doped carbon dots from biomass Highland barley for the detection of Hg^{2+}. *Sensors*. 2019, vol. 19(14), 3169.

90. Ding, H., Ji, Y., Wei, J. S., Gao, Q. Y., Zhou, Z. Y., Xiong, H. M. Facile synthesis of red-emitting carbon dots from pulp-free lemon juice for bioimaging. *Journal of Materials Chemistry B*. 2017, vol. 5(26), 5272–5277.

91. Dong, Y., Wang, R., Li, H., Shao, J., Chi, Y., Lin, X., Chen, G. Polyamine-functionalized carbon quantum dots for chemical sensing. *Carbon*. 2012, vol. 50(8), 2810–2815.

92. Li, X., Zhang, S., Kulinich, S. A., Liu, Y., Zeng, H. Engineering surface states of carbon dots to achieve controllable luminescence for solid-luminescent composites and sensitive Be^{2+} detection. *Scientific Reports*. 2014, vol. 4(1).

93. Bao, L., Zhang, Z. L., Tian, Z. Q., Zhang, L., Liu, C., Lin, Y., Qi, B., Pang, D. W. Electrochemical tuning of luminescent carbon nanodots: from preparation to luminescence mechanism. *Advanced Materials*. 2011, vol. 23(48), 5801–5806.

CHAPTER 12

Nonlinear Optical Properties of Inorganic Polymers

SIJO FRANCIS[1*], REMYA VIJAYAN[2], EBEY P. KOSHY[1], and BEENA MATHEW[2]

[1]*Department of Chemistry, St. Joseph's College, Moolamattom, Kerala, India*

[2]*School of Chemical Sciences, Mahatma Gandhi University, Kottayam, Kerala, India*

Corresponding author. E-mail: srsijofrancis@gmail.com

ABSTRACT

Nonlinear optical materials have a wide range of applications in industry and electronics because of their versatile mechanical properties. Inorganic polymers are macromolecules having noncarbon skeletal structure and inorganic or organic side chain. Coordination polymers, cross-linked polymers, metal-organic frameworks, thin films, and hybrid materials have additional optoelectronic qualities.

12.1 INTRODUCTION

Polymers are chains compounds composed of repeating monomer units interlinked by covalent bonds. Inorganic polymers are macromolecules have noncarbon skeletal structure and inorganic or organic side chain. The major repeating atoms include oxygen, nitrogen, phosphorous, and silicon. These polymers have multifaceted characteristics and some of them are described briefly here. Polyphosphazenes have phosphorus and nitrogen in their backbone and their major applications are as flame-resistant materials and in the medical field. Conjugated phosphole materials that are derived from phosphorous materials have electronic properties. Icosahedral boron clusters have extreme stabilities due to their geometric features and used to prepare hybrid materials and coordination polymers. Dendrimers when modified by adding phosphorus or silicon moieties acquire biomedical qualities. Ferrocene molecules can be incorporated in a polymer backbone or side chain of polysilanes. Polystannanes are easy to process and have high stability in addition to their peculiar semiconductivity. Metallopolymers are constituted by incorporating metals into the polymer backbone has nonlinear optical capacities. Metal-organic frameworks are a peculiar class of metallopolymers and have ample nonlinear optical characteristic [1]. Polythiazyl has formula $(SN)_x$ and has superconducting properties and is obtained by spontaneous and room temperature polymerization of S_2N_2. Quartz and mica are also inorganic polymers having Si–O bonds. Quartz has a crystalline structure with SiO_4 tetrahedral arrangement, whereas mica has a noncrystalline character. Polysilanes show peculiar optical and electronic properties since silicon atoms in the chain is less electronegative than carbon atoms [2].

12.1.1 BASIC THEORY

When light falls on certain materials, the frequency of incident light changes. When infrared radiation falls on some crystals it emits green light due to frequency variation. According to the classical theory of Raman Effect, when a molecule is placed in a static electric field, it gets polarized. The induced polarization has two components with respect to electric field, a linear and a higher order nonlinear part as shown below

$$P = P_{linear} + P_{nonlinear}$$

The nonlinear part leads to the production of new frequency components and second harmonic generation. Nonlinear-optical processes are generally N-wave-mixing process and in frequency doubling is a three-wave mixing process. Phase matching is a term used in nonlinear optics for the cumulative effect of conservation of energy and momentum. It is observed that phase matching take places at the time of N-wave mixing processes.

12.1.2 MAJOR OPTICAL PROPERTIES

Two major optical properties are of at most importance. They are luminescence and frequency doubling.

12.1.2.1   LUMINESCENCE

Luminescence is a nonthermal process involving the emission of light in the UV, visible, or IR regions. The origin of this phenomenon is the electronically excited states of molecules giving light with a higher wavelength and they are of two types; fluorescence and phosphorescence. They differ in two senses. First, the former is derived from singlet excited state and latter is from the triplet excited state. Second, the latter has a longer lifetime. Another type of luminescence is electroluminescence as a consequence of radiative recombination of electrons and holes in a semiconductor material. LEDs are the best electroluminescent devices [3]. Upconversion photoluminescence is a nonlinear excitation of light with low energy and produces a higher energy light photons. Common upconversion materials are the trivalent lanthanide (Ln^{3+}) ions such as Er^{3+}, Tm^{3+}, Ho^{3+}, Pr^{3+}, and Nd^{3+}. Upconversion nanomaterials have interdisciplinary optical applications [4]. Luminescent colloidal carbon dots derived from precursor fluorescent nanoparticles are also known [5]. Semiconductor nanocrystals added by some dopants can drastically improve the existing properties and introduce new nonlinear properties [6].

12.1.2.2   SECOND-HARMONIC GENERATION (SHG)

Second-harmonic generation (SHG) is the creation of a wave with twice the frequency of input radiation in a nonlinear optical (NLO) material. It is also called frequency doubling. Frequency doubling is a nonlinear-optical phenomenon where a specific wavelength of light is converted to its half value and thus the frequency gets doubled. SHG is a nonlinear ultrasonic method for detecting and monitoring microstructural changes occurring in metals. The measurement of second harmonic generation has the unique capacity to detect cracking, aging, and others of transportation industries, and defense systems [7]. A good ultraviolet nonlinear optical material should have a relatively larger SHG coefficient, moderate birefringence, a wide UV transparency range, and high stability [8]. The second harmonic generation in aluminum nitride microring generator involves two optical modes, the fundamental and second-harmonic frequency [9]. SHG can be quenched by the highly symmetric plasmonic materials [10].

12.1.2.3   OTHER NLO PROPERTIES

NLO properties occurred when materials are exposed by high energy laser light. A number of effects have occurred simultaneously. Some of them are discussed below.

12.1.2.3.1   Kerr Effect

It is an electrooptical phenomenon. When light propagates through NLO crystals and glasses, a change in refractive index occurs by the electric fields, and the magnitude of this effect depends on the noncollinear property of the antiferromagnets [11]. The nonlinear refractive index shown by a single layer of graphene that was supported on quartz showed a property that is independent on the thermal nonlinearities and sample characteristics [12]. The linear electrooptic effect in quartz with low intensity is called Pockels effect.

12.1.2.3.2   Two-Photon Excited Fluorescence

It is an imaging technique used in biological systems where scanning of 1 mm thickness is possible. The two photons used for excitation and have lower energy than the emitted radiations. Nonlinear microscopy is based on this multiphoton effect and the generated signals depend solely on the excitation intensity and are utilized in vivo imaging of biological samples. This dual-wavelength excitation technique is used to study and inhibit the metabolism of malignant cancerous cells [13]. Absorption of double photons in the NIR region stimulates fluorescent dyes providing microscopic analysis, a substitute technique to confocal microscopy. The chromophore group in indole is the amino acid tryptophan, which is commonly used as a fluorescent marker. The femtosecond excitation by double pulses produces polarized fluorescence in indole [14].

12.1.2.3.3   Multiphoton Absorption

The simultaneous absorption of multiple photons by a material is called multiphoton absorption. Multiphoton absorption is a prominent NLO effect and exhibited by dyes, conjugated polymers, certain solid-state materials and metal-organic frameworks (MOFs). MOFs consist of metal ion/metal cluster bonded to organic molecules (linkers) and the intrinsic NLO properties of MOFs can be tailored by choosing appropriate chromophore linkers. This property finds applications in telecommunications, photonics, defense, and biomedicines [15]. Semiconductor zinc oxide powder showed a simultaneous three-photon absorption by femtosecond laser radiation in the ultraviolet range [16]. Graphene–ZnO nanocomposites showed a controllable five-photon absorption, which is highly useful in eye protection devices [17]. Theoretically, multiphoton absorption is possible in optical solitons (fiber) with anticubic nonlinearity [18]. Zero-dimensional lead halide perovskite nanocrystals exhibited superior nonlinearities like multiphoton absorption coefficients and strong multiphoton absorption cross-sections in three and four-photon absorptions applicable in photonics and ultrafast optical switching devices [19].

12.1.2.3.4 *Triplet–Triplet Annihilation*

In organic upconversion, low-energy light is converted to high-energy light either by two-photon absorption or triplet–triplet annihilation. Triplet–triplet annihilation requires a triplet sensitizer and triplet acceptor. The sensitizer absorbs light and then undergoes intersystem crossing and transfers its triplet energy to the triplet state of the acceptor. The two nearby triplet acceptors may endure the triplet–triplet annihilation and produce a special fluorescence [20]. This is a strategy for harvesting light of low-energy from the solar spectrum using photovoltaic technologies [21]. Triplet–triplet annihilation is used in dye-sensitized solar cells for solar energy conversion with high efficiencies [22].

12.1.2.3.5 *Stimulated Emission*

Semiconductor colloidal nanoplatelets offer advantageous optoelectronic properties. In CdSe the spontaneous and stimulated light emission showed a decrease in intensity with increasing lateral size of the particles [23]. Inorganic colloidal nanocrystals $CsPbX_3$ is a member perovskite family showed strong nonlinear absorption and emissions. Generally, halide perovskite materials have intense performance in photo harvesting and photoemissions. Stimulated emissions should be stable and tunable with wavelength. Here green stimulated emission from the colloidal nanocrystals upon three-photon pumping is observed. These nanocrystals offer new corridors in nonlinear photonics and nonlinear optical devices [24].

12.1.2.3.6 *Photorefractive Effect*

It is a nonlinear property of certain materials. Some crystals change their refractive index when light falls on it. This property is widely exploited in holography where alternate bright and dark fringes are observed. When electrons are excited from the conduction band, they give rise to bright fringes and if the recombination of electrons with holes in the conduction band, dark fringes may results. The ferroelectric crystal lithium niobate has a photorefractive effect. $LiNbO_3$ has linear and nonlinear properties and on In-doping give additional resistance optical damage [25]. On doping hafnium niobate crystals induces a decrease of photorefraction results in an increase in photoconductivity [26]. Iron doping on lithium niobate crystals also causes changes in refractive index due to photorefraction [27]. Photorefractive optical nonlinearity of n lithium niobate crystals depends on the nonlinear response of the medium [28]. The photorefractive effect ferroelectric liquid crystal doped with photoconductive material is also studied. Light scattering is less in smectic liquid crystals compared to nematic liquid crystals [29]. The photorefractive effect of photoconductive ferroelectric liquid crystals containing photoconductive chiral dopants was also established [30]. Terthiophene compounds with chiral photoconductive dopants also exhibit photorefractivity that can be useful in the formation of a dynamic hologram [31].

12.1.2.3.7 *Third-Harmonic Generation*

Z-scan technique is used for the characterization of third-order NLO properties of materials. The Z-scan technique is particularly useful for the simultaneous measurement of both nonlinear refractive index and nonlinear absorption coefficient [32]. The name Z-scan is derived from the fact that laser beam focused through the z-axis. Z-scan depends on the molecular reorientations, scattering, local heating and absorption by excited states. Z-scan is measured in femtosecond timescale. The excited state of organic molecules has picosecond to nanosecond lifetime. Z-scan setup gives nonlinear refractive index and two-photon absorption coefficients simultaneously.

12.2 CLASSIFICATION OF NONLINEAR POLYMERS

NLOs are mainly of three types—inorganic, organic, and hybrid materials.

12.2.1 INORGANIC NLOS

Inorganic NLO materials can be used as solid-state lasers in the ultraviolet and deep-UV regions [33]. Inorganic metal cyanurates with planar $(C_3N_3O_3)^{3-}$ anion as a fundamental building block is an efficient NLO material [34]. Low temperature synthesized $Sr_3(O_3C_3N_3)_2$ crystal with noncenterosymmetric space group has frequency doubling capacity [35]. Ferrocenyl moieties cause enhancement of NLO response in the V-shaped triple-hybrid decaborane derivative compounds [36].

12.2.2 ORGANIC NLOS

NLO properties of organic chromophores have a crucial role in the industry. Second-order nonlinear optical properties of ionic organic crystals very much depend on their molecular structure. Presence of π conjugation, properties of donor–acceptor systems, a suitable combination of anions and their counter ions, type of crystal packing, and others affect the NLO properties. For example, the second-order NLO effect of the best organic NLO crystal 4-N,N-dimethylamino-4,-N,-methyl-stilbazolium tosylate depends largely on the orientation of the chromophores. The presence of heterocyclics like thiophene, furan, and pyrrole should improve the second-order optical nonlinearity of organic molecules [37]. In multiple porphyrins of zinc, the extension of electron delocalization positively affects the third-order nonlinear optical properties, absorption coefficients, and refractive indices [38]. Solvation effects also have an important influence on the nonlinear optical properties organic materials such as 1,1 dicyano, 6-(di-butyl amine) hexatriene [39]. Theoretically, nonlinear optical properties of open-shell systems have greater larger NLO properties than the closed-shell NLO systems [40].

12.2.3 HYBRID MATERIALS

Hybrid materials have multifunctional and improved properties to be utilized in energy and bioimaging. By combining organic or organometallic species the mechanical and optical properties are modified at the macro- and nanoscale dimensions. The crystalline organic–inorganic perovskites are the examples of light-emitting hybrid materials. Sol–gel synthetic strategy is adopted here and by embedding organic dyes, quantum dots, or trivalent lanthanide complexes to hybrid hosts by some sort of supramolecular host–guest interactions, the stability, and flexibility of light-emitting centers can be improved. Molecular self-assembly in crystals are important for NLO applications. In hybrid materials NLO response improved by adding covalent bonded dyes and through hydrogen bonding. The association of NLO dyes into inorganic or hybrid matrices are a practical way to generate NLO materials. Some spacer molecules containing silane functional groups group also impact the NLO properties. The phenomenon cross-linking improves the NLO properties of dye-containing hybrid matrixes. The introduction of metal centers has a positive impact on the NLO response of organic dyes. The carbon nanotubes based and graphene-based nonlinear scattering are also important. Two-photon absorption in picosecond is observed in graphene-based materials. Hybrid composites of graphene with ZnO or CdS nanoparticles in PMMA glasses exhibit applications of third-harmonic generation in nanosecond and picoseconds regimes [41]. Sol–gel-derived MOFs have fine tune the optical response because of their high solubility and covalent bonding. Hybrid organic–inorganic perovskites have unique optoelectronic properties [42]. NLO switching due to order–disorder transformation of the organic cation and the displacement followed by reorientation of the inorganic component was established [43].

12.2.4   METAL-ORGANIC FRAMEWORKS (MOFS)

MOFs are hybrid materials formed by the assembly of metal ions and polydentate ligands. The NLO response of MOFs depends on the metal ion and the ligand. The mixing of the energy levels of ligands and metal ions may happen. d-p mixing and generation of new charge-transfer states metal-to-ligand charge transfer) and ligand-to-metal charge transfer happened when the transition metals are used. The luminescence effects of heavy transition metals are generally due to spin–orbit coupling and the well-known f–f transitions. Two-photon transitions from long-lived excited states of ligand levels may also result. To get an optical response of MOFs multiphoton absorbing ligand molecules are to be selected. NLO chromophores are classified into dipolar, quadrupolar, and octupolar molecules. Extended π conjugation is a prerequisite for two-photon molecular chromophores. Example of such compounds includes aromatic compounds like triphenylamine that has many arms. Intramolecular charge transfer also supports two-photon chromophore activities. Intramolecular two-photon excited emission in MOFs is facilitated by the large separation between the donor and acceptor sites. And the presence of electron-withdrawing groups like the nitro group and the absence of electron donor groups are helpful in the efficient charge transfer. Carboxylic group is the most used functional groups for linking in MOFs. π-stacked aromatic units are widely enhancing the NLO response of intermolecular two-photon transitions [44].

12.2.4.1 MICROWAVE-ASSISTED SOLVOTHERMAL STRATEGY

Some MOFs can be prepared by microwave-assisted solvothermal strategy and has excellent photoluminescence properties. Microwave-assisted synthesis is known from its high energy efficiency and reduction in time. The major classification of light emission characteristics of MOFs include luminescence based on highly conjugated ligands, the lanthanide metal center and that due to charge-transfer. The optical response of the MOFs can be fruitfully modified toward the applications in the visible region [45].

12.2.4.2 SOLID-STATE HIGH-TEMPERATURE PROCEDURE

The commercial NLO materials are generally explored in the IR region. The bandgap of NLO material in the IR region can be widened by incorporating an electropositive element into the framework. Applications of new NLO MOFs in laser sources are commonly based on its second-harmonic generation. Nonlinear optical properties of metal sulfides prepared in solid-state by applying high-temperature were found in the literature. The NLO applications of $Ba_2Ga_8MS_{16}$ (M = Si and Ge) in the mid and far-IR transparent regions with large bandgaps. These metal sulfides have three-dimensional framework structure and have noncentrosymmetric space. They show intense SHG signals by high polarization because of their peculiar stacking arrangement of alternate mixed and pure tetrahedral [46]. Two-dimensional semiconductor materials that obtained by embedding monolayer MoS within microcavity have shown strong nonlinear optical properties [47]. Sr_2CeO_4 nanocrystals observe luminescence decay profiles decreases with increasing the grain size [48]. Borate halides (Pb_2BO_3I) prepared hydrothermal method has high-performance nonlinear optical properties [49].

12.2.5   THIN-FILM NLO MATERIALS

Optical properties of the thin film of Violet 1-doped polyvinyl alcohol when measured using Z-scan technique showed excellent nonlinear refractive index and nonlinear absorption coefficient adequate for optical limiting applications [50]. Thermally evaporated copper tetra *tert*-butyl phthalocyanine thin films exhibited linear and nonlinear optical properties (third order NLO) in femtosecond Z-scan by two-photon absorption. Phthalocyanine compounds have organic nonlinearities generally that make it suited for applications like passive Q-switching and fast optical switching [51]. Cd-doped CuO nanoparticles on thin films in the polyvinyl alcohol matrix have characteristic Z-scan properties, enhanced nonlinear refractive index, and nonlinear absorption coefficient have been obtained [52]. The third-order nonlinear optical properties of $CaCu_3Ti_4O_{12}$ thin film by means of the Z-scan technique using femtosecond laser pulse were prepared by feasible RF magnetron sputtering method [53]. Surface-anchored metal-organic frameworks of thin films display special refractive index (n) is high when measured by liquid phase epitaxy and found applications microelectronic and sensing devices. The highly porous crystalline MOFs based on Cu^{2+} are highly tunable materials and may have laminar triclinic crystal structures [54].

12.2.6   POLYMER NLOS

Macroscopic organic materials have nonlinear optical properties dependent on the polarization reaction of molecular electrons. Polymers have low dielectric constant. The absorption coefficient of NLO polymers is 10^{-4} times less than that of semiconductor and inorganic NLOs. Polymer NLOs has advanced stability and can be used in optical switching purposes [55]. Polymers are generally important in near-IR (NIR) and mid-IR applications. Chalcogenide hybrid inorganic/organic polymers show Z-scan characteristics for optics and photonics applications. NLO properties of sulfur-based polymers can be utilized in infrared spectrometers and cameras and gas sensors [56].

12.2.7 NANOCOMPOSITES NLOS

Metal nanoparticles and nanocomposites have unique optical nonlinear properties. The nanocomposite thin films generated by silver nanoparticles incorporated in a matrix aliphatic urethane acrylate polymer matrix shows nonlinear optical properties in the standard Z-scan using pico- and femto-second scanning pulses. The polymer matrix provides a platform for the desired NLO properties of the metal nanoparticles. The surface plasmon resonance properties of the noble metal nanoparticles significantly enhance the NLO behavior and photonic applications. NLO responses of the prepared thin film composite material can be used in photonic applications [57]. The NLO properties of gold-decorated graphene nanocomposites adversely affected by the changes in the electronic structure of graphene due to rehybridization of metal d-orbitals and graphene p-orbitals [58]. Graphene oxide-silver nanocomposite has third-order nonlinear susceptibility than GO and is attributed to the complex energy band structures of the nanocomposite. Thus GO-Ag nanocomposite can safely be used as a good optical limiter for military purposes [59]. Polymer matrix has a limiting influence on the nonlinear optical properties of polymer/ZnO nanocomposites. PMMA, poly(vinylidene fluoride), PVA and polystyrene are the polymers selected. The optical limiting threshold was established using Z-scan technique and the study shows the concentration of ZnO also affect NLO response. PMMA/ZnO at high concentration of ZnO has a low optical limiting threshold compared to other composites studied [60]. Carboxylic acid-functionalized multiwalled carbon nanotubes at three concentrations of carbon nanotubes and Z-scan in three different intensities of the laser beam show an increase in the high order of nonlinear refractive coefficient as the beam intensity increases [61].

12.3 APPLICATIONS OF INORGANIC POLYMER NLOS

Polymer NLOs can be used for the separation and storage of materials. Poly(2-chloroaniline) is useful for ammonia gas sensing applications since the electrical resistance depends on the concentration of different ammonia gas concentrations [62]. NLO materials have applications in optical fiber communication, ultrafast switches, sensors, optical computing, laser amplifiers, and lanthanide-doped nanocrystals have luminescence applications [63]. Based on nonlinear optical properties of materials a number of sensing systems and communications systems are operating. Metalloporphyrins and metallophthalocyanines are used for optical power limiting purposes [64]. Te-based glasses with WO_3 were used for radioprotection application in the medical field [65]. Light-emitting diodes are a practical application of NLOs.

12.4   CONCLUSION

The interaction of light with a nonlinear medium mainly in inorganic single crystals, inorganic thin films, and hybrid materials are discussed here. Crystals that have no center of symmetry are traditionally known as nonlinear materials. Inorganic polymer materials have specific optoelectronic and electronic properties. Polymers can be suitably tailored for specific NLO functionalities. Generally, organic compounds displayed greater NLO response than inorganic materials. Nonlinear optical materials found applications in photonics, lasing, optoelectronic devices, and so forth.

KEYWORDS

- **single crystals**
- **thin films**
- **hybrid materials**
- **second harmonic generation**
- **Z-scan**
- **inorganic polymers**

REFERENCES

1. Caminade, Anne-Marie, Evamarie Hey-Hawkins, and Ian Manners. "Smart inorganic polymers." *Chemical Society Reviews* 45, 19 (2016): 5144–5146.
2. Kensuke Naka. "Inorganic polymers: overview." In: Kobayashi S., Müllen K. (eds) Encyclopedia of Polymeric Nanomaterials. Springer, Berlin, Heidelberg. (2015), 995–1000.
3. Parola, Stephane, et al. "Optical properties of hybrid organic-inorganic materials and their applications." *Advanced Functional Materials* 26, 36 (2016): 6506–6544.
4. Nadort, Annemarie, Jiangbo Zhao, and Ewa M. Goldys. "Lanthanide upconversion luminescence at the nanoscale: fundamentals and optical properties." *Nanoscale* 8, 27 (2016): 13099–13130.

5. Reckmeier, C. J., et al. "Luminescent colloidal carbon dots: optical properties and effects of doping." *Optics Express* 24, 2 (2016): A312–A340.
6. Adhikari, Samrat Das, et al. "Luminescence, plasmonic, and magnetic properties of doped semiconductor nanocrystals." (2017).
7. Matlack, K. H., et al. "Review of second harmonic generation measurement techniques for material state determination in metals." *Journal of Nondestructive Evaluation* 34, 1 (2015): 273.
8. Zou, Guohong, et al. "ACdCO 3 F (A= K and Rb): new noncentrosymmetric materials with remarkably strong second-harmonic generation (SHG) responses enhanced via π-interaction." *RSC Advances* 5, 103 (2015): 84754–84761.
9. Guo, Xiang, Chang-Ling Zou, and Hong X. Tang. "Second-harmonic generation in aluminum nitride microrings with 2500%/W conversion efficiency." *Optica* 3, 10 (2016): 1126–1131.
10. Celebrano, Michele, et al. "Mode matching in multiresonant plasmonic nanoantennas for enhanced second harmonic generation." *Nature Nanotechnology* 10, 5 (2015): 412.
11. Sivadas, Nikhil, Satoshi Okamoto, and Di Xiao. "Gate-controllable magneto-optic Kerr effect in layered collinear antiferromagnets." *Physical Review Letters* 117, 26 (2016): 267203.
12. Dremetsika, Evdokia, et al. "Measuring the nonlinear refractive index of graphene using the optical Kerr effect method." *Optics Letters* 41, 14 (2016): 3281–3284.
13. Hou, Jue, et al. "Correlating two-photon excited fluorescence imaging of breast cancer cellular redox state with seahorse flux analysis of normalized cellular oxygen consumption." *Journal of Biomedical Optics* 21, 6 (2016): 060503.
14. Herbrich, Sebastian, et al. "Two-color two-photon excited fluorescence of indole: determination of wavelength-dependent molecular parameters." *The Journal of Chemical Physics* 142, 2 (2015): 024310.
15. Medishetty, Raghavender, et al. "Multi-photon absorption in metal–organic frameworks." *Angewandte Chemie International Edition* 56, 46 (2017): 14743–14748.
16. Dominguez, Christian Tolentino, et al. "Multi-photon excited coherent random laser emission in ZnO powders." *Nanoscale* 7, 1 (2015): 317–323.
17. Tong, Qing, et al. "Nonlinear optical and multi-photon absorption properties in graphene–ZnO nanocomposites." *Nanotechnology* 29, 16 (2018): 165706.
18. Khan, Salam, et al. "Stochastic perturbation of optical solitons having anti-cubic nonlinearity with bandpass filters and multi-photon absorption." *Optik* 178 (2019): 1120–1124.
19. Krishnakanth, K. N., et al. "Broadband ultrafast nonlinear optical studies revealing exciting multi-photon absorption coefficients in phase pure zero-dimensional Cs 4 PbBr 6 perovskite films." *Nanoscale* 11, 3 (2019): 945–954.
20. Ye, Changqing, et al. "Photon upconversion: from two-photon absorption (TPA) to triplet–triplet annihilation (TTA)." *Physical Chemistry Chemical Physics* 18, 16 (2016): 10818–10835.
21. Hoseinkhani, S., et al. "Achieving the photon up-conversion thermodynamic yield upper limit by sensitized triplet–triplet annihilation." *Physical Chemistry Chemical Physics* 17, 6 (2015): 4020–4024.
22. Simpson, Catherine, et al. "An intermediate band dye-sensitised solar cell using triplet–triplet annihilation." *Physical Chemistry Chemical Physics* 17, 38 (2015): 24826–24830.

23. Olutas, Murat, et al. "Lateral size-dependent spontaneous and stimulated emission properties in colloidal CdSe nanoplatelets." *ACS Nano* 9, 5 (2015): 5041–5050.

24. Wang, Yue, et al. "Nonlinear absorption and low-threshold multiphoton pumped stimulated emission from all-inorganic perovskite nanocrystals." *Nano Letters* 16, 1 (2016): 448–453.

25. Kong, Yongfa, Jinke Wen, and Huafu Wang. "New doped lithium niobate crystal with high resistance to photorefraction—$LiNbO_3$: *Applied Physics Letters* 66, 3 (1995): 280–281.

26. Razzari, Luca, et al. "Photorefractivity of Hafnium-doped congruent lithium–niobate crystals." *Applied Physics Letters* 86, 13 (2005): 131914.

27. Beyer, O., et al. "Photorefractive effect in iron-doped lithium niobate crystals induced by femtosecond pulses of 1.5 μm wavelength." *Applied Physics Letters* 88, 5 (2006): 051120.

28. Perin, A. S., V. M. Shandarov, and V. Yu Ryabchenok. "Photonic waveguide structures in photorefractive lithium niobate with pyroelectric mechanism of nonlinear response." *Physics of Wave Phenomena* 24, 1 (2016): 7–10.

29. Sasaki, Takeo, et al. "Spontaneous polarization-vector-reorientation photorefractive effect in ferroelectric liquid crystals." *Applied Physics Letters* 78, 26 (2001): 4112–4114.

30. Sasaki, Takeo, et al. "Real-time dynamic hologram in photorefractive ferroelectric liquid crystal with two-beam coupling gain coefficient of over 800 cm^{-1} and response time of 8 ms." *Applied Physics Letters* 102, 6 (2013): 31.

31. Sasaki, Takeo, Satoshi Kajikawa, and Yumiko Naka. "Dynamic amplification of light signals in photorefractive ferroelectric liquid crystalline mixtures." *Faraday Discussions* 174 (2014): 203–218.

32. Badran, Hussain A., et al. "Determination of optical constants and nonlinear optical coefficients of Violet 1-doped polyvinyl alcohol thin film." *Pramana* 86, 1 (2016): 135–145.

33. Halasyamani, P. Shiv, and Weiguo Zhang. "Inorganic materials for UV and deep-UV nonlinear-optical applications." (2017): 12077–12085.

34. Liang, Fei, et al. "Molecular construction using (C3N3O3) 3–anions: analysis and prospect for inorganic metal cyanurates nonlinear optical materials." *Crystal Growth & Design* 17, 7 (2017): 4015–4020.

35. Kalmutzki, Markus Johannes, et al. "Formation, structure, and frequency-doubling effect of a modification of strontium cyanurate (α-SCY)." *Inorganic Chemistry* 56, 6 (2017): 3357–3362.

36. Muhammad, Shabbir. "Quantum chemical design of triple hybrid organic, inorganic and organometallic materials: an efficient two-dimensional second-order nonlinear optical material." *Materials Chemistry and Physics* 220 (2018): 286–292.

37. Liu, Xiu, et al. "Molecular structures and second-order nonlinear optical properties of ionic organic crystal materials." *Crystals* 6, 12 (2016): 158.

38. de Torres, Miriam, et al. "Extended π-conjugated ruthenium zinc–porphyrin complexes with enhanced nonlinear-optical properties." *Chemical Communications* 51, 14 (2015): 2855–2858.

39. Chen, Guanhua, Daqi Lu, and William A. Goddard III. "Valence-bond charge-transfer solvation model for nonlinear optical properties of organic molecules in polar solvents." *The Journal of Chemical Physics* 101, 7 (1994): 5860–5864.

40. Nakano, Masayoshi and Benoît Champagne. "Nonlinear optical properties in open-shell molecular systems." *Wiley Interdisciplinary Reviews: Computational Molecular Science* 6, 2 (2016): 198–210.
41. Parola, Stephane, et al. "Optical properties of hybrid organic-inorganic materials and their applications." *Advanced Functional Materials* 26, 36 (2016): 6506–6544.
42. Chen, Daqin, and Xiao Chen. "Luminescent perovskite quantum dots: synthesis, microstructures, optical properties and applications." *Journal of Materials Chemistry C* 7, 6 (2019): 1413–1446.
43. Chen, Tianliang, et al. "An organic–inorganic hybrid co-crystal complex as a high-performance solid-state nonlinear optical switch." *Journal of Materials Chemistry C* 4, 2 (2016): 266–271.
44. Medishetty, Raghavender, et al. "Nonlinear optical properties, upconversion and lasing in metal–organic frameworks." *Chemical Society Reviews* 46, 16 (2017): 4976–5004.
45. Liang, Weibin, Ravichandar Babarao, and Deanna M. D'Alessandro. "Microwave-assisted solvothermal synthesis and optical properties of tagged MIL-140A metal–organic frameworks." *Inorganic Chemistry* 52, 22 (2013): 12878–12880.
46. Liu, Bin-Wen, et al. "Syntheses, structures, and nonlinear-optical properties of metal sulfides $Ba2Ga8MS16$ (M= Si, Ge)." *Inorganic Chemistry* 54, 3 (2015): 976–981.
47. Day, Jared K., et al. "Microcavity enhanced second harmonic generation in 2D MoS 2." *Optical Materials Express* 6, 7 (2016): 2360–2365.
48. Stefanski, M., et al. "Size and temperature dependence of optical properties of Eu^{3+}: Sr_2CeO_4 nanocrystals for their application in luminescence thermometry." *Materials Research Bulletin* 76 (2016): 133–139.
49. Yu, Hongwei, et al. "Pb2BO3I: A borate iodide with the largest second-harmonic generation (SHG) response in the $KBe_2BO_3F_2$ (KBBF) family of nonlinear optical (NLO) materials." *Angewandte Chemie International Edition* 57, 21 (2018): 6100–6103.
50. Badran, Hussain A., et al. "Determination of optical constants and nonlinear optical coefficients of Violet 1-doped polyvinyl alcohol thin film." *Pramana* 86, 1 (2016): 135–145.
51. Kumar, KV Anil, et al. "Wavelength dependent nonlinear optical switching in electron beam irradiated CuTTBPc thin film." *RSC Advances* 6, 26 (2016): 22083–22089.
52. Tamgadge, Y. S., et al. "Thermally stimulated third-order optical nonlinearity in Cd-doped CuO–PVA thin films under CW laser illumination." *Applied Physics B* 120, 2 (2015): 373–381.
53. Yin, Congfei, et al. "Preparation and characterization of RF magnetron sputtered CuO/CaTi4O9 thin films with enhanced third-order nonlinear optical properties." *Materials Characterization* 126 (2017): 96–103.
54. Redel, Engelbert, et al. "On the dielectric and optical properties of surface-anchored metal-organic frameworks: a study on epitaxially grown thin films." *Applied Physics Letters* 103, 9 (2013): 091903.
55. Ulrich, Donald R. "Nonlinear optical polymer systems and devices." *Molecular Crystals and Liquid Crystals* 160, 1 (1988): 1–31.
56. Babaeian, Masoud, et al. "Nonlinear optical properties of chalcogenide hybrid inorganic/organic polymers (CHIPs) are using the Z-scan technique." *Optical Materials Express* 8, 9 (2018): 2510–2519.

57. Misra, Nilanjal, et al. "Nonlinear optical studies of inorganic nanoparticles–polymer nanocomposite coatings fabricated by electron beam curing." *Optics & Laser Technology* 79 (2016): 24–31.

58. Pradhan, Prabin, et al. "Optical limiting and nonlinear optical properties of gold-decorated graphene nanocomposites." *Optical Materials* 39 (2015): 182–187.

59. Biswas, S., et al. "Enhanced nonlinear optical properties of graphene oxide–silver nanocomposites measured by Z-scan technique." *RSC Advances* 6, 13 (2016): 10319–10325.

60. Shanshool, Haider Mohammed, et al. "Influence of polymer matrix on nonlinear optical properties and optical limiting threshold of polymer-ZnO nanocomposites." *Journal of Materials Science: Materials in Electronics* 27, 9 (2016): 9503–9513.

61. Sousani, Abbas, et al. "Synthesis of nanocomposites based on carbon nanotube/smart copolymer with nonlinear optical properties." *Optical Materials* 67 (2017): 172–179.

62. Pandule, Sudam S., Mahadeo R. Patil, and Rangappa S. Keri. "Properties and ammonia gas sensing applications of different inorganic acid-doped poly(2-chloroanilines)." *Polymer Bulletin* 75, 10 (2018): 4469–4483.

63. Fischer, Stefan, et al. "Precise tuning of surface quenching for luminescence enhancement in core–shell lanthanide-doped nanocrystals." *Nano Letters* 16, 11 (2016): 7241–7247.

64. Dini, Danilo, Mario JF Calvete, and Michael Hanack. "Nonlinear optical materials for the smart filtering of optical radiation." *Chemical Reviews* 116, 22 (2016): 13043–13233.

65. Sayyed, M. I., Saleem I. Qashou, and Z. Y. Khattari. "Radiation shielding competence of newly developed TeO_2-WO_3 glasses." *Journal of Alloys and Compounds* 696 (2017): 632–638.

CHAPTER 13

Nonlinear Optical Properties of Metal Nanoparticles and Their Composites Synthesized by Different Methods

REMYA VIJAYAN[1*], SIJO FRANCIS[2], and BEENA MATHEW[1]

[1]*School of Chemical Sciences, Mahatma Gandhi University, Kottayam, Kerala, India*

[2]*Department of Chemistry, St. Joseph's College, Moolamattom, Kerala, India*

Corresponding author. E-mail: remyavijayan88@gmail.com

ABSTRACT

The optical properties of metal nanoparticles (MNPs) have a huge significance in physical chemistry from Faraday's explorations of colloidal gold in the mid-1800s. It is possible to change the electronic and optical properties of the MNPs during material design, hence the MNPs possess an immense technological significance. Different methods have been used to synthesize MNPs with a broad collection of sizes, shapes, and dielectric environments. In this chapter, the nonlinear optical characteristics of different nanoparticles have been discussed and we have summarized several processes for the preparation of their composite materials. The nonlinear optical properties of these materials were studied by the Z-scan method.

13.1 INTRODUCTION

Nonlinear optics is a branch of modern optics and it has a good resemblance with the laser physics. The nonlinear optics explains different nonlinear effects that arise during the interaction between laser and matter [1, 2]. The optical characters of the material system are altered during the interaction of a laser with nonlinear optical materials [3]. The laser light is extremely strong to generate nonlinear optical phenomena. The invention of the second-harmonic generation by Franken et al. in 1961 resulted in the origin of the research field in nonlinear optics [4]. Most of the nonlinear optical phenomena observed in the presence of laser radiation, but some types of nonlinear optical effects like Pockel's and Kerr's electrooptic effects were identified prior to the invention of the laser [5]. The important aim of the nonlinear optics is to find out the occurrence of novel phenomena during the interaction of strong laser radiations and materials and study the explanations of these processes and also discover its promising applications. Nonlinear optics has enormous significance and extensive scientific values. The optical materials with large optical response can be applied in various photonic and optoelectronic applications.

The development of nanoscience and nanotechnology has given a good opportunity for nonlinear optics. A large number of nanomaterials exhibit significant nonlinear optical properties; this resulted in the production of several nano and nano-scale optoelectronic and photonic devices. Nonlinear optics has immense scientific importance and are widely applied in information and image processing, laser technology, and optical computing, biosensors, and imaging [6, 7]. Nanomaterials having vast nonlinear properties can be applied as contrast materials in nonlinear microscopy, photocatalysis, and optical limiting applications [8, 9]. Z-scanning technique is an important method to study of nonlinear optical properties. It was invented by Sheik–Bahae et al. in 1989 [10]. Here the test sample is needed to move along the direction of the optical axis, hence the method is known as the Z-scan method. It is an easy and simple method for evaluating the nonlinear optical properties and gives the idea about the magnitude of the optical linearity and its sign. In 1990, Sheik–Bahae et al. used this method to determine the nonlinear absorption coefficient of a measurement material [11].

13.2   NONLINEAR OPTICAL PROPERTIES OF METAL NANOPARTICLES

The metal nanoparticles (MNPs) have gained a significant consideration as potential nonlinear optical materials. Among the MNPs, gold and silver nanoparticles (NPs) have been attained more interest due to their broad surface plasmon resonance (SPR) absorption peak observed in the visible region of the electromagnetic spectrum. The SPR peak of the MNPs is affected by the size, shape, and dielectric parameters of the surrounding medium. The MNPs can be utilized for the manufacture of photonic devices due to the high stability, large third-order susceptibility, and ultrafast responses. Silver, gold, and copper MNPs show SPR peak in the visible region while the platinum, palladium, and chromium nanoparticles exhibit SPR band in the wavelengths shorter than 300 nm. There are various methods used to prepare composite materials with MNPs showing nonlinear optical properties. The optical properties of these composite materials studied using lasers with the operating frequencies in the region of the SPR of the nanoparticles. Because of the fast response and large value of nonlinear optical parameters, the composite based on MNPs gained huge attention. Under light illumination, the SPR of MNPs results in high linear absorption and stimulates nonlinear optical effects in the same spectral region. The composite material with a high concentration of MNPs shows large nonlinear optical effects. Thus these composite materials are very useful in optical device applications in optical computing, optical correlators, and phase conjugators [12]. The optical properties of some of the MNPs are discussed below.

13.2.1 NONLINEAR OPTICAL STUDIES OF GOLD NANOPARTICLES (AUNPS)

Several reports are available in the literature describing the nonlinear optical properties of gold nanoparticles. Ricard et al. reported the first result on the nonlinear optical effects of gold nanoparticles in 1985 [13]. They have synthesized gold nanoparticles with an average diameter of 10 nm and the third-order nonlinear susceptibility of 1.5×10^{-9} esu at 530 nm. The intensifications to the nonlinearities of the electrons in gold particles were also examined.

Compared to other composite materials, those prepared with gold nanoparticles embedded in a dielectric matrix are more important, due to its strong SPR absorption peak in the visible region [14]. These SPR peaks are controlled by the nanoparticle's surroundings and its shape. The exceptional linear and nonlinear optical properties of these materials increase their applicability in optical limiters [15], cancer treatment [16], optical switching [17], and others.

A report on the influence of the gold nanoparticles size on the nonlinear optical properties was reported by Sánchezdena et al. [18]. For this, they were prepared the nanoparticles with different diameters (5.1, 13.4, and 14.2 nm) and using the ion implantation method embedded them in the sapphire matrix. They established that the gold nanoparticles displayed a negative nonlinear absorption under 532-nm, 26-ps pulses, which increases with size. Using the nanosphere lithography (NSL), Yoon et al. synthesized the triangular gold nanoparticle arrays with four larger sizes on SiO_2 substrates. The sizes of the nanoparticles obtained were 37, 70, 140, and 190 nm. Their SPR peaks respectively located at 552, 566, 580, and 606 nm. A shift of the absorption peak toward the longer wavelength observed with increasing particle diameter [19].

In a study, the optical properties of gold nanoparticles embedded into various matrices (Al_2O_3, ZnO, and SiO_2) were analyzed using the Z-scan method. Moreover, the other parameters like the nonlinear absorption coefficient, nonlinear refractive index, and the real and imaginary parts of the third-order nonlinear susceptibility were also estimated [20]. Here, the annealing atmosphere alters the absorption coefficient when the wavelength was between 300 nm and 800 nm. The samples annealed in Ar exhibit a higher absorption coefficient than the sample annealed in air. SPR band of gold nanoparticles in absorption spectra appeared at 524 nm, 540 nm, and 500 nm respectively for Al_2O_3, ZnO, and SiO_2 [21–23]. The absorption coefficients obtained at the wavelength of 532 nm respectively were 3.3×10^{-16}, -9×10^{-15}, and 2.2×10^{-16} [24]. These results were much smaller than that of composite materials. Hence the nonlinear optical properties of the composite materials were originated due to the presence of gold nanoparticles.

13.2.2   NONLINEAR OPTICAL STUDIES OF SILVER NANOPARTICLES (AGNPS)

A considerable overlap among the interband absorption and the plasmon resonance absorption happens in the case of gold nanoparticles, this results in the considerable diminishes of the efficiency of plasmon excitation. But the plasmon excitation in silver nanoparticles is more effective than that in gold nanoparticles. It is due to the position of the interband transition absorption and SPR peak. In silver nanoparticles, the interband transition absorption located at about 320 nm is far away from its SPR wavelength of 400 nm. The separation of these two bands helps the individual analyze the nonlinear optical effects that originated from inters band transitions and

SPR. There are several research groups examined the nonlinear absorption of silver nanoparticles.

Gurudas et al. analyzed the picoseconds optical nonlinearity in silver nanodots at 532 nm. It was synthesized by using pulsed laser deposition [25]. Using the Z-scan technique, the nonlinear properties of these nanoparticle films were studied. The occurrence of various-sized and various-shaped nanoparticles in the samples was proved by broad SPR absorption.

In another study, Zheng et al. examined the shape-dependent nonlinear optical properties of nanostructured silver nanoplates, nanowires, and nanoparticles suspensions and their silica gel glass composites at both 532 and 1064 nm by using Z-scan technique [26, 27]. At these wavelengths, the nonlinear properties of nanoparticles depend on its shape.

Unnikrishnan and coworkers examined the nonlinear optical absorption in silver nanosol at different wavelengths (456, 477, and 532 nm) by open aperture Z-scan method [28]. These wavelengths were respectively located within the SPR band, on the border of the SPR band, and outside the SPR band. At higher input excitation, they are changes their behavior from saturable absorption (SA) to reverse saturable absorption (RSA). This type of behavior was also shown by silver nanoparticles in ZrO_2 [29] and silver nanoparticles in PMMA [30] under nanosecond laser pulse at 532 nm. However, at resonant wavelength, Ganeev and Ryasnyansky have examined the nonlinear optical absorption of silver nanoparticles and found that silver nanoparticles display either SA for 1.2-ps pulsed laser or RSA for 8-ns pulsed laser. Thus the nonlinear optical properties of materials depend on the wavelength and pulse width [31].

The noble metallic nanostructures embedded on a transparent dielectric matrix are widely used in optical signal processing devices because of their large nonlinear optical properties originated due to SPR and the quantum size effect [32, 33]. The more exceptional performance of the material is derived from its quantum size effect. Compared to other noble MNPs, the silver nanoparticle experiences a lower intrinsic loss of plasmonic energy at visible frequencies that gives rise to SPR [34]. The surface frequency of the MNPs depends on the shape of the nanoparticles and their dielectric environment. Which in turn affect the optical properties of nanoparticles [35, 36]. Sakho et al. prepared the noncovalent functionalized reduced graphene oxide (NF-RGO) and NF-RGO/Ag-NPs (NF-RGO with various concentrations of silver nanoparticles) and the open aperture Z-scan technique was used to study the nonlinear optics.

13.2.3   NONLINEAR OPTICAL STUDIES OF PALLADIUM (PD) AND PLATINUM (PT) NANOPARTICLES

The gold and silver nanoparticles show a strong SPR absorption peak in the visible region. Hence they have attained numerous attention and nonlinear optical properties of these nanoparticles are widely examined. However, the SPR of transition MNPs is positioned in the ultraviolet region and its nonlinear optical can be affected by other nonlinear processes. Palladium (Pd) and Platinum (Pt) nanoparticles are exhibit remarkable nonlinear optical properties like SA [37], RSA [38], and two-photon or multiphoton processes [39, 40]. The platinum nanoparticles are used different applications like optical limiting [41] and mode-locking [42]. The platinum nanoparticles exhibit SPR peak at 215 nm, situated away from an excitation wavelength of 532 nm [43], and observed SA at lower fluences [44].

13.2.4   NONLINEAR OPTICAL STUDIES OF SILICON-BASED NANOMATERIALS

There are a large number of reports are available on the nonlinear optical characters of silicon materials examined by different characterization methods. Silicon-based optical material has very significant applications in high-speed signal processing and no-chip communications [45]. And also used for the preparation of receivers, modulators, filters in optical communication [46], in real-time A-to-D converters. The silicon photonic devices can be easily produced by the mature silicon processing technology and which gives a low-cost, large-volume electronic circuit production [47].

In a study, a nanocrystalline Si/SiO_2 multilayer was synthesized and their cross-sectional microstructures were examined with the help of transmission microscopy and Raman spectroscopy. The photoluminescence band of this nanocrystalline material was centered at 870 nm. By Z-scanning method with the excitation of two laser pulses were employed for the study of multilayer nonlinear optical effect. At room temperature, the photoluminescence was observed at 870 nm. Here, the emission was originated by the recombination of photoexcited carriers via the interface states with an energy level inside the gap and it is revealed by the large Stokes shift between the linear absorption edge and emission band [48].

Zhang et al. reported the appearance of nonlinear absorption and nonlinear refraction in multilayers of Si/SiO_2 in femtosecond excitation of 800 nm. It has occurred through the transition process from the amorphous to nanocrys-talline phases. If the samples contain a large amount of amorphous Si phases, then the two-photon absorption process dominates. But the phonon-assisted one-photon transition process between the valence band and interface states dominates nonlinear optical properties in nanocrystalline-Si/SiO$_2$ multilayer. This indicates that this material can be used for the fabrication of sensitive photonic devices like optical switches and Q-switch lasers [49]. The silicon nanocrystals are extensively used in photonics. It can be used as nonlinear material in a variety of devices, like in bistable optical cavities, in waveguide optical mode monitors, and in wavelength shifters. In quantum random number production it is exploited as an entropy resource [50].

13.3   DIFFERENT METHODS USED FOR THE SYNTHESIS OF METAL NANOPARTICLES AND THEIR COMPOSITE MATERIALS

In the following session, we have discussed some of the various techniques used for the production of MNPs and their composite materials.

13.3.1   VACUUM ELECTRON-BEAM COEVAPORATION METHOD

The vacuum electron beam evaporation is a widely accepted powerful method for the preparation of nanoparticles and its composites. The source material is firstly evaporated and subsequently, thin-film coatings are formed on the surface of the substrate. This reaction is carried out at a temperature of above 3500 °C and gives a pure and thin film of nanomaterials [51].

Semiconductor nanocrystals or quantum dots are achieving numerous considerations because of their atom like behavior, which is the reason behind their higher optical and electrical properties [52]. Among the semi-conductor nanocrystals, the indirect bandgap semiconductors like silicon (Si) and germanium (Ge) are widely used in nanophotonics and nanoelectronics, hence these nanostructures are broadly studied. Also, these nanostructures are very much compatible with conventional integrated circuit technology [53, 54]. The optical and electrical properties of semiconductor nanostruc-tures embedded in dielectrics have gained a specific consideration. This is because of the following reason. Carriers in a semiconductor nanocrystal can three-dimensionally confined if the crystallite size of these semiconductors is lesser than the exciton Bohr radius. Hence the electronic state is supposed to display zero-dimensional features [55, 56]. The quantum confinement effect of Ge nanostructures is more obvious than that of Si nanostructures. This is due to the large Bohr radius of bulk Ge (24.3 nm) compared to that of Si (4.9 nm).

In a report, Wan et al. synthesized germanium (Ge) nanostructures embedded in Al_2O_3 dielectric by a vacuum electron beam evaporation method [57]. The amorphous film with uniform thickness and sharp interfaces were formed. In this study, they had noticed a strong blue-shift of the absorption band and a large third-order nonlinear optical susceptibility of the amorphous Ge nanocrystals deriving from the quantum confinement effect. Also, negative photoconductivity was observed in metal-insulator-semiconductor structures containing Ge nanocrystals. The photogenerated electrons charge the Ge nanocrystals and it screens the bias voltage. As a result current at a given voltage was decreased. Because of the quantum confinement effect, the bandgap of the semiconductor was increased which resulted in the blue-shift of the absorption band edge and optical luminescence [58, 59]. Additionally, the third-order optical nonlinearity was increased due to this effect. The optical absorption analysis was carried out at room temperature in the visible to near-infrared range. Compared to other dielectrics, Al_2O_3 has several advantages like a high dielectric constant and high bandgap which help to survive high-temperature processing steps [60].

13.3.2   NANOSPHERE LITHOGRAPHY

NSL is an easy technique employed for the production of periodic particle array (PPA) surfaces with nanometer-scale properties. In 1995, Hulteen et al. developed NSL from natural lithography and they have demonstrated that it is a powerful method for the production of periodic particle arrays with adjustable shape and magnitude, and for the analysis of the optical characters of nanoparticles [61].

They have prepared a variety of PPA surfaces using identical single-layer and double-layer NSL masks by self-assembly of polymer nanospheres with diameter $D=5264$ nm and with changing both the substrate material S and the particle material M. Here, substrate material was an insulator, semiconductor, or metal and the particle material was metal, inorganic ionic insulator, or an organic p-electron semiconductor. The mica, Si (100), Si (111), or Cu (100) were the substrate materials, and the particle materials used were M as Ag, CaF_2, and cobalt phthalocyanine (CoPc).

13.3.3   ORGANOMETALLIC PYROLYSIS METHOD

The organometallic pyrolysis method is widely used for the preparation of nanoparticles and its composite materials. Salah et al. synthesized Au and Au–CdSe nanocomposite with different nanoparticles size and concentrations by the organometallic pyrolysis method [62]. Using the open aperture Z-scan technique at near-resonant excitation using 532 nm of 6 ns Q-switched Nd-YAG laser, the nonlinear properties of the samples were investigated. Dependence of nonlinear optical properties on laser excitation energy is examined for different nanoparticles size and concentrations. As the nanoparticle's size increases, the location of the excitonic peak changed to 613 from 538 nm. The increase was observed in nonlinear absorption coefficients are attributed to the resonance between surface plasmon of gold nanoparticles and quantum dots exciton band.

13.3.4   MELT-QUENCHING TECHNIQUE

Melt-quenching technique is widely used for the preparation of glasses containing MNPs and their composites. The other methods used for the preparation of glasses containing MNPs are sol–gel method, ion implantation, sputtering, ion exchange, and others [63, 64]. These materials are extensively used in the optical switching, ultrafast imaging, signal processing, optical telecommunications because of its special features like strong resorption of the surface plasmon, large third-order optical nonlinearities for a large local-field enhancement factor, and ultrafast electron response within a few of picoseconds [65,66].

Lin et al. introduced an easy method for the formation of bismuth (Bi) nanoparticles in glasses [67]. The atomic state Bi was prepared by the reduction of Bi ions by Al. Later the atomic or molecular Bi combined together. After subsequent heat treatment, Bi grows and forms nanocrystals. As the heat treatment temperature increase, the average size of Bi nanoparticles increases. With increasing of heat treatment temperature, the fundamental absorption edge displays a redshift due to the size effect and multiple scattering of nanoparticles. The Z-scan technique was used to examine the nonlinear optical natures of the Bi nanoparticle composite glasses and they have concluded that Bi nanoparticle doped glasses exploited as potential material for optical switching.

13.3.5   COLLOIDAL CHEMICAL SYNTHESIS

The nanoparticles and nanocomposites synthesized by the chemical route have some advantages. The chemical route gives an opportunity to manage the size, shape, and distribution of nanoparticles. Most importantly by this method, we can enhance the crystallinity of the nanoparticles by changing the concentration of the reagents and their combination at various temperatures. The polymer nanocomposites (PNCs) are generally prepared by using the chemical route and these materials have a wide variety of applications. It is a multiphase material containing a dispersion of nanosized particles within the polymer matrix.

CdSe is one of the promising semiconducting materials among the II–VI semiconductors and it is used in solar cells, thin-film transistors, memory devices, gamma-ray detectors, and optoelectronics [68, 69]. Tripathi and coworkers proposed a colloidal chemical synthesis technique for the preparation of Ag-CdSe hybrid PNC by mixing the silver colloids with CdSe PNC [70]. The effect of different Ag concentrations on linear and nonlinear optical properties of the PNC films has been examined by means of the Z-scan method. The large free carrier absorption originating from the local field enhancement effect of silver causes the increase in the nonlinearity of CdSe PNC films with increasing silver content. The consequence of the quantum confinement and thermo-optical effects is the occurrence of large nonlinearity effects. Therefore, the Ag–CdSe hybrid PNCs with huge nonlinear performance can be used in nonlinear photonic devices.

13.4   CONCLUSION

Nonlinear optics was originated soon after the innovation of the laser in 1960 and its research field has been constantly emerging. A nonlinear optics study includes several types of nonlinear effects that happen during the interactions between a laser and matter. To date, nonlinear optics gained a wide range of applications. In this chapter, we have discussed some of the nonlinear characters of some MNPs and their composites along with their application. Also, we have briefly explained some important methods for the preparation of different MNPs and nanocomposites with their nonlinear optical behavior. The investigation on nonlinear optical properties of MNPs is very essential for the invention of new optoelectronic elements and device

KEYWORDS

- **nonlinear optical properties**
- **metal nanoparticles**
- **Z-scan**
- **nanosphere lithography**
- **melt-quenching technique**
- **organometallic pyrolysis method**

REFERENCES

1. Boyd, R., Masters, M., "Book review: nonlinear optics, third edition." *Journal of Biomedical Optics*, 2009, 14(2), 029902.
2. Marder, Seth R., John E. Sohn, and Galen D. Stucky. *Materials for Nonlinear Optics Chemical Perspectives*. No. ACS-SYMPOSIUM-SER-455. American Chemical Society Washington DC, 1991.
3. Shanon, Zahraa S., and Raad Sh. "Study of the Nonlinear Optical Properties of Lithium Triborate Crystal by Using Z-Scan." *International Journal of Science and Research (IJSR)* (2016): 1683.
4. Franken, PA, et al. "Generation of optical harmonics." *Physical Review Letters* 7.4 (1961): 118.
5. Schroer, Christian G., and Bruno Lengeler. "X-ray optics." *Springer Handbook of Lasers and Optics*. Springer, Berlin, Heidelberg, 2012. 1461–1474.
6. Ju, Seongmin, et al. "Nonlinear optical properties of zinc doped germano-silicate glass optical fiber." *Journal of Nonlinear Optical Physics & Materials* 19.04 (2010): 791–799.
7. Wang, Yong, et al. "Four-wave mixing microscopy of nanostructures." *Advances in Optics and Photonics* 3.1 (2011): 1–52.
8. Ren, Pengrong, Huiqing Fan, and Xin Wang. "Electrospun nanofibers of ZnO/BaTiO3 heterostructures with enhanced photocatalytic activity." *Catalysis Communications* 25 (2012): 32–35.
9. Krishnan, Shiji, et al. "Two-photon assisted excited state absorption in multiferroic YCrO3 nanoparticles." *Chemical Physics Letters* 529 (2012): 59–63.
10. Sheik-Bahae, Mansoor, Ali A. Said, and Eric W. Van Stryland. "High-sensitivity, single-beam n 2 measurements." *Optics Letters* 14.17 (1989): 955–957.
11. Sheik-Bahae, M., David J. Hagan, and Eric W. Van Stryland. "Dispersion and band-gap scaling of the electronic Kerr effect in solids associated with two-photon absorption." *Physical Review Letters* 65.1 (1990): 96.
12. Tripathi, S. K., et al. "Third-order nonlinear optical response of Ag–CdSe/PVA hybrid nanocomposite." *Applied Physics A* 120.3 (2015): 1047–1057.
13. Ricard, D., Ph Roussignol, and Chr Flytzanis. "Surface-mediated enhancement of optical phase conjugation in metal colloids." *Optics Letters* 10.10 (1985): 511–513.

14. Wang, Kai, et al. "Intensity-dependent reversal of nonlinearity sign in a gold nanoparticle array." *Optics Letters* 35.10 (2010): 1560–1562.

15. Philip, Reji, et al. "Picosecond optical nonlinearity in monolayer-protected gold, silver, and gold-silver alloy nanoclusters." *Physical Review B* 62.19 (2000): 13160.

16. Huang, Xiaohua, Svetlana Neretina, and Mostafa A. El-Sayed. "Gold nanorods: from synthesis and properties to biological and biomedical applications." *Advanced Materials* 21.48 (2009): 4880–4910.

17. Gibbs Hyatt, M. *Optical Bistability: Controlling Light with Light.* Academic Press, 1985.

18. Sánchez-Dena, O., et al. "Size-and shape-dependent nonlinear optical response of Au nanoparticles embedded in sapphire." *Optical Materials Express* 4.1 (2014): 92–100.

19. Jun, H-S., et al. "3rd order nonlinear optical properties of Au: SiO_2 nanocomposite films with varying Au particle size." *Physica Status Solidi (a)* 203.6 (2006): 1211–1216.

20. Sánchez-Dena, O., et al. "Size-and shape-dependent nonlinear optical response of Au nanoparticles embedded in sapphire." *Optical Materials Express* 4.1 (2014): 92–100.

21. Dakka, Anass, et al. "Optical properties of $Ag–TiO_2$ nanocermet films prepared by cosputtering and multilayer deposition techniques." *Applied Optics* 39.16 (2000): 2745–2753.

22. Pal, U., et al. "Preparation of Au/ZnO nanocomposites by radio frequency co-sputtering." *Solar Energy Materials and Solar Cells* 70.3 (2001): 363–368.

23. Pinçon, N., et al. "Third-order nonlinear optical response of Au: SiO2 thin films: Influence of gold nanoparticle concentration and morphologic parameters." *The European Physical Journal D-Atomic, Molecular, Optical and Plasma Physics* 19.3 (2002): 395–402.

24. Zhang, X. J., W. Ji, and S. H. Tang. "Determination of optical nonlinearities and carrier lifetime in ZnO." *JOSA B* 14.8 (1997): 1951–1955.

25. Gurudas U, Brooks E, Bubb DM, Heiroth S, Lippert T, Wokaun A. Saturable and reverse saturable absorption in silver nanodots at 532 nm using picosecond laser pulses. *Journal of Applied Physics*. 2008;104 (7): 073107–073125.

26. Zheng, Chan, et al. "Shape dependence of nonlinear optical behaviors of nanostructured silver and their silica gel glass composites." *Applied Physics Letters* 93.14 (2008): 143108.

27. Zheng, C., et al. "Observation of nonlinear saturable and reverse-saturable absorption in silver nanowires and their silica gel glass composite." *Applied Physics B* 101.4 (2010): 835–840.

28. Unnikrishnan, K. P., et al. "Nonlinear optical absorption in silver nanosol." *Journal of Physics D: Applied Physics* 36.11 (2003): 1242.

29. Anija, M., et al. "Nonlinear light transmission through oxide-protected Au and Ag nanoparticles: an investigation in the nanosecond domain." *Chemical Physics Letters* 380.1–2 (2003): 223–229.

30. Deng, Yan, et al. "Nonlinear optical properties of silver colloidal solution by in situ synthesis technique." *Current Applied Physics* 8.1 (2008): 13–17.

31. Ganeev, R. A., and A. I. Ryasnyansky. "Nonlinear optical characteristics of nanoparticles in suspensions and solid matrices." *Applied Physics B* 84.1–2 (2006): 295–302.

32. Hou, Wenbo, and Stephen B. Cronin. "A review of surface plasmon resonance-enhanced photocatalysis." *Advanced Functional Materials* 23.13 (2013): 1612–1619.

33. Antosiewicz, Tomasz J., and S. Peter Apell. "Plasmonic glasses: optical properties of amorphous metal-dielectric composites." *Optics Express* 22.2 (2014): 2031–2042.
34. Li, Rang, et al. "Giant enhancement of nonlinear optical response in Nd: YAG single crystals by embedded silver nanoparticles." *ACS Omega* 2.4 (2017): 1279–1286.
35. Hua, Yi, et al. "Shape-dependent nonlinear optical properties of anisotropic gold nanoparticles." *The Journal of Physical Chemistry Letters* 6.24 (2015): 4904–4908.
36. Sato, Rodrigo, et al. "Experimental investigation of nonlinear optical properties of Ag nanoparticles: Effects of size quantization." *Physical Review B* 90.12 (2014): 125417.
37. Ganeev, R. A., et al. "Low-and high-order nonlinear optical properties of Au, Pt, Pd, and Ru nanoparticles." *Journal of Applied Physics* 103.6 (2008): 063102.
38. Qu, Shiliang, et al. "A theoretical and experimental study on optical limiting in platinum nanoparticles." *Optics Communications* 203.3–6 (2002): 283–288.
39. Papagiannouli, I., et al. "Third-order optical nonlinearities of PVP/Pd nanohybrids." *Optical Materials* 72 (2017): 226–232.
40. Iliopoulos, Konstantinos, et al. "Preparation and nonlinear optical response of novel palladium-containing micellar nanohybrids." *Optical Materials* 33.8 (2011): 1342–1349.
41. Chehrghani, A., and M. J. Torkamany. "Nonlinear optical properties of laser synthesized Pt nanoparticles: saturable and reverse saturable absorption." *Laser Physics* 24.1 (2013): 015901.
42. Ganeev, R. A., R. I. Tugushev, and T. Usmanov. "Application of the nonlinear optical properties of platinum nanoparticles for the mode locking of Nd: glass laser." *Applied Physics B* 94.4 (2009): 647–651.
43. Henglein, A., B. G. Ershov, and M. Malow. "Absorption spectrum and some chemical reactions of colloidal platinum in aqueous solution." *The Journal of Physical Chemistry* 99.38 (1995): 14129–14136.
44. Gao, Yachen, et al. "Saturable absorption and reverse saturable absorption in platinum nanoparticles." *Optics Communications* 251.4–6 (2005): 429–433.
45. Tsang, H. K., and Y. Liu. "Nonlinear optical properties of silicon waveguides." *Semiconductor Science and Technology* 23.6 (2008): 064007.
46. Ji, Hua, et al. "1.28-Tb/s demultiplexing of an OTDM DPSK data signal using a silicon waveguide." *IEEE Photonics Technology Letters* 22.23 (2010): 1762–1764.
47. Miller, David AB. "Optical interconnects to electronic chips." *Applied Optics* 49.25 (2010): F59-F70.
48. Zhang, Pei, et al. "Interface state-related linear and nonlinear optical properties of nanocrystalline Si/SiO$_2$ multilayers." *Applied Surface Science* 292 (2014): 262–266.
49. Zhang, Pei, et al. "Tunable nonlinear optical properties in nanocrystalline Si/SiO2 multilayers under femtosecond excitation." *Nanoscale Research Letters* 9.1 (2014): 28.
50. Bisadi, Z., et al. "Silicon nanocrystals for nonlinear optics and secure communications." *Physica Status Solidi (a)* 212.12 (2015): 2659–2671.
51. Wan, Q., et al. "Resonant tunneling of Si nanocrystals embedded in Al 2 O 3 matrix synthesized by vacuum electron-beam co-evaporation." *Applied Physics Letters* 81.3 (2002): 538–540.
52. Pavesi, L. "Dal Negro, C. Mazzoleni, G. Franzo, and F. Priolo." *Nature* 408 (2000): 440.
53. Canham, Leigh T. "Silicon quantum wire array fabrication by electrochemical and chemical dissolution of wafers." *Applied Physics Letters* 57.10 (1990): 1046–1048.
54. Pavesi, Lorenzo, et al. "Optical gain in silicon nanocrystals." *Nature* 408.6811 (2000): 440–444.

55. Zhuravlev, K. S., A. M. Gilinsky, and A. Yu Kobitsky. "Mechanism of photoluminescence of Si nanocrystals fabricated in a SiO_2 matrix." *Applied Physics Letters* 73.20 (1998): 2962–2964.

56. Choi, Suk-Ho, and R. G. Elliman. "Reversible charging effects in SiO_2 films containing Si nanocrystals." *Applied Physics Letters* 75.7 (1999): 968–970.

57. Wan, Q., T. H. Wang, and C. L. Lin. "Third-order optical nonlinearity and negative photoconductivity of Ge nanocrystals in Al2O3 dielectric." *Nanotechnology* 14.11 (2003): L15.

58. Brenner, M. P., et al. "Brenner et al. Reply." *Physical Review Letters* 80.16 (1998): 3668.

59. Ledoux, G., et al. "Photoluminescence of size-separated silicon nanocrystals: Confirmation of quantum confinement." *Applied Physics Letters* 80.25 (2002): 4834–4836.

60. Roy Chowdhuri, A., et al. "Metalorganic chemical vapor deposition of aluminum oxide on Si: evidence of interface SiO_2 formation." *Applied Physics Letters* 80.22 (2002): 4241–4243.

61. Hulteen, John C., and Richard P. Van Duyne. "Nanosphere lithography: A materials general fabrication process for periodic particle array surfaces." *Journal of Vacuum Science & Technology A: Vacuum, Surfaces, and Films* 13.3 (1995): 1553–1558.

62. Salah, Abeer, et al. "Effects of nanoparticles size and concentration and laser power on nonlinear optical properties of Au and Au–CdSe nanocrystals." *Applied Surface Science* 353 (2015): 112–117.

63. Mattei, Giovanni, Paolo Mazzoldi, and Harry Bernas. "Metal nanoclusters for optical properties." *Materials Science with Ion Beams*. Springer, Berlin, Heidelberg, 2009. 287–316.

64. Uchida, K., et al. "Optical nonlinearities of a high concentration of small metal particles dispersed in glass: copper and silver particles." *JOSA B* 11.7 (1994): 1236–1243.

65. Garcia, Hernando, Ramki Kalyanaraman, and Radhakrishna Sureshkumar. "Nonlinear optical properties of multi-metal nanocomposites in a glass matrix." *Journal of Physics B: Atomic, Molecular and Optical Physics* 42.17 (2009): 175401.

66. Wang, Y. H., et al. "Nonlinear optical properties of Cu nanoclusters by ion implantation in silicate glass." *Optics Communications* 283.3 (2010): 486–489.

67. Lin, Geng, et al. "Linear and nonlinear optical properties of glasses doped with Bi nanoparticles." *Journal of Non-Crystalline Solids* 357.11–13 (2011): 2312–2315.

68. Cheng, Wei-Jen, Ding-Jhih Chen, and Chaur-Jeng Wang. "High-temperature corrosion of Cr–Mo steel in molten $LiNO_3$–$NaNO_3$–KNO_3 eutectic salt for thermal energy storage." *Solar Energy Materials and Solar Cells* 132 (2015): 563–569.

69. Kaur, Ramneek, and S. K. Tripathi. "Study of conductivity switching mechanism of CdSe/PVP nanocomposite for memory device application." *Microelectronic Engineering* 133 (2015): 59–65.

70. Tripathi, S. K., et al. "Third-order nonlinear optical response of Ag–CdSe/PVA hybrid nanocomposite." *Applied Physics A* 120.3 (2015): 1047–1057.

CHAPTER 14

Phosphors: A Promising Optical Material for Multifunctional Applications

VIJI VIDYADHARAN[1*], KAMAL P. MANI[2*], CYRIAC JOSEPH[3], and M. KAILASNATH[2]

[1]*Department of Optoelectronics, University of Kerala, Thiruvananthapuram 695581, Kerala, India*

[2]*International School of Photonics, Cochin University of Science and Technology, Kochi 682022, Kerala, India*

[3]*School of Pure & Applied Physics, Mahatma Gandhi University, Kottayam 686560, Kerala, India*

Corresponding author. E-mail: vijiv.opto@keralauniversity.ac.in; kamalspap@gmail.com

ABSTRACT

The solid luminescent materials that convert energy into electromagnetic radiation, usually in the visible region, are termed as phosphor materials. The emission of electromagnetic radiation or light is called luminescence, which is the basic mechanism behind phosphors or luminescent materials. Based on various excitation sources, the luminescence has many classifications like photoluminescence, chemiluminescence, bioluminescence, radioluminescence, and electroluminescence. Phosphors are generally made from host lattices, in which a small amount of impurities or dopants or activators are added intentionally. These impurities can be transition metal ions, rare earth (RE) ions, or complex ions. Also, there is a vast area of host materials are available as silicates, oxides, titanates, molybdates, tungstates, and others. An efficient phosphor material for luminescence applications, we should choose suitable host material and activator ion. For this, there are a number of phosphor synthesis methods available, which affect the physical and luminescence properties of the phosphor material. Also, the phosphor materials find many applications in various fields and some of them include lighting applications, cathode ray tubes, scintillators, display devices, long persistent phosphors, light emitting diodes, organic fluorescent pigments, and others. This is the theme of this chapter.

In this chapter, we discuss about the phosphors, various luminescence and their basic mechanisms in Sections 14.1 and 14.2. Sections 14.3, 14.4, and 14.5 briefly introduce the importance and properties of activators (dopants), codopants, and host materials used in the phosphor material, respectively. Various phosphor synthesis methods and their advantages are explained in Section 14.6. Finally, Section 14.7 deals with various applications of phosphor materials and important optical properties of some of RE-doped oxide phosphors are also included.

14.1   PHOSPHORS

Phosphors are solid luminescent materials that convert energy into electromagnetic radiation, usually in the visible region. In other words, phosphors are the materials that emit photons when excited by an external energy source. This emission of light is termed as luminescence, originated from the latin word "luminis" means light. Generally, luminescence can be divided into two: fluorescence and phosphorescence. The word fluorescence came from the Latin word "fluor" means flow and phosphorescence from the Greek words "phos" (means light) and "phoros" (means carrying). In terms of decay time, we can define fluorescence and phosphorescence as for fluorescence the decay time is less than 10 ms (usually $\approx 10^{-8}$–10^{-9} s) and for phosphorescence the decay time is greater than 0.1 s (usually 0.1 s to few hours). Fluorescence is very fast with a spin allowed transition ($\Delta s = 0$), whereas phosphorescence is very slow with a spin-forbidden transition ($\Delta s = 1$). However, in terms of modern usage, light emission from a material during the time it is exposed to exciting radiation is termed as fluorescence, whereas the emission after the excitation ceased off is termed as phosphorescence.

Simply, phosphors emit energy from an excited electron as light. This excitation of an electron can be caused by absorption of energy from an external source such as another electron, a photon or an electric field. An excited electron occupies a quantum state whose energy is above the minimum energy ground state. In the case of solids, semiconductors, and insulators, the electronic ground state is referred to the levels in the valence band, which is completely filled with electrons. The excited quantum state often lies in the conduction band, which is empty and separated from the valence band by an energy bandgap. So minimum energy equal to the bandgap energy is necessary to excite an electron in a semiconductor or insulator, and the energy released during the de-excitation of the electrons is often nearly equal to the bandgap. Under normal conditions, electrons are forbidden to have energies between the valence band and the conduction band. During the de-excitation process electrons lose the excess energy before coming to rest at the lowest energy in the conduction band. This excess energy in the form of photons is termed as luminescence.

14.2 CLASSIFICATION OF LUMINESCENCE ON THE BASIS OF NATURE OF EXCITATION

14.2.1 PHOTOLUMINESCENCE

Photoluminescence is the spontaneous light emission from any form of matter under optical excitation. Usually, the emission is observed by the excitation with electromagnetic radiation often Ultraviolet, visible light, or infrared. In other words, the process of photon excitation followed by photon emission is known as photoluminescence. The fluorescent lamps for the household and general lighting devices are based on photoluminescence.

14.2.2 ELECTROLUMINESCENCE

It is the luminescence caused by electric current or a strong electric field. There is another type of electroluminescence known as injection luminescence. Here the electrons are injected from an external supply across a p–n junction. On applying a direct voltage across the junction, such that the electrons flow to the p region, luminescence is produced by electron–hole recombination in that region. Light-emitting diodes (LEDs) that are commonly used for display and lighting applications are based on this principle. Electroluminescent devices are fabricated using organic or inorganic electroluminescent materials. Generally, semiconductors of wide bandgap are used as active materials in order to allow the exit of the light. Powder phosphor-based electroluminescent panels are frequently used as backlights in liquid crystal displays.

14.2.3   X-RAY LUMINESCENCE

It is the luminescence from materials that absorbs x-rays and converts the absorbed energy efficiently into UV or visible emission.

14.2.4   CATHODOLUMINESCENCE

It is an optical and electromagnetic phenomenon in which a beam of electrons impact on a luminescent material, such as a phosphor, causing the emission of photons in the visible spectrum. It is the reverse process of the photoelectric effect, in which electron emission is induced by irradiation with photons. The cathodoluminescence in a material depends on its composition, lattice structure, and imposed stress or strain on the structure of the material. The familiar example is the TV screen that uses a cathode ray tube (CRT). In CRTs, sulfide-based phosphors like zinc sulfide or cadmium sulfide are used.

14.2.5   RADIOLUMINESCENCE

It is the phenomenon by which light is produced in a material by the bombardment of ionized radiation such as beta particles. In radioluminescent light sources, a radioactive substance is mixed with phosphor materials to produce light of a particular color. Radioluminescence can see around high power radiation sources like nuclear reactors and radioisotopes. It is used as a light source for night illumination, signboards, or other applications where light must be produced for long periods without external energy sources.

14.2.6   MECHANOLUMINESCENCE

It is also known as triboluminescence or fractoluminescence. An inorganic and organic material subjected to mechanical stress emits light is called mechanoluminescence. It has been observed that all piezoelectric crystals exhibit triboluminescence. The spectra of mechanoluminescence are similar to those of photoluminescence in many substances.

14.2.7   CHEMILUMINESCENCE

It is the luminescence where the energy is supplied by chemical reactions. But not all the chemical molecules are capable of luminescence. Oxidation of white phosphorous in the air is the best-known example of chemiluminescence.

14.2.8   BIOLUMINESCENCE

Luminescence caused by chemical reactions in living things is a form of chemiluminescence. Biochemical reactions inside the cells of some living organisms produce electronically excited states of the biomolecules, which results in luminescence. Fireflies, some bacteria and fungi, angelfish, many sea creatures, and others are the examples of luminescence living things.

14.2.9   THERMOLUMINESCENCE

It is the re-emission of previously absorbed electromagnetic or other ionizing radiation, from certain crystalline materials such as minerals upon heating. The major applications of thermoluminescent materials are in dosimeter for a dating application. $CaSO_4$, LiF, and CaF_2 are some of the examples of common thermoluminescent materials [2].

14.3   BASIC MECHANISMS OF PHOTOLUMINESCENCE

The basic mechanism observed in photoluminescent materials is the emission of light as a result of the absorption of light energy; in other words emission of light when excited by photons. UV–Vis–IR part of the electromagnetic radiations are commonly used as the excitation energy source. We know for all solids there is an energy gap between the valence band and the conduction band. So the electrons in the valence band are forbidden to reach the conduction band under normal conditions. But a photon that has energy greater than the bandgap energy can be absorbed by the valence electron and can reach the conduction band. This process is termed as photoexcitation. During this process, the electron has an excess energy that it loses before coming to the lowest energy level in the conduction band. This excess energy is released or emitted from the material and the process is known as emission. Figure 14.1 shows the photoluminescence mechanism.

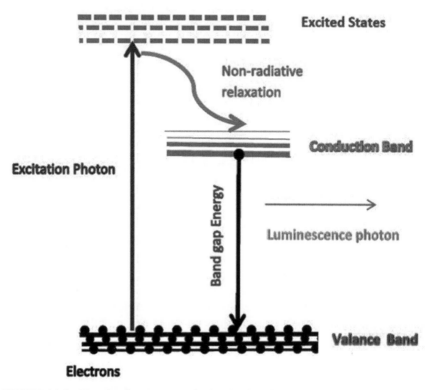

FIGURE 14.1   Photoluminescence mechanism in phosphors.

This emission process can occur as radiative transition, nonradiative transition, multiphonon relaxation, cross-relaxation, or upconversion [3]. The visible emission from a luminescent material is a process of returning to the ground state by radiative emission. In this case, efficiency is the ratio of the number of photons emitted to the number of photons absorbed. But the radiative emission can also occur by the assistance of nonradiative transitions such as multiphonon relaxation, and energy transfer between different ions or same ions. The energy transfer between the same ions is known as cross-relaxation. The energy absorbed by the luminescent materials, which is not emitted as radiation, is dissipated to the crystal lattice, is known as nonradiative transition.

14.4 ACTIVATORS/DOPANTS

Phosphors are often made from the combination of a crystalline host lattice and a small amount of certain impurities, the activator ion, or dopants. Generally, the host matrix should be nonabsorbing to the radiation source used for the excitation process. The activator ion absorbs energy and gets excited. The luminescence occurs when this excited activator ion de-excites to the ground state by releasing energy in the form of photons. The proper choice of host lattice and activator ions is essential to obtain an efficient phosphor material. The activator ions used in the phosphor materials can be transition metals, complex ions, or rare-earth ions.

14.4.1 TRANSITION METALS

Transition metal ions are widely used as luminescent centers in commercial phosphors. They have an electronic configuration of $1s^2\, 2s^2\, 2p^6\, 3s^2\, 3p^6\, 3d^n\, 4s^2$, where n means the number of 3d electrons ($1 < n < 10$). These 3d valence electrons are responsible for the optical transitions. Also these 3d orbitals are not shielded from the outer shells, which leads to broad spectral bands ($s > 0$) or sharp spectral lines ($s<0$) because of the strong electron-lattice coupling effect [4]. Ti^{3+}, V^{4+}, V^{3+}, V^{2+}, Cr^{4+}, Cr^{3+}, Cr, Mn^{5+}, Mn^{4+}, Mn^{3+}, Mn^{2+}, Fe^{3+}, Fe^{2+}, Fe^+, Co^{3+}, Co^{2+}, Co^+, Ni^{3+}, Ni^{2+}, Ni^+, Cu^{2+}, and others are the most commonly used transition metal activators in phosphors.

14.4.2 COMPLEX IONS

Complex ions are an important class of luminescent centers and are widely used luminescence centers in practical phosphors. Molybdate (MoO_4^{2-}), vanadate (VO_4^{3-}), and tungstate (WO_4^{2-}) are the examples of complex ions and the materials that contain these are known as scheelite compounds [4]. The luminescence center of MoO_4^{2-} complex ion consists of a central Mo metal ion, coordinated by four O^{2-} ions in tetrahedral symmetry. The intrinsic luminescence of these complex ions is often observed in the blue to green spectral regions due to the spin-forbidden $^3T_1 \rightarrow {}^1A_1$ transition. Also, scheelite phosphors exhibit efficient absorption of blue or near-ultraviolet light due to strong broad charge-transfer band enabling them to have an efficient energy transfer from the complex ions to rare-earth ions, which finds potential application in white LEDs.

14.4.3 RARE EARTH IONS

The rare earth (RE) elements mainly include 17 elements consisting of the 15 lanthanides from lanthanum to lutetium, scandium, and yttrium. The electronic configuration of a lanthanide ion can be written as follows:

$1s^2 \, 2s^2 \, 2p^6 \, 3s^2 \, 3p^6 \, 3d^{10} \, 4s^2 \, 4p^6 \, 4d^{10} \, 4f^n \, 5s^2 \, 5p^6 \, 6s^2$, where n is an integer from 2 to 14.

The partially filled 4f electronic energy levels of lanthanide ions are characteristic of each ion. The levels are not affected much by the environment because 4f electrons are well shielded from external electric fields by the outer $5s^2$ and $5p^6$ electrons. This feature is in strong contrast with transition metal ions, whose 3d elements located in an outer orbit, which is heavily affected by the environmental or crystal electric field. When incorporated in crystalline or amorphous host materials, the RE ions exist as +3 or +2 ions. The triply charged RE ions exhibit narrowband intra 4f luminescence in a wide variety of host materials. Also, some of the RE ions, like Eu^{2+} and Sm^{2+}, exhibit luminescence [4, 5]. Because of the shielding of 4f electrons, the positions of the RE electronic levels are much more influenced by spin–orbit interactions than by the applied crystal field. The characteristic energy levels of 4f electrons of trivalent lanthanide ions have been thoroughly investigated by Dieke and his coworkers and are known as Dieke diagram [5]. The Dieke diagram is applicable to RE^{3+} ions in almost any environment because the maximum variation of the energy levels is, at most, of the order of several hundred cm^{-1}. There are mainly two mechanisms for light emission in RE-doped phosphor materials. They are 4f–4f transitions and 4f–5d transitions.

4f–4f transitions: In the case of 4f–4f transitions electrons are transferred between different energy levels of the 4f orbitals of the same RE ion. But these transitions are forbidden since the parity selection rule says that electronic transitions between energy levels with the same parity cannot occur [6]. But their 4f–4f transitions can occur because the parity selection rule is relaxed due to a perturbation such as electron–vibration coupling and uneven crystal field effect from the host lattice [4]. In the case of RE ions the 4f electrons are shielded from the external fields by 5s and 5p electrons, so the crystal field effect from the host is very small. Thus, RE^{3+} ion is uniquely characterized by its energy levels, relatively independent of host matrices. So RE^{3+} ions show narrow f–f absorption and emission.

4f–5d transitions: There are a number of RE^{3+} ions like Ce^{3+}, Pr^{3+}, Tb^{3+}, Er^{3+}, and Eu^{3+} emit through an intra-band 4f–4f transitions or a 5d–4f transitions. The 4f–5d transitions are dipole allowed and unlike the 4f state, the 5d states can diffuse and overlap with ligand orbitals. This is due to the strong dependence of 4f–5d transition in the host material. Thus with increasing host crystal field the emission will change from line emission in the UV region to band emission in the visible region.

RE-doped phosphor materials have a very important role in optoelectronics and are widely used in solid-state lighting applications [7–10]. The RE-doped phosphors have high efficiency and color tunability, which can be varied by doping different RE ions with different compositions [7].

14.5   CO-DOPANTS

In addition to the activator ions or dopants, there is one more component that is deliberately added in to the phosphor material is known as co-dopant. This co-dopant can be sensitizers or charge compensators.

14.5.1 SENSITIZER

In some cases, the exciting radiation is not absorbed by the activator ion. So we need to add another ion in the host lattice called sensitizer. Now the sensitizer ion absorbs the exciting radiation and transfers it to the activator ion. Thus in the case of phosphor materials, luminescence can be achieved by the direct excitation of the activator ion or by the indirect excitation via a sensitizing ion. In this case, the sensitizer and activators are termed as donor and acceptor, respectively. For this, the emission range of the donor/sensitizer atom should match with the absorption range of the acceptor/activator ion. RE ions like Ce^{3+} and Yb^{3+} can be used as sensitizer ions [4].

14.5.2   CHARGE COMPENSATORS

The most common role of the codopant, which is acting as a charge compensator, is to compensate the imbalance charge between the host and the dopants. This results in the reduction of induced defects that can reduce the efficiency of the phosphor material. In some cases, these charge compensators also help to improve the luminescence properties. For acting as a charge compensator, the size and valency of the codopant should probably match with the activator ions or the cations of the host material. Li^+, K^+, Na^+, and others are some of the examples of commonly used charge compensators [1, 4].

14.6   HOST MATERIALS

A large number of phosphor materials are available for the applications of white LEDs. Oxides, nitrides, oxynitrides, sulfides, fluorides, phosphates, selenides, aluminates, silicates, and orthosilicates are the most commonly used phosphor host materials. Some of the examples for phosphor host materials are YVO_4, $CaWO_4$, $SrCaMoO_4$, $CaTiO_3$, GaN, $Sr_2Si_5N_8$, ZnS, $(Sr_3M)_4F$ (M= Ca, Sr, B), $BaMgAl_{10}O_{17}$, $(Sr, Mg)_3(PO_4)_2$, SrSSe, M_3SiO_5, and M_2SiO_4 (M = Ca, Sr, Ba) [1, 4, 11–14]. The main requirements for the host lattice of phosphors include chemical stability, strong covalent chemical bonds, and large crystal field splitting [4]. Among the above-mentioned host materials, oxides have high chemical and thermal stability that is suitable for a high efficient phosphor [4].

14.7   SYNTHESIS TECHNIQUES OF PHOSPHOR MATERIALS

In recent years, there have been significant advances in the chemical synthesis of advanced phosphor and ceramic materials for a wide variety of applications. Suppose if there are three or four elements required to produce a particular phosphor composition it is very difficult to maintain the quality as well as phase purity. For example, solid-state reactions are highly desirable and favorable to obtain for low cost and large scale production of phosphors. However, due to the use of multiple precursors, several sintering and grinding steps may be required for a full reaction. If we want to prepare small, dispersed, and uniformly shaped particles with narrow size distribution we have to adopt sol gel or wet chemical methods but they lack in high cost, complex chemical steps, and low production volume. So the prerequisite for the end application actually determines the choice of a synthesis method. There are a number of methods that are currently in use to synthesize the phosphor materials and are namely solid-state reaction method, sol–gel method, coprecipitation method, hydrothermal synthesis method, combustion method, wet chemical method, microwave synthesis method, spray pyrolysis, green synthesis method, and others [4, 11, 15].

14.7.1 SOLID-STATE REACTION METHOD

Solid-state reaction technique is the most commonly used method for the preparation of polycrystalline materials. The high-temperature solid-state reaction method is straightforward and suitable for mass production. For the preparation of polycrystalline phosphor materials, the high-temperature solid-state reaction method is commonly used. This method uses a mixture of solid starting materials, not any solvents. At room temperature, the materials of the solid mixture do not react with each other. They will react only at high temperatures, often at 1000–1500 °C. The reaction conditions, starting materials, and their structural properties, the surface area of the solids, their reactivity, thermodynamics of the reaction, and others are the factors affecting the feasibility of a solid-state reaction technique [16]. Initially, the stoichiometric amount of the solid reactants were weighed and mixed. The natural mixing of the solid reactants is done by using agate mortar and pestle. The fine mixing process will help to improve the reaction rate of the reactants. It is necessary to choose a suitable container that is chemically inert and can withstand high temperatures. Usually, platinum or gold crucibles were used for this purpose. Then the mixture taken in the platinum/gold crucible is fired at an appropriately high temperature. The product obtained is crushed and finely ground for further characterizations.

The advantages of solid-state reaction technique include low production cost, produces more amount of products, environmentally friendly, no need for purification steps since there are no solvents used, and others. But there are some disadvantages also for solid-state reaction method, which include less homogeneity of the product and the presence of defects/impurities.

14.7.2 SOL–GEL METHOD

Sol–gel method is one of the most important techniques for the synthesis of various functional materials because it offers many advantages such as higher uniformity in particle size distribution, nonagglomeration, and fine powders. The surface area of powders produced from sol–gel is very high and higher surface area phosphors have better thermal conductivity when excited with high-energy photons that are favorable to the lifetime of UV GaN-based WLEDs [17]. The sol–gel process involves the evolution of inorganic networks through the formation of a colloidal suspension (sol) and gelation of the sol to form a network in a continuous liquid phase (gel). The precursors used for the synthesis are usually a metal or metalloid element surrounded by various reactive ligands. The initial phase of the synthesis involves hydrolysis, which is the processing of starting materials to form a dispersible oxide and thereby forming a sol in contact with water or dilute acid. The second phase of the synthesis is the condensation of the hydrolyzed sol often referred as gelation. Gelation is the formation of an oxide- or alcohol-bridged network by a polycondensation reaction that results in the increase in the viscosity of the solution. Aging of the gel is called syneresis during which the polycondensation reactions are continued until the gel is transformed into a solid mass accompanied by contraction of the gel network and expulsion of solvent from gel pores. Ostwald ripening/coarsening and phase transformations may occur concurrently with syneresis. Drying of the gel involves the removal of water and other volatile liquids from the gel network. The gel can be either cross-linked or noncross-linked depending on the stoichiometry of the reactants. The high viscosity of the gel causes low cation mobility that prevents the different mixed cation from segregation. So during calcination, there is a little segregation of various cations takes place. Calcination of the gel in the air or other gases causes breakdown of the gel and subsequently, the cations are oxidized to form crystallites of mixed cation oxides. The major advantages of the sol–gel route are that it uses a relatively low temperature and it can produce very fine high purity product. This method is a simple, economic, and effective way to produce high-quality phosphors [18].

14.7.3   CO-PRECIPITATION METHOD

Co-precipitation method is considered as a simple, cost-effective, and one of the most convenient strategies for the preparation of phosphors. In a typical synthesis procedure, metal precursor solutions are prepared either using suitable metal salts or by the dissolution of metal oxides in an acidic solution. The prepared precursors are mixed in prefixed concentration ratios, to promote local supersaturation, primary nucleation and successive stages of crystallization. One of the essential prerequisite for the occurrence of co-precipitation is that the material should have a low solubility in the solvent selected for the synthesis process. The phosphor precursors are precipitated out of the solution above a critical size. By employing the centrifugation process, phosphor precursors are collected. The collected precipitates are washed thoroughly and annealed at higher temperatures to enhance the crystallinity of the samples [19].

14.7.4   HYDROTHERMAL METHOD

Among many synthesis methods, the hydrothermal method is one of the best solution routes and is widely applied to generate different kinds of nanophosphor materials. Hydrothermal method is convenient, lower processing temperature, environmental friendly solution processing for the preparation of phosphors with desired form, sizes, and shapes. Hydro-thermal uses homogeneous or heterogeneous reaction that takes place in the presence of a solvent (aqueous/nonaqueous) above the room tempera-ture at a pressure greater than 1 atmospheric pressure in a closed system [20–22]. They have distinct advantages like high-crystallized samples with a narrow particles size distribution, controlled morphology, and high purity without postannealing at high treatment. In the hydrothermal process, the reaction takes place in a sealed vessel, in which water can be brought to temperatures well above its boiling point by the increase of autogenous pressure resulting from heating. The basic principle is that small crystals will homogeneously nucleate and grow from solution when subjected to high temperature and pressure. During the nucleation and growth process, water is both a catalyst and occasionally a solid-state phase component. We can control the product properties by varying temperature, treatment time, and concentration of the solution Hence, hydrothermal synthesis is a promising route for producing nanocrystalline phosphors with low energy waste and high efficiency.

The hydrothermal process is environmentally benign, inexpensive, and allows for the reduction of free energies for various equilibria compared to other synthesis methods. The main advantage of this process is the low crystallization temperature with the one-step process and avoiding thermal treatments. Also, it is a soft chemical route with versatility in producing a large class of nanoscale materials in the form of powders. The nanostructure of bulk and thin films grown using this approach may bring new applications to materials for biosensors, implants, or substrates for growing biological cells. A great variety of phosphor materials like metal oxides, hydroxides, phosphates, silicates, and carbonates, both as particles and nanostructures, could be synthesized using the hydrothermal method.

14.7.5   COMBUSTION METHOD

Combustion method is generally used to synthesis complex oxide phosphors materials in powder form. This process makes use of the exothermic reaction between an oxidizer such as metal nitrates and organic fuel. Commonly used organic fuels are urea (H_2NCONH_2), carbohydrazide ($CO(NHNH_2)_2$), or glycine ($C_2H_5NO_2$). The combustion reaction usually takes palace below 5000 °C inside a muffle furnace or on a hot plate. In a typical reaction, the precursor (a mixture of water, metal nitrates, and fuel) on heating decomposes, dehydrates, and blowout into a flame. The resultant product is a voluminous, fluffy foamy powder, which occupies almost the entire volume of the reaction vessel. The key behind this method is that the chemical energy released from the exothermic reaction between the metal nitrates and fuel can rapidly heat the system to high temperatures without an external heat source [23, 24]. Phosphors prepared by these methods are usually homogeneous, contain fewer impurities, and have higher surface area than powders prepared by conventional solid-state methods. These methods mainly rely on certain parameters such as nature of the fuel, fuel to oxidizer ratio, ignition temperature, and water content of the precursor mixture. One of the major advantages of this method is that large-scale production at relatively low temperatures but with a disadvantage of highly agglomerated particles and cannot be dispersed in solvents.

14.7.6   WET CHEMICAL METHOD

The wet chemical method is also known as evaporation method. This method has the capability to produce phosphors powders between temperature 80 °C and 100 °C. Being a low-temperature synthesis method is very cost-effective, power-saving, and high homogeneity as compared to other conventional methods [25]. It is an ideal technique used to produce fine, chemically homogeneous, and single-phase powders. The molecular motions and hence the chemical reaction may proceed very rapidly in the liquid state. Polycrystalline mixed compound phosphors are prepared using this method. For that aqueous solutions of constituent hosts with stoichiometric ratios were taken and then evaporated till the mixture becomes anhydrous. The compound, obtained in its powder form, is evaporated at 80 °C for 8 h. The dried samples were then slowly cooled to room temperature. The resultant polycrystalline mass was crushed to fine particles in a crucible and used in further studies.

14.7.7   MICROWAVE SYNTHESIS METHOD

The conventional heating methods often results in-homogeneous heating via energy transfer between molecules. The difficulties of conventional heating methods can be overcome by using microwave-assisted volumetric heating. In microwave heating, the molecules/atoms of dielectric materials: liquids or solids, interact with microwaves and undergo rapid translational and rotational movements. The continuous reorientation leads to the friction between molecules, which subsequently generate heat. The interaction between molecule and microwave is actually the interaction between the electric dipoles present in the materials and the applied electric field that offered by microwave. Hence, microwave heating is also known as dielectric heating. The fast and homogeneous heating effects of microwave irradiation generates a more rapid and simultaneous environment for the formation of nanoparticles [26, 27]. The frequencies allotted for microwave-assisted heating are 918 MHz and 2.45 GHz and the latter frequency is the most often used one. Some of the advantages of microwave-assisted heating are the short reaction time, high energy efficiency, and the ability to induce the formation of particles with small size, narrow size distribution, and high purity. In the past few years, microwave-assisted heating has been applied in the soft chemical synthesis of various nanocrystalline materials and presents a promising trend in its future development.

14.7.8   SPRAY PYROLYSIS METHOD

Spray pyrolysis is an easy method to produce fine size and spherical-shaped phosphor particles. It has some advantages such as low-cost precursor materials and high production rate. Oxide powders with homogeneous particle size less than 100 nm may be produced by this method, which allows for continuous operation and offers the opportunity to tailor materials on a nanoscale. For the spray pyrolysis synthesis, liquid precursors as a starting solution, such as nitrate, acetate and chloride are prepared by dissolving in water or alcohol as precursor solvent [28]. The basic principle of this method is that when a droplet of the spray solution reaches the hot substrate, fine powders would be deposited due to the pyrolytic decomposition of the solution. The droplets, which are atomized from a starting solution, are supplied to a series of furnaces. And then the aerosol droplets experience evaporation of the solvent, diffusion of solute, drying and precipitation, the reaction between precursor and surrounding gas. The precipitate undergoes pyrolysis, or sintering inside the furnace at a higher temperature to form the final product. For phosphor particles, two furnace systems in series are required to enhance the quality such as crystallinity and morphology of the final product. The first furnace promotes the formation of microporous particle, and the second furnace further densifies the microporous particle and increases the crystallinity. In spray pyrolysis, the final composition of phosphor particles would be determined by the starting precursor solution composition. The morphology and particle size can be easily controlled to some degree by the choice of precursors, concentration, droplet size, and the residence time in the furnace.

Although the spray pyrolysis method is a relatively simple and low-cost method to produce phosphor particles with a high production rate in a single continuous process, it also has the limitations to obtain nanophosphor particles with high luminescence efficiency. In this process, hollow and highly porous phosphor particles are generated. In the case of phosphors porosity act as structural defects and can lead to a decrease in luminescence efficiency of the prepared phosphor particles. Moreover, in order to produce nanosized particles by spray pyrolysis, it is required to have lower overall concentration of starting precursor solution or by introducing smaller droplet size into the reactors, as a result, the production rate in decrease.

14.7.9   GREEN SYNTHESIS METHOD

Eco-friendly synthesis (green synthesis) of nanophosphors is an expanding research area due to their nontoxic nature and also the use of environmentally benign precursors in the synthesis process that do not release any harmful by-products. Eco- friendly synthetic approaches have more advantages over the conventional physicochemical methods of nanoparticle synthesis [29–34]. Biosynthetic routes can provide improved defined size and morphology of nanostructures compared to that of physicochemical methods of production and no need to use high energy, temperature, pressure for synthesis procedure, and also this method can be scaled up for large-scale synthesis [35, 36]. By using green synthesis method we can do the synthesis in one step using biological organisms such as bacteria, actinobacteria, yeasts, molds, and algae and plants or their products. Molecules in plants and microorganisms, such as proteins, enzymes, amines, alkaloids, and pigments perform nanoparticle synthesis by reduction [35, 36].

14.8 APPLICATIONS OF PHOSPHOR MATERIALS

The electricity usage and environmental issues lead to the search for the new applications of materials science especially the application of luminescence. In the present era, the developing applications and applied research helped us to improve or overcome the drawbacks of the preceding. For example, the discovery of LEDs helped us to overcome the drawbacks of the conventional lighting devices like the incandescent lamp and fluorescent lamps [37–39]. The luminescent materials or phosphors find many applications and also there are a lot of research is going on to the search for new applications of it. Some of the important applications include lighting applications, CRTs, scintillators, vacuum fluorescent displays, field emission displays, long persistent phosphors, LEDs and diode lasers, electroluminescent materials, plasma display panels, organic fluorescent pigments, luminous paints, and others [1, 4, 38–44].

14.8.1 AGRICULTURAL FIELD

Light conversion films to cover the greenhouse and LED plant lamp (LED-PL) are the most important applications of phosphors in the agriculture field [45]. For using as a light conversion agent, the phosphors should be able to excite by UV (290–350 nm) or green-greenish yellow (510–580 nm) and emit blue (400–480), or red (600–680 nm) at the same time. This can match with the absorption range of chlorophyll and will increase the efficiency of the plant photosynthesis, leading to increase yield and quality of crops.

14.8.2 CATHODE RAY TUBES

Phosphors for CRTs are mainly used in color television sets and giant screens to computers. In color television sets, a combination of red, green, and blue phosphors are used. Examples of phosphors currently used in color television picture tube include Y_2O_2S: Eu^{3+} (red), ZnS: Cu, Al (green) and ZnS: Ag (blue) [46]. For this application, phosphors should have some important characteristics like high color purity of emission color, high emission quantum efficiency, long lifetime, and stability of quality.

14.8.3   LONG PERSISTENT PHOSPHORS

Due to the unique mechanism of photoelectron storage and release, long persistent luminescence (long-lasting afterglow/phosphorescence) plays an important role in the areas of spectroscopy, photochemistry, and photonics. Currently, more research has focused on the morphology, operational wavelength, and persistent duration of long-lasting persistent phosphors. The major applications of long persistent phosphors include bioimaging, photocatalysts, optical sensors, detectors, and photonic devices [47].

14.8.4   SCINTILLATORS

Scintillators play an important role in radiation detection and high energy physics. The material in which fluorescence is caused by incident radiation such as α, β, or γ rays is termed as scintillators and are used in X-ray security, particle detectors, nuclear cameras, computed tomography, space telescopes, and gas explorations. The important applications of scintillators in the medical field include X-ray imaging, computed tomography, positron emission tomography detectors, and others [48, 49]. There is a wide variety of inorganic and organic scintillating materials that are available in the form of crystals, ceramics, glasses, gases, liquids, and plastics. Some of the currently using scintillating materials are CaF_2: Eu^{2+}, CeF_3, $CdWO_4$, ZnS: Ag, $LuAlO_3$: Ce^{3+}, Gd_2O_2S:Pr^{3+}: Ce, Anthracene, and others [50–55].

14.8.5   VACUUM FLUORESCENT DISPLAYS

Vacuum fluorescent display (VFD) is a practical display device operates in the same way as the cathode ray tube. But the phosphors used in VFDs are usually called low energy electron excited phosphors. Some of the required characteristics of phosphors for the application of vacuum fluorescent display include thermal stability, essentially low emission threshold voltage, long lifetime under low energy electron beam excitation, efficient luminescence, low level of release of harmful substances, and others [46]. Examples of phosphor materials currently used for vacuum fluorescent displays are $SrTiO_3$: Pr^{3+}, $ZnGa_2O_4$, $ZnGa_2O_4$: Mn, GaN: Zn, and others [46, 56, 57].

14.8.6 LIGHTING APPLICATIONS

For the past decades, incandescent bulbs and fluorescent or compact fluorescent lamps were used for the lighting applications. But in the case of incandescent bulbs, only 10% of input electricity is converted into visible light and the remaining 90% electricity is emitted as heat. Also, the fluorescent lamps and compact fluorescent lamps use mercury, which is highly toxic and harmful to the environment. By using suitable luminescent phosphors we can overcome these issues, since they have high absorption efficiency, good color rendering index, long-term stability, high quantum efficiency, low cost, and environmental friendly. The lighting applications include white LEDs, phosphor-converted white LEDs, traffic signals, aviation lighting, signboards, street lighting, backlighting general illumination, and others.

14.8.6.1 PHOSPHOR-CONVERTED WHITE LEDS

By using phosphor-converted white LEDs (pc-WLEDs) we can solve these disadvantages of single/multi-LED chip produced white light. pc-WLEDs show the characteristics of high efficiency, long lifetime, and energy saving [37]. Mainly the near UV excited blue, green, and red phosphor materials produce warm white light with high color rendering properties [58]. Single or multiple phosphor materials can be used to produce white light. But the multiple phosphors-based white LEDs have high color rendering index and tunable color temperature [37]. Thus for phosphor-converted white LEDs, phosphor materials play an important role by downconversion process. As a result, downconversion phosphors got more importance in the white LED applications. For this purpose novel white, red, green, and blue-emitting phosphor materials should be needed. Now we can look into some of the phosphor materials and their optical properties that can find application in phosphor-converted white LEDs.

14.8.6.2 $SR_{0.5}CA_{0.5}TIO_3$: X PR^{3+} PHOSPHOR

They show strong reddish-orange emission corresponding to $^1D_2 \rightarrow ^3H_4$ transition at 611 nm under UV excitation [59]. We can see that the luminescence bands of Pr^{3+} from ^3PJ ($J = 0$, 1, 2) levels are completely quenched and only sensitized luminescence band from the $^1D_2$ levels are observed. The presence of the low-lying intervalence charge transfer state, its energy position depends on the nature of the host lattice, causes the quenching of $^3P_0$ emission. The sample, $Sr_{0.5}Ca_{0.5}TiO_3$: 0.1 Pr^{3+} shows high emission intensity with color purity of 91.7%. Also the high value of stimulated emission cross-section and quantum efficiency and low value of nonradiative relaxation rate.

14.8.6.3 $Sr_{0.5}Ca_{0.5}TiO_3$: X Dy^{3+} PHOSPHOR

They show strong blue and yellow emissions corresponding to $^4F_{9/2} \rightarrow {}^6H_{15/2}$ and $^4F_{9/2} \rightarrow {}^6H_{13/2}$ transitions respectively, whereas under 386 nm excitation weak red emission corresponding to $^4F_{9/2} \rightarrow {}^6H_{11/2}$ transition was also reported [60]. The CIE color coordinates of the prepared phosphor samples lie in the near-white light region with lower color purity. Changes in correlated color temperature and yellow/blue ratio of $Sr_{0.5}Ca_{0.5}TiO_3$: Dy^{3+} phosphor samples with varying Dy^{3+} concentration and different excitation wavelength is also reported.

14.8.6.4   Tb2(MoO4)3 PHOSPHOR

Terbium molybdate nanophosphors prepared by the conventional sol–gel method show green emission under UV excitation [61]. The emission is due to the activation of Tb^{3+} ion through sensitizing MoO_4^{2-} groups in the host lattices and sensitizing Tb^{3+} ions directly via the f–f transitions. $Tb_2(MoO_4)_3$ is an example of host sensitized phosphor material and also a stable, efficient, and potential green phosphor for future SSL devices.

14.8.6.5   Tb2-xSmx(MoO4)3 NANOPHOSPHOR

Due to different energy transfer schemes, $Tb_{2-x}Smx(MoO_4)_3$ phosphor shows the characteristic emissions of Sm^{3+} and Tb^{3+} ions upon excitation at 290, 376, and 485 nm [62]. The nonradiative nature of the $Tb^{3+} \rightarrow Sm^{3+}$ energy transfer is confirmed by the fast nonradiative decay of the Tb^{3+} emission. The efficient energy transfer process from $Tb^{3+} \rightarrow Sm^{3+}$ occurs via the dipole–dipole and produce multicolor emission in the prepared phosphors.

14.8.6.6   Tb2-xEux(MoO4)3 NANOPHOSPHOR

In $Tb_{2-x}Eux(MoO_4)_3$, the energy transfer from the MoO_4^{2-} group to Tb^{3+}/Eu^{3+} ions and from Tb^{3+} ions to Eu^{3+} ions cause the Eu^{3+} ions to show their characteristic strong emissions on exciting the charge transfer band at 290 nm [63]. The emission color of the phosphor can be tuned from green to white and finally to red by changing the doping concentrations of the Eu^{3+}, implying the as-prepared multicolor tunable phosphors could be a promising candidate for n-UV pumped pc-LEDs.

ACKNOWLEDGMENTS

The author, Dr Viji Vidyadharan, is thankful to Science and Engineering Research Board (SERB), Department of Science and Technology, Government of India for National Postdoctoral Research Fellowship (PDF/2016/002564/PMS). Also Dr Kamal P Mani acknowledges the financial support from the University Grants Commission (UGC), Govt. of India, New Delhi toward Dr D. S Kothari Postdoctoral fellowship program (No. F.4-2/2006 (BSR)/PH/18-19/0012).

KEYWORDS

- **phosphor**
- **luminescence**
- **photoluminescence**
- **activators**
- **rare earth ions**
- **synthesis methods**
- **phosphor applications**

REFERENCES

1. Ropp, R. C., Luminescence and the Solid State, Second Edition, Elsevier B.V., The Netherlands, 2004.
2. Kortov, V., Radiation Measurements 42 (2007) 576–581.
3. Weber, M. J. Phys. Rev. B 8 (1973) 54–64.

4. Shionoya, S., Yen, W. M., Yamamoto, H., Phosphor Handbook, second ed. CRC Press, Boca Raton, London, New York, 2006.
5. Dieke, G. H., Spectroscopic observations on maser materials—Advances in Quantum Electronics, Ed. By Singer, J. R., Columbia University Press, New York, 1961.
6. Haris, D. C., Bertolucci, M. D., Symmetry and Spectroscopy, Oxford University press, 1978.
7. Nazarov, M., Noh, D. Y., J. Rare Earth 28 (2011) 1–11.
8. S.Ye, Xiao, F., Pan, Y. X., Y. Y. Ma, Zhang, Q. Y., Mater. Sci. Eng. R 71 (2010) 1–34.
9. Blasse, G., J. Less-Common Met. 112 (1985) 79–82.
10. Sommer, C., Hartmann, P., Pachler, P., Hoschopf, H., Wenzl, F. P., J. Alloy. Compd. 520 (2012) 146–152.
11. Nazarov, M., Noh, D. Y., New Generation of Europium and Terbium Activated Phosphors from Syntheses to Applications, Pan Stanford Publishing, 2011.
12. Blasse, G., Grabmaier, B. C., Luminescent Materials, Springer-Verlag, Berlin, 1994.
13. Dutta, D. P., Tyagi, A. K., Solid State Phenom. 155 (2009) 113–143.
14. Shinde, K. N., Dhoble, S. J., Swart, H. C., Park, K. Phosphate Phosphors for Solid State Lighting, Springer-Verlag, Berlin, 2012.
15. (a) Kirby, K. W., Mater. Res, Bull. 23 (1988) 881–890; (b) Kakihana, M., J. Sol-Gel Sci. Technol. 6 (1996) 7–55.
16. West, A. R., Solid State Chemistry and its applications, Wiley and Sons, New York, 1987.
17. Rao, R. P., J. Electrochem. Soc. 143 (1996) 189.
18. Cho, Y. S., Burdick, V. L., Amarakoon, V. R. W., J. Am. Cerm. Sc. 82 (1999) 1416.
19. Qin, X., Liu, X., Huang, W., Bettinelli, M., Liu, X. Chem. Rev. 117 (2017) 4488–4527.
20. Byrappa, K., Adschiri, T., Prog. Cryst. Growth Chem. 53 (2007) 117.
21. Chen, X. P., Huang, X. Y., Zhang, Q. Y., J. Appl. Phys. 106 (2009) 063518.
22. Chen, X. P., Zhang, Q. Y., Yang, C. H., Chen, D. D., Zhao, C. Spectrochim. ActaA 74 (2009) 441.
23. Huang, X. Y., Yu, D. C., Zhang, Q. Y., J. Appl. Phys. 106 (2009) 113521.
24. Huang, X. Y., Zhang, Q. Y., J. Appl. Phys. 105 (2009) 053521.
25. Bhake, A. M., Nair, G., Zade, G. D., Dhoble, S. J., Luminescence, 31, 8 (2016) 1468–1473.
26. Ekthammathat, N., Thongtem, T., Anukorn, P., Thongtem, S., Rare Met. 30 (2011) 572–576.
27. Ramírez, D. P., Domínguez-Crespo, M. A., Torres-Huerta, A. M., Rosales, H. D., Meneses, E. R., Rodríguez, E., J. Alloy. Compd. 643 (2015) S209.
28. Okuyama, K., Lenggoro, I. W., Chem. Eng. Sci. 58 (2003) 537–547.
29. Dahoumane, S. A., Yéprémian, C., Djédiat, C., J. Nanopart. Res.18 (2016) 79.
30. Sinha, S., I. Pan, Chanda, P., Sen, S. K.. J. Appl. Biosci. 19 (2009)1113–1130.
31. Azizi, S., Ahmad, M. B., Namvar, F., Mohamad, R., Mater. Lett. 116 (2014) 275–277.
32. Salam, H. A., Sivaraj, R., R. Venckatesh, Mater. Lett. 131 (2014) 16–18.
33. Agarwal, H., S. Venkat Kumar, Rajeshkumar, S., Resource Effic. Technol. 3 (2017) 406–413.
34. Dobrucka, R., Dugaszewska, J., Saudi J. Biol. Sci. (2016) 23, 517–523.
35. Mérillon, J. M., Ramawat, K. G. (eds.), Fungal Metabolites, Reference Series in Phytochemistry, DOI 10.1007/978-3-319-25001-48.

36. Wongpreecha, J., Polpanich, D., Suteewong, T., C. Kaewsaneha, P. Tangboriboonrat, Carbohydr. Polym. 199 (2018) 641–648.
37. Humphreys, C. J., MRS Bull. 33 (2008) 459–470.
38. Krames, M. R., Shchekin, O. B., Mach, R. M., Mucller, G. O., Zhou, L., Harbers, G., Craford, M. G., J. Display Technol. 3 (2007) 160–175.
39. Steigerwald, D. A., Batt, J. C., Collins, D., Fletcher, R. M., Holcomb, M. O., Ludowise, M. J., Martin, P. S., Rudaz, S. L., IEEE J. Sel. Top. Quantum Electron. 8 (2002) 310–320.
40. Chein, M. C., Tien, C. H., Proc. of SPIE 8486 (2012) 84860R 1–2.
41. Pan, Y. L., Boutou, V., Chang, R. K., Ozden, I., Davitt, K., Nurmikko, A. V., Opt. Lett. 28 (2003) 1707–1709.
42. Xu, H., Zhang, J., Davitt, K. M., Song, Y. K., Nurmiko, A. V., J. Phys. D Appl. Phys. 41 (2008) 094013 1–13
43. Zukauskas, A., Shur, M. S., Gaska, R., Introduction to Solid State Lighting, Wiley, New York, 2002.
44. Goins, G. D., Yorio, N. C., Sanwo, M. M., Brown, C. S., J. Exp. Bot. 48 (1997) 1407–1413.
45. Ru-Shi Liu, Phosphors, Up-conversion Nano Particles, Quantum dots and Their Applications, Volume 2, Springer Singapore, pp-119–137, 2016.
46. Shionoya, S., Yen, W. M., Yamamoto, H., Practical Applications of Phosphors, CRC Press, Boca Raton, 2007.
47. Li, Y., Gecevicius, M., Qiu, J. Chem. Soc. Rev. 45 (2016) 2090–2136.
48. Lecoq, P. Nucl. Instrum. Methods Phys. Res. Sect. A 809 (2016) 130–139.
49. We., K., Hei. D., Weng, X., Tan, X., Liu, J. Appl. Radiat. Isot. 156 (2020) 108992.
50. Nakamura, F., T. Kato, Okada, G., Kawaguchi, N., Fukuda, K., Yanagida, T., Ceram. Int. 43 (2017) 604–609.
51. Belli, P., Cerulli, R., Dai, C. J., Danevich, F. A., Incicchitti, A., Kobychev, V. V., Ponkratenko, O. A., Prosperi, D., Tretyak, V. I., Zdesenko, Y. G., Nucl. Instrum. Methods Phys. Res. Sect. A 498 (2003) 352–361.
52. Fazzini, T., Bizzeti, P. G., Maurenzig, P. R., C. Stramaccioni, Danevich, F. A., Kobychev, V. V., Tretyab, V. I., Zdensenko, Y. G., Nucl. Instrum. Methods Phys. Res. Sect. A 410 (1998) 213–219.
53. Stedman, R., Rev. Sci. Instrum. 31 (1960) 1156.
54. Petrosyan, A. G., Shirinyan, G. O., Pedrini, C., Durjardin, C., Ovanesyan, K. L., Manucharyan, R. G., Butaeva, T. I., Derzyan, M. V., Cryst. Res. Technol. 3 (1998) 241–248.
55. Blahuta, S., Viana, B., Bessiere, A.. Mattmann, E., B. LaCourse, Opt. Mater. 33 (2011) 1514–1518.
56. Itoh, S., H. Tok, Tamura, K., Kataoka, F., Jpn. J. Appl. Phys. 38 (1999) 6387–6391.
57. Toki, T., Kataoka, H., Itoh, S., Technical Digest of Japan Display'92, 1992, 421–423.
58. De Sousa, P. C., Serra, O. A., J. Lumin. 129 (2009) 1664–1668.
59. Vidyadharan, V., Remya, M. P., Joseph, C., Unnikrishanan, N. V., Biju, P. R., Mater. Chem. Phys. 170 (2016) 38–43.
60. Vidyadharan, V., Sreeja, E., Jose, S. K., Joseph, C., Unnikrishanan, N. V., Biju, P. R., Lumin. 31 (2016) 202–209.
61. Mani, K. P., Vimal, G., Biju, P. R., Unnikrishanan, N. V., Ittyachan, M. A., Joseph, C., J, Mater. Sci. Mater. Electron. 27 (2016) 966–975.

62. Mani, K. P., Vimal, G., Biju, P. R., Unnikrishanan, N. V., Ittyachan, M. A., Joseph, C., Opt. Mater. 42 (2015) 237–244.
63. Mani, K. P., Vimal, G., Biju, P. R., Unnikrishanan, N. V., Ittyachan, M. A., Joseph, C., J. Mol. Struct. 1105 (2016) 279–285.

PART IV
Molecular Physics and Diffusion

CHAPTER 15

Theoretical Values of Diffusion Coefficients of Electrolytes in Aqueous Solutions: Important Parameters with Application in Fundamental and Technological Areas

ANA C. F. RIBEIRO,[1*] LUIS M. P. VERÍSSIMO,[1] ARTUR J. M. VALENTE,[1] S. C. G. S. ANDRADE,[1] A. M. T. D. P. V. CABRAL,[2] and M. A. ESTESO[3,4]

[1]Department of Chemistry, University of Coimbra, 3004-535 Coimbra, Portugal

[2]Faculty of Pharmacy, University of Coimbra, 3000-295 Coimbra, Portugal

[3]U.D. Physical Chemistry, University of Alcala, 28805 Alcala de Henares, Madrid, Spain

[4]Catholic University of "Santa Teresa de Jesus de Avila," Los Canteros street, 05005 Avila, Spain

[*]Corresponding author. E-mail: anacfrib@ci.uc.pt

ABSTRACT

Mutual diffusion coefficients of different electrolytes, in aqueous solutions, at 298.15 K, have been estimated by using Onsager–Fuoss and Pikal theories for over 300 electrolytes. These theories are described and discussed. The limitations of these theories are dependent on different types of electrolytes, that is, symmetrical univalent (1:1), symmetrical polyvalent (basically 2:2), and nonsymmetrical polyvalent. We believe that the compiled diffusion coefficients are of utmost importance for fundamental and practical applications.

15.1   INTRODUCTION

The development of technology and science request precise data concerning the transport and thermodynamic properties of electrolyte solutions. For example, for food technology, corrosion, wastewater, and soil treatment and remediation, those relevant parameters needed to be known for a better and accurate assessment of the processes. Among them, diffusion property data of the electrolytes in aqueous solutions are of great interest not only for fundamental purposes, helping to understand the nature of the structure of aqueous electrolyte solutions, but also for many technical fields such biomedical and pharmaceutical applications [1–7].

Diffusion is an irreversible process that can be defined as the transport of substance through a medium, occurring in any state of matter, driven by a concentration gradient (strictly, by a chemical potential). That mass transport phenomenon is a pure diffusion (i.e., neither convection nor migration are present), which occurs without visible movement of the solution, and in addition, there is no macroscopic flow of the denser parts of the solution relatively to the lighter ones.

The scarcity of diffusion coefficients in the scientific literature, arising from the difficulty of their accurate experimental measurement, associated to their industrial and research need, well justifies efforts in their estimation. Thus for estimative purposes, and as an initial approach to the experimental mutual diffusion coefficients, the respective theoretical values can be estimated by using Onsager–Fuoss and Pikal models [8, 9]. Estimations from these theories are appropriate for electrolytes in aqueous dilute solutions (i.e., for $c \leq 0.010$ mol dm^{-3}), as has been shown for other similar systems, where the theoretical data are consistent with experimental results (deviations are, in general, $\leq 3\%$) [8, 9]. Concerning more concentrated solutions ($c \leq 1.000$ mol dm^{-3}), no definite conclusions is possible. Thus a complete dataset of diffusion coefficients for electrolytes are presented here.

15.2   NERNST–HARTLEY, ONSAGER–FUOSS, AND PIKAL THEORIES: ITS APPLICABILITY AND LIMITATIONS

The calculation of binary diffusion coefficients (D) of solutions of different electrolytes can be made on the basis of the Nernst, Onsager–Fuoss, and Pikal models [1–9], suggesting that the D is a product of both kinetic (molar mobility coefficient of a diffusing substance, U_m) and thermodynamic factors

($c\partial\mu/\partial c$, where μ represents the chemical potential) or the gradient of the free energy (Eq. (15.1)).

Assuming that each ion of the diffusing electrolyte can be regarded as moving under the influence of two forces: (1) a gradient of the chemical potential for those ionic species; and (2) an electrical field produced by the motion of oppositely charged ions, we come up to the Nernst–Hartley equation (15.1) [1]

$$D = D^0 \, [1 + (d \, \ln\gamma_\pm/d \, \ln c)] \tag{15.1}$$

and

$$D^0 = [(v_1 + v_2) \, \lambda_1^0 \lambda_2^0/(v_1 \, |Z_1| \, (\lambda_1^0 + \lambda_2^0))] \, (R \, T/F^2)$$

where D^0 is the Nernst limiting value of the diffusion coefficient, λ^0 are the limiting ionic conductivities of the ions (subscripts 1 and 2 are for cation and anion, respectively), Z is the algebraic valence of the ion, v is the number of ions formed upon complete dissociation of one solute "molecule," T is the absolute temperature, R and F are the gas and Faraday constants, respectively, and $\gamma_\pm$ is the mean molar activity coefficient.

Onsager and Fuoss [8] improved Eq. (15.1) by taking into account the electrophoretic effects (Eq. (15.2)):

$$D = (D^0 + \Sigma\Delta_n)[1 + (d \, \ln\gamma_\pm/d \, \ln c)] \tag{15.2}$$

The electrophoretic term, Δ_n, in (Eq. (15.2)) can be described through the following equation:

$$\Delta_n = k_B T A_n \, (Z_1^n t_2^0 + Z_2^n t_1^0)^2/(a^n \, |Z_1 Z_2|) \tag{15.3}$$

where k_B is the Boltzmann constant, A_n are functions of the dielectric constant, of the viscosity of the solvent, of the temperature, and of the dimensionless concentration-dependent quantity (κa), being the reciprocal of average radius of the ionic atmosphere; and t_1^0 and t_2^0 are the limiting transport numbers of the cation and anion, respectively.

Having in mind that the expression for the electrophoretic effect has been derived on the basis of the expansion of the exponential Boltzmann function, once that function must be consistent with the Poisson equation, we would have to take into account the electrophoretic term of the first order ($n = 1$). However, for symmetrical electrolytes it has been demonstrated that we can consider the second term without a significant error [8]. In fact, we see that for symmetrical uni-univalent both theories give similar results and they are consistent with experimental results [8]. Concerning nonsymmetrical electrolytes, we can consider only the first electrophoretic term.

Thus the experimental data can be compared with the calculated D on the basis of Eqs. (15.4) and (15.5)

$$D = (D^0 + \Delta_1 + \Delta_2) [1 + c\, (d\, \ln\gamma_{\pm}/dc)] \tag{15.4}$$

$$D = (D^0 + \Delta_1) [1 + c\, (d\, \ln\gamma_{\pm}/dc)] \tag{15.5}$$

For symmetrical and nonsymmetrical electrolytes, respectively.

The theory of mutual diffusion of binary electrolytes, developed by Pikal [9], has been based on the Onsager–Fuoss equation, but new terms resulting from the application of the Boltzmann exponential function for the study of diffusion have been taken into consideration.

Thus the electrophoretic correction appears as the sum of two terms:

$$\Delta v_j = \Delta v_j^L + \Delta v_j^S \tag{15.6}$$

where Δv_j^L represents the effect of long-range electrostatic interactions, and Δv_j^S take into account short-range electrostatic interactions.

Designating by $M = 10^{12}L/c$ the solute thermodynamic mobility, where L is the thermodynamic diffusion coefficient, ΔM can be represented by Eq. (15.7)

$$(1/M) = (1/M^0)\, (1 - (\Delta M/M^0)) \tag{15.7}$$

where M^0 is the value of M for infinitesimal concentration, and

$$M = \Delta M^{0F} + \Delta M_1 + \Delta M_2 + \Delta M_A + \Delta M_{H1} + \Delta M_{H2} + \Delta M_{H3} \tag{15.8}$$

The first term on the right-hand side of Eq. (15.8), ΔM^{0F}, represents the Onsager–Fuoss term for the effect of the concentration in the solute thermodynamic mobility, M; the second term, ΔM_1, is a consequence of the approximation applied on the ionic thermodynamic force; and the other terms result from the Boltzmann exponential function.

The relation between the solute thermodynamic mobility and the mutual diffusion coefficient is given by

$$D = (L/c)\ 10^3 RTv\ [1 + c\ (\mathrm{d}\ \ln\gamma_\pm/\mathrm{d}c)] \tag{15.9}$$

From Eqs. (15.9) and (15.7) a version of Pikal's equation (Eq. (15.10)) is obtained

$$D = \frac{1}{\dfrac{1}{M^0}\left(1 - \dfrac{\Delta M}{M^0}\right)}\ (10^3 RTv)\ [1 + c\ (\mathrm{d}\ \ln\gamma_\pm/\mathrm{d}c)] \tag{15.10}$$

Concerning symmetrical but polyvalent electrolytes, we can well see that Pikal's theory (Eq. (15.10)) is a better approximation than the Onsager–Fuoss (Eq. (15.2)). In polyvalent nonsymmetrical electrolytes, the agreement between experimental data and Pikal calculations is not so good, eventually because of the full use of Boltzmann's exponential in Pikal's development.

15.3   RESULTS

Based on the Onsager–Fuoss and Pikal equations described above a set of data for different types of electrolytes have been computed. Tables 15.1 and 15.2 summarize the dependence of D on the concentration of electrolytes in aqueous solutions at 298.15 K [10]. These tables also include, for each electrolyte, the corresponding equivalent conductivity at infinitesimal concentration, L^0, and the corresponding minimum approach distance of the ions, a.

TABLE 15.1   Dependence of D on the Concentration of Electrolytes in Aqueous Solutions

Electrolyte	AgClO$_3$		AgClO$_4$		AgF		AgNO$_2$	
	$L^0 = 126.5$		$L^0 = 129.2$		$L^0 = 117.3$		$L^0 = 133.9$	
	$a = 1.75$		$a = 2.91$		$a = 3.0$		$a = 2.8$	
Conc.	Dof	Dpikal	Dof	Dpikal	Dof	Dpikal	Dof	Dpikal
0.000	1.683	1.683	1.717	1.717	1.556	1.557	1.772	1.772
0.001	1.657	1.656	1.690	1.689	1.533	1.532	1.745	1.743
0.002	1.647	1.647	1.681	1.679	1.524	1.523	1.735	1.733
0.003	1.640	1.640	1.675	1.672	1.519	1.517	1.728	1.726
0.004	1.635	1.635	1.669	1.667	1.514	1.512	1.723	1.720
0.005	1.630	1.630	1.664	1.662	1.510	1.507	1.718	1.715
0.006	1.626	1.626	1.660	1.658	1.506	1.504	1.713	1.711
0.007	1.622	1.623	1.657	1.654	1.503	1.500	1.710	1.707
0.008	1.621	1.620	1.653	1.651	1.500	1.498	1.706	1.703

TABLE 15.1   *(Continued)*

0.009	1.618	1.617	1.651	1.648	1.498	1.495	1.703	1.700
0.010	1.616	1.614	1.648	1.645	1.495	1.492	1.700	1.697
0.020	1.595	1.595	1.629	1.625	1.479	1.474	1.680	1.676
0.030	1.582	1.582	1.618	1.611	1.469	1.462	1.668	1.661
0.040	1.572	1.572	1.610	1.601	1.463	1.453	1.660	1.651
0.050	1.564	1.564	1.605	1.592	1.458	1.445	1.654	1.642
0.060	1.559	1.557	1.601	1.585	1.455	1.438	1.649	1.634
0.070	1.554	1.550	1.599	1.578	1.453	1.432	1.646	1.627
0.080	1.550	1.544	1.596	1.572	1.451	1.426	1.644	1.620
0.090	1.547	1.531	1.594	1.566	1.449	1.420	1.642	1.614
0.100	1.545	1.526	1.592	1.560	1.448	1.415	1.640	1.609

Electrolyte	$AgNO_3$		Ag_2SO_4		$AlBr_3$		$AlCl_3$	
	$L^0 = 133.32$		$L^0 = 141.7$		$L^0 = 141.4$		$L^0 = 139.3$	
	$a = 2.45$		$a = 6.3$		$a = 6.0$		$a = 6.0$	
Conc.	Dof	Dpikal	Dof	Dpikal	Dof	Dpikal	Dof	Dpikal
0.000	1.766	1.766	1.392	1.392	1.240	1.240	1.225	1.225
0.001	1.738	1.737	1.327	1.321	1.158	1.074	1.144	1.062
0.002	1.728	1.726	1.309	1.298	1.145	0.992	1.130	0.982
0.003	1.721	1.718	1.298	1.280	1.140	0.921	1.125	0.912
0.004	1.714	1.712	1.290	1.265	1.133	0.835	1.119	0.828
0.005	1.709	1.706	1.284	1.251	1.129	0.746	1.115	0.740
0.006	1.704	1.701	1.278	1.238	1.126	0.658	1.112	0.654
0.007	1.700	1.697	1.274	1.236	1.125	0.575	1.110	0.572
0.008	1.696	1.693	1.270	1.224	1.124	0.508	1.109	0.506
0.009	1.692	1.689	1.267	1.213	1.124	0.440	1.109	0.438
0.010	1.689	1.686	1.264	1.201	1.124	0.380	1.109	0.379
0.020	1.663	1.658	1.250	1.061	1.124	0.105	1.109	0.105
0.030	1.645	1.638	1.244	0.889	1.130	0.041	1.115	0.041
0.040	1.631	1.621	1.242	0.716	1.139	0.020	1.124	0.020
0.050	1.619	1.606	1.242	0.563	1.150	0.012	1.135	0.012
0.060	1.608	1.592	1.243	0.451	1.152	0.007	1.137	0.007
0.070	1.599	1.578	1.246	0.352	1.156	0.005	1.141	0.005
0.080	1.590	1.566	1.250	0.277	1.160	0.004	1.145	0.004
0.090	1.581	1.553	1.251	0.221	1.164	0.003	1.149	0.003
0.100	1.573	1.541	1.253	0.178	1.169	0.002	1.153	0.002
0.100	1.573	1.541	1.253	0.178	1.169	0.002	1.153	0.002

TABLE 15.1   *(Continued)*

Electrolyte	$Al(ClO_4)_3$		$Al(NO_3)_3$		$Al_2(SO_4)_3$		$BaBr_2$	
	$L^0 = 130.3$ $a = 6.3$		$L^0 = 134.42$ $a = 6.0$		$L^0 = 142.8$ $a = 6.5$		$L^0 = 142.1$ $a = 4.0$	
Conc.	Dof	Dpikal	Dof	Dpikal	Dof	Dpikal	Dof	Dpikal
0.000	1.155	1.155	1.188	1.188	0.781	0.781	1.403	1.404
0.001	1.077	1.006	1.109	1.033	0.719	0.655	1.341	1.322
0.002	1.065	0.939	1.095	0.957	0.711	0.532	1.322	1.295
0.003	1.058	0.864	1.090	0.891	0.709	0.397	1.310	1.275
0.004	1.052	0.783	1.084	0.810	0.711	0.295	1.301	1.257
0.005	1.048	0.699	1.080	0.726	0.709	0.215	1.293	1.242
0.006	1.046	0.625	1.077	0.643	0.709	0.158	1.288	1.227
0.007	1.045	0.546	1.075	0.564	0.709	0.119	1.283	1.215
0.008	1.044	0.474	1.074	0.499	0.710	0.091	1.279	1.202
0.009	1.044	0.410	1.074	0.433	0.711	0.072	1.275	1.189
0.010	1.046	0.354	1.074	0.376	0.712	0.057	1.272	1.176
0.020	1.043	0.097	1.074	0.105	0.724	0.011	1.252	1.062
0.030	1.049	0.038	1.079	0.041	0.729	0.004	1.242	0.919
0.040	1.062	0.019	1.088	0.020	0.735	0.002	1.238	0.768
0.050	1.064	0.011	1.098	0.012	0.741	0.001	1.239	0.630
0.060	1.067	0.007	1.100	0.007	0.747	0.001	1.236	0.511
0.070	1.071	0.005	1.104	0.005	0.752	0.001	1.235	0.421
0.080	1.075	0.003	1.107	0.004	0.758	0.000	1.235	0.341
0.090	1.079	0.002	1.112	0.003	0.762	0.000	1.236	0.277
0.100	1.083	0.002	1.116	0.002	0.767	0.000	1.238	0.227

TABLE 15.1   *(Continued)*

Electrolyte	$Ba(BrO_3)_2$		$Ba(CHO_2)_2$		$Ba(C_2H_3O_2)_2$		BaC_2O_4	
	$L^0 = 119.5$		$L^0 = 118.8$		$L^0 = 104.7$		$L^0 = 112.5$	
	a = 4.3		a = 4.3		a = 4.8		a = 3.5	
Conc.	Dof	Dpikal	Dof	Dpikal	Dof	Dpikal	Dof	Dpikal
0.000	1.188	1.188	1.180	1.180	0.996	0.996	0.736	0.736
0.001	1.135	1.125	1.127	1.118	0.951	0.948	0.629	0.723
0.002	1.119	1.104	1.111	1.097	0.937	0.933	0.597	0.675
0.003	1.108	1.089	1.101	1.082	0.929	0.921	0.578	0.681
0.004	1.101	1.076	1.093	1.069	0.922	0.912	0.565	0.687
0.005	1.095	1.064	1.087	1.058	0.917	0.900	0.556	0.692
0.006	1.090	1.054	1.082	1.047	0.914	0.892	0.550	0.695
0.007	1.086	1.044	1.079	1.037	0.910	0.884	0.546	0.696
0.008	1.083	1.040	1.075	1.034	0.908	0.877	0.544	0.695
0.009	1.080	1.031	1.072	1.025	0.906	0.870	0.543	0.692
0.010	1.077	1.022	1.070	1.016	0.903	0.862	0.544	0.687
0.020	1.060	0.917	1.053	0.912	0.889	0.792	0.591	0.500
0.030	1.052	0.802	1.045	0.798	0.884	0.698	0.681	0.437
0.040	1.050	0.693	1.042	0.690	0.882	0.595	0.800	0.389
0.050	1.049	0.574	1.042	0.572	0.880	0.494	0.942	0.389
0.060	1.048	0.469	1.040	0.468	0.880	0.405	1.087	0.369
0.070	1.047	0.381	1.040	0.380	0.880	0.335	1.244	0.351
0.080	1.048	0.310	1.041	0.310	0.881	0.273	1.408	0.335
0.090	1.049	0.253	1.042	0.253	0.882	0.223	1.580	0.321
0.100	1.051	0.208	1.043	0.209	0.884	0.184	1.759	0.308

TABLE 15.1   *(Continued)*

Electrolyte	BaCl$_2$		Ba(ClO$_3$)$_2$		Ba(ClO$_4$)$_2$		Ba$_2$Fe(CN)$_6$	
	$L^0 = 139.3$		$L^0 = 127.6$		$L^0 = 131.0$		$L^0 = 171.8$	
	a = 4.0		a = 4.3		a = 4.3		a = 4.5	
Conc.	Dof	Dpikal	Dof	Dpikal	Dof	Dpikal	Dof	Dpikal
0.000	1.385	1.385	1.274	1.274	1.307	1.307	0.800	0.800
0.001	1.324	1.307	1.217	1.204	1.249	1.235	0.727	0.792
0.002	1.306	1.280	1.200	1.180	1.231	1.210	0.720	−0.296
0.003	1.294	1.260	1.189	1.163	1.220	1.192	0.716	−0.232
0.004	1.285	1.243	1.181	1.148	1.212	1.177	0.711	−0.218
0.005	1.278	1.228	1.175	1.135	1.205	1.163	0.708	−0.999
0.006	1.273	1.213	1.170	1.123	1.200	1.150	0.706	−6.543
0.007	1.268	1.200	1.166	1.111	1.196	1.138	0.706	0.949
0.008	1.265	1.196	1.162	1.107	1.192	1.134	0.706	0.376
0.009	1.262	1.184	1.159	1.096	1.189	1.123	0.707	0.215
0.010	1.259	1.171	1.157	1.086	1.186	1.113	0.711	0.126
0.020	1.240	1.036	1.138	0.966	1.168	0.989	0.708	0.020
0.030	1.233	0.889	1.131	0.835	1.160	0.855	0.717	0.007
0.040	1.231	0.752	1.128	0.713	1.157	0.729	0.735	0.004
0.050	1.231	0.609	1.128	0.583	1.156	0.595	0.736	0.002
0.060	1.231	0.488	1.126	0.470	1.155	0.480	0.739	0.001
0.070	1.231	0.389	1.126	0.378	1.154	0.386	0.743	0.001
0.080	1.233	0.312	1.126	0.305	1.155	0.311	0.747	0.001
0.090	1.235	0.253	1.128	0.247	1.156	0.252	0.752	0.000
0.100	1.238	0.206	1.130	0.203	1.158	0.206	0.757	0.000

TABLE 15.1   *(Continued)*

Electrolyte	BaI_2		$Ba(IO_3)_2$		$Ba(NO_2)_2$		$Ba(NO_3)_2$	
	$L^0 = 140.6$		$L^0 = 104.7$		$L^0 = 135.7$		$L^0 = 135.12$	
	$a = 4.0$		$a = 4.6$		$a = 4.0$		$a = 4.0$	
Conc.	Dof	Dpikal	Dof	Dpikal	Dof	Dpikal	Dof	Dpikal
0.000	1.391	1.391	0.996	0.996	1.350	1.350	1.344	1.345
0.001	1.330	1.311	0.951	0.947	1.290	1.273	1.285	1.268
0.002	1.311	1.285	0.937	0.931	1.272	1.248	1.267	1.243
0.003	1.298	1.264	0.928	0.921	1.260	1.229	1.255	1.224
0.004	1.289	1.247	0.922	0.912	1.251	1.212	1.246	1.208
0.005	1.282	1.232	0.917	0.904	1.244	1.198	1.239	1.194
0.006	1.276	1.218	0.913	0.897	1.238	1.185	1.233	1.181
0.007	1.272	1.205	0.909	0.889	1.233	1.173	1.229	1.169
0.008	1.268	1.192	0.907	0.882	1.230	1.161	1.225	1.157
0.009	1.264	1.180	0.904	0.870	1.226	1.149	1.222	1.145
0.010	1.261	1.167	0.903	0.863	1.223	1.137	1.219	1.133
0.020	1.241	1.055	0.887	0.786	1.204	1.030	1.199	1.027
0.030	1.231	0.914	0.882	0.706	1.194	0.896	1.189	0.894
0.040	1.227	0.764	0.880	0.607	1.190	0.752	1.185	0.751
0.050	1.228	0.628	0.878	0.509	1.191	0.621	1.187	0.620
0.060	1.226	0.510	0.877	0.420	1.189	0.506	1.184	0.506
0.070	1.225	0.421	0.877	0.345	1.188	0.419	1.183	0.419
0.080	1.225	0.340	0.878	0.283	1.188	0.340	1.183	0.340
0.090	1.225	0.277	0.879	0.236	1.188	0.277	1.184	0.277
0.100	1.227	0.227	0.880	0.196	1.190	0.228	1.185	0.228

TABLE 15.1 *(Continued)*

Electrolyte	Ba(OH)$_2$		Ba(SCN)$_2$		BaS$_2$O$_3$		BeCl$_2$	
	$L^0 = 261.3$		$L^0 = 130.2$		$L^0 = 151.1$		$L^0 = 121.3$	
	a = 4.3		a = 4.3		a = 4.5		a = 5.5	
Conc.	Dof	Dpikal	Dof	Dpikal	Dof	Dpikal	Dof	Dpikal
0.000	1.923	1.924	1.299	1.299	0.981	0.981	1.130	1.130
0.001	1.836	1.789	1.242	1.228	0.833	0.849	1.084	1.058
0.002	1.809	1.740	1.224	1.204	0.789	0.812	1.072	1.032
0.003	1.792	1.704	1.213	1.186	0.761	0.766	1.064	1.017
0.004	1.780	1.673	1.205	1.171	0.741	0.744	1.059	1.000
0.005	1.770	1.645	1.198	1.157	0.728	0.725	1.056	0.985
0.006	1.762	1.619	1.193	1.144	0.718	0.709	1.053	0.970
0.007	1.756	1.594	1.189	1.132	0.712	0.695	1.052	0.955
0.008	1.751	1.586	1.185	1.128	0.707	0.683	1.049	0.940
0.009	1.746	1.564	1.182	1.118	0.704	0.672	1.047	0.925
0.010	1.743	1.542	1.180	1.107	0.703	0.662	1.045	0.910
0.020	1.714	1.311	1.161	0.985	0.749	0.608	1.037	0.733
0.030	1.702	1.071	1.153	0.852	0.859	0.585	1.037	0.561
0.040	1.698	0.864	1.150	0.727	0.997	0.566	1.035	0.402
0.050	1.697	0.667	1.149	0.594	1.156	0.546	1.036	0.286
0.060	1.694	0.513	1.148	0.479	1.332	0.524	1.037	0.207
0.070	1.693	0.397	1.147	0.385	1.520	0.501	1.040	0.153
0.080	1.693	0.310	1.148	0.311	1.719	0.524	1.042	0.116
0.090	1.695	0.246	1.149	0.252	1.926	0.498	1.045	0.089
0.100	1.697	0.198	1.151	0.207	2.140	0.473	1.049	0.070

TABLE 15.1   *(Continued)*

Electrolyte	Be(NO₃)₂		BeSO₄		BeSeO₄		CaBr₂	
	$L^0 = 116.42$		$L^0 = 124.8$		$L^0 = 120.7$		$L^0 = 137.9$	
	a = 5.5		a = 6.0		a = 6.0		a = 4.5	
Conc.	Dof	Dpikal	Dof	Dpikal	Dof	Dpikal	Dof	Dpikal
0.000	1.102	1.103	0.766	0.766	0.751	0.751	1.351	1.351
0.001	1.057	1.033	0.652	0.647	0.639	0.636	1.292	1.272
0.002	1.045	1.008	0.617	0.608	0.605	0.598	1.274	1.245
0.003	1.038	0.994	0.595	0.581	0.584	0.571	1.263	1.224
0.004	1.033	0.978	0.580	0.558	0.569	0.549	1.255	1.212
0.005	1.030	0.964	0.569	0.540	0.558	0.531	1.249	1.197
0.006	1.027	0.950	0.560	0.524	0.550	0.515	1.244	1.183
0.007	1.026	0.935	0.554	0.509	0.544	0.501	1.240	1.170
0.008	1.023	0.921	0.550	0.496	0.540	0.488	1.237	1.157
0.009	1.021	0.907	0.547	0.490	0.537	0.482	1.234	1.144
0.010	1.019	0.892	0.546	0.480	0.536	0.473	1.232	1.130
0.020	1.012	0.722	0.581	0.406	0.570	0.401	1.214	0.983
0.030	1.011	0.555	0.659	0.348	0.648	0.346	1.207	0.840
0.040	1.010	0.400	0.764	0.298	0.750	0.299	1.207	0.679
0.050	1.011	0.286	0.886	0.256	0.870	0.259	1.204	0.536
0.060	1.012	0.207	1.020	0.233	1.002	0.238	1.204	0.420
0.070	1.015	0.153	1.165	0.202	1.145	0.208	1.204	0.329
0.080	1.017	0.116	1.312	0.178	1.289	0.184	1.205	0.260
0.090	1.020	0.090	1.465	0.157	1.439	0.164	1.207	0.208
0.100	1.023	0.071	1.623	0.141	1.594	0.147	1.209	0.168

TABLE 15.1 *(Continued)*

Electrolyte	$Ca(C_2H_3O_2)_2$		CaC_2O_4		$CaCl_2$		$Ca(ClO_3)_2$	
	$L^0 = 100.5$		$L^0 = 108.3$		$L^0 = 135.8$		$L^0 = 124.1$	
	a = 5.3		a = 5.3		a = 4.9		a = 4.8	
Conc.	Dof	Dpikal	Dof	Dpikal	Dof	Dpikal	Dof	Dpikal
0.000	0.969	0.969	0.714	0.714	1.335	1.261	1.237	1.237
0.001	0.926	0.921	0.605	0.614	1.278	1.261	1.183	1.170
0.002	0.914	0.905	0.573	0.582	1.262	1.235	1.167	1.147
0.003	0.906	0.894	0.553	0.558	1.251	1.217	1.157	1.131
0.004	0.901	0.884	0.539	0.531	1.244	1.197	1.150	1.117
0.005	0.896	0.879	0.529	0.515	1.239	1.182	1.144	1.100
0.006	0.893	0.871	0.522	0.500	1.235	1.168	1.140	1.087
0.007	0.891	0.863	0.516	0.487	1.231	1.155	1.137	1.075
0.008	0.888	0.856	0.512	0.476	1.230	1.141	1.134	1.064
0.009	0.886	0.848	0.510	0.465	1.226	1.128	1.132	1.052
0.010	0.884	0.841	0.509	0.456	1.224	1.115	1.129	1.040
0.020	0.872	0.752	0.542	0.388	1.211	0.982	1.113	0.926
0.030	0.870	0.642	0.617	0.377	1.210	0.811	1.108	0.778
0.040	0.867	0.536	0.715	0.350	1.211	0.643	1.107	0.628
0.050	0.867	0.429	0.828	0.325	1.213	0.498	1.106	0.494
0.060	0.867	0.339	0.954	0.302	1.217	0.391	1.105	0.386
0.070	0.868	0.268	1.088	0.282	1.221	0.302	1.106	0.307
0.080	0.870	0.214	1.229	0.263	1.226	0.236	1.108	0.242
0.090	0.872	0.172	1.380	0.246	1.232	0.187	1.110	0.193
0.100	0.874	0.139	1.527	0.231	1.238	0.151	1.112	0.156

TABLE 15.1   *(Continued)*

Electrolyte	$Ca(ClO_4)_2$		$Ca_3[Fe(CN)_6]_2$		CaI_2		$Ca(NO_2)_2$	
	$L^0 = 126.8$		$L^0 = 158.4$		$L^0 = 136.4$		$L^0 = 131.5$	
	a = 4.8		a = 5.0		a = 4.5		a = 4.5	
Conc.	Dof	Dpikal	Dof	Dpikal	Dof	Dpikal	Dof	Dpikal
0.000	1.261	1.261	0.824	0.824	1.339	1.340	1.301	1.301
0.001	1.206	1.192	0.737	0.966	1.281	1.262	1.244	1.227
0.002	1.190	1.169	0.724	0.719	1.264	1.235	1.227	1.201
0.003	1.180	1.152	0.715	0.525	1.253	1.214	1.217	1.182
0.004	1.173	1.137	0.710	0.448	1.245	1.202	1.209	1.170
0.005	1.167	1.120	0.707	0.369	1.239	1.188	1.203	1.157
0.006	1.163	1.107	0.706	0.306	1.234	1.174	1.198	1.144
0.007	1.159	1.094	0.707	0.268	1.230	1.161	1.194	1.131
0.008	1.156	1.082	0.705	0.233	1.226	1.148	1.191	1.119
0.009	1.154	1.070	0.703	0.200	1.224	1.135	1.188	1.107
0.010	1.151	1.058	0.702	0.174	1.222	1.122	1.187	1.095
0.020	1.135	0.940	0.706	0.056	1.204	0.977	1.169	0.956
0.030	1.130	0.787	0.717	0.026	1.197	0.836	1.162	0.822
0.040	1.129	0.633	0.720	0.014	1.197	0.677	1.162	0.668
0.050	1.127	0.497	0.725	0.008	1.194	0.535	1.160	0.530
0.060	1.127	0.387	0.730	0.005	1.194	0.419	1.159	0.417
0.070	1.128	0.308	0.735	0.004	1.194	0.329	1.159	0.328
0.080	1.130	0.242	0.740	0.003	1.195	0.260	1.161	0.260
0.090	1.132	0.193	0.746	0.002	1.197	0.208	1.162	0.209
0.100	1.134	0.155	0.751	0.002	1.199	0.168	1.165	0.169

TABLE 15.1   *(Continued)*

Electrolyte	$Ca(NO_3)_2$		$Ca(OH)_2$		$Ca(SCN)_2$		$CaSO_4$	
	$L^0 = 130.92$		$L^0 = 257.1$		$L^0 = 126.0$		$L^0 = 139.3$	
	a = 4.5		a = 4.8		a = 4.8		a = 5.0	
Conc.	Dof	Dpikal	Dof	Dpikal	Dof	Dpikal	Dof	Dpikal
0.000	1.296	1.296	1.826	1.826	1.254	1.254	0.907	0.908
0.001	1.240	1.223	1.743	1.698	1.199	1.186	0.771	0.776
0.002	1.223	1.197	1.719	1.652	1.183	1.163	0.729	0.735
0.003	1.212	1.178	1.704	1.618	1.173	1.146	0.703	0.706
0.004	1.204	1.166	1.693	1.589	1.166	1.131	0.685	0.684
0.005	1.198	1.153	1.685	1.559	1.160	1.114	0.673	0.665
0.006	1.194	1.140	1.678	1.534	1.156	1.101	0.664	0.648
0.007	1.190	1.128	1.673	1.510	1.152	1.089	0.657	0.627
0.008	1.187	1.115	1.669	1.486	1.150	1.077	0.652	0.616
0.009	1.184	1.103	1.666	1.463	1.148	1.065	0.649	0.605
0.010	1.182	1.091	1.662	1.440	1.145	1.053	0.648	0.596
0.020	1.164	0.954	1.637	1.217	1.129	0.936	0.690	0.536
0.030	1.158	0.820	1.629	0.957	1.124	0.784	0.787	0.502
0.040	1.158	0.667	1.627	0.723	1.123	0.631	0.913	0.511
0.050	1.156	0.530	1.624	0.538	1.121	0.496	1.058	0.488
0.060	1.155	0.417	1.623	0.402	1.121	0.387	1.219	0.462
0.070	1.155	0.328	1.624	0.308	1.122	0.308	1.390	0.436
0.080	1.156	0.260	1.626	0.237	1.123	0.242	1.572	0.410
0.090	1.158	0.209	1.628	0.185	1.126	0.193	1.761	0.386
0.100	1.160	0.169	1.631	0.147	1.128	0.155	1.961	0.363

TABLE 15.1   *(Continued)*

Electrolyte	CaS_2O_3		$CdBr_2$		$CdCl_2$		$Cd(ClO_3)_2$	
	$L^0 = 164.9$		$L^0 = 132.4$		$L^0 = 130.3$		$L^0 = 118.6$	
	a = 5.0		a = 4.0		a = 4.0		a = 4.3	
Conc.	Dof	Dpikal	Dof	Dpikal	Dof	Dpikal	Dof	Dpikal
0.000	0.942	0.943	1.277	1.277	1.263	1.263	1.237	1.237
0.001	0.800	0.804	1.223	1.198	1.209	1.186	1.183	1.170
0.002	0.757	0.761	1.207	1.173	1.193	1.160	1.167	1.147
0.003	0.730	0.731	1.196	1.152	1.183	1.140	1.157	1.131
0.004	0.711	0.707	1.189	1.135	1.175	1.123	1.150	1.117
0.005	0.698	0.688	1.183	1.119	1.170	1.108	1.144	1.100
0.006	0.689	0.671	1.178	1.104	1.165	1.094	1.140	1.087
0.007	0.682	0.650	1.174	1.091	1.161	1.080	1.137	1.075
0.008	0.676	0.638	1.171	1.077	1.158	1.067	1.134	1.064
0.009	0.673	0.627	1.169	1.064	1.156	1.054	1.132	1.052
0.010	0.672	0.617	1.167	1.051	1.154	1.041	1.129	1.040
0.020	0.715	0.556	1.151	0.930	1.139	0.923	1.113	0.926
0.030	0.816	0.519	1.144	0.782	1.132	0.777	1.108	0.778
0.040	0.946	0.523	1.142	0.633	1.129	0.630	1.107	0.628
0.050	1.096	0.495	1.145	0.505	1.132	0.503	1.106	0.494
0.060	1.263	0.464	1.143	0.399	1.130	0.399	1.105	0.386
0.070	1.441	0.434	1.143	0.322	1.130	0.322	1.106	0.307
0.080	1.628	0.404	1.143	0.256	1.130	0.256	1.108	0.242
0.090	1.824	0.377	1.144	0.206	1.132	0.206	1.110	0.193
0.100	2.031	0.351	1.146	0.167	1.133	0.167	1.112	0.156

TABLE 15.1   *(Continued)*

Electrolyte	$Cd(ClO_4)_2$		CdI_2		$Cd(NO_2)_2$		$Cd(NO_3)_2$	
	$L^0 = 121.3$		$L^0 = 130.9$		$L^0 = 126.0$		$L^0 = 125.42$	
	a = 4.3		a = 4.0		a = 4.0		a = 4.0	
Conc.	Dof	Dpikal	Dof	Dpikal	Dof	Dpikal	Dof	Dpikal
0.000	1.196	1.197	1.267	1.267	1.232	1.232	1.228	1.228
0.001	1.146	1.127	1.213	1.189	1.180	1.158	1.176	1.154
0.002	1.131	1.103	1.197	1.164	1.164	1.134	1.160	1.130
0.003	1.121	1.085	1.187	1.144	1.154	1.115	1.150	1.111
0.004	1.114	1.070	1.179	1.127	1.147	1.099	1.143	1.095
0.005	1.109	1.055	1.174	1.111	1.142	1.084	1.138	1.081
0.006	1.105	1.042	1.169	1.097	1.137	1.070	1.133	1.067
0.007	1.101	1.030	1.165	1.083	1.133	1.058	1.129	1.055
0.008	1.099	1.026	1.162	1.070	1.130	1.045	1.127	1.042
0.009	1.096	1.014	1.160	1.057	1.128	1.033	1.124	1.030
0.010	1.095	1.003	1.157	1.044	1.126	1.020	1.122	1.017
0.020	1.080	0.871	1.142	0.925	1.111	0.907	1.107	0.905
0.030	1.074	0.730	1.135	0.779	1.104	0.767	1.100	0.765
0.040	1.073	0.604	1.133	0.631	1.102	0.624	1.098	0.623
0.050	1.074	0.478	1.136	0.504	1.105	0.500	1.101	0.499
0.060	1.073	0.375	1.134	0.399	1.103	0.397	1.099	0.397
0.070	1.073	0.295	1.134	0.322	1.103	0.322	1.099	0.322
0.080	1.074	0.233	1.134	0.256	1.103	0.256	1.099	0.256
0.090	1.076	0.187	1.135	0.206	1.104	0.206	1.100	0.206
0.100	1.078	0.151	1.137	0.167	1.106	0.168	1.102	0.168

TABLE 15.1   *(Continued)*

Electrolyte	CdSO$_4$		CdSeO$_4$		CeCl$_3$		Ce(NO$_3$)$_3$	
	$L^0 = 133.8$		$L^0 = 129.7$		$L^0 = 143.3$		$L^0 = 138.42$	
	a = 3.6		a = 4.5		a = 6.0		a = 6.0	
Conc.	Dof	Dpikal	Dof	Dpikal	Dof	Dpikal	Dof	Dpikal
0.000	0.857	0.857	0.839	0.839	1.266	1.266	1.227	1.227
0.001	0.734	0.772	0.715	0.731	1.179	1.103	1.142	1.072
0.002	0.697	0.764	0.678	0.701	1.164	1.025	1.127	0.997
0.003	0.674	0.764	0.655	0.660	1.157	0.956	1.120	0.932
0.004	0.659	0.765	0.639	0.641	1.150	0.873	1.113	0.853
0.005	0.648	0.766	0.628	0.625	1.145	0.786	1.108	0.770
0.006	0.641	0.767	0.620	0.611	1.142	0.699	1.105	0.687
0.007	0.636	0.766	0.615	0.598	1.140	0.616	1.102	0.606
0.008	0.634	0.765	0.612	0.587	1.138	0.547	1.101	0.540
0.009	0.633	0.661	0.609	0.577	1.138	0.477	1.100	0.471
0.010	0.634	0.654	0.608	0.567	1.137	0.415	1.100	0.411
0.020	0.685	0.594	0.651	0.510	1.136	0.118	1.098	0.118
0.030	0.790	0.539	0.749	0.479	1.141	0.046	1.103	0.046
0.040	0.927	0.516	0.869	0.454	1.149	0.023	1.111	0.023
0.050	1.087	0.491	1.009	0.429	1.160	0.013	1.121	0.013
0.060	1.255	0.467	1.163	0.406	1.162	0.008	1.123	0.008
0.070	1.436	0.444	1.328	0.383	1.166	0.006	1.127	0.006
0.080	1.627	0.423	1.503	0.397	1.170	0.004	1.131	0.004
0.090	1.826	0.403	1.685	0.373	1.174	0.003	1.135	0.003
0.100	2.032	0.384	1.873	0.351	1.179	0.002	1.139	0.002

TABLE 15.1 *(Continued)*

Electrolyte	Ce(SO$_4$)$_2$		CoBr$_2$		Co(C$_2$H$_3$O$_2$)$_2$		CoCl$_2$	
	$L^0 = 146.8$		$L^0 = 132.4$		$L^0 = 95.0$		$L^0 = 130.3$	
	a = 7.5		a = 4.5		a = 5.3		a = 4.5	
Conc.	Dof	Dpikal	Dof	Dpikal	Dof	Dpikal	Dof	Dpikal
0.000	0.727	0.727	1.277	1.277	0.931	0.931	1.263	1.263
0.001	0.782	0.633	1.222	1.200	0.890	0.884	1.209	1.187
0.002	0.802	0.389	1.207	1.173	0.879	0.868	1.193	1.160
0.003	0.818	0.227	1.197	1.152	0.872	0.856	1.183	1.140
0.004	0.831	0.136	1.189	1.140	0.867	0.846	1.176	1.128
0.005	0.836	0.087	1.184	1.125	0.863	0.841	1.171	1.114
0.006	0.841	0.058	1.180	1.111	0.860	0.833	1.167	1.100
0.007	0.846	0.041	1.176	1.097	0.859	0.826	1.163	1.086
0.008	0.850	0.030	1.173	1.084	0.856	0.818	1.160	1.073
0.009	0.854	0.023	1.171	1.070	0.854	0.810	1.158	1.060
0.010	0.858	0.018	1.170	1.057	0.852	0.802	1.157	1.047
0.020	0.876	0.003	1.154	0.905	0.842	0.707	1.141	0.898
0.030	0.876	0.001	1.148	0.760	0.841	0.591	1.136	0.755
0.040	0.877	0.001	1.150	0.602	0.839	0.482	1.137	0.599
0.050	0.878	0.000	1.147	0.466	0.838	0.377	1.135	0.465
0.060	0.880	0.000	1.147	0.360	0.839	0.293	1.134	0.359
0.070	0.881	0.000	1.147	0.279	0.840	0.228	1.135	0.279
0.080	0.883	0.000	1.149	0.218	0.842	0.179	1.136	0.219
0.090	0.884	0.000	1.151	0.173	0.844	0.143	1.138	0.174
0.100	0.886	0.000	1.153	0.140	0.846	0.115	1.141	0.140

TABLE 15.1   *(Continued)*

Electrolyte	$Co(ClO_3)_2$		$Co(ClO_4)_2$		CoI_2		$Co(NH_3)_6Cl_2$	
	$L^0 = 118.6$ a = 4.8		$L^0 = 121.3$ a = 4.8		$L^0 = 130.9$ a = 4.5		$L^0 = 182.0$ a = 3.5	
Conc.	Dof	Dpikal	Dof	Dpikal	Dof	Dpikal	Dof	Dpikal
0.000	1.174	1.175	1.196	1.197	1.267	1.267	1.180	1.180
0.001	1.125	1.109	1.145	1.129	1.213	1.191	0.998	1.085
0.002	1.110	1.087	1.131	1.106	1.197	1.164	0.943	1.014
0.003	1.101	1.070	1.122	1.089	1.187	1.144	0.908	0.999
0.004	1.095	1.056	1.115	1.074	1.180	1.132	0.885	0.989
0.005	1.090	1.039	1.110	1.056	1.175	1.117	0.868	0.982
0.006	1.086	1.026	1.106	1.043	1.170	1.103	0.856	0.975
0.007	1.083	1.014	1.103	1.031	1.167	1.090	0.848	0.969
0.008	1.081	1.002	1.101	1.018	1.164	1.076	0.842	0.963
0.009	1.079	0.990	1.099	1.005	1.162	1.063	0.840	0.957
0.010	1.077	0.977	1.097	0.993	1.161	1.050	0.839	0.950
0.020	1.063	0.858	1.083	0.869	1.145	0.900	0.897	0.801
0.030	1.059	0.706	1.079	0.713	1.139	0.757	1.027	0.761
0.040	1.059	0.557	1.079	0.561	1.141	0.600	1.198	0.727
0.050	1.058	0.431	1.077	0.432	1.138	0.466	1.400	0.752
0.060	1.058	0.331	1.078	0.332	1.138	0.359	1.613	0.740
0.070	1.059	0.260	1.079	0.260	1.138	0.279	1.841	0.725
0.080	1.061	0.203	1.081	0.203	1.140	0.219	2.081	0.708
0.090	1.063	0.161	1.083	0.160	1.142	0.174	2.331	0.691
0.100	1.065	0.129	1.085	0.129	1.144	0.140	2.591	0.672

TABLE 15.1 *(Continued)*

Electrolyte	$Co(NO_3)_2$		$CoSO_4$		$CrBr_3$		$CrCl_3$	
	$L^0 = 125.42$		$L^0 = 133.8$		$L^0 = 145.4$		$L^0 = 143.3$	
	a = 4.5		a = 5.0		a = 6.0		a = 6.0	
Conc.	Dof	Dpikal	Dof	Dpikal	Dof	Dpikal	Dof	Dpikal
0.000	1.228	1.228	0.857	0.857	1.282	1.282	1.266	1.266
0.001	1.176	1.156	0.729	0.733	1.194	1.116	1.179	1.103
0.002	1.160	1.131	0.691	0.695	1.179	1.036	1.164	1.025
0.003	1.151	1.111	0.666	0.667	1.172	0.966	1.157	0.956
0.004	1.144	1.100	0.650	0.646	1.165	0.881	1.150	0.873
0.005	1.139	1.086	0.638	0.628	1.160	0.792	1.145	0.786
0.006	1.134	1.073	0.630	0.612	1.157	0.704	1.142	0.699
0.007	1.131	1.060	0.624	0.592	1.155	0.620	1.140	0.616
0.008	1.128	1.048	0.619	0.580	1.153	0.550	1.138	0.547
0.009	1.126	1.035	0.616	0.570	1.153	0.479	1.138	0.477
0.010	1.125	1.023	0.615	0.560	1.153	0.417	1.137	0.415
0.020	1.110	0.880	0.656	0.498	1.151	0.118	1.136	0.118
0.030	1.105	0.744	0.749	0.458	1.156	0.046	1.141	0.046
0.040	1.106	0.593	0.869	0.458	1.165	0.023	1.149	0.023
0.050	1.103	0.462	1.008	0.428	1.176	0.013	1.160	0.013
0.060	1.103	0.358	1.161	0.397	1.178	0.008	1.162	0.008
0.070	1.104	0.279	1.325	0.368	1.182	0.006	1.166	0.006
0.080	1.105	0.219	1.498	0.341	1.186	0.004	1.170	0.004
0.090	1.107	0.174	1.679	0.316	1.190	0.003	1.174	0.003
0.100	1.109	0.140	1.870	0.294	1.195	0.002	1.179	0.002

TABLE 15.1 *(Continued)*

Electrolyte	$Cr(ClO_4)_3$		$Cr(NO_3)_3$		$Cr_2(SO_4)_3$		CsBr	
	$L^0 = 134.3$		$L^0 = 138.42$		$L^0 = 146.8$		$L^0 = 155.2$	
	a = 6.3		a = 6.0		a = 6.5		a = 3.15	
Conc.	Dof	Dpikal	Dof	Dpikal	Dof	Dpikal	Dof	Dpikal
0.000	1.192	1.192	1.227	1.227	0.808	0.808	2.065	2.066
0.001	1.109	1.042	1.142	1.072	0.741	0.683	2.034	2.032
0.002	1.094	0.977	1.127	0.997	0.731	0.562	2.022	2.020
0.003	1.087	0.904	1.120	0.932	0.729	0.427	2.014	2.012
0.004	1.080	0.825	1.113	0.853	0.730	0.322	2.007	2.005
0.005	1.075	0.742	1.108	0.770	0.728	0.238	2.001	1.999
0.006	1.072	0.667	1.105	0.687	0.727	0.178	1.997	1.994
0.007	1.070	0.587	1.102	0.606	0.727	0.135	1.992	1.989
0.008	1.069	0.513	1.101	0.540	0.728	0.104	1.988	1.985
0.009	1.069	0.446	1.100	0.471	0.729	0.082	1.985	1.982
0.010	1.071	0.388	1.100	0.411	0.730	0.066	1.981	1.978
0.020	1.066	0.109	1.098	0.118	0.742	0.013	1.958	1.953
0.030	1.072	0.043	1.103	0.046	0.747	0.005	1.944	1.937
0.040	1.085	0.021	1.111	0.023	0.753	0.002	1.935	1.925
0.050	1.087	0.012	1.121	0.013	0.759	0.001	1.929	1.915
0.060	1.090	0.008	1.123	0.008	0.765	0.001	1.924	1.906
0.070	1.094	0.005	1.127	0.006	0.771	0.001	1.920	1.898
0.080	1.098	0.004	1.131	0.004	0.776	0.000	1.917	1.891
0.090	1.102	0.003	1.135	0.003	0.781	0.000	1.914	1.884
0.100	1.106	0.002	1.139	0.002	0.785	0.000	1.912	1.878

TABLE 15.1 *(Continued)*

Electrolyte	CsBrO$_3$		CsCHO$_2$		CsC$_2$H$_3$O$_2$		CsCl	
	$L^0 = 132.6$		$L^0 = 131.9$		$L^0 = 117.8$		$L^0 = 153.1$	
	a = 3.21		a = 3.0		a = 3.5		a = 3.0	
Conc.	Dof	Dpikal	Dof	Dpikal	Dof	Dpikal	Dof	Dpikal
0.000	1.721	1.721	1.708	1.708	1.423	1.423	2.043	2.044
0.001	1.694	1.693	1.681	1.680	1.400	1.398	2.012	2.010
0.002	1.684	1.683	1.672	1.670	1.392	1.390	2.000	1.998
0.003	1.678	1.675	1.665	1.663	1.386	1.383	1.992	1.989
0.004	1.672	1.670	1.660	1.657	1.381	1.378	1.985	1.982
0.005	1.667	1.665	1.655	1.652	1.377	1.374	1.979	1.976
0.006	1.663	1.661	1.651	1.648	1.373	1.371	1.974	1.971
0.007	1.660	1.657	1.647	1.644	1.370	1.367	1.969	1.966
0.008	1.656	1.653	1.644	1.641	1.367	1.364	1.965	1.962
0.009	1.653	1.650	1.641	1.638	1.365	1.362	1.961	1.958
0.010	1.651	1.648	1.638	1.635	1.363	1.359	1.957	1.954
0.020	1.632	1.627	1.619	1.614	1.347	1.340	1.932	1.927
0.030	1.621	1.613	1.607	1.600	1.338	1.327	1.915	1.908
0.040	1.614	1.602	1.600	1.589	1.332	1.317	1.903	1.893
0.050	1.608	1.593	1.594	1.580	1.328	1.308	1.894	1.881
0.060	1.605	1.585	1.590	1.572	1.325	1.300	1.886	1.869
0.070	1.602	1.578	1.588	1.564	1.322	1.292	1.880	1.859
0.080	1.599	1.571	1.585	1.558	1.320	1.285	1.874	1.850
0.090	1.597	1.564	1.583	1.551	1.318	1.278	1.869	1.841
0.100	1.596	1.558	1.581	1.545	1.317	1.283	1.865	1.832

TABLE 15.1 *(Continued)*

Electrolyte	$CsClO_3$		$CsClO_4$		CsF		CsI	
	$L^0 = 141.4$		$L^0 = 144.1$		$L^0 = 132.2$		$L^0 = 153.7$	
	a = 2.29		a = 1.61		a = 5.21		a = 3.16	
Conc.	Dof	Dpikal	Dof	Dpikal	Dof	Dpikal	Dof	Dpikal
0.000	1.868	1.868	1.910	1.910	1.713	1.714	2.046	2.046
0.001	1.839	1.838	1.880	1.879	1.688	1.686	2.014	2.013
0.002	1.828	1.827	1.868	1.868	1.679	1.677	2.003	2.001
0.003	1.820	1.819	1.860	1.860	1.672	1.670	1.995	1.992
0.004	1.814	1.812	1.854	1.853	1.667	1.665	1.988	1.986
0.005	1.810	1.807	1.848	1.848	1.663	1.661	1.982	1.980
0.006	1.805	1.802	1.843	1.843	1.660	1.658	1.978	1.975
0.007	1.801	1.798	1.839	1.839	1.657	1.654	1.973	1.970
0.008	1.797	1.794	1.835	1.835	1.654	1.652	1.969	1.966
0.009	1.794	1.791	1.834	1.832	1.652	1.649	1.966	1.963
0.010	1.791	1.788	1.831	1.829	1.649	1.647	1.962	1.959
0.020	1.768	1.764	1.806	1.806	1.636	1.630	1.940	1.935
0.030	1.753	1.748	1.790	1.790	1.628	1.619	1.926	1.919
0.040	1.743	1.735	1.777	1.778	1.624	1.611	1.917	1.906
0.050	1.735	1.725	1.768	1.768	1.621	1.604	1.911	1.896
0.060	1.730	1.715	1.760	1.759	1.619	1.597	1.906	1.888
0.070	1.725	1.707	1.754	1.751	1.618	1.590	1.902	1.880
0.080	1.722	1.699	1.749	1.744	1.618	1.583	1.899	1.873
0.090	1.719	1.692	1.745	1.737	1.619	1.576	1.896	1.866
0.100	1.717	1.685	1.741	1.731	1.619	1.569	1.894	1.860

TABLE 15.1 *(Continued)*

Electrolyte	$CsIO_3$		$CsNO_2$		$CsNO_3$		CsOH	
	$L^0 = 117.8$		$L^0 = 148.8$		$L^0 = 148.22$		$L^0 = 274.4$	
	$a = 2.4$		$a = 2.53$		$a = 2.13$		$a = 3.0$	
Conc.	Dof	Dpikal	Dof	Dpikal	Dof	Dpikal	Dof	Dpikal
0.000	1.423	1.423	1.978	1.979	1.970	1.971	2.944	2.945
0.001	1.399	1.398	1.948	1.946	1.940	1.938	2.893	2.891
0.002	1.391	1.389	1.936	1.935	1.928	1.927	2.874	2.871
0.003	1.384	1.383	1.928	1.927	1.920	1.918	2.861	2.857
0.004	1.380	1.378	1.923	1.920	1.913	1.911	2.849	2.846
0.005	1.377	1.373	1.917	1.914	1.907	1.906	2.840	2.836
0.006	1.373	1.370	1.912	1.909	1.904	1.901	2.831	2.827
0.007	1.370	1.366	1.908	1.905	1.899	1.896	2.824	2.819
0.008	1.367	1.363	1.904	1.901	1.895	1.892	2.817	2.813
0.009	1.364	1.360	1.900	1.897	1.892	1.888	2.811	2.806
0.010	1.361	1.358	1.897	1.894	1.888	1.885	2.805	2.800
0.020	1.344	1.338	1.873	1.869	1.863	1.860	2.765	2.757
0.030	1.333	1.323	1.858	1.852	1.848	1.843	2.739	2.729
0.040	1.325	1.312	1.848	1.839	1.836	1.829	2.721	2.707
0.050	1.319	1.302	1.840	1.829	1.828	1.818	2.708	2.689
0.060	1.315	1.296	1.834	1.820	1.821	1.809	2.697	2.675
0.070	1.312	1.289	1.830	1.812	1.816	1.800	2.689	2.662
0.080	1.309	1.282	1.826	1.805	1.811	1.792	2.682	2.650
0.090	1.307	1.276	1.824	1.798	1.808	1.785	2.676	2.639
0.100	1.306	1.270	1.822	1.792	1.805	1.778	2.671	2.629

TABLE 15.1   *(Continued)*

Electrolyte	Cs_2SO_4		$CuBr_2$		$Cu(C_2H_3O_2)_2$		$CuCl_2$	
	$L^0 = 156.6$		$L^0 = 135.0$		$L^0 = 97.6$		$L^0 = 132.9$	
	$a = 3.3$		$a = 4.5$		$a = 5.3$		$a = 4.5$	
Conc.	Dof	Dpikal	Dof	Dpikal	Dof	Dpikal	Dof	Dpikal
0.000	1.567	1.567	1.312	1.313	0.949	0.950	1.298	1.298
0.001	1.490	1.481	1.256	1.235	0.908	0.902	1.242	1.221
0.002	1.464	1.445	1.239	1.207	0.896	0.886	1.225	1.195
0.003	1.447	1.418	1.229	1.187	0.889	0.874	1.215	1.174
0.004	1.434	1.396	1.221	1.175	0.883	0.865	1.207	1.162
0.005	1.424	1.375	1.215	1.160	0.879	0.860	1.201	1.148
0.006	1.415	1.365	1.211	1.146	0.876	0.852	1.197	1.134
0.007	1.407	1.348	1.207	1.132	0.874	0.844	1.193	1.121
0.008	1.401	1.331	1.204	1.119	0.872	0.836	1.190	1.108
0.009	1.395	1.315	1.202	1.106	0.869	0.829	1.188	1.095
0.010	1.390	1.299	1.200	1.092	0.867	0.821	1.186	1.082
0.020	1.354	1.147	1.183	0.943	0.857	0.729	1.169	0.935
0.030	1.333	1.032	1.177	0.799	0.855	0.616	1.163	0.794
0.040	1.318	0.889	1.177	0.639	0.853	0.508	1.164	0.636
0.050	1.309	0.754	1.175	0.499	0.852	0.402	1.162	0.498
0.060	1.302	0.633	1.174	0.388	0.853	0.315	1.161	0.387
0.070	1.294	0.531	1.175	0.302	0.854	0.247	1.161	0.302
0.080	1.287	0.445	1.176	0.238	0.856	0.195	1.163	0.238
0.090	1.281	0.373	1.178	0.189	0.858	0.156	1.165	0.190
0.100	1.276	0.325	1.180	0.153	0.860	0.126	1.167	0.153

TABLE 15.1 *(Continued)*

Electrolyte	$Cu(ClO_3)_2$		$Cu(ClO_4)_2$		$Cu(NO_3)_2$		$CuSO_4$	
	$L^0 = 121.2$		$L^0 = 123.9$		$L^0 = 128.02$		$L^0 = 136.4$	
	a = 4.8		a = 4.8		a = 4.8		a = 5.0	
Conc.	Dof	Dpikal	Dof	Dpikal	Dof	Dpikal	Dof	Dpikal
0.000	1.205	1.205	1.228	1.228	1.261	1.261	0.881	0.882
0.001	1.153	1.139	1.175	1.160	1.207	1.190	0.749	0.754
0.002	1.138	1.116	1.160	1.136	1.191	1.166	0.709	0.714
0.003	1.128	1.100	1.150	1.119	1.181	1.147	0.684	0.686
0.004	1.121	1.085	1.143	1.105	1.174	1.132	0.667	0.664
0.005	1.116	1.068	1.138	1.087	1.168	1.114	0.655	0.646
0.006	1.112	1.056	1.134	1.074	1.164	1.100	0.647	0.629
0.007	1.109	1.044	1.130	1.062	1.161	1.087	0.640	0.609
0.008	1.106	1.032	1.128	1.049	1.158	1.074	0.635	0.597
0.009	1.105	1.020	1.126	1.037	1.156	1.061	0.632	0.587
0.010	1.102	1.008	1.123	1.024	1.154	1.048	0.631	0.578
0.020	1.087	0.891	1.108	0.904	1.138	0.921	0.672	0.517
0.030	1.083	0.741	1.104	0.749	1.134	0.760	0.768	0.480
0.040	1.082	0.591	1.103	0.595	1.133	0.602	0.890	0.484
0.050	1.081	0.461	1.102	0.463	1.132	0.466	1.032	0.457
0.060	1.081	0.357	1.102	0.358	1.132	0.359	1.189	0.429
0.070	1.082	0.282	1.103	0.283	1.133	0.283	1.357	0.401
0.080	1.084	0.221	1.104	0.221	1.134	0.221	1.534	0.374
0.090	1.086	0.176	1.107	0.175	1.137	0.175	1.718	0.349
0.100	1.088	0.141	1.109	0.141	1.139	0.140	1.914	0.326

TABLE 15.1　*(Continued)*

Electrolyte	FeBr$_2$		FeCl$_2$		FeCl$_3$		Fe(ClO$_4$)$_2$	
	$L^0 = 131.9$ a = 4.5		$L^0 = 129.8$ a = 4.5		$L^0 = 144.3$ a = 6.0		$L^0 = 120.8$ a = 4.8	
Conc.	Dof	Dpikal	Dof	Dpikal	Dof	Dpikal	Dof	Dpikal
0.000	1.270	1.270	1.256	1.256	1.276	1.276	1.190	1.190
0.001	1.216	1.193	1.202	1.181	1.188	1.113	1.140	1.123
0.002	1.200	1.166	1.187	1.154	1.172	1.035	1.125	1.100
0.003	1.190	1.145	1.177	1.133	1.165	0.967	1.116	1.083
0.004	1.183	1.133	1.170	1.122	1.157	0.884	1.110	1.068
0.005	1.178	1.118	1.165	1.107	1.152	0.797	1.105	1.050
0.006	1.173	1.104	1.161	1.093	1.149	0.710	1.101	1.037
0.007	1.170	1.090	1.157	1.080	1.147	0.627	1.098	1.024
0.008	1.167	1.077	1.155	1.066	1.145	0.558	1.096	1.012
0.009	1.165	1.063	1.152	1.053	1.144	0.487	1.094	0.999
0.010	1.164	1.050	1.151	1.040	1.144	0.424	1.091	0.987
0.020	1.148	0.898	1.135	0.891	1.142	0.121	1.078	0.863
0.030	1.143	0.753	1.130	0.748	1.147	0.048	1.074	0.706
0.040	1.144	0.595	1.132	0.592	1.156	0.024	1.074	0.555
0.050	1.142	0.460	1.129	0.459	1.166	0.014	1.073	0.427
0.060	1.141	0.354	1.129	0.354	1.168	0.009	1.073	0.327
0.070	1.142	0.274	1.130	0.274	1.172	0.006	1.074	0.256
0.080	1.144	0.215	1.131	0.215	1.176	0.004	1.076	0.200
0.090	1.146	0.170	1.133	0.171	1.180	0.003	1.078	0.158
0.100	1.148	0.137	1.135	0.137	1.185	0.002	1.081	0.126

TABLE 15.1 *(Continued)*

Electrolyte	$Fe(ClO_4)_3$		$Fe(NO_3)_2$		$Fe(NO_3)_3$		$FeSO_4$	
	$L^0 = 135.3$ $a = 4.8$		$L^0 = 124.92$ $a = 4.5$		$L^0 = 139.42$ $a = 6.0$		$L^0 = 133.3$ $a = 5.0$	
Conc.	Dof	Dpikal	Dof	Dpikal	Dof	Dpikal	Dof	Dpikal
0.000	1.201	1.201	1.221	1.222	1.236	1.237	0.853	0.853
0.001	1.116	1.050	1.170	1.150	1.150	1.081	0.744	0.748
0.002	1.099	0.982	1.154	1.124	1.134	1.007	0.712	0.716
0.003	1.090	0.919	1.145	1.105	1.127	0.942	0.692	0.693
0.004	1.085	0.853	1.138	1.094	1.120	0.863	0.677	0.673
0.005	1.081	0.786	1.133	1.080	1.115	0.781	0.665	0.654
0.006	1.076	0.733	1.129	1.067	1.111	0.697	0.657	0.637
0.007	1.073	0.666	1.126	1.054	1.109	0.617	0.648	0.614
0.008	1.070	0.602	1.123	1.041	1.108	0.550	0.640	0.599
0.009	1.067	0.541	1.121	1.029	1.107	0.481	0.633	0.585
0.010	1.066	0.484	1.120	1.016	1.106	0.420	0.627	0.571
0.020	1.065	0.164	1.104	0.874	1.104	0.121	0.594	0.450
0.030	1.063	0.069	1.099	0.737	1.109	0.048	0.580	0.353
0.040	1.067	0.035	1.101	0.586	1.117	0.024	0.571	0.299
0.050	1.073	0.020	1.098	0.456	1.126	0.014	0.567	0.239
0.060	1.081	0.013	1.098	0.353	1.129	0.009	0.566	0.192
0.070	1.095	0.009	1.099	0.274	1.132	0.006	0.567	0.156
0.080	1.097	0.006	1.100	0.215	1.136	0.004	0.568	0.128
0.090	1.100	0.005	1.102	0.171	1.141	0.003	0.571	0.106
0.100	1.103	0.004	1.104	0.138	1.145	0.002	0.575	0.089

TABLE 15.1   *(Continued)*

Electrolyte	H_3AsO_4		HBr		$HCHO_2$		$HC_2H_3O_2$	
	$L^0 = 383.7$		$L^0 = 428.1$		$L^0 = 404.8$		$L^0 = 390.7$	
	a = 6.7		a = 6.0		a = 6.3		a = 6.8	
Conc.	Dof	Dpikal	Dof	Dpikal	Dof	Dpikal	Dof	Dpikal
0.000	1.650	1.650	3.410	3.410	2.534	2.535	1.954	1.954
0.001	1.610	1.609	3.349	3.346	2.485	2.482	1.911	1.909
0.002	1.596	1.594	3.328	3.324	2.467	2.464	1.896	1.894
0.003	1.587	1.584	3.312	3.309	2.455	2.452	1.886	1.883
0.004	1.579	1.576	3.301	3.297	2.445	2.442	1.877	1.874
0.005	1.573	1.570	3.291	3.287	2.437	2.433	1.871	1.867
0.006	1.567	1.564	3.282	3.278	2.430	2.426	1.865	1.861
0.007	1.563	1.559	3.275	3.271	2.424	2.420	1.860	1.856
0.008	1.559	1.554	3.269	3.264	2.419	2.414	1.855	1.851
0.009	1.555	1.550	3.263	3.258	2.414	2.409	1.851	1.846
0.010	1.551	1.546	3.258	3.252	2.410	2.404	1.848	1.842
0.020	1.528	1.517	3.224	3.213	2.382	2.371	1.824	1.812
0.030	1.514	1.496	3.204	3.189	2.365	2.349	1.810	1.791
0.040	1.505	1.478	3.192	3.170	2.355	2.330	1.801	1.773
0.050	1.499	1.460	3.184	3.154	2.348	2.314	1.794	1.755
0.060	1.494	1.442	3.179	3.139	2.343	2.297	1.790	1.737
0.070	1.490	1.423	3.176	3.124	2.340	2.280	1.786	1.718
0.080	1.487	1.402	3.173	3.108	2.337	2.262	1.784	1.697
0.090	1.485	1.380	3.172	3.092	2.335	2.243	1.782	1.675
0.100	1.483	1.357	3.171	3.075	2.334	2.223	1.780	1.651

TABLE 15.1   *(Continued)*

Electrolyte	HCN		H_2CO_3		HCl		$HClO_3$	
	$L^0 = 427.7$		$L^0 = 419.0$		$L^0 = 426.0$		$L^0 = 414.3$	
	a = 6.0		a = 6.8		a = 4.08		a = 6.3	
Conc.	Dof	Dpikal	Dof	Dpikal	Dof	Dpikal	Dof	Dpikal
0.000	3.395	3.396	2.309	2.310	3.335	3.335	2.903	2.904
0.001	3.335	3.332	2.170	2.146	3.273	3.270	2.849	2.847
0.002	3.314	3.310	2.130	2.085	3.250	3.247	2.830	2.827
0.003	3.298	3.295	2.105	2.053	3.234	3.230	2.816	2.813
0.004	3.287	3.283	2.087	2.015	3.221	3.216	2.806	2.803
0.005	3.277	3.273	2.072	1.978	3.210	3.205	2.797	2.794
0.006	3.269	3.264	2.060	1.942	3.200	3.195	2.790	2.786
0.007	3.261	3.257	2.049	1.906	3.191	3.186	2.784	2.779
0.008	3.255	3.250	2.040	1.870	3.184	3.178	2.778	2.773
0.009	3.249	3.244	2.032	1.832	3.177	3.170	2.773	2.768
0.010	3.244	3.239	2.025	1.794	3.171	3.163	2.768	2.763
0.020	3.210	3.200	1.986	1.367	3.126	3.118	2.738	2.728
0.030	3.191	3.175	1.964	0.958	3.100	3.087	2.721	2.705
0.040	3.178	3.156	1.953	0.675	3.081	3.064	2.710	2.686
0.050	3.170	3.140	1.947	0.462	3.067	3.045	2.703	2.670
0.060	3.165	3.125	1.943	0.324	3.056	3.029	2.699	2.654
0.070	3.162	3.110	1.945	0.234	3.048	3.015	2.695	2.637
0.080	3.159	3.094	1.942	0.175	3.042	3.001	2.693	2.620
0.090	3.158	3.078	1.940	0.134	3.037	2.988	2.691	2.602
0.100	3.157	3.061	1.939	0.104	3.033	2.975	2.691	2.583

TABLE 15.1   *(Continued)*

Electrolyte	HClO$_4$		H$_2$CrO$_4$		HF		HI	
	$L^0 = 417.0$		$L^0 = 434.7$		$L^0 = 405.1$		$L^0 = 426.6$	
	a = 6.3		a = 6.5		a = 6.3		a = 4.5	
Conc.	Dof	Dpikal	Dof	Dpikal	Dof	Dpikal	Dof	Dpikal
0.000	3.005	3.005	2.730	2.731	2.546	2.547	2.546	2.547
0.001	2.950	2.947	2.572	2.551	2.496	2.494	2.495	2.492
0.002	2.930	2.927	2.527	2.485	2.479	2.476	2.476	2.473
0.003	2.916	2.913	2.498	2.435	2.466	2.463	2.463	2.460
0.004	2.905	2.902	2.477	2.390	2.457	2.453	2.452	2.448
0.005	2.897	2.893	2.461	2.374	2.449	2.445	2.443	2.439
0.006	2.889	2.885	2.446	2.339	2.442	2.438	2.435	2.431
0.007	2.883	2.878	2.434	2.303	2.436	2.432	2.428	2.424
0.008	2.877	2.872	2.423	2.268	2.431	2.426	2.422	2.417
0.009	2.872	2.867	2.414	2.231	2.426	2.421	2.416	2.411
0.010	2.867	2.862	2.406	2.194	2.422	2.416	2.411	2.406
0.020	2.836	2.826	2.362	1.769	2.393	2.382	2.376	2.367
0.030	2.819	2.803	2.338	1.322	2.377	2.360	2.355	2.340
0.040	2.808	2.784	2.325	0.949	2.366	2.342	2.339	2.320
0.050	2.801	2.768	2.319	0.698	2.359	2.325	2.328	2.303
0.060	2.797	2.752	2.316	0.503	2.355	2.309	2.319	2.287
0.070	2.793	2.736	2.315	0.370	2.351	2.292	2.313	2.273
0.080	2.791	2.719	2.317	0.279	2.348	2.274	2.307	2.259
0.090	2.790	2.701	2.316	0.215	2.347	2.255	2.303	2.245
0.100	2.789	2.682	2.315	0.170	2.346	2.234	2.300	2.230

TABLE 15.1 (*Continued*)

Electrolyte	HIO$_3$		HNO$_3$		H$_3$PO$_4$		H$_2$SO$_4$	
	$L^0 = 390.7$		$L^0 = 421.12$		$L^0 = 385.7$		$L^0 = 429.5$	
	a = 6.6		a = 6.0		a = 6.6		a = 6.5	
Conc.	Dof	Dpikal	Dof	Dpikal	Dof	Dpikal	Dof	Dpikal
0.000	1.954	1.954	3.158	3.158	1.738	1.738	2.594	2.595
0.001	1.911	1.909	3.100	3.098	1.697	1.696	2.442	2.420
0.002	1.896	1.894	3.080	3.077	1.683	1.681	2.398	2.356
0.003	1.885	1.883	3.065	3.062	1.673	1.671	2.370	2.306
0.004	1.877	1.874	3.054	3.051	1.665	1.662	2.350	2.262
0.005	1.870	1.867	3.045	3.041	1.659	1.655	2.335	2.246
0.006	1.864	1.861	3.037	3.033	1.653	1.649	2.320	2.211
0.007	1.859	1.855	3.030	3.026	1.648	1.644	2.309	2.176
0.008	1.855	1.850	3.024	3.019	1.644	1.639	2.298	2.140
0.009	1.851	1.846	3.018	3.013	1.640	1.635	2.290	2.104
0.010	1.847	1.841	3.013	3.008	1.637	1.631	2.282	2.067
0.020	1.822	1.811	2.980	2.970	1.613	1.602	2.239	1.645
0.030	1.808	1.790	2.962	2.946	1.599	1.581	2.215	1.212
0.040	1.798	1.772	2.950	2.927	1.590	1.563	2.202	0.860
0.050	1.792	1.754	2.942	2.911	1.583	1.545	2.196	0.627
0.060	1.787	1.737	2.937	2.896	1.578	1.527	2.192	0.449
0.070	1.783	1.718	2.933	2.880	1.574	1.508	2.191	0.329
0.080	1.781	1.698	2.931	2.864	1.571	1.488	2.193	0.248
0.090	1.779	1.677	2.929	2.848	1.569	1.467	2.191	0.191
0.100	1.777	1.654	2.928	2.830	1.567	1.443	2.191	0.150

TABLE 15.1 *(Continued)*

Electrolyte	Hg(CN)$_2$		HgCl$_2$		Hg$_2$Cl$_2$		KBr	
	$L^0 = 141.6$ $a = 4.0$		$L^0 = 139.9$ $a = 4.0$		$L^0 = 144.9$ $a = 4.0$		$L^0 = 151.9$ $a = 3.67$	
Conc.	Dof	Dpikal	Dof	Dpikal	Dof	Dpikal	Dof	Dpikal
0.000	1.399	1.399	1.385	1.385	1.442	1.443	2.020	2.020
0.001	1.337	1.318	1.324	1.305	1.378	1.361	1.989	1.987
0.002	1.318	1.291	1.305	1.279	1.357	1.334	1.978	1.976
0.003	1.306	1.271	1.293	1.259	1.344	1.314	1.970	1.968
0.004	1.297	1.253	1.284	1.242	1.334	1.297	1.963	1.961
0.005	1.289	1.238	1.277	1.227	1.327	1.282	1.958	1.956
0.006	1.284	1.224	1.271	1.213	1.320	1.268	1.953	1.951
0.007	1.279	1.211	1.266	1.200	1.315	1.256	1.949	1.947
0.008	1.275	1.198	1.262	1.187	1.310	1.243	1.946	1.943
0.009	1.271	1.185	1.259	1.175	1.307	1.231	1.942	1.939
0.010	1.268	1.173	1.256	1.163	1.303	1.219	1.939	1.936
0.020	1.248	1.059	1.236	1.051	1.281	1.110	1.918	1.913
0.030	1.238	0.917	1.226	0.911	1.270	0.974	1.906	1.898
0.040	1.234	0.765	1.222	0.762	1.265	0.826	1.899	1.887
0.050	1.235	0.629	1.223	0.626	1.265	0.688	1.893	1.877
0.060	1.233	0.510	1.220	0.509	1.262	0.566	1.889	1.869
0.070	1.232	0.420	1.219	0.419	1.261	0.471	1.886	1.868
0.080	1.232	0.340	1.219	0.339	1.261	0.385	1.883	1.862
0.090	1.232	0.276	1.220	0.276	1.261	0.315	1.882	1.856
0.100	1.234	0.227	1.222	0.227	1.263	0.260	1.881	1.851

TABLE 15.1 *(Continued)*

Electrolyte	$KBrO_3$		$KCHO_2$		$KC_2H_3O_2$		K_2CO_3	
	$L^0 = 129.3$		$L^0 = 128.6$		$L^0 = 114.5$		$L^0 = 142.8$	
	$a = 2.57$		$a = 3.3$		$a = 3.8$		$a = 3.8$	
Conc.	Dof	Dpikal	Dof	Dpikal	Dof	Dpikal	Dof	Dpikal
0.000	1.689	1.689	1.677	1.677	1.401	1.401	1.424	1.425
0.001	1.662	1.661	1.651	1.649	1.379	1.377	1.354	1.344
0.002	1.653	1.651	1.641	1.640	1.371	1.369	1.332	1.311
0.003	1.646	1.644	1.635	1.633	1.365	1.363	1.317	1.286
0.004	1.641	1.639	1.630	1.627	1.360	1.358	1.306	1.264
0.005	1.636	1.634	1.625	1.623	1.356	1.354	1.297	1.243
0.006	1.632	1.630	1.621	1.619	1.353	1.351	1.290	1.235
0.007	1.629	1.626	1.618	1.615	1.350	1.347	1.284	1.218
0.008	1.625	1.623	1.615	1.612	1.348	1.345	1.279	1.202
0.009	1.622	1.620	1.612	1.609	1.345	1.342	1.274	1.186
0.010	1.620	1.617	1.609	1.606	1.343	1.340	1.270	1.170
0.020	1.600	1.595	1.591	1.586	1.329	1.322	1.244	1.019
0.030	1.588	1.581	1.581	1.573	1.320	1.310	1.229	0.906
0.040	1.579	1.570	1.574	1.563	1.315	1.300	1.221	0.767
0.050	1.573	1.561	1.570	1.554	1.311	1.297	1.217	0.641
0.060	1.569	1.553	1.566	1.546	1.308	1.290	1.216	0.530
0.070	1.565	1.546	1.563	1.539	1.306	1.284	1.212	0.439
0.080	1.563	1.539	1.561	1.532	1.304	1.278	1.211	0.364
0.090	1.562	1.533	1.559	1.526	1.303	1.272	1.210	0.304
0.100	1.560	1.528	1.558	1.520	1.303	1.266	1.210	0.263

TABLE 15.1   *(Continued)*

Electrolyte	$K_2C_2O_4$		KCl		$KClO_3$		$KClO_4$	
	$L^0 = 122.3$		$L^0 = 149.8$		$L^0 = 138.1$		$L^0 = 140.8$	
	a = 3.8		a = 3.8		a = 3.3		a = 4.95	
Conc.	Dof	Dpikal	Dof	Dpikal	Dof	Dpikal	Dof	Dpikal
0.000	1.171	1.171	1.993	1.993	1.830	1.831	1.870	1.870
0.001	1.112	1.097	1.963	1.961	1.803	1.801	1.843	1.841
0.002	1.093	1.063	1.951	1.950	1.792	1.791	1.833	1.831
0.003	1.081	1.037	1.944	1.942	1.785	1.783	1.827	1.825
0.004	1.073	1.013	1.938	1.935	1.780	1.777	1.822	1.820
0.005	1.066	0.991	1.932	1.930	1.775	1.772	1.818	1.816
0.006	1.060	0.983	1.928	1.925	1.770	1.768	1.814	1.812
0.007	1.056	0.965	1.924	1.921	1.766	1.764	1.812	1.809
0.008	1.052	0.947	1.920	1.917	1.763	1.760	1.809	1.807
0.009	1.048	0.929	1.917	1.914	1.760	1.757	1.807	1.804
0.010	1.045	0.911	1.914	1.911	1.757	1.754	1.805	1.802
0.020	1.026	0.745	1.893	1.888	1.738	1.733	1.795	1.789
0.030	1.015	0.630	1.881	1.873	1.726	1.719	1.791	1.783
0.040	1.009	0.501	1.872	1.862	1.719	1.708	1.791	1.779
0.050	1.007	0.395	1.867	1.852	1.714	1.699	1.792	1.777
0.060	1.007	0.312	1.862	1.843	1.710	1.691	1.795	1.775
0.070	1.004	0.248	1.858	1.836	1.706	1.683	1.799	1.773
0.080	1.002	0.199	1.855	1.828	1.704	1.677	1.804	1.772
0.090	1.001	0.162	1.852	1.822	1.702	1.670	1.809	1.770
0.100	1.001	0.138	1.850	1.815	1.701	1.664	1.815	1.768

TABLE 15.1 *(Continued)*

Electrolyte	K_2CrO_4		KF		$K_3Fe(CN)_6$		$K_4Fe(CN)_6$	
	$L^0 = 158.5$		$L^0 = 128.9$		$L^0 = 172.4$		$L^0 = 181.6$	
	a = 3.5		a = 2.96		a = 3.5		a = 3.5	
Conc.	Dof	Dpikal	Dof	Dpikal	Dof	Dpikal	Dof	Dpikal
0.000	1.574	1.574	1.682	1.682	1.497	1.497	1.456	1.456
0.001	1.497	1.488	1.656	1.654	1.362	1.439	1.274	1.826
0.002	1.472	1.459	1.646	1.645	1.324	1.449	1.231	2.819
0.003	1.454	1.435	1.640	1.638	1.301	1.481	1.207	4.621
0.004	1.441	1.415	1.635	1.632	1.283	1.509	1.191	6.429
0.005	1.431	1.397	1.630	1.627	1.271	1.529	1.183	3.529
0.006	1.422	1.380	1.626	1.623	1.260	1.537	1.167	2.323
0.007	1.414	1.355	1.622	1.620	1.252	1.529	1.154	1.439
0.008	1.408	1.339	1.619	1.616	1.248	1.506	1.143	0.919
0.009	1.402	1.323	1.616	1.613	1.240	0.655	1.134	0.617
0.010	1.397	1.307	1.613	1.610	1.232	0.601	1.126	0.434
0.020	1.365	1.188	1.595	1.590	1.190	0.272	1.098	0.051
0.030	1.345	1.056	1.584	1.576	1.175	0.155	1.072	0.016
0.040	1.334	0.929	1.576	1.566	1.168	0.090	1.065	0.008
0.050	1.327	0.807	1.571	1.557	1.160	0.057	1.067	0.004
0.060	1.323	0.728	1.567	1.549	1.157	0.038	1.074	0.003
0.070	1.322	0.625	1.565	1.542	1.157	0.027	1.085	0.002
0.080	1.319	0.536	1.562	1.535	1.159	0.021	1.108	0.001
0.090	1.317	0.459	1.560	1.529	1.162	0.016	1.110	0.001
0.100	1.316	0.394	1.558	1.523	1.166	0.012	1.113	0.001

TABLE 15.1 *(Continued)*

Electrolyte	KH_2AsO_4		$KHCO_3$		KH_2PO_4		K_2HPO_4	
	$L^0 = 107.5$		$L^0 = 118.0$		$L^0 = 109.5$		$L^0 = 130.5$	
	a = 3.6		a = 3.8		a = 3.6		a = 3.5	
Conc.	Dof	Dpikal	Dof	Dpikal	Dof	Dpikal	Dof	Dpikal
0.000	1.238	1.238	1.476	1.476	1.287	1.287	1.282	1.282
0.001	1.216	1.215	1.452	1.451	1.265	1.264	1.218	1.204
0.002	1.209	1.207	1.444	1.442	1.257	1.255	1.198	1.176
0.003	1.204	1.201	1.438	1.436	1.252	1.250	1.185	1.152
0.004	1.199	1.197	1.433	1.431	1.247	1.245	1.175	1.131
0.005	1.196	1.193	1.429	1.427	1.244	1.241	1.167	1.111
0.006	1.192	1.189	1.426	1.424	1.240	1.238	1.161	1.092
0.007	1.190	1.186	1.423	1.420	1.238	1.235	1.155	1.066
0.008	1.187	1.184	1.420	1.417	1.235	1.232	1.151	1.047
0.009	1.185	1.181	1.418	1.415	1.233	1.229	1.146	1.029
0.010	1.183	1.179	1.416	1.412	1.231	1.227	1.143	1.012
0.020	1.169	1.161	1.401	1.394	1.216	1.209	1.122	0.875
0.030	1.161	1.149	1.392	1.382	1.208	1.197	1.107	0.731
0.040	1.155	1.139	1.387	1.372	1.203	1.187	1.099	0.603
0.050	1.152	1.130	1.383	1.369	1.199	1.178	1.095	0.493
0.060	1.148	1.121	1.380	1.363	1.196	1.170	1.093	0.426
0.070	1.146	1.113	1.378	1.356	1.193	1.162	1.094	0.348
0.080	1.144	1.116	1.376	1.351	1.192	1.165	1.091	0.286
0.090	1.143	1.110	1.375	1.345	1.190	1.159	1.090	0.236
0.100	1.142	1.104	1.374	1.339	1.189	1.152	1.089	0.196
0.200	1.143	1.032	1.380	1.273	1.192	1.083	1.101	0.048
0.300	1.150	0.940	1.391	1.185	1.200	0.993	1.126	0.019
0.400	1.161	0.834	1.407	1.080	1.211	0.888	1.141	0.009
0.500	1.174	0.726	1.424	0.968	1.225	0.780	1.159	0.005
0.600	1.188	0.624	1.442	0.857	1.240	0.676	1.178	0.004
0.700	1.203	0.533	1.462	0.754	1.256	0.583	1.197	0.002
0.800	1.219	0.455	1.480	0.660	1.272	0.501	1.217	0.002
0.900	1.233	0.389	1.497	0.578	1.287	0.431	1.237	0.001
0.100	1.142	1.104	1.374	1.339	1.189	1.152	1.089	0.196

TABLE 15.1   *(Continued)*

Electrolyte	KHS		KI		KIO$_3$		KIO$_4$	
	$L^0 = 138.5$ a = 3.3		$L^0 = 150.4$ a = 3.88		$L^0 = 114.5$ a = 2.77		$L^0 = 128.0$ a = 2.31	
Conc.	Dof	Dpikal	Dof	Dpikal	Dof	Dpikal	Dof	Dpikal
0.000	1.836	1.837	2.001	2.001	1.401	1.401	1.666	1.666
0.001	1.808	1.807	1.970	1.969	1.378	1.377	1.640	1.639
0.002	1.798	1.796	1.960	1.958	1.370	1.368	1.630	1.629
0.003	1.791	1.789	1.952	1.950	1.364	1.362	1.623	1.622
0.004	1.785	1.783	1.945	1.943	1.360	1.357	1.618	1.617
0.005	1.780	1.778	1.940	1.938	1.356	1.353	1.615	1.612
0.006	1.776	1.774	1.936	1.933	1.352	1.349	1.610	1.608
0.007	1.772	1.770	1.932	1.929	1.349	1.346	1.607	1.604
0.008	1.769	1.766	1.928	1.925	1.346	1.343	1.603	1.600
0.009	1.766	1.763	1.925	1.922	1.344	1.340	1.600	1.597
0.010	1.763	1.760	1.922	1.919	1.342	1.338	1.598	1.595
0.020	1.743	1.738	1.902	1.897	1.325	1.319	1.578	1.573
0.030	1.732	1.724	1.891	1.882	1.315	1.306	1.565	1.559
0.040	1.724	1.713	1.884	1.874	1.308	1.296	1.556	1.548
0.050	1.719	1.704	1.878	1.866	1.303	1.287	1.550	1.538
0.060	1.715	1.696	1.874	1.859	1.300	1.280	1.545	1.529
0.070	1.712	1.689	1.871	1.853	1.297	1.272	1.541	1.521
0.080	1.709	1.682	1.869	1.847	1.296	1.266	1.538	1.514
0.090	1.708	1.676	1.868	1.841	1.294	1.259	1.536	1.507
0.100	1.706	1.670	1.867	1.836	1.292	1.253	1.534	1.505

TABLE 15.1   *(Continued)*

Electrolyte	KMnO$_4$		KNO$_2$		KNO$_3$		KOH	
	$L^0 = 136.3$		$L^0 = 145.5$		$L^0 = 144.92$		$L^0 = 271.1$	
	a = 3.3		a = 3.0		a = 3.5		a = 3.3	
Conc.	Dof	Dpikal	Dof	Dpikal	Dof	Dpikal	Dof	Dpikal
0.000	1.803	1.803	1.936	1.937	1.929	1.929	2.852	2.853
0.001	1.775	1.774	1.907	1.905	1.899	1.897	2.803	2.800
0.002	1.765	1.764	1.896	1.894	1.887	1.886	2.784	2.781
0.003	1.759	1.756	1.888	1.886	1.880	1.877	2.771	2.767
0.004	1.753	1.750	1.882	1.879	1.873	1.870	2.760	2.756
0.005	1.748	1.745	1.877	1.874	1.867	1.864	2.751	2.747
0.006	1.744	1.741	1.872	1.869	1.861	1.859	2.743	2.739
0.007	1.740	1.737	1.868	1.865	1.857	1.854	2.736	2.731
0.008	1.737	1.734	1.864	1.861	1.852	1.850	2.729	2.725
0.009	1.734	1.731	1.861	1.858	1.848	1.846	2.723	2.719
0.010	1.731	1.728	1.858	1.855	1.845	1.842	2.718	2.713
0.020	1.712	1.707	1.836	1.831	1.817	1.812	2.680	2.672
0.030	1.700	1.693	1.823	1.816	1.797	1.790	2.657	2.645
0.040	1.693	1.682	1.814	1.804	1.782	1.772	2.640	2.625
0.050	1.688	1.673	1.808	1.794	1.769	1.755	2.628	2.608
0.060	1.684	1.665	1.803	1.786	1.758	1.740	2.619	2.594
0.070	1.681	1.658	1.800	1.779	1.747	1.725	2.611	2.582
0.080	1.678	1.651	1.797	1.772	1.736	1.711	2.604	2.570
0.090	1.677	1.645	1.794	1.765	1.727	1.698	2.599	2.560
0.100	1.675	1.639	1.792	1.759	1.718	1.684	2.595	2.550

TABLE 15.1 *(Continued)*

Electrolyte	KSCN		K_2SO_3		K_2SO_4		K_2SeO_4	
	$L^0 = 140.0$		$L^0 = 145.5$		$L^0 = 153.3$		$L^0 = 149.2$	
	a = 3.3		a = 3.8		a = 3.0		a = 3.5	
Conc.	Dof	Dpikal	Dof	Dpikal	Dof	Dpikal	Dof	Dpikal
0.000	1.859	1.859	1.452	1.453	1.528	1.528	1.489	1.489
0.001	1.830	1.829	1.381	1.372	1.453	1.446	1.416	1.406
0.002	1.820	1.818	1.358	1.338	1.429	1.410	1.393	1.377
0.003	1.813	1.811	1.343	1.313	1.412	1.384	1.377	1.354
0.004	1.807	1.805	1.332	1.291	1.398	1.362	1.365	1.334
0.005	1.802	1.799	1.323	1.271	1.388	1.348	1.355	1.315
0.006	1.798	1.795	1.315	1.262	1.379	1.330	1.347	1.298
0.007	1.794	1.791	1.309	1.246	1.371	1.313	1.340	1.272
0.008	1.790	1.788	1.304	1.230	1.364	1.297	1.334	1.255
0.009	1.787	1.784	1.299	1.214	1.358	1.280	1.329	1.239
0.010	1.784	1.781	1.295	1.198	1.353	1.265	1.324	1.223
0.020	1.764	1.760	1.267	1.050	1.320	1.115	1.295	1.098
0.030	1.753	1.745	1.252	0.938	1.300	1.009	1.277	0.960
0.040	1.745	1.734	1.244	0.800	1.286	0.884	1.267	0.830
0.050	1.740	1.725	1.239	0.672	1.277	0.762	1.261	0.709
0.060	1.736	1.717	1.238	0.559	1.271	0.661	1.258	0.631
0.070	1.733	1.710	1.235	0.465	1.267	0.572	1.257	0.534
0.080	1.730	1.703	1.233	0.388	1.265	0.520	1.254	0.451
0.090	1.728	1.697	1.232	0.324	1.267	0.451	1.252	0.382
0.100	1.727	1.691	1.232	0.282	1.264	0.391	1.251	0.325

TABLE 15.1 *(Continued)*

Electrolyte	$LaBr_3$		$La(C_2H_3O_2)_3$		$LaCl_3$		$La(ClO_4)_3$	
	$L^0 = 148.1$		$L^0 = 110.7$		$L^0 = 146.0$		$L^0 = 137.0$	
	a = 6.0		a = 6.8		a = 5.75		a = 6.3	
Conc.	Dof	Dpikal	Dof	Dpikal	Dof	Dpikal	Dof	Dpikal
0.000	1.310	1.310	0.916	0.916	1.293	1.294	1.167	1.167
0.001	1.218	1.143	0.847	0.820	1.192	1.116	1.088	1.018
0.002	1.201	1.064	0.834	0.775	1.166	1.039	1.074	0.952
0.003	1.194	0.995	0.825	0.730	1.152	0.958	1.068	0.877
0.004	1.186	0.911	0.820	0.678	1.137	0.874	1.061	0.797
0.005	1.180	0.823	0.816	0.621	1.125	0.800	1.057	0.713
0.006	1.177	0.734	0.813	0.561	1.114	0.715	1.055	0.639
0.007	1.174	0.649	0.811	0.509	1.106	0.632	1.053	0.560
0.008	1.173	0.579	0.810	0.452	1.098	0.556	1.052	0.487
0.009	1.172	0.506	0.810	0.399	1.091	0.486	1.052	0.422
0.010	1.171	0.441	0.809	0.351	1.085	0.424	1.055	0.365
0.020	1.169	0.127	0.805	0.104	1.035	0.121	1.051	0.101
0.030	1.173	0.050	0.809	0.041	1.001	0.047	1.057	0.039
0.040	1.182	0.025	0.814	0.021	0.977	0.023	1.070	0.020
0.050	1.192	0.014	0.816	0.012	0.962	0.013	1.072	0.011
0.060	1.195	0.009	0.819	0.008	0.942	0.008	1.075	0.007
0.070	1.199	0.006	0.822	0.005	0.927	0.005	1.078	0.005
0.080	1.203	0.004	0.826	0.004	0.914	0.004	1.082	0.003
0.090	1.207	0.003	0.829	0.003	0.904	0.003	1.087	0.003
0.100	1.212	0.003	0.832	0.002	0.897	0.002	1.091	0.002

TABLE 15.1   *(Continued)*

Electrolyte	La(IO$_3$)$_3$		La(NO$_3$)$_3$		La$_2$(SO$_4$)$_3$		LiBr	
	$L^0 = 110.7$ a = 6.7		$L^0 = 141.12$ a = 6.0		$L^0 = 149.5$ a = 6.5		$L^0 = 117.08$ a = 4.76	
Conc.	Dof	Dpikal	Dof	Dpikal	Dof	Dpikal	Dof	Dpikal
0.000	0.916	0.916	1.252	1.252	0.825	0.825	1.379	1.379
0.001	0.847	0.820	1.163	1.097	0.755	0.700	1.356	1.355
0.002	0.833	0.775	1.147	1.023	0.744	0.582	1.348	1.347
0.003	0.825	0.731	1.139	0.959	0.741	0.447	1.342	1.341
0.004	0.819	0.680	1.131	0.881	0.742	0.341	1.338	1.336
0.005	0.815	0.623	1.126	0.798	0.740	0.254	1.334	1.332
0.006	0.813	0.565	1.122	0.715	0.739	0.191	1.331	1.329
0.007	0.811	0.506	1.120	0.634	0.739	0.145	1.328	1.326
0.008	0.809	0.457	1.118	0.567	0.740	0.113	1.326	1.323
0.009	0.810	0.404	1.117	0.497	0.740	0.089	1.324	1.321
0.010	0.808	0.356	1.117	0.435	0.742	0.072	1.322	1.319
0.020	0.804	0.107	1.114	0.127	0.753	0.015	1.308	1.303
0.030	0.808	0.043	1.118	0.050	0.758	0.005	1.301	1.292
0.040	0.813	0.021	1.126	0.025	0.764	0.003	1.296	1.284
0.050	0.815	0.012	1.136	0.014	0.771	0.002	1.293	1.276
0.060	0.818	0.008	1.138	0.009	0.777	0.001	1.291	1.269
0.070	0.821	0.005	1.142	0.006	0.782	0.001	1.289	1.262
0.080	0.825	0.004	1.146	0.004	0.788	0.000	1.288	1.255
0.090	0.828	0.003	1.150	0.003	0.793	0.000	1.288	1.248
0.100	0.831	0.002	1.155	0.003	0.797	0.000	1.287	1.240

TABLE 15.1   *(Continued)*

Electrolyte	LiBrO$_3$		LiCHO$_2$		LiC$_2$H$_3$O$_2$		LiCl	
	$L^0 = 94.48$		$L^0 = 93.78$		$L^0 = 79.68$		$L^0 = 114.98$	
	a = 4.8		a = 4.8		a = 5.3		a = 4.25	
Conc.	Dof	Dpikal	Dof	Dpikal	Dof	Dpikal	Dof	Dpikal
0.000	1.216	1.216	1.210	1.210	1.060	1.060	1.319	1.319
0.001	1.198	1.197	1.192	1.191	1.045	1.044	1.297	1.295
0.002	1.192	1.190	1.186	1.184	1.040	1.039	1.289	1.287
0.003	1.187	1.186	1.181	1.180	1.036	1.035	1.284	1.282
0.004	1.184	1.182	1.178	1.176	1.033	1.032	1.279	1.277
0.005	1.181	1.179	1.175	1.173	1.031	1.030	1.276	1.274
0.006	1.178	1.177	1.172	1.171	1.029	1.028	1.273	1.270
0.007	1.176	1.174	1.170	1.168	1.028	1.026	1.270	1.268
0.008	1.174	1.172	1.168	1.166	1.026	1.025	1.268	1.266
0.009	1.173	1.171	1.167	1.165	1.025	1.023	1.266	1.263
0.010	1.171	1.169	1.165	1.163	1.024	1.022	1.264	1.261
0.020	1.162	1.157	1.156	1.151	1.017	1.013	1.253	1.247
0.030	1.157	1.149	1.151	1.143	1.014	1.006	1.247	1.239
0.040	1.153	1.142	1.148	1.137	1.012	1.000	1.244	1.232
0.050	1.151	1.136	1.146	1.131	1.011	0.995	1.242	1.227
0.060	1.150	1.131	1.144	1.125	1.010	0.990	1.242	1.222
0.070	1.149	1.125	1.144	1.119	1.010	0.984	1.242	1.217
0.080	1.149	1.119	1.143	1.114	1.011	0.978	1.243	1.213
0.090	1.149	1.113	1.144	1.107	1.012	0.971	1.244	1.208
0.100	1.150	1.106	1.144	1.101	1.012	0.965	1.245	1.203

TABLE 15.1 *(Continued)*

Electrolyte	LiClO$_3$		LiClO$_4$		Li$_2$CrO$_4$		LiF	
	$L^0 = 103.28$		$L^0 = 105.98$		$L^0 = 123.68$		$L^0 = 94.08$	
	a = 4.8		a = 5.0		a = 5.0		a = 3.72	
Conc.	Dof	Dpikal	Dof	Dpikal	Dof	Dpikal	Dof	Dpikal
0.000	1.288	1.288	1.264	1.264	1.061	1.062	1.213	1.213
0.001	1.268	1.267	1.244	1.243	1.012	1.009	1.194	1.193
0.002	1.261	1.260	1.237	1.236	0.997	0.992	1.188	1.186
0.003	1.256	1.254	1.232	1.231	0.987	0.977	1.183	1.182
0.004	1.252	1.250	1.229	1.227	0.980	0.966	1.180	1.178
0.005	1.249	1.247	1.226	1.224	0.974	0.957	1.177	1.175
0.006	1.246	1.244	1.223	1.221	0.970	0.948	1.174	1.172
0.007	1.244	1.242	1.221	1.219	0.966	0.940	1.172	1.169
0.008	1.242	1.239	1.220	1.217	0.963	0.932	1.170	1.167
0.009	1.240	1.237	1.218	1.215	0.960	0.930	1.168	1.165
0.010	1.238	1.236	1.217	1.214	0.957	0.923	1.166	1.163
0.020	1.227	1.222	1.209	1.203	0.941	0.853	1.155	1.150
0.030	1.221	1.213	1.206	1.197	0.935	0.774	1.149	1.141
0.040	1.217	1.206	1.206	1.193	0.932	0.688	1.145	1.133
0.050	1.215	1.199	1.206	1.189	0.930	0.620	1.143	1.126
0.060	1.213	1.193	1.208	1.186	0.929	0.536	1.140	1.125
0.070	1.212	1.186	1.210	1.182	0.930	0.458	1.139	1.120
0.080	1.211	1.180	1.213	1.178	0.931	0.390	1.138	1.116
0.090	1.211	1.173	1.216	1.174	0.932	0.332	1.137	1.111
0.100	1.212	1.167	1.220	1.169	0.934	0.283	1.137	1.106

TABLE 15.1   *(Continued)*

Electrolyte	LiI		LiIO$_3$		LiNO$_2$		LiNO$_3$	
	$L^0 = 115.58$		$L^0 = 79.68$		$L^0 = 110.68$		$L^0 = 110.1$	
	a = 5.88		a = 5.3		a = 3.93		a = 2.64	
Conc.	Dof	Dpikal	Dof	Dpikal	Dof	Dpikal	Dof	Dpikal
0.000	1.370	1.370	1.060	1.060	1.340	1.340	1.336	1.336
0.001	1.348	1.347	1.045	1.044	1.318	1.317	1.314	1.313
0.002	1.340	1.339	1.040	1.039	1.310	1.308	1.306	1.304
0.003	1.335	1.333	1.036	1.035	1.305	1.303	1.300	1.299
0.004	1.331	1.329	1.033	1.032	1.300	1.298	1.296	1.294
0.005	1.327	1.325	1.031	1.030	1.297	1.294	1.292	1.290
0.006	1.325	1.322	1.029	1.028	1.293	1.291	1.289	1.286
0.007	1.322	1.320	1.028	1.026	1.291	1.288	1.286	1.283
0.008	1.320	1.317	1.026	1.025	1.288	1.285	1.283	1.280
0.009	1.318	1.315	1.025	1.023	1.286	1.283	1.281	1.277
0.010	1.316	1.313	1.024	1.022	1.284	1.281	1.279	1.275
0.020	1.305	1.299	1.017	1.013	1.270	1.264	1.263	1.257
0.030	1.299	1.289	1.014	1.006	1.262	1.255	1.253	1.245
0.040	1.295	1.280	1.012	1.000	1.258	1.247	1.246	1.236
0.050	1.293	1.273	1.011	0.995	1.254	1.240	1.242	1.228
0.060	1.292	1.265	1.010	0.990	1.251	1.233	1.238	1.221
0.070	1.292	1.256	1.010	0.984	1.249	1.227	1.235	1.214
0.080	1.291	1.247	1.011	0.978	1.247	1.221	1.234	1.208
0.090	1.291	1.238	1.012	0.971	1.246	1.216	1.233	1.202
0.100	1.291	1.228	1.012	0.965	1.246	1.210	1.231	1.197

TABLE 15.1 *(Continued)*

Electrolyte	Li_2SO_4		$MgBr_2$		$Mg(BrO_3)_2$		$Mg(C_2H_3O_2)_2$	
	$L^0 = 118.48$		$L^0 = 131.46$		$L^0 = 108.86$		$L^0 = 94.06$	
	$a = 3.9$		$a = 5.5$		$a = 5.8$		$a = 6.3$	
Conc.	Dof	Dpikal	Dof	Dpikal	Dof	Dpikal	Dof	Dpikal
0.000	1.003	1.003	1.264	1.264	1.086	1.086	0.924	0.924
0.001	0.956	0.952	1.210	1.188	1.040	1.028	0.884	0.878
0.002	0.941	0.934	1.195	1.161	1.028	1.009	0.874	0.864
0.003	0.931	0.920	1.186	1.146	1.020	0.995	0.868	0.853
0.004	0.923	0.912	1.180	1.129	1.015	0.982	0.863	0.843
0.005	0.917	0.902	1.175	1.114	1.011	0.971	0.860	0.834
0.006	0.912	0.893	1.172	1.100	1.009	0.959	0.857	0.826
0.007	0.908	0.885	1.169	1.085	1.006	0.948	0.855	0.823
0.008	0.904	0.876	1.166	1.071	1.003	0.937	0.853	0.815
0.009	0.901	0.869	1.163	1.056	1.001	0.925	0.851	0.806
0.010	0.898	0.863	1.161	1.041	0.999	0.914	0.850	0.797
0.020	0.878	0.790	1.150	0.868	0.990	0.786	0.844	0.683
0.030	0.868	0.734	1.148	0.690	0.988	0.621	0.841	0.542
0.040	0.861	0.662	1.146	0.513	0.987	0.466	0.841	0.412
0.050	0.858	0.588	1.146	0.376	0.988	0.346	0.842	0.307
0.060	0.855	0.520	1.147	0.277	0.989	0.257	0.843	0.233
0.070	0.852	0.457	1.150	0.207	0.992	0.194	0.846	0.176
0.080	0.850	0.414	1.152	0.158	0.994	0.149	0.849	0.135
0.090	0.849	0.361	1.155	0.123	0.997	0.117	0.851	0.105
0.100	0.848	0.315	1.158	0.098	1.001	0.093	0.852	0.084

TABLE 15.1 *(Continued)*

Electrolyte	MgC_2O_4		$MgCl_2$		$Mg(ClO_4)_2$		$MgCrO_4$	
	$L^0 = 101.86$		$L^0 = 129.36$		$L^0 = 120.36$		$L^0 = 138.06$	
	a = 6.3		a = 5.5		a = 5.8		a = 6.0	
Conc.	Dof	Dpikal	Dof	Dpikal	Dof	Dpikal	Dof	Dpikal
0.000	0.677	0.677	1.249	1.249	1.185	1.185	0.870	0.870
0.001	0.574	0.576	1.196	1.176	1.135	1.118	0.738	0.735
0.002	0.543	0.542	1.181	1.151	1.121	1.096	0.698	0.691
0.003	0.523	0.517	1.172	1.131	1.113	1.079	0.672	0.660
0.004	0.510	0.497	1.165	1.112	1.107	1.065	0.655	0.636
0.005	0.500	0.479	1.160	1.096	1.103	1.051	0.642	0.615
0.006	0.492	0.467	1.157	1.081	1.100	1.038	0.632	0.598
0.007	0.486	0.455	1.154	1.067	1.097	1.025	0.625	0.583
0.008	0.482	0.444	1.151	1.053	1.094	1.011	0.620	0.569
0.009	0.480	0.434	1.149	1.039	1.092	0.998	0.617	0.563
0.010	0.479	0.425	1.147	1.025	1.090	0.984	0.615	0.553
0.020	0.507	0.362	1.133	0.886	1.080	0.835	0.652	0.484
0.030	0.575	0.320	1.130	0.714	1.078	0.648	0.740	0.432
0.040	0.666	0.311	1.131	0.551	1.077	0.479	0.857	0.385
0.050	0.771	0.282	1.130	0.418	1.078	0.350	0.993	0.342
0.060	0.890	0.257	1.131	0.322	1.080	0.257	1.142	0.321
0.070	1.011	0.235	1.133	0.246	1.082	0.192	1.305	0.286
0.080	1.138	0.216	1.135	0.190	1.085	0.147	1.470	0.256
0.090	1.269	0.199	1.138	0.150	1.088	0.115	1.641	0.230
0.100	1.405	0.185	1.142	0.119	1.092	0.091	1.817	0.208

TABLE 15.1 *(Continued)*

Electrolyte	$Mg_2Fe(CN)_6$		MgI_2		$Mg(IO_3)_2$		$Mg(NO_2)_2$	
	$L^0 = 161.16$		$L^0 = 129.96$		$L^0 = 94.06$		$L^0 = 125.06$	
	$a = 7.5$		$a = 5.5$		$a = 6.3$		$a = 5.5$	
Conc.	Dof	Dpikal	Dof	Dpikal	Dof	Dpikal	Dof	Dpikal
0.000	0.711	0.711	1.254	1.254	0.924	0.924	1.220	1.220
0.001	0.728	0.624	1.201	1.179	0.884	0.878	1.168	1.148
0.002	0.736	0.450	1.186	1.153	0.874	0.864	1.154	1.123
0.003	0.743	0.336	1.177	1.137	0.868	0.853	1.145	1.109
0.004	0.750	0.242	1.171	1.121	0.863	0.843	1.139	1.093
0.005	0.752	0.182	1.166	1.106	0.860	0.834	1.135	1.079
0.006	0.754	0.135	1.163	1.092	0.857	0.826	1.131	1.066
0.007	0.756	0.104	1.160	1.078	0.855	0.823	1.129	1.052
0.008	0.758	0.080	1.157	1.064	0.853	0.815	1.126	1.039
0.009	0.760	0.065	1.154	1.049	0.851	0.806	1.123	1.025
0.010	0.762	0.052	1.152	1.034	0.850	0.797	1.121	1.011
0.020	0.772	0.011	1.141	0.863	0.844	0.683	1.110	0.848
0.030	0.772	0.004	1.139	0.687	0.841	0.542	1.108	0.678
0.040	0.773	0.002	1.137	0.512	0.841	0.412	1.106	0.508
0.050	0.774	0.001	1.137	0.375	0.842	0.307	1.107	0.374
0.060	0.776	0.001	1.138	0.277	0.843	0.233	1.108	0.277
0.070	0.777	0.000	1.141	0.208	0.846	0.176	1.110	0.208
0.080	0.779	0.000	1.143	0.159	0.849	0.135	1.113	0.159
0.090	0.780	0.000	1.146	0.123	0.851	0.105	1.116	0.124
0.100	0.781	0.000	1.150	0.098	0.852	0.084	1.119	0.098

TABLE 15.1 *(Continued)*

Electrolyte	$Mg(NO_3)_2$		$MgSO_4$		MgS_2O_3		$MnBr_2$	
	$L^0 = 124.48$		$L^0 = 132.86$		$L^0 = 140.46$		$L^0 = 131.9$	
	$a = 5.5$		$a = 3.9$		$a = 6.0$		$a = 4.5$	
Conc.	Dof	Dpikal	Dof	Dpikal	Dof	Dpikal	Dof	Dpikal
0.000	1.216	1.216	0.848	0.848	0.879	0.879	1.270	1.270
0.001	1.164	1.145	0.725	0.762	0.746	0.743	1.216	1.193
0.002	1.150	1.120	0.688	0.751	0.705	0.698	1.200	1.166
0.003	1.141	1.105	0.665	0.696	0.679	0.666	1.190	1.145
0.004	1.135	1.090	0.650	0.683	0.662	0.642	1.183	1.133
0.005	1.131	1.076	0.639	0.672	0.648	0.621	1.178	1.118
0.006	1.127	1.062	0.632	0.661	0.638	0.604	1.173	1.104
0.007	1.125	1.049	0.627	0.652	0.631	0.588	1.170	1.090
0.008	1.122	1.036	0.624	0.643	0.626	0.574	1.167	1.077
0.009	1.119	1.022	0.623	0.634	0.623	0.568	1.165	1.063
0.010	1.117	1.008	0.624	0.626	0.621	0.558	1.164	1.050
0.020	1.106	0.846	0.671	0.541	0.659	0.487	1.148	0.898
0.030	1.105	0.677	0.772	0.504	0.747	0.434	1.143	0.753
0.040	1.103	0.507	0.907	0.473	0.865	0.385	1.144	0.595
0.050	1.103	0.374	1.055	0.445	1.003	0.341	1.142	0.460
0.060	1.104	0.277	1.217	0.460	1.154	0.318	1.141	0.354
0.070	1.106	0.208	1.391	0.437	1.318	0.282	1.142	0.274
0.080	1.109	0.159	1.575	0.415	1.484	0.252	1.144	0.215
0.090	1.112	0.124	1.767	0.393	1.657	0.226	1.146	0.170
0.100	1.115	0.099	1.966	0.372	1.836	0.204	1.148	0.137

TABLE 15.1 *(Continued)*

Electrolyte	MnCl$_2$		Mn(ClO$_4$)$_2$		Mn(NO$_3$)$_2$		MnSO$_4$	
	$L^0 = 129.8$		$L^0 = 120.8$		$L^0 = 124.92$		$L^0 = 133.3$	
	a = 4.5		a = 4.8		a = 4.5		a = 5.0	
Conc.	Dof	Dpikal	Dof	Dpikal	Dof	Dpikal	Dof	Dpikal
0.000	1.256	1.256	1.190	1.190	1.221	1.222	0.853	0.853
0.001	1.202	1.181	1.140	1.123	1.170	1.150	0.726	0.729
0.002	1.187	1.154	1.125	1.100	1.154	1.124	0.687	0.691
0.003	1.177	1.133	1.116	1.083	1.145	1.105	0.663	0.664
0.004	1.170	1.122	1.110	1.068	1.138	1.094	0.646	0.642
0.005	1.165	1.107	1.105	1.050	1.133	1.080	0.635	0.624
0.006	1.161	1.093	1.101	1.037	1.129	1.067	0.627	0.608
0.007	1.157	1.080	1.098	1.024	1.126	1.054	0.620	0.588
0.008	1.155	1.066	1.096	1.012	1.123	1.041	0.616	0.577
0.009	1.152	1.053	1.094	0.999	1.121	1.029	0.613	0.566
0.010	1.151	1.040	1.091	0.987	1.120	1.016	0.612	0.557
0.020	1.135	0.891	1.078	0.863	1.104	0.874	0.653	0.495
0.030	1.130	0.748	1.074	0.706	1.099	0.737	0.746	0.454
0.040	1.132	0.592	1.074	0.555	1.101	0.586	0.865	0.452
0.050	1.129	0.459	1.073	0.427	1.098	0.456	1.003	0.422
0.060	1.129	0.354	1.073	0.327	1.098	0.353	1.156	0.391
0.070	1.130	0.274	1.074	0.256	1.099	0.274	1.319	0.362
0.080	1.131	0.215	1.076	0.200	1.100	0.215	1.491	0.335
0.090	1.133	0.171	1.078	0.158	1.102	0.171	1.671	0.310
0.100	1.135	0.137	1.081	0.126	1.104	0.138	1.862	0.288

TABLE 15.1   *(Continued)*

Electrolyte	NH_4Br		NH_4CHO_2		$NH_4C_2H_3O_2$		$(NH_4)_2C_2O_4$	
	$L^0 = 152.1$		$L^0 = 128.8$		$L^0 = 114.7$		$L^0 = 122.5$	
	a = 2.8		a = 3.0		a = 3.5		a = 3.5	
Conc.	Dof	Dpikal	Dof	Dpikal	Dof	Dpikal	Dof	Dpikal
0.000	2.023	2.023	1.679	1.679	1.403	1.403	1.172	1.173
0.001	1.991	1.990	1.653	1.651	1.380	1.379	1.113	1.096
0.002	1.980	1.978	1.643	1.641	1.372	1.370	1.095	1.068
0.003	1.972	1.969	1.637	1.634	1.366	1.364	1.083	1.044
0.004	1.965	1.962	1.631	1.629	1.361	1.359	1.074	1.022
0.005	1.960	1.957	1.627	1.624	1.358	1.355	1.067	1.001
0.006	1.955	1.952	1.623	1.620	1.354	1.351	1.061	0.981
0.007	1.950	1.947	1.619	1.616	1.351	1.348	1.056	0.954
0.008	1.946	1.943	1.616	1.613	1.348	1.345	1.052	0.935
0.009	1.943	1.939	1.613	1.610	1.346	1.343	1.049	0.916
0.010	1.939	1.936	1.610	1.607	1.344	1.340	1.046	0.898
0.020	1.916	1.911	1.592	1.587	1.329	1.322	1.028	0.758
0.030	1.901	1.894	1.581	1.573	1.320	1.310	1.015	0.614
0.040	1.891	1.882	1.573	1.563	1.314	1.300	1.008	0.493
0.050	1.884	1.872	1.568	1.554	1.311	1.291	1.005	0.394
0.060	1.879	1.863	1.564	1.546	1.307	1.283	1.003	0.336
0.070	1.875	1.855	1.562	1.539	1.305	1.276	1.005	0.270
0.080	1.872	1.848	1.559	1.532	1.303	1.268	1.002	0.218
0.090	1.869	1.841	1.557	1.526	1.301	1.261	1.001	0.178
0.100	1.866	1.835	1.555	1.520	1.300	1.267	1.000	0.147

TABLE 15.1 *(Continued)*

Electrolyte	NH_4Cl		NH_4ClO_4		$(NH_4)_2CrO_4$		NH_4F	
	$L^0 = 150.0$ a = 2.8		$L^0 = 141.0$ a = 3.0		$L^0 = 122.5$ a = 3.3		$L^0 = 129.1$ a = 3.0	
Conc.	Dof	Dpikal	Dof	Dpikal	Dof	Dpikal	Dof	Dpikal
0.000	1.996	1.996	1.873	1.873	1.172	1.173	1.684	1.684
0.001	1.965	1.964	1.844	1.843	1.114	1.096	1.658	1.656
0.002	1.954	1.952	1.833	1.832	1.095	1.065	1.648	1.647
0.003	1.946	1.944	1.827	1.824	1.083	1.039	1.642	1.640
0.004	1.940	1.937	1.820	1.818	1.074	1.015	1.636	1.634
0.005	1.934	1.931	1.815	1.813	1.067	0.993	1.632	1.629
0.006	1.929	1.926	1.811	1.808	1.061	0.972	1.628	1.625
0.007	1.925	1.922	1.807	1.804	1.056	0.966	1.624	1.621
0.008	1.921	1.918	1.803	1.800	1.052	0.947	1.621	1.618
0.009	1.917	1.914	1.800	1.797	1.048	0.929	1.618	1.615
0.010	1.914	1.911	1.797	1.794	1.045	0.911	1.615	1.612
0.020	1.891	1.886	1.776	1.771	1.029	0.736	1.597	1.592
0.030	1.876	1.870	1.764	1.756	1.015	0.594	1.586	1.578
0.040	1.867	1.857	1.755	1.745	1.008	0.508	1.578	1.568
0.050	1.860	1.847	1.749	1.736	1.004	0.409	1.573	1.559
0.060	1.854	1.839	1.745	1.727	1.002	0.331	1.569	1.551
0.070	1.850	1.831	1.742	1.720	1.001	0.268	1.567	1.544
0.080	1.848	1.824	1.739	1.713	1.003	0.219	1.564	1.537
0.090	1.845	1.817	1.736	1.707	1.001	0.181	1.562	1.531
0.100	1.842	1.811	1.734	1.701	1.000	0.158	1.560	1.525

TABLE 15.1   *(Continued)*

Electrolyte	NH$_4$HCO$_3$		NH$_4$H$_2$PO$_4$		(NH$_4$)$_2$HPO$_4$		NH$_4$HSO$_3$	
	$L^0 = 118.2$		$L^0 = 109.7$		$L^0 = 130.7$		$L^0 = 123.7$	
	a = 3.4		a = 3.4		a = 3.3		a = 3.4	
Conc.	Dof	Dpikal	Dof	Dpikal	Dof	Dpikal	Dof	Dpikal
0.000	1.477	1.478	1.288	1.288	1.283	1.284	1.586	1.586
0.001	1.454	1.452	1.266	1.265	1.220	1.206	1.561	1.560
0.002	1.445	1.444	1.258	1.256	1.200	1.174	1.552	1.551
0.003	1.440	1.437	1.253	1.251	1.187	1.149	1.546	1.544
0.004	1.435	1.432	1.248	1.246	1.177	1.126	1.541	1.539
0.005	1.431	1.428	1.245	1.242	1.168	1.105	1.537	1.534
0.006	1.427	1.424	1.241	1.238	1.162	1.084	1.533	1.531
0.007	1.424	1.421	1.238	1.235	1.156	1.078	1.530	1.527
0.008	1.421	1.418	1.236	1.233	1.151	1.061	1.527	1.524
0.009	1.419	1.415	1.234	1.230	1.147	1.043	1.524	1.521
0.010	1.416	1.413	1.231	1.228	1.143	1.026	1.522	1.519
0.020	1.400	1.394	1.217	1.210	1.123	0.854	1.505	1.500
0.030	1.391	1.382	1.208	1.197	1.107	0.710	1.495	1.487
0.040	1.385	1.372	1.203	1.187	1.098	0.618	1.489	1.476
0.050	1.381	1.363	1.199	1.178	1.093	0.509	1.485	1.468
0.060	1.378	1.355	1.196	1.170	1.091	0.420	1.482	1.460
0.070	1.375	1.348	1.193	1.162	1.090	0.346	1.479	1.453
0.080	1.373	1.341	1.191	1.155	1.091	0.286	1.477	1.446
0.090	1.372	1.334	1.189	1.148	1.089	0.239	1.475	1.440
0.100	1.371	1.328	1.188	1.140	1.088	0.210	1.474	1.433

TABLE 15.1   *(Continued)*

Electrolyte	NH_4I		NH_4IO_3		NH_4NO_3		NH_4OH	
	$L^0 = 150.6$		$L^0 = 114.7$		$L^0 = 145.12$		$L^0 = 271.3$	
	a = 2.8		a = 3.4		a = 2.3		a = 3.0	
Conc.	Dof	Dpikal	Dof	Dpikal	Dof	Dpikal	Dof	Dpikal
0.000	2.004	2.004	1.403	1.403	1.931	1.931	2.858	2.858
0.001	1.973	1.971	1.380	1.379	1.901	1.900	2.808	2.806
0.002	1.961	1.959	1.372	1.370	1.890	1.889	2.789	2.786
0.003	1.954	1.951	1.366	1.364	1.882	1.880	2.776	2.772
0.004	1.947	1.944	1.361	1.359	1.875	1.874	2.765	2.761
0.005	1.941	1.938	1.357	1.355	1.871	1.868	2.755	2.751
0.006	1.936	1.933	1.354	1.351	1.866	1.863	2.747	2.743
0.007	1.932	1.929	1.351	1.348	1.862	1.859	2.740	2.736
0.008	1.928	1.925	1.348	1.345	1.858	1.855	2.733	2.729
0.009	1.924	1.921	1.346	1.343	1.854	1.852	2.727	2.722
0.010	1.921	1.918	1.344	1.340	1.851	1.848	2.722	2.717
0.020	1.898	1.893	1.328	1.322	1.827	1.824	2.682	2.674
0.030	1.884	1.877	1.319	1.309	1.813	1.807	2.657	2.646
0.040	1.874	1.864	1.314	1.299	1.802	1.795	2.639	2.625
0.050	1.867	1.854	1.310	1.291	1.794	1.784	2.626	2.608
0.060	1.861	1.846	1.307	1.283	1.788	1.774	2.616	2.593
0.070	1.857	1.838	1.304	1.275	1.783	1.766	2.608	2.580
0.080	1.855	1.831	1.302	1.268	1.779	1.758	2.601	2.568
0.090	1.852	1.824	1.300	1.261	1.777	1.751	2.595	2.557
0.100	1.849	1.818	1.299	1.254	1.774	1.744	2.590	2.547

TABLE 15.1 *(Continued)*

Electrolyte	NH$_4$SCN		(NH$_4$)$_2$SO$_4$		NaBr		NaBrO$_3$	
	$L^0 = 140.2$ $a = 3.5$		$L^0 = 153.5$ $a = 3.3$		$L^0 = 128.5$ $a = 3.58$		$L^0 = 105.9$ $a = 3.9$	
Conc.	Dof	Dpikal	Dof	Dpikal	Dof	Dpikal	Dof	Dpikal
0.000	1.861	1.861	1.530	1.530	1.627	1.628	1.405	1.406
0.001	1.833	1.831	1.455	1.445	1.602	1.600	1.385	1.383
0.002	1.823	1.821	1.430	1.414	1.593	1.591	1.378	1.376
0.003	1.815	1.813	1.414	1.389	1.586	1.584	1.372	1.370
0.004	1.809	1.807	1.401	1.368	1.581	1.579	1.368	1.366
0.005	1.804	1.802	1.391	1.348	1.576	1.574	1.365	1.363
0.006	1.800	1.798	1.382	1.329	1.573	1.570	1.362	1.360
0.007	1.796	1.794	1.375	1.322	1.569	1.567	1.359	1.357
0.008	1.793	1.790	1.368	1.306	1.566	1.563	1.357	1.354
0.009	1.790	1.787	1.362	1.291	1.563	1.560	1.355	1.352
0.010	1.787	1.784	1.357	1.276	1.561	1.558	1.353	1.350
0.020	1.768	1.763	1.327	1.119	1.544	1.538	1.340	1.335
0.030	1.757	1.749	1.307	0.982	1.534	1.525	1.333	1.327
0.040	1.749	1.738	1.294	0.889	1.528	1.515	1.329	1.320
0.050	1.745	1.729	1.287	0.770	1.523	1.507	1.326	1.314
0.060	1.740	1.721	1.283	0.664	1.520	1.499	1.324	1.309
0.070	1.737	1.714	1.280	0.570	1.517	1.492	1.322	1.304
0.080	1.735	1.707	1.280	0.488	1.515	1.492	1.321	1.299
0.090	1.733	1.700	1.277	0.418	1.513	1.487	1.321	1.294
0.100	1.732	1.704	1.276	0.376	1.512	1.482	1.320	1.289

TABLE 15.1   *(Continued)*

Electrolyte	NaCHO$_2$		Na$_2$CO$_3$		Na$_2$C$_2$O$_4$		NaCl	
	$L^0 = 105.2$		$L^0 = 119.4$		$L^0 = 98.9$		$L^0 = 126.4$	
	a = 3.9		a = 4.4		a = 4.4		a = 4.0	
Conc.	Dof	Dpikal	Dof	Dpikal	Dof	Dpikal	Dof	Dpikal
0.000	1.397	1.397	1.161	1.161	0.987	0.987	1.611	1.611
0.001	1.376	1.375	1.107	1.101	0.941	0.931	1.585	1.584
0.002	1.369	1.368	1.091	1.079	0.928	0.910	1.577	1.575
0.003	1.364	1.362	1.080	1.063	0.919	0.892	1.570	1.568
0.004	1.360	1.358	1.072	1.048	0.913	0.876	1.565	1.563
0.005	1.356	1.355	1.065	1.035	0.908	0.861	1.561	1.559
0.006	1.354	1.351	1.060	1.030	0.904	0.857	1.557	1.555
0.007	1.351	1.349	1.056	1.019	0.901	0.844	1.554	1.552
0.008	1.349	1.346	1.053	1.008	0.898	0.831	1.551	1.549
0.009	1.347	1.344	1.050	0.997	0.896	0.818	1.549	1.546
0.010	1.345	1.342	1.048	0.987	0.896	0.806	1.546	1.543
0.020	1.332	1.327	1.028	0.873	0.880	0.671	1.531	1.525
0.030	1.325	1.319	1.020	0.760	0.874	0.544	1.522	1.515
0.040	1.322	1.312	1.018	0.670	0.875	0.453	1.517	1.507
0.050	1.318	1.306	1.014	0.563	0.872	0.356	1.513	1.500
0.060	1.316	1.301	1.013	0.468	0.870	0.279	1.510	1.493
0.070	1.314	1.296	1.012	0.387	0.870	0.220	1.508	1.487
0.080	1.314	1.291	1.012	0.320	0.870	0.175	1.506	1.482
0.090	1.313	1.286	1.013	0.266	0.871	0.141	1.506	1.476
0.100	1.313	1.281	1.015	0.222	0.872	0.115	1.505	1.471

TABLE 15.1 *(Continued)*

Electrolyte	NaClO$_3$		NaClO$_4$		Na$_2$CrO$_4$		NaF	
	$L^0 = 114.7$ a = 3.23		$L^0 = 117.4$ a = 3.4		$L^0 = 135.1$ a = 4.1		$L^0 = 105.5$ a = 3.34	
Conc.	Dof	Dpikal	Dof	Dpikal	Dof	Dpikal	Dof	Dpikal
0.000	1.502	1.503	1.529	1.529	1.259	1.259	1.401	1.401
0.001	1.479	1.478	1.506	1.504	1.199	1.194	1.380	1.379
0.002	1.471	1.470	1.497	1.496	1.180	1.171	1.372	1.371
0.003	1.466	1.464	1.492	1.489	1.167	1.155	1.367	1.365
0.004	1.461	1.459	1.487	1.485	1.158	1.140	1.363	1.361
0.005	1.457	1.455	1.483	1.480	1.150	1.128	1.360	1.357
0.006	1.453	1.451	1.479	1.477	1.144	1.117	1.356	1.354
0.007	1.450	1.448	1.476	1.474	1.139	1.106	1.354	1.351
0.008	1.448	1.445	1.473	1.471	1.134	1.095	1.351	1.349
0.009	1.445	1.443	1.471	1.468	1.131	1.085	1.349	1.347
0.010	1.443	1.440	1.468	1.466	1.127	1.075	1.347	1.345
0.020	1.427	1.423	1.453	1.448	1.104	0.994	1.334	1.329
0.030	1.418	1.411	1.444	1.436	1.092	0.893	1.326	1.319
0.040	1.413	1.402	1.438	1.427	1.087	0.795	1.321	1.310
0.050	1.409	1.395	1.434	1.419	1.085	0.698	1.318	1.304
0.060	1.406	1.388	1.431	1.412	1.082	0.627	1.316	1.297
0.070	1.404	1.382	1.429	1.406	1.080	0.541	1.313	1.291
0.080	1.402	1.376	1.427	1.400	1.080	0.465	1.312	1.286
0.090	1.400	1.370	1.426	1.394	1.080	0.399	1.311	1.280
0.100	1.399	1.365	1.425	1.388	1.080	0.343	1.310	1.275

TABLE 15.1   *(Continued)*

Electrolyte	$Na_4Fe(CN)_6$		NaH_2AsO_4		$NaHCO_3$		NaH_2PO_4	
	$L^0 = 158.2$		$L^0 = 84.1$		$L^0 = 94.6$		$L^0 = 86.1$	
	a = 4.6		a = 4.6		a = 4.2		a = 4.2	
Conc.	Dof	Dpikal	Dof	Dpikal	Dof	Dpikal	Dof	Dpikal
0.000	1.139	1.139	1.078	1.079	1.255	1.255	1.115	1.115
0.001	0.991	1.161	1.062	1.061	1.236	1.235	1.099	1.098
0.002	0.960	1.236	1.057	1.055	1.230	1.229	1.093	1.092
0.003	0.944	1.182	1.053	1.051	1.226	1.224	1.089	1.087
0.004	0.928	1.000	1.049	1.048	1.222	1.220	1.086	1.084
0.005	0.917	0.773	1.047	1.045	1.219	1.217	1.083	1.081
0.006	0.908	0.573	1.045	1.043	1.216	1.215	1.081	1.079
0.007	0.902	0.418	1.043	1.040	1.214	1.212	1.079	1.077
0.008	0.897	0.312	1.041	1.038	1.212	1.210	1.077	1.075
0.009	0.893	0.237	1.039	1.037	1.211	1.208	1.075	1.073
0.010	0.890	0.183	1.038	1.036	1.209	1.207	1.074	1.072
0.020	0.873	0.031	1.029	1.025	1.199	1.195	1.065	1.060
0.030	0.873	0.009	1.024	1.017	1.193	1.187	1.060	1.053
0.040	0.880	0.005	1.022	1.011	1.190	1.180	1.057	1.047
0.050	0.892	0.003	1.019	1.005	1.188	1.175	1.055	1.041
0.060	0.895	0.002	1.018	1.000	1.186	1.170	1.053	1.036
0.070	0.899	0.001	1.017	0.995	1.185	1.165	1.052	1.031
0.080	0.905	0.001	1.016	0.990	1.184	1.160	1.052	1.026
0.090	0.911	0.001	1.016	0.984	1.184	1.155	1.052	1.021
0.100	0.917	0.000	1.016	0.979	1.185	1.149	1.052	1.015

TABLE 15.1   *(Continued)*

Electrolyte	Na$_2$HPO$_4$		NaI		NaIO$_3$		NaMnO$_4$	
	$L^0 = 107.1$		$L^0 = 127.0$		$L^0 = 91.1$		$L^0 = 112.9$	
	a = 4.1		a = 4.23		a = 4.2		a = 3.8	
Conc.	Dof	Dpikal	Dof	Dpikal	Dof	Dpikal	Dof	Dpikal
0.000	1.065	1.065	1.615	1.615	1.200	1.201	1.484	1.484
0.001	1.016	1.007	1.590	1.589	1.183	1.182	1.461	1.460
0.002	1.001	0.986	1.581	1.579	1.177	1.175	1.454	1.452
0.003	0.991	0.969	1.575	1.573	1.173	1.171	1.448	1.446
0.004	0.984	0.954	1.570	1.568	1.169	1.167	1.443	1.441
0.005	0.978	0.940	1.566	1.563	1.166	1.164	1.440	1.438
0.006	0.973	0.927	1.562	1.560	1.164	1.162	1.436	1.434
0.007	0.970	0.914	1.559	1.556	1.162	1.160	1.433	1.431
0.008	0.967	0.901	1.556	1.553	1.160	1.158	1.431	1.429
0.009	0.964	0.889	1.553	1.551	1.158	1.156	1.429	1.426
0.010	0.962	0.876	1.551	1.549	1.157	1.154	1.427	1.424
0.020	0.945	0.775	1.536	1.531	1.147	1.143	1.413	1.408
0.030	0.937	0.652	1.527	1.520	1.142	1.135	1.405	1.397
0.040	0.934	0.541	1.522	1.511	1.139	1.129	1.400	1.388
0.050	0.934	0.443	1.518	1.504	1.136	1.123	1.397	1.384
0.060	0.932	0.377	1.515	1.498	1.135	1.118	1.394	1.379
0.070	0.931	0.307	1.513	1.492	1.134	1.113	1.392	1.374
0.080	0.931	0.250	1.512	1.486	1.133	1.108	1.391	1.368
0.090	0.931	0.205	1.511	1.480	1.133	1.103	1.390	1.364
0.100	0.932	0.170	1.511	1.474	1.133	1.098	1.390	1.359

TABLE 15.1   *(Continued)*

Electrolyte	Na_2MoO_4		$NaNO_2$		$NaNO_3$		NaOH	
	$L^0 = 124.6$		$L^0 = 122.1$		$L^0 = 121.52$		$L^0 = 247.7$	
	a = 4.4		a = 3.95		a = 2.98		a = 3.9	
Conc.	Dof	Dpikal	Dof	Dpikal	Dof	Dpikal	Dof	Dpikal
0.000	1.196	1.196	1.573	1.573	1.568	1.568	2.128	2.128
0.001	1.140	1.134	1.549	1.547	1.543	1.542	2.087	2.085
0.002	1.123	1.113	1.540	1.538	1.534	1.533	2.072	2.069
0.003	1.112	1.096	1.534	1.532	1.529	1.526	2.061	2.058
0.004	1.103	1.082	1.529	1.527	1.524	1.521	2.053	2.049
0.005	1.097	1.069	1.525	1.523	1.519	1.517	2.045	2.041
0.006	1.091	1.063	1.521	1.519	1.515	1.513	2.039	2.035
0.007	1.087	1.053	1.518	1.516	1.512	1.509	2.033	2.029
0.008	1.083	1.042	1.516	1.513	1.509	1.506	2.028	2.023
0.009	1.080	1.032	1.513	1.510	1.506	1.503	2.024	2.018
0.010	1.078	1.022	1.511	1.508	1.504	1.501	2.019	2.014
0.020	1.057	0.913	1.496	1.490	1.487	1.482	1.990	1.980
0.030	1.048	0.805	1.487	1.480	1.476	1.469	1.972	1.962
0.040	1.046	0.717	1.482	1.472	1.469	1.459	1.960	1.946
0.050	1.042	0.611	1.478	1.465	1.465	1.451	1.951	1.932
0.060	1.040	0.514	1.475	1.459	1.461	1.444	1.943	1.920
0.070	1.040	0.430	1.473	1.453	1.459	1.437	1.937	1.909
0.080	1.040	0.359	1.471	1.448	1.456	1.431	1.933	1.899
0.090	1.041	0.301	1.471	1.442	1.455	1.425	1.929	1.889
0.100	1.042	0.253	1.470	1.437	1.453	1.420	1.926	1.879

TABLE 15.1   *(Continued)*

Electrolyte	NaSCN		Na$_2$SO$_3$		Na$_2$SO$_4$		Na$_2$S$_2$O$_3$	
	$L^0 = 116.6$		$L^0 = 122.1$		$L^0 = 129.9$		$L^0 = 137.5$	
	a = 3.9		a = 4.4		a = 3.5		a = 4.1	
Conc.	Dof	Dpikal	Dof	Dpikal	Dof	Dpikal	Dof	Dpikal
0.000	1.521	1.522	1.180	1.180	1.229	1.229	1.272	1.272
0.001	1.498	1.497	1.125	1.119	1.171	1.165	1.211	1.206
0.002	1.490	1.488	1.108	1.097	1.151	1.144	1.192	1.184
0.003	1.484	1.482	1.097	1.080	1.138	1.128	1.179	1.167
0.004	1.480	1.478	1.088	1.066	1.128	1.113	1.169	1.152
0.005	1.476	1.474	1.082	1.053	1.120	1.100	1.162	1.141
0.006	1.472	1.470	1.077	1.048	1.113	1.088	1.155	1.129
0.007	1.469	1.467	1.072	1.037	1.108	1.068	1.150	1.118
0.008	1.467	1.464	1.069	1.026	1.103	1.056	1.145	1.108
0.009	1.464	1.462	1.066	1.016	1.098	1.044	1.141	1.097
0.010	1.462	1.459	1.064	1.006	1.094	1.033	1.138	1.087
0.020	1.448	1.443	1.043	0.894	1.068	0.945	1.114	1.008
0.030	1.440	1.433	1.035	0.784	1.050	0.847	1.102	0.908
0.040	1.435	1.425	1.033	0.695	1.039	0.754	1.096	0.812
0.050	1.431	1.419	1.029	0.588	1.031	0.662	1.094	0.716
0.060	1.428	1.413	1.027	0.492	1.025	0.602	1.091	0.646
0.070	1.426	1.407	1.027	0.409	1.022	0.522	1.090	0.560
0.080	1.425	1.402	1.027	0.341	1.016	0.452	1.089	0.483
0.090	1.424	1.397	1.028	0.284	1.012	0.390	1.089	0.416
0.100	1.424	1.392	1.029	0.238	1.008	0.337	1.090	0.358

TABLE 15.1   *(Continued)*

Electrolyte	Na_2WO_4		$NdBr_3$		$NdCl_3$		$Nd(ClO_4)_3$	
	$L^0 = 119.5$		$L^0 = 142.7$		$L^0 = 140.6$		$L^0 = 131.6$	
	a = 4.6		a = 6.0		a = 6.0		a = 6.3	
Conc.	Dof	Dpikal	Dof	Dpikal	Dof	Dpikal	Dof	Dpikal
0.000	1.162	1.162	1.254	1.254	1.239	1.239	1.167	1.167
0.001	1.108	1.102	1.170	1.088	1.156	1.076	1.088	1.018
0.002	1.091	1.080	1.156	1.007	1.141	0.996	1.074	0.952
0.003	1.081	1.067	1.151	0.936	1.136	0.927	1.068	0.877
0.004	1.073	1.054	1.144	0.850	1.129	0.843	1.061	0.797
0.005	1.067	1.041	1.139	0.761	1.125	0.755	1.057	0.713
0.006	1.062	1.030	1.136	0.673	1.122	0.669	1.055	0.639
0.007	1.058	1.019	1.135	0.590	1.120	0.586	1.053	0.560
0.008	1.054	1.008	1.134	0.522	1.119	0.519	1.052	0.487
0.009	1.052	0.992	1.133	0.453	1.119	0.451	1.052	0.422
0.010	1.049	0.981	1.133	0.392	1.119	0.391	1.055	0.365
0.020	1.031	0.870	1.133	0.109	1.118	0.109	1.051	0.101
0.030	1.023	0.776	1.139	0.043	1.124	0.043	1.057	0.039
0.040	1.021	0.659	1.148	0.021	1.133	0.021	1.070	0.020
0.050	1.018	0.549	1.158	0.012	1.143	0.012	1.072	0.011
0.060	1.017	0.453	1.161	0.008	1.146	0.008	1.075	0.007
0.070	1.016	0.372	1.165	0.005	1.149	0.005	1.078	0.005
0.080	1.017	0.306	1.169	0.004	1.153	0.004	1.082	0.003
0.090	1.018	0.262	1.173	0.003	1.157	0.003	1.087	0.003
0.100	1.020	0.218	1.177	0.002	1.162	0.002	1.091	0.002

TABLE 15.1   *(Continued)*

Electrolyte	NiBr$_2$		Ni(C$_2$H$_3$O$_2$)$_2$		NiCl$_2$		Ni(ClO$_3$)$_2$	
	$L^0 = 132.4$		$L^0 = 95.0$		$L^0 = 130.3$		$L^0 = 118.6$	
	a = 4.5		a = 5.3		a = 4.5		a = 4.8	
Conc.	Dof	Dpikal	Dof	Dpikal	Dof	Dpikal	Dof	Dpikal
0.000	1.277	1.277	0.931	0.931	1.263	1.263	1.174	1.175
0.001	1.222	1.200	0.890	0.884	1.209	1.187	1.125	1.109
0.002	1.207	1.173	0.879	0.868	1.193	1.160	1.110	1.087
0.003	1.197	1.152	0.872	0.856	1.183	1.140	1.101	1.070
0.004	1.189	1.140	0.867	0.846	1.176	1.128	1.095	1.056
0.005	1.184	1.125	0.863	0.841	1.171	1.114	1.090	1.039
0.006	1.180	1.111	0.860	0.833	1.167	1.100	1.086	1.026
0.007	1.176	1.097	0.859	0.826	1.163	1.086	1.083	1.014
0.008	1.173	1.084	0.856	0.818	1.160	1.073	1.081	1.002
0.009	1.171	1.070	0.854	0.810	1.158	1.060	1.079	0.990
0.010	1.170	1.057	0.852	0.802	1.157	1.047	1.077	0.977
0.020	1.154	0.905	0.842	0.707	1.141	0.898	1.063	0.858
0.030	1.148	0.760	0.841	0.591	1.136	0.755	1.059	0.706
0.040	1.150	0.602	0.839	0.482	1.137	0.599	1.059	0.557
0.050	1.147	0.466	0.838	0.377	1.135	0.465	1.058	0.431
0.060	1.147	0.360	0.839	0.293	1.134	0.359	1.058	0.331
0.070	1.147	0.279	0.840	0.228	1.135	0.279	1.059	0.260
0.080	1.149	0.218	0.842	0.179	1.136	0.219	1.061	0.203
0.090	1.151	0.173	0.844	0.143	1.138	0.174	1.063	0.161
0.100	1.153	0.140	0.846	0.115	1.141	0.140	1.065	0.129

TABLE 15.1   *(Continued)*

Electrolyte	$Ni(ClO_4)_2$		$Ni(NO_3)_2$		$NiSO_4$		$PbBr_2$	
	$L^0 = 121.3$		$L^0 = 125.42$		$L^0 = 133.8$		$L^0 = 148.4$	
	$a = 4.8$		$a = 4.5$		$a = 5.0$		$a = 3.8$	
Conc.	Dof	Dpikal	Dof	Dpikal	Dof	Dpikal	Dof	Dpikal
0.000	1.196	1.197	1.228	1.228	0.857	0.857	1.477	1.477
0.001	1.145	1.129	1.176	1.156	0.729	0.733	1.410	1.395
0.002	1.131	1.106	1.160	1.131	0.691	0.695	1.389	1.363
0.003	1.122	1.089	1.151	1.111	0.666	0.667	1.375	1.342
0.004	1.115	1.074	1.144	1.100	0.650	0.646	1.365	1.323
0.005	1.110	1.056	1.139	1.086	0.638	0.628	1.357	1.307
0.006	1.106	1.043	1.134	1.073	0.630	0.612	1.350	1.298
0.007	1.103	1.031	1.131	1.060	0.624	0.592	1.344	1.285
0.008	1.101	1.018	1.128	1.048	0.619	0.580	1.339	1.272
0.009	1.099	1.005	1.126	1.035	0.616	0.570	1.335	1.260
0.010	1.097	0.993	1.125	1.023	0.615	0.560	1.332	1.248
0.020	1.083	0.869	1.110	0.880	0.656	0.498	1.308	1.129
0.030	1.079	0.713	1.105	0.744	0.749	0.458	1.295	1.013
0.040	1.079	0.561	1.106	0.593	0.869	0.458	1.289	0.872
0.050	1.077	0.432	1.103	0.462	1.008	0.428	1.286	0.734
0.060	1.078	0.332	1.103	0.358	1.161	0.397	1.287	0.608
0.070	1.079	0.260	1.104	0.279	1.325	0.368	1.285	0.502
0.080	1.081	0.203	1.105	0.219	1.498	0.341	1.284	0.414
0.090	1.083	0.160	1.107	0.174	1.679	0.316	1.284	0.343
0.100	1.085	0.129	1.109	0.140	1.870	0.294	1.285	0.289

TABLE 15.1   *(Continued)*

Electrolyte	$Pb(C_2H_3O_2)_2$		$PbCl_2$		$Pb(ClO_4)_2$		PbI_2	
	$L^0 = 111.0$		$L^0 = 146.3$		$L^0 = 137.3$		$L^0 = 146.9$	
	a = 4.5		a = 4.5		a = 4.0		a = 3.8	
Conc.	Dof	Dpikal	Dof	Dpikal	Dof	Dpikal	Dof	Dpikal
0.000	1.032	1.033	1.458	1.458	1.370	1.370	1.463	1.463
0.001	0.984	0.982	1.392	1.378	1.308	1.295	1.397	1.383
0.002	0.969	0.965	1.371	1.347	1.289	1.271	1.376	1.351
0.003	0.960	0.953	1.358	1.325	1.276	1.252	1.363	1.330
0.004	0.953	0.946	1.347	1.307	1.266	1.237	1.352	1.312
0.005	0.947	0.938	1.339	1.292	1.259	1.224	1.344	1.296
0.006	0.942	0.931	1.332	1.283	1.253	1.211	1.337	1.287
0.007	0.939	0.924	1.327	1.270	1.248	1.200	1.332	1.274
0.008	0.936	0.917	1.322	1.258	1.243	1.189	1.327	1.262
0.009	0.933	0.910	1.318	1.246	1.240	1.178	1.323	1.250
0.010	0.931	0.904	1.314	1.234	1.236	1.168	1.319	1.238
0.020	0.914	0.827	1.291	1.117	1.215	1.071	1.296	1.121
0.030	0.906	0.755	1.278	1.004	1.204	0.949	1.283	1.007
0.040	0.904	0.663	1.272	0.865	1.199	0.814	1.277	0.867
0.050	0.901	0.569	1.270	0.730	1.199	0.687	1.274	0.731
0.060	0.900	0.481	1.270	0.606	1.196	0.571	1.275	0.606
0.070	0.900	0.403	1.268	0.501	1.194	0.480	1.273	0.501
0.080	0.900	0.336	1.267	0.414	1.194	0.395	1.272	0.414
0.090	0.901	0.281	1.267	0.343	1.195	0.326	1.272	0.343
0.100	0.902	0.236	1.268	0.289	1.196	0.271	1.273	0.289

TABLE 15.1 *(Continued)*

Electrolyte	$Pb(NO_3)_2$		$PrBr_3$		$PrCl_3$		$Pr(ClO_4)_3$	
	$L^0 = 141.42$		$L^0 = 143.8$		$L^0 = 141.7$		$L^0 = 132.7$	
	a = 3.8		a = 6.0		a = 6.0		a = 6.3	
Conc.	Dof	Dpikal	Dof	Dpikal	Dof	Dpikal	Dof	Dpikal
0.000	1.412	1.412	1.266	1.266	1.250	1.250	1.177	1.177
0.001	1.348	1.335	1.180	1.100	1.165	1.087	1.096	1.028
0.002	1.328	1.306	1.165	1.019	1.151	1.008	1.082	0.962
0.003	1.314	1.285	1.160	0.948	1.145	0.939	1.075	0.888
0.004	1.304	1.269	1.153	0.863	1.138	0.855	1.069	0.808
0.005	1.297	1.254	1.148	0.774	1.133	0.768	1.065	0.725
0.006	1.290	1.245	1.145	0.686	1.130	0.681	1.062	0.651
0.007	1.285	1.233	1.143	0.602	1.128	0.599	1.060	0.571
0.008	1.280	1.221	1.142	0.534	1.127	0.531	1.059	0.497
0.009	1.276	1.210	1.141	0.464	1.126	0.461	1.059	0.432
0.010	1.273	1.199	1.141	0.402	1.126	0.401	1.061	0.374
0.020	1.250	1.089	1.140	0.113	1.125	0.113	1.057	0.104
0.030	1.238	0.982	1.146	0.044	1.131	0.044	1.063	0.041
0.040	1.231	0.850	1.155	0.022	1.140	0.022	1.076	0.020
0.050	1.229	0.719	1.165	0.013	1.150	0.013	1.078	0.012
0.060	1.229	0.599	1.168	0.008	1.152	0.008	1.081	0.007
0.070	1.227	0.498	1.172	0.005	1.156	0.005	1.085	0.005
0.080	1.226	0.412	1.176	0.004	1.160	0.004	1.089	0.004
0.090	1.226	0.342	1.180	0.003	1.164	0.003	1.093	0.003
0.100	1.227	0.290	1.185	0.002	1.169	0.002	1.097	0.002

TABLE 15.1   *(Continued)*

Electrolyte	$Pr(NO_3)_3$		RbBr		$RbBrO_3$		$RbC_2H_3O_2$	
	$L^0 = 136.82$		$L^0 = 155.9$		$L^0 = 133.3$		$L^0 = 118.5$	
	a = 6.0		a = 3.4		a = 2.7		a = 3.5	
Conc.	Dof	Dpikal	Dof	Dpikal	Dof	Dpikal	Dof	Dpikal
0.000	1.212	1.212	2.075	2.075	1.727	1.728	1.428	1.428
0.001	1.129	1.057	2.043	2.042	1.700	1.699	1.404	1.403
0.002	1.114	0.982	2.031	2.030	1.690	1.689	1.396	1.394
0.003	1.108	0.916	2.023	2.021	1.683	1.681	1.390	1.388
0.004	1.101	0.836	2.017	2.014	1.678	1.676	1.385	1.383
0.005	1.097	0.753	2.011	2.008	1.673	1.670	1.381	1.378
0.006	1.094	0.669	2.006	2.003	1.669	1.666	1.377	1.375
0.007	1.092	0.590	2.002	1.999	1.665	1.662	1.374	1.371
0.008	1.091	0.524	1.998	1.995	1.662	1.659	1.372	1.368
0.009	1.090	0.456	1.994	1.991	1.659	1.655	1.369	1.366
0.010	1.090	0.397	1.991	1.988	1.656	1.652	1.367	1.363
0.020	1.088	0.113	1.969	1.964	1.636	1.631	1.351	1.344
0.030	1.093	0.044	1.956	1.948	1.623	1.616	1.342	1.331
0.040	1.102	0.022	1.947	1.936	1.615	1.605	1.336	1.321
0.050	1.112	0.013	1.941	1.926	1.609	1.595	1.332	1.312
0.060	1.114	0.008	1.936	1.917	1.604	1.587	1.328	1.304
0.070	1.118	0.005	1.932	1.909	1.601	1.580	1.326	1.296
0.080	1.122	0.004	1.930	1.902	1.599	1.573	1.323	1.288
0.090	1.126	0.003	1.927	1.896	1.596	1.567	1.322	1.281
0.100	1.130	0.002	1.926	1.889	1.594	1.561	1.321	1.287

TABLE 15.1 *(Continued)*

Electrolyte	RbCl		RbClO$_3$		RbClO$_4$		RbF	
	$L^0 = 153.8$		$L^0 = 142.1$		$L^0 = 144.8$		$L^0 = 132.9$	
	a = 3.6		a = 2.48		a = 1.86		a = 4.01	
Conc.	Dof	Dpikal	Dof	Dpikal	Dof	Dpikal	Dof	Dpikal
0.000	2.054	2.054	1.876	1.876	1.918	1.918	1.720	1.720
0.001	2.022	2.021	1.847	1.845	1.888	1.887	1.694	1.692
0.002	2.011	2.009	1.836	1.834	1.876	1.876	1.684	1.682
0.003	2.002	2.000	1.828	1.826	1.868	1.868	1.677	1.675
0.004	1.996	1.993	1.823	1.820	1.862	1.861	1.672	1.670
0.005	1.990	1.987	1.817	1.815	1.856	1.856	1.668	1.665
0.006	1.985	1.982	1.813	1.810	1.851	1.851	1.664	1.661
0.007	1.981	1.978	1.809	1.806	1.849	1.847	1.660	1.658
0.008	1.977	1.974	1.805	1.802	1.845	1.843	1.657	1.654
0.009	1.973	1.970	1.801	1.799	1.841	1.840	1.654	1.651
0.010	1.970	1.967	1.798	1.795	1.838	1.837	1.652	1.649
0.020	1.947	1.941	1.776	1.772	1.813	1.813	1.635	1.631
0.030	1.933	1.924	1.762	1.756	1.797	1.793	1.626	1.619
0.040	1.923	1.911	1.752	1.743	1.786	1.780	1.620	1.610
0.050	1.916	1.900	1.745	1.733	1.777	1.769	1.616	1.602
0.060	1.909	1.890	1.739	1.724	1.770	1.759	1.612	1.595
0.070	1.904	1.880	1.735	1.717	1.764	1.751	1.610	1.589
0.080	1.900	1.879	1.732	1.710	1.759	1.743	1.608	1.583
0.090	1.896	1.872	1.729	1.703	1.756	1.736	1.608	1.578
0.100	1.894	1.865	1.728	1.697	1.753	1.730	1.607	1.572

TABLE 15.1   *(Continued)*

Electrolyte	RbI		RbIO$_3$		RbNO$_2$		RbNO$_3$	
	$L^0 = 154.4$		$L^0 = 118.5$		$L^0 = 149.5$		$L^0 = 148.92$	
	a = 3.51		a = 2.61		a = 7.95		a = 2.37	
Conc.	Dof	Dpikal	Dof	Dpikal	Dof	Dpikal	Dof	Dpikal
0.000	2.055	2.055	1.428	1.428	1.987	1.988	1.979	1.979
0.001	2.024	2.022	1.404	1.403	1.959	1.957	1.948	1.947
0.002	2.013	2.010	1.395	1.394	1.950	1.948	1.936	1.935
0.003	2.004	2.002	1.389	1.387	1.943	1.942	1.928	1.927
0.004	1.998	1.995	1.385	1.382	1.939	1.937	1.922	1.920
0.005	1.992	1.989	1.381	1.378	1.935	1.933	1.917	1.914
0.006	1.987	1.985	1.377	1.374	1.932	1.929	1.912	1.909
0.007	1.983	1.980	1.374	1.370	1.929	1.926	1.908	1.905
0.008	1.979	1.976	1.371	1.367	1.927	1.924	1.904	1.901
0.009	1.976	1.973	1.368	1.364	1.925	1.922	1.900	1.897
0.010	1.972	1.969	1.366	1.362	1.923	1.920	1.897	1.894
0.020	1.951	1.946	1.348	1.342	1.913	1.905	1.872	1.869
0.030	1.938	1.930	1.337	1.328	1.908	1.896	1.857	1.852
0.040	1.930	1.918	1.330	1.318	1.907	1.887	1.846	1.838
0.050	1.924	1.908	1.324	1.308	1.906	1.879	1.838	1.827
0.060	1.919	1.900	1.320	1.300	1.907	1.869	1.832	1.818
0.070	1.916	1.892	1.317	1.293	1.908	1.859	1.827	1.810
0.080	1.913	1.885	1.315	1.286	1.909	1.848	1.823	1.803
0.090	1.911	1.879	1.314	1.279	1.911	1.835	1.821	1.797
0.100	1.910	1.882	1.312	1.273	1.913	1.821	1.818	1.791

TABLE 15.1 *(Continued)*

Electrolyte	RbOH		Rb$_2$SO$_4$		SmCl$_3$		Sm(ClO$_4$)$_3$	
	$L^0 = 275.1$		$L^0 = 157.3$		$L^0 = 142.1$		$L^0 = 133.1$	
	a = 3.0		a = 3.3		a = 6.0		a = 6.3	
Conc.	Dof	Dpikal	Dof	Dpikal	Dof	Dpikal	Dof	Dpikal
0.000	2.964	2.964	1.570	1.570	1.254	1.254	1.181	1.181
0.001	2.913	2.910	1.493	1.482	1.169	1.091	1.099	1.032
0.002	2.893	2.890	1.467	1.449	1.154	1.012	1.085	0.966
0.003	2.880	2.876	1.450	1.423	1.148	0.943	1.078	0.892
0.004	2.868	2.864	1.437	1.401	1.141	0.860	1.072	0.812
0.005	2.859	2.854	1.426	1.380	1.136	0.772	1.067	0.729
0.006	2.850	2.846	1.417	1.361	1.133	0.686	1.065	0.655
0.007	2.843	2.838	1.409	1.353	1.131	0.603	1.063	0.575
0.008	2.836	2.831	1.403	1.337	1.130	0.535	1.062	0.501
0.009	2.830	2.825	1.397	1.321	1.129	0.465	1.061	0.435
0.010	2.824	2.819	1.392	1.305	1.129	0.404	1.064	0.377
0.020	2.783	2.776	1.360	1.142	1.128	0.114	1.059	0.105
0.030	2.758	2.747	1.339	1.000	1.133	0.045	1.065	0.041
0.040	2.739	2.725	1.326	0.902	1.142	0.022	1.078	0.020
0.050	2.726	2.708	1.319	0.780	1.152	0.013	1.080	0.012
0.060	2.715	2.693	1.314	0.671	1.155	0.008	1.083	0.007
0.070	2.707	2.680	1.311	0.574	1.158	0.005	1.087	0.005
0.080	2.700	2.668	1.311	0.491	1.163	0.004	1.091	0.004
0.090	2.694	2.657	1.308	0.420	1.167	0.003	1.095	0.003
0.100	2.689	2.647	1.307	0.377	1.171	0.002	1.100	0.002

TABLE 15.1   *(Continued)*

Electrolyte	$Sm(NO_3)_3$		$SrBr_2$		$Sr(BrO_3)_2$		$Sr(CHO_2)_2$	
	$L^0 = 137.22$		$L^0 = 137.9$		$L^0 = 115.3$		$L^0 = 114.6$	
	$a = 6.0$		$a = 4.0$		$a = 4.3$		$a = 4.3$	
Conc.	Dof	Dpikal	Dof	Dpikal	Dof	Dpikal	Dof	Dpikal
0.000	1.216	1.216	1.351	1.351	1.150	1.150	1.142	1.143
0.001	1.132	1.061	1.292	1.271	1.100	1.088	1.092	1.082
0.002	1.117	0.986	1.274	1.244	1.084	1.068	1.077	1.061
0.003	1.111	0.920	1.263	1.224	1.075	1.052	1.068	1.046
0.004	1.104	0.840	1.254	1.206	1.068	1.039	1.061	1.033
0.005	1.100	0.757	1.247	1.191	1.062	1.028	1.055	1.022
0.006	1.096	0.674	1.242	1.176	1.058	1.017	1.051	1.011
0.007	1.094	0.594	1.238	1.163	1.054	1.006	1.047	1.001
0.008	1.093	0.528	1.234	1.150	1.051	1.003	1.044	0.997
0.009	1.093	0.460	1.231	1.137	1.049	0.994	1.042	0.988
0.010	1.092	0.400	1.228	1.124	1.046	0.984	1.040	0.979
0.020	1.091	0.114	1.210	1.007	1.031	0.876	1.024	0.871
0.030	1.096	0.045	1.201	0.862	1.024	0.757	1.017	0.754
0.040	1.104	0.022	1.198	0.711	1.022	0.646	1.015	0.644
0.050	1.114	0.013	1.200	0.577	1.022	0.528	1.015	0.527
0.060	1.117	0.008	1.198	0.463	1.021	0.426	1.014	0.425
0.070	1.120	0.006	1.197	0.378	1.020	0.343	1.014	0.342
0.080	1.124	0.004	1.197	0.303	1.021	0.276	1.014	0.276
0.090	1.128	0.003	1.198	0.245	1.022	0.224	1.016	0.224
0.100	1.132	0.002	1.200	0.200	1.024	0.184	1.017	0.184

TABLE 15.1 *(Continued)*

Electrolyte	Sr(C$_2$H$_3$O$_2$)$_2$		SrCl$_2$		Sr(ClO$_3$)$_2$		Sr(ClO$_4$)$_2$	
	$L^0 = 100.5$		$L^0 = 135.8$		$L^0 = 124.1$		$L^0 = 126.8$	
	a = 4.8		a = 4.89		a = 4.3		a = 4.3	
Conc.	Dof	Dpikal	Dof	Dpikal	Dof	Dpikal	Dof	Dpikal
0.000	0.969	0.969	1.334	1.334	1.237	1.237	1.261	1.261
0.001	0.926	0.922	1.277	1.259	1.183	1.168	1.206	1.190
0.002	0.913	0.907	1.260	1.234	1.167	1.145	1.190	1.166
0.003	0.905	0.896	1.250	1.216	1.156	1.127	1.179	1.148
0.004	0.899	0.887	1.243	1.196	1.149	1.112	1.171	1.133
0.005	0.895	0.875	1.238	1.182	1.143	1.099	1.165	1.119
0.006	0.891	0.866	1.234	1.168	1.138	1.087	1.161	1.106
0.007	0.888	0.859	1.231	1.155	1.134	1.075	1.157	1.094
0.008	0.886	0.851	1.228	1.141	1.131	1.071	1.153	1.090
0.009	0.884	0.843	1.227	1.128	1.128	1.060	1.151	1.079
0.010	0.882	0.836	1.224	1.115	1.126	1.049	1.148	1.067
0.020	0.869	0.762	1.214	0.986	1.109	0.927	1.131	0.940
0.030	0.864	0.664	1.214	0.818	1.102	0.793	1.124	0.803
0.040	0.863	0.557	1.219	0.651	1.100	0.670	1.122	0.676
0.050	0.862	0.456	1.223	0.507	1.100	0.542	1.122	0.545
0.060	0.861	0.369	1.229	0.400	1.099	0.433	1.121	0.434
0.070	0.862	0.302	1.236	0.310	1.099	0.345	1.121	0.346
0.080	0.863	0.244	1.243	0.243	1.100	0.276	1.122	0.276
0.090	0.864	0.198	1.252	0.194	1.101	0.223	1.123	0.223
0.100	0.866	0.162	1.260	0.156	1.103	0.182	1.125	0.182

TABLE 15.1   *(Continued)*

Electrolyte	$Sr_2Fe(CN)_6$		SrI_2		$Sr(NO_2)_2$		$Sr(NO_3)_2$	
	$L^0 = 167.6$ $a = 5.0$		$L^0 = 136.4$ $a = 4.0$		$L^0 = 131.5$ $a = 4.0$		$L^0 = 130.92$ $a = 4.0$	
Conc.	Dof	Dpikal	Dof	Dpikal	Dof	Dpikal	Dof	Dpikal
0.000	0.766	0.766	1.339	1.340	1.301	1.301	1.296	1.296
0.001	0.837	−0.793	1.281	1.260	1.244	1.225	1.240	1.221
0.002	0.875	−1.507	1.264	1.235	1.227	1.201	1.223	1.197
0.003	0.895	−1.323	1.252	1.214	1.216	1.182	1.212	1.178
0.004	0.910	2.134	1.244	1.197	1.208	1.165	1.203	1.161
0.005	0.924	0.995	1.237	1.182	1.201	1.151	1.197	1.147
0.006	0.936	0.561	1.232	1.168	1.196	1.137	1.192	1.134
0.007	0.947	0.357	1.227	1.155	1.192	1.125	1.188	1.122
0.008	0.961	0.246	1.224	1.142	1.188	1.113	1.184	1.109
0.009	0.966	0.178	1.221	1.129	1.185	1.101	1.181	1.097
0.010	0.971	0.138	1.218	1.116	1.183	1.089	1.179	1.085
0.020	1.003	0.026	1.200	1.001	1.165	0.979	1.161	0.977
0.030	1.025	0.009	1.191	0.858	1.157	0.842	1.153	0.840
0.040	1.035	0.005	1.188	0.708	1.154	0.698	1.149	0.697
0.050	1.034	0.003	1.190	0.575	1.155	0.569	1.151	0.569
0.060	1.033	0.002	1.188	0.462	1.153	0.459	1.149	0.459
0.070	1.032	0.001	1.187	0.378	1.152	0.376	1.148	0.376
0.080	1.032	0.001	1.187	0.303	1.153	0.303	1.148	0.303
0.090	1.031	0.001	1.188	0.245	1.154	0.246	1.149	0.246
0.100	1.031	0.000	1.190	0.201	1.155	0.201	1.151	0.201

TABLE 15.1 *(Continued)*

Electrolyte	$SrSO_4$		SrS_2O_3		$TlC_2H_3O_2$		$TlCl$	
	$L^0 = 139.3$		$L^0 = 146.9$		$L^0 = 115.9$		$L^0 = 151.2$	
	a = 4.5		a = 4.5		a = 3.5		a = 2.8	
Conc.	Dof	Dpikal	Dof	Dpikal	Dof	Dpikal	Dof	Dpikal
0.000	0.907	0.908	0.942	0.943	1.411	1.411	2.012	2.013
0.001	0.772	0.788	0.802	0.816	1.388	1.386	1.981	1.980
0.002	0.731	0.755	0.759	0.781	1.380	1.378	1.970	1.968
0.003	0.706	0.712	0.732	0.737	1.374	1.372	1.962	1.959
0.004	0.688	0.691	0.714	0.715	1.369	1.367	1.955	1.953
0.005	0.676	0.674	0.701	0.697	1.365	1.363	1.950	1.947
0.006	0.667	0.659	0.692	0.681	1.362	1.359	1.945	1.942
0.007	0.661	0.646	0.686	0.668	1.359	1.356	1.940	1.937
0.008	0.658	0.634	0.682	0.656	1.356	1.353	1.936	1.933
0.009	0.655	0.624	0.679	0.645	1.353	1.350	1.933	1.930
0.010	0.654	0.614	0.678	0.635	1.351	1.348	1.929	1.926
0.020	0.698	0.559	0.723	0.579	1.336	1.329	1.906	1.901
0.030	0.802	0.532	0.830	0.551	1.327	1.317	1.892	1.885
0.040	0.930	0.511	0.963	0.527	1.321	1.307	1.882	1.873
0.050	1.079	0.490	1.117	0.502	1.318	1.298	1.875	1.862
0.060	1.244	0.468	1.288	0.476	1.314	1.290	1.869	1.853
0.070	1.420	0.446	1.470	0.451	1.311	1.282	1.865	1.846
0.080	1.606	0.468	1.662	0.464	1.309	1.275	1.863	1.839
0.090	1.799	0.444	1.863	0.437	1.308	1.268	1.860	1.832
0.100	2.000	0.422	2.070	0.411	1.307	1.273	1.857	1.826

TABLE 15.1   *(Continued)*

Electrolyte	$TlClO_4$		TlF		$TlNO_2$		$TlNO_3$	
	$L^0 = 142.2$		$L^0 = 130.3$		$L^0 = 146.9$		$L^0 = 146.32$	
	a = 3.0		a = 3.0		a = 2.8		a = 2.8	
Conc.	Dof	Dpikal	Dof	Dpikal	Dof	Dpikal	Dof	Dpikal
0.000	1.887	1.888	1.695	1.696	1.954	1.955	1.946	1.947
0.001	1.858	1.857	1.669	1.668	1.924	1.923	1.916	1.915
0.002	1.848	1.846	1.659	1.658	1.913	1.911	1.905	1.904
0.003	1.841	1.838	1.653	1.651	1.906	1.903	1.898	1.895
0.004	1.834	1.832	1.648	1.645	1.899	1.897	1.892	1.889
0.005	1.829	1.826	1.643	1.640	1.894	1.891	1.886	1.883
0.006	1.825	1.822	1.639	1.636	1.889	1.886	1.881	1.879
0.007	1.821	1.818	1.635	1.632	1.885	1.882	1.877	1.874
0.008	1.817	1.814	1.632	1.629	1.881	1.878	1.873	1.870
0.009	1.814	1.811	1.629	1.626	1.878	1.875	1.870	1.867
0.010	1.811	1.808	1.626	1.623	1.874	1.871	1.867	1.864
0.020	1.790	1.785	1.607	1.603	1.852	1.847	1.844	1.840
0.030	1.777	1.770	1.596	1.589	1.838	1.831	1.830	1.824
0.040	1.768	1.758	1.589	1.578	1.828	1.819	1.821	1.812
0.050	1.762	1.749	1.583	1.569	1.822	1.809	1.814	1.802
0.060	1.758	1.740	1.580	1.561	1.817	1.801	1.809	1.793
0.070	1.755	1.733	1.577	1.554	1.813	1.793	1.805	1.786
0.080	1.752	1.726	1.574	1.547	1.810	1.786	1.803	1.779
0.090	1.749	1.720	1.572	1.541	1.807	1.780	1.800	1.772
0.100	1.747	1.714	1.570	1.535	1.805	1.774	1.798	1.766

TABLE 15.1   *(Continued)*

Electrolyte	Tl_2SO_4		YCl_3		$Y(NO_3)_3$		$ZnBr_2$	
	$L^0 = 154.7$		$L^0 = 141.5$		$L^0 = 136.62$		$L^0 = 131.9$	
	a = 3.3		a = 6.0		a = 6.0		a = 4.5	
Conc.	Dof	Dpikal	Dof	Dpikal	Dof	Dpikal	Dof	Dpikal
0.000	1.543	1.543	1.248	1.248	1.210	1.210	1.270	1.270
0.001	1.467	1.457	1.164	1.085	1.127	1.055	1.216	1.193
0.002	1.442	1.425	1.149	1.006	1.113	0.980	1.200	1.166
0.003	1.425	1.400	1.143	0.937	1.107	0.914	1.190	1.145
0.004	1.412	1.378	1.136	0.853	1.100	0.834	1.183	1.133
0.005	1.402	1.358	1.132	0.766	1.095	0.750	1.178	1.118
0.006	1.393	1.340	1.129	0.679	1.092	0.667	1.173	1.104
0.007	1.386	1.332	1.127	0.596	1.090	0.587	1.170	1.090
0.008	1.379	1.316	1.125	0.529	1.089	0.522	1.167	1.077
0.009	1.373	1.301	1.125	0.460	1.089	0.454	1.165	1.063
0.010	1.368	1.285	1.125	0.399	1.089	0.395	1.164	1.050
0.020	1.338	1.127	1.124	0.112	1.087	0.112	1.148	0.898
0.030	1.317	0.988	1.129	0.044	1.092	0.044	1.143	0.753
0.040	1.305	0.893	1.138	0.022	1.101	0.022	1.144	0.595
0.050	1.297	0.773	1.149	0.013	1.111	0.013	1.142	0.460
0.060	1.293	0.666	1.151	0.008	1.113	0.008	1.141	0.354
0.070	1.290	0.571	1.155	0.005	1.117	0.005	1.142	0.274
0.080	1.290	0.489	1.159	0.004	1.120	0.004	1.144	0.215
0.090	1.287	0.419	1.163	0.003	1.125	0.003	1.146	0.170
0.100	1.286	0.377	1.167	0.002	1.129	0.002	1.148	0.137

TABLE 15.1 *(Continued)*

Electrolyte	$ZnCl_2$		$Zn(ClO_3)_2$		$Zn(ClO_4)_2$		ZnF_2	
	$L^0 = 129.8$ a = 4.5		$L^0 = 118.1$ a = 4.8		$L^0 = 120.8$ a = 4.8		$L^0 = 108.9$ a = 4.8	
Conc.	Dof	Dpikal	Dof	Dpikal	Dof	Dpikal	Dof	Dpikal
0.000	1.256	1.256	1.169	1.169	1.190	1.190	1.087	1.087
0.001	1.202	1.181	1.119	1.103	1.140	1.123	1.041	1.029
0.002	1.187	1.154	1.105	1.081	1.125	1.100	1.027	1.009
0.003	1.177	1.133	1.096	1.064	1.116	1.083	1.019	0.995
0.004	1.170	1.122	1.089	1.050	1.110	1.068	1.013	0.982
0.005	1.165	1.107	1.085	1.033	1.105	1.050	1.009	0.967
0.006	1.161	1.093	1.081	1.020	1.101	1.037	1.005	0.956
0.007	1.157	1.080	1.078	1.008	1.098	1.024	1.002	0.945
0.008	1.155	1.066	1.076	0.996	1.096	1.012	1.000	0.934
0.009	1.152	1.053	1.074	0.984	1.094	0.999	0.999	0.924
0.010	1.151	1.040	1.072	0.971	1.091	0.987	0.996	0.913
0.020	1.135	0.891	1.058	0.851	1.078	0.863	0.984	0.807
0.030	1.130	0.748	1.054	0.699	1.074	0.706	0.980	0.671
0.040	1.132	0.592	1.054	0.551	1.074	0.555	0.980	0.535
0.050	1.129	0.459	1.053	0.425	1.073	0.427	0.979	0.418
0.060	1.129	0.354	1.053	0.326	1.073	0.327	0.979	0.324
0.070	1.130	0.274	1.054	0.256	1.074	0.256	0.980	0.256
0.080	1.131	0.215	1.056	0.200	1.076	0.200	0.982	0.201
0.090	1.133	0.171	1.058	0.158	1.078	0.158	0.984	0.160
0.100	1.135	0.137	1.061	0.127	1.081	0.126	0.986	0.128

TABLE 15.1 *(Continued)*

Electrolyte	ZnI_2		$Zn(NO_3)_2$		$Zn(SCN)_2$		$ZnSO_4$	
	$L^0 = 130.4$		$L^0 = 124.92$		$L^0 = 120.0$		$L^0 = 133.3$	
	a = 4.5		a = 4.5		a = 4.8		a = 3.64	
Conc.	Dof	Dpikal	Dof	Dpikal	Dof	Dpikal	Dof	Dpikal
0.000	1.260	1.260	1.221	1.222	1.184	1.184	0.853	0.853
0.001	1.206	1.184	1.170	1.150	1.134	1.117	0.730	0.768
0.002	1.191	1.157	1.154	1.124	1.119	1.095	0.693	0.760
0.003	1.181	1.137	1.145	1.105	1.110	1.077	0.670	0.758
0.004	1.174	1.125	1.138	1.094	1.104	1.063	0.655	0.759
0.005	1.168	1.110	1.133	1.080	1.099	1.045	0.644	0.760
0.006	1.164	1.096	1.129	1.067	1.095	1.032	0.637	0.760
0.007	1.161	1.083	1.126	1.054	1.092	1.020	0.633	0.759
0.008	1.158	1.069	1.123	1.041	1.090	1.007	0.630	0.661
0.009	1.156	1.056	1.121	1.029	1.088	0.995	0.629	0.654
0.010	1.155	1.043	1.120	1.016	1.086	0.982	0.630	0.648
0.020	1.139	0.893	1.104	0.874	1.072	0.859	0.681	0.585
0.030	1.134	0.749	1.099	0.737	1.068	0.704	0.785	0.529
0.040	1.135	0.593	1.101	0.586	1.068	0.553	0.920	0.506
0.050	1.133	0.459	1.098	0.456	1.067	0.426	1.079	0.481
0.060	1.133	0.354	1.098	0.353	1.067	0.327	1.246	0.456
0.070	1.133	0.274	1.099	0.274	1.068	0.256	1.425	0.433
0.080	1.135	0.215	1.100	0.215	1.070	0.200	1.614	0.411
0.090	1.137	0.171	1.102	0.171	1.072	0.158	1.812	0.391
0.100	1.139	0.137	1.104	0.138	1.075	0.126	2.016	0.414

TABLE 15.1   *(Continued)*

Electrolyte	$ZnSeO_4$	
	$L^0 = 129.2$	
	$a = 50$	
Conc.	Dof	Dpikal
0.000	0.834	0.835
0.001	0.710	0.715
0.002	0.673	0.677
0.003	0.649	0.651
0.004	0.633	0.630
0.005	0.622	0.612
0.006	0.614	0.597
0.007	0.608	0.577
0.008	0.603	0.565
0.009	0.601	0.555
0.010	0.600	0.546
0.020	0.640	0.485
0.030	0.731	0.446
0.040	0.848	0.448
0.050	0.983	0.420
0.060	1.133	0.392
0.070	1.293	0.365
0.080	1.462	0.339
0.090	1.638	0.316
0.100	1.825	0.294

The units for conductivities equivalent to infinitesimal concentration, L^0, are cm^{-2} Ω^{-1} equiv^{-1}, for a, the minimum approach distance of the ions, 10^{-8} cm; Conc. It means $c/(\text{mol dm}^{-3})$. Dof means Dof/$(10^{-9} \text{ m}^2 \text{ s}^{-1})$ the diffusion coefficient calculated by Onsager–Fuoss theory (Eq. (15.1)) and Dpikal means Dpikal/$(10^{-9} \text{ m}^2 \text{ s}^{-1})$ as calculated by Pikal equation (Eq. (15.2)).

TABLE 15.2 Table of Selected Electrolytes

No.	Electrolyte	λ_1^0	λ_2^0	z_1	z_2	v_1	v_2	a
001	AgClO$_3$	61.9	64.6	1	−1	1	1	1.75**
002	AgClO$_4$	61.9	67.3	1	−1	1	1	2.91**
003	AgF	61.9	55.4	1	−1	1	1	3.0*
004	AgNO$_2$	61.9	72.0	1	−1	1	1	2.8*
005	AgNO$_3$	61.9	71.42	1	−1	1	1	2.45**
006	Ag$_2$SO$_4$	61.9	79.8	1	−2	2	1	6.3*
007	AlBr$_3$	63.0	78.4	3	−1	1	3	6.0*
008	AlCl$_3$	63.0	76.3	3	−1	1	3	6.0*
009	Al(ClO$_4$)$_3$	63.0	67.3	3	−1	1	3	6.3*
010	Al(NO$_3$)$_3$	63.0	71.42	3	−1	1	3	6.0*
011	Al$_2$(SO$_4$)$_3$	63.0	79.8	3	−2	2	3	6.5*
012	BaBr$_2$	63.7	78.4	2	−1	1	2	4.0*
013	Ba(BrO$_3$)$_2$	63.7	55.8	2	−1	1	2	4.3*
014	Ba(CHO$_2$)$_2$	63.7	55.1#	2	−1	1	2	4.3*
015	Ba(C$_2$H$_3$O$_2$)$_2$	63.7	41.0	2	−1	1	2	4.8*
016	BaC$_2$O$_4$	63.7	48.8#	2	−2	1	1	3.5*
017	BaCl$_2$	63.0	76.3	2	−1	1	2	4.0*
018	Ba(ClO$_3$)$_2$	63.0	64.6	2	−1	1	2	4.3*
019	Ba(ClO$_4$)$_2$	63.7	67.3	2	−1	1	2	4.3*
020	Ba$_2$Fe(CN)$_6$	63.7	108.1#	2	−4	2	1	4.5*
021	BaI$_2$	63.7	76.9	2	−1	1	2	4.0*
022	Ba(IO$_3$)$_2$	63.7	41.0	2	−1	1	2	4.6*
023	Ba(NO$_2$)$_2$	63.7	72.0	2	−1	1	2	4.0*
024	Ba(NO$_3$)$_2$	63.7	71.42	2	−1	1	2	4.0*
025	Ba(OH)$_2$	63.7	197.6	2	−1	1	2	4.3*
026	Ba(SCN)$_2$	63.7	66.5	2	−1	1	2	4.3*
027	BaS$_2$O$_3$	63.7	87.4	2	−2	1	1	4.5*
028	BeCl$_2$	45.0	76.3	2	−1	1	2	5.5*

TABLE 15.2   *(Continued)*

No.	Electrolyte	λ_1^0	$\lambda_2^{\ 0}$	z_1	z_2	v_1	v_2	a
029	$Be(NO_3)_2$	45.0	71.42	2	−1	1	2	5.5*
030	$BeSO_4$	45.0	79.8	2	−2	1	1	6.0*
031	$BeSeO_4$	45.0	75.7	2	−2	1	1	6.0*
032	$CaBr_2$	59.5	78.4	2	−1	1	2	4.5*
033	$Ca(C_2H_3O_2)_2$	59.5	41.0	2	−1	1	2	5.3*
034	CaC_2O_4	59.5	48.8#	2	−2	1	1	5.3*
035	$CaCl_2$	59.5	76.3	2	−1	1	2	4.9***
036	$Ca(ClO_3)_2$	59.5	64.6	2	−1	1	2	4.8*
037	$Ca(ClO_4)_2$	59.5	67.3	2	−1	1	2	4.8*
038	$Ca_3[Fe(CN)_6]_2$	59.5	98.9#	2	−3	3	2	5.0*
039	CaI_2	59.5	76.9	2	−1	1	2	4.5*
040	$Ca(NO_2)_2$	59.5	72.0	2	−1	1	2	4.5*
041	$Ca(NO_3)_2$	59.5	71.42	2	−1	1	2	4.5*
042	$Ca(OH)_2$	59.5	197.6	2	−1	1	2	4.8*
043	$Ca(SCN)_2$	59.5	66.5	2	−1	1	2	4.8*
044	$CaSO_4$	59.5	79.8	2	−2	1	1	5.0*
045	CaS_2O_3	59.5	87.4	2	−2	1	1	5.0*
046	$CdBr_2$	54.0	78.4	2	−1	1	2	4.0*
047	$CdCl_2$	54.0	76.3	2	−1	1	2	4.0*
048	$Cd(ClO_3)_2$	54.0	64.6	2	−1	1	2	4.3*
049	$Cd(ClO_4)_2$	54.0	67.3	2	−1	1	2	4.3*
050	CdI_2	54.0	76.9	2	−1	1	2	4.0*
051	$Cd(NO_2)_2$	54.0	72.0	2	−1	1	2	4.0*
052	$Cd(NO_3)_2$	54.0	71.42	2	−1	1	2	4.0*
053	$CdSO_4$	54.0	79.8	2	−2	1	1	3.6****
054	$CdSeO_4$	54.0	75.7	2	−2	1	1	4.5*
055	$CeCl_3$	67.0	76.3	3	−1	1	3	6.0*
056	$Ce(NO_3)_3$	67.0	71.42	3	−1	1	3	6.0*
057	$Ce(SO_4)_2$	67.0	79.8	4	−2	1	2	7.5*
058	$CoBr_2$	54.0	78.4	2	−1	1	2	4.5*
059	$Co(C_2H_3O_2)_2$	54.0	41.0	2	−1	1	2	5.3*
060	$CoCl_2$	54.0	76.3	2	−1	1	2	4.5*
061	$Co(ClO_3)_2$	54.0	64.6	2	−1	1	2	4.8*
062	$Co(ClO_4)_2$	54.0	67.3	2	−1	1	2	4.8*

TABLE 15.2 *(Continued)*

No.	Electrolyte	λ_1^0	λ_2^0	z_1	z_2	v_1	v_2	a
063	CoI_2	54.0	76.9	2	−1	1	2	4.5*
064	$Co(NH_3)_6Cl_2$	105.7#	76.3	2	−2	1	1	3.5*
065	$Co(NO_3)_2$	54.0	71.42	2	−1	1	2	4.5*
066	$CoSO_4$	54.0	79.8	2	−2	1	1	5.0*
067	$CrBr_3$	67.0	78.4	3	−1	1	3	6.0*
068	$CrCl_3$	67.0	76.3	3	−1	1	3	6.0*
069	$Cr(ClO_4)_3$	67.0	67.3	3	−1	1	3	6.3*
070	$Cr(NO_3)_3$	67.0	71.42	3	−1	1	3	6.0*
071	$Cr_2(SO_4)_3$	67.0	79.8	3	−2	2	3	6.5*
072	CsBr	76.8	78.4	1	−1	1	1	3.15**
073	$CsBrO_3$	76.8	55.8	1	−1	1	1	3.21**
074	$CsCHO_2$	76.8	55.1#	1	−1	1	1	3.0*
075	$CsC_2H_3O_2$	76.8	41.0	1	−1	1	1	3.5*
076	CsCl	76.8	76.3	1	−1	1	1	3.0**
077	$CsClO_3$	76.8	64.6	1	−1	1	1	2.29**
078	$CsClO_4$	76.8	67.3	1	−1	1	1	1.61**
079	CsF	76.8	55.4	1	−1	1	1	5.21**
080	CsI	76.8	76.9	1	−1	1	1	3.16**
081	$CsIO_3$	76.8	41.0	1	−1	1	1	2.4**
082	$CsNO_2$	76.8	72.0	1	−1	1	1	2.53**
083	$CsNO_3$	76.8	71.42	1	−1	1	1	2.13**
084	CsOH	76.8	197.6	1	−1	1	1	3.0*
085	Cs_2SO_4	76.8	79.8	1	−2	2	1	3.3*
086	$CuBr_2$	56.6	78.4	2	−1	1	2	4.5*
087	$Cu(C_2H_3O_2)_2$	56.6	41.0	2	−1	1	2	5.3*
088	$CuCl_2$	56.6	76.3	2	−1	1	2	4.5*
089	$Cu(ClO_3)_2$	56.6	64.6	2	−1	1	2	4.8*
090	$Cu(ClO_4)_2$	56.6	67.3	2	−1	1	2	4.8*
091	$Cu(NO_3)_2$	56.6	71.42	2	−1	1	2	4.8*
092	$CuSO_4$	56.6	79.8	2	−2	1	1	5.0*
093	$FeBr_2$	53.5	78.4	2	−1	1	2	4.5*
094	$FeCl_2$	53.5	76.3	2	−1	1	2	4.5*
095	$FeCl_3$	68.0	76.3	3	−1	1	3	6.0*
096	$Fe(ClO_4)_2$	53.5	67.3	2	−1	1	2	4.8*

TABLE 15.2   *(Continued)*

No.	Electrolyte	λ_1^0	λ_2^0	z_1	z_2	v_1	v_2	a
097	$Fe(ClO_4)_3$	68.0	67.3	3	−1	1	3	4.8*
098	$Fe(NO_3)_2$	53.5	71.42	2	−1	1	2	4.5*
099	$Fe(NO_3)_3$	68.0	71.42	3	−1	1	3	6.0*
100	$FeSO_4$	53.5	79.8	2	−2	1	1	5.0*
101	H_3AsO_4	349.7	34.0	1	−1	1	1	6.7*
102	HBr	349.7	78.4	1	−1	1	1	6.0*
103	$HCHO_2$	349.7	55.1#	1	−1	1	1	6.3*
104	$HC_2H_3O_2$	349.7	41.0	1	−1	1	1	6.8*
105	HCN	349.7	78.0	1	−1	1	1	6.0*
106	H_2CO_3	349.7	69.3	1	−2	2	1	6.8*
107	HCl	349.7	76.3	1	−1	1	1	4.08**
108	$HClO_3$	349.7	64.6	1	−1	1	1	6.3*
109	$HClO_4$	349.7	67.3	1	−1	1	1	6.3*
110	H_2CrO_4	349.7	85.0	1	−2	2	1	6.5*
111	HF	349.7	55.4	1	−1	1	1	6.3*
112	HI	349.7	76.9	1	−1	1	1	4.5*
113	HIO_3	349.7	41.0	1	−1	1	1	6.6*
114	HNO_3	349.7	71.42	1	−1	1	1	6.0*
115	H_3PO_4	349.7	36.0	1	−1	1	1	6.6*
116	H_2SO_4	349.7	79.8	1	−2	2	1	6.5*
117	$Hg(CN)_2$	63.6	78.0	2	−1	1	2	4.0*
118	$HgCl_2$	63.6	76.3	2	−1	1	2	4.0*
119	Hg_2Cl_2	68.6	76.3	2	−1	1	2	4.0*
120	KBr	73.50	78.4	1	−1	1	1	3.67**
121	$KBrO_3$	73.50	55.8	1	−1	1	1	2.57**
122	$KCHO_2$	73.50	55.1#	1	−1	1	1	3.3*
123	$KC_2H_3O_2$	73.50	41.0	1	−1	1	1	3.8*
124	K_2CO_3	73.50	69.3	1	−2	2	1	3.8*
125	$K_2C_2O_4$	73.50	48.8#	1	−2	2	1	3.8*
126	KCl	73.50	76.3	1	−1	1	1	3.8***
127	$KClO_3$	73.50	64.6	1	−1	1	1	3.3*
128	$KClO_4$	73.50	67.3	1	−1	1	1	4.95***
129	K_2CrO_4	73.50	85.0	1	−2	2	1	3.5*
130	KF	73.50	55.4	1	−1	1	1	2.96**

TABLE 15.2 *(Continued)*

No.	Electrolyte	λ_1^0	λ_2^0	z_1	z_2	v_1	v_2	a
131	$K_3Fe(CN)_6$	73.50	98.9#	1	−3	3	1	3.5*
132	$K_4Fe(CN)_6$	73.50	108.1#	1	−4	4	1	3.5*
133	KH_2AsO_4	73.50	34.0	1	−1	1	1	3.6*
134	$KHCO_3$	73.50	44.5	1	−1	1	1	3.8*
135	KH_2PO_4	73.50	36.0	1	−1	1	1	3.6*
136	K_2HPO_4	73.50	57.0	1	−2	2	1	3.5*
137	KHS	73.50	65.0	1	−1	1	1	3.3*
138	KI	73.50	76.9	1	−1	1	1	3.88**
139	KIO_3	73.50	41.0	1	−1	1	1	2.77**
140	KIO_4	73.50	54.5	1	−1	1	1	2.31**
141	$KMnO_4$	73.50	62.8	1	−1	1	1	3.3*
142	KNO_2	73.50	72.0	1	−1	1	1	3.0*
143	KNO_3	73.50	71.42	1	−1	1	1	3.5***
144	KOH	73.50	197.6	1	−1	1	1	3.3*
145	KSCN	73.50	66.5	1	−1	1	1	3.3*
146	K_2SO_3	73.50	72.0	1	−2	2	1	3.8*
147	K_2SO_4	73.50	79.8	1	−2	2	1	3.0*
148	K_2SeO_4	73.50	75.7	1	−2	2	1	3.5*
149	$LaBr_3$	69.7	78.4	3	−1	1	3	6.0*
150	$La(C_2H_3O_2)_3$	69.7	41.0	3	−1	1	3	6.8*
151	$LaCl_3$	69.7	76.3	3	−1	1	3	5.75***
152	$La(ClO_4)_3$	69.7	67.3	3	−1	1	3	6.3*
153	$La(IO_3)_3$	69.7	41.0	3	−1	1	3	6.7*
154	$La(NO_3)_3$	69.7	71.42	3	−1	1	3	6.0*
155	$La_2(SO_4)_3$	69.7	79.8	3	−2	2	3	6.5*
156	LiBr	38.68	78.4	1	−1	1	1	4.76**
157	$LiBrO_3$	38.68	55.8	1	−1	1	1	4.8*
158	$LiCHO_2$	38.68	55.1#	1	−1	1	1	4.8*
159	$LiC_2H_3O_2$	38.68	41.0	1	−1	1	1	5.3*
160	LiCl	38.68	76.3	1	−1	1	1	4.25***
161	$LiClO_3$	38.68	64.6	1	−1	1	1	4.8*
162	$LiClO_4$	38.68	67.3	1	−1	1	1	5.0***
163	Li_2CrO_4	38.68	85.0	1	−2	2	1	5.0*
164	LiF	38.68	55.4	1	−1	1	1	3.72**

TABLE 15.2   *(Continued)*

No.	Electrolyte	λ_1^0	λ_2^0	z_1	z_2	v_1	v_2	a
165	LiI	38.68	76.9	1	−1	1	1	5.88**
166	LiIO$_3$	38.68	41.0	1	−1	1	1	5.3*
167	LiNO$_2$	38.68	72.0	1	−1	1	1	3.93**
168	LiNO$_3$	38.68	71.42	1	−1	1	1	2.64**
169	Li$_2$SO$_4$	38.68	79.8	1	−2	2	1	3.9***
170	MgBr$_2$	53.06	78.4	2	−1	1	2	5.5*
171	Mg(BrO$_3$)$_2$	53.06	55.8	2	−1	1	2	5.8*
172	Mg(C$_2$H$_3$O$_2$)$_2$	53.06	41.0	2	−1	1	2	6.3*
173	MgC$_2$O$_4$	53.06	48.8#	2	−2	1	1	6.3*
174	MgCl$_2$	53.06	76.3	2	−1	1	2	5.5*
175	Mg(ClO$_4$)$_2$	53.06	67.3	2	−1	1	2	5.8*
176	MgCrO$_4$	53.06	85.0	2	−2	1	1	6.0*
177	Mg$_2$Fe(CN)$_6$	53.06	108.1#	2	−4	2	1	7.5*
178	MgI$_2$	53.06	76.9	2	−1	1	2	5.5*
179	Mg(IO$_3$)$_2$	53.06	41.0	2	−1	1	2	6.3*
180	Mg(NO$_2$)$_2$	53.06	72.0	2	−1	1	2	5.5*
181	Mg(NO$_3$)$_2$	53.06	71.42	2	−1	1	2	5.5*
182	MgSO$_4$	53.06	79.8	2	−2	1	1	3.9***
183	MgS$_2$O$_3$	53.06	87.4	2	−2	1	1	6.0*
184	MnBr$_2$	53.5	78.4	2	−1	1	2	4.5*
185	MnCl$_2$	53.5	76.3	2	−1	1	2	4.5*
186	Mn(ClO$_4$)$_2$	53.5	67.3	2	−1	1	2	4.8*
187	Mn(NO$_3$)$_2$	53.5	71.42	2	−1	1	2	4.5*
188	MnSO$_4$	53.5	79.8	2	−2	1	1	5.0***
189	NH$_4$Br	73.7	78.4	1	−1	1	1	2.8*
190	NH$_4$CHO$_2$	73.7	55.1#	1	−1	1	1	3.0*
191	NH$_4$C$_2$H$_3$O$_2$	73.7	41.0	1	−1	1	1	3.5*
192	(NH$_4$)$_2$C$_2$O$_4$	73.7	48.8#	1	−2	2	1	3.5*
193	NH$_4$Cl	73.7	76.3	1	−1	1	1	2.8*
194	NH$_4$ClO$_4$	73.7	67.3	1	−1	1	1	3.0*
195	(NH$_4$)$_2$CrO$_4$	73.7	48.8#	1	−2	2	1	3.3*
196	NH$_4$F	73.7	55.4	1	−1	1	1	3.0*
197	NH$_4$HCO$_3$	73.7	44.5	1	−1	1	1	3.4*
198	NH$_4$H$_2$PO$_4$	73.7	36.0	1	−1	1	1	3.4*

TABLE 15.2 *(Continued)*

No.	Electrolyte	λ_1^0	λ_2^0	z_1	z_2	v_1	v_2	a
199	$(NH_4)_2HPO_4$	73.7	57.0	1	−2	2	1	3.3*
200	NH_4HSO_3	73.7	50.0	1	−1	1	1	3.4*
201	NH_4I	73.7	76.9	1	−1	1	1	2.8*
202	NH_4IO_3	73.7	41.0	1	−1	1	1	3.4*
203	NH_4NO_3	73.7	71.42	1	−1	1	1	2.3*
204	NH_4OH	73.7	197.6	1	−1	1	1	3.0*
205	NH_4SCN	73.7	66.5	1	−1	1	1	3.5*
206	$(NH_4)_2SO_4$	73.7	79.8	1	−2	2	1	3.3*
207	$NaBr$	50.1	78.4	1	−1	1	1	3.58**
208	$NaBrO_3$	50.1	55.8	1	−1	1	1	3.9*
209	$NaCHO_2$	50.1	55.1#	1	−1	1	1	3.9*
210	Na_2CO_3	50.1	69.3	1	−2	2	1	4.4*
211	$Na_2C_2O_4$	50.1	48.8#	1	−2	2	1	4.4*
212	$NaCl$	50.1	76.3	1	−1	1	1	4.0***
213	$NaClO_3$	50.1	64.6	1	−1	1	1	3.23**
214	$NaClO_4$	50.1	67.3	1	−1	1	1	3.4**
215	Na_2CrO_4	50.1	85.0	1	−2	2	1	4.1*
216	NaF	50.1	55.4	1	−1	1	1	3.34**
217	$Na_4Fe(CN)_6$	50.1	108.1#	1	−4	4	1	4.6*
218	NaH_2AsO_4	50.1	34.0	1	−1	1	1	4.2*
219	$NaHCO_3$	50.1	44.5	1	−1	1	1	4.2*
220	NaH_2PO_4	50.1	36.0	1	−1	1	1	4.2*
221	Na_2HPO_4	50.1	57.0	1	−2	2	1	4.1*
222	NaI	50.1	76.9	1	−1	1	1	4.23**
223	$NaIO_3$	50.1	41.0	1	−1	1	1	4.2*
224	$NaMnO_4$	50.1	62.8	1	−1	1	1	3.8*
225	Na_2MoO_4	50.1	74.5	1	−2	2	1	4.4*
226	$NaNO_2$	50.1	72.0	1	−1	1	1	3.95**
227	$NaNO_3$	50.1	71.42	1	−1	1	1	2.98**
228	$NaOH$	50.1	197.6	1	−1	1	1	3.9*
229	$NaSCN$	50.1	66.5	1	−1	1	1	3.9*
230	Na_2SO_3	50.1	72.0	1	−2	2	1	4.4*
231	Na_2SO_4	50.1	79.8	1	−2	2	1	3.5****
232	$Na_2S_2O_3$	50.1	87.4	1	−2	2	1	4.1*

TABLE 15.2 *(Continued)*

No.	Electrolyte	λ_1^0	λ_2^0	z_1	z_2	v_1	v_2	a
233	Na_2WO_4	50.1	69.4	1	−2	2	1	4.6*
234	$NdBr_3$	64.3	78.4	3	−1	1	3	6.0*
235	$NdCl_3$	64.3	76.3	3	−1	1	3	6.0*
236	$Nd(ClO_4)_3$	64.3	67.3	3	−1	1	3	6.3*
237	$NiBr_2$	54.0	78.4	2	−1	1	2	4.5*
238	$Ni(C_2H_3O_2)_2$	54.0	41.0	2	−1	1	2	5.3*
239	$NiCl_2$	54.0	76.3	2	−1	1	2	4.5*
240	$Ni(ClO_3)_2$	54.0	64.6	2	−1	1	2	4.8*
241	$Ni(ClO_4)_2$	54.0	67.3	2	−1	1	2	4.8*
242	$Ni(NO_3)_2$	54.0	71.42	2	−1	1	2	4.5*
243	$NiSO_4$	54.0	79.8	2	−2	1	1	5.0*
244	$PbBr_2$	70.0	78.4	2	−1	1	2	3.8*
245	$Pb(C_2H_3O_2)_2$	70.0	41.0	2	−1	1	2	4.5*
246	$PbCl_2$	70.0	76.3	2	−1	1	2	3.8*
247	$Pb(ClO_4)_2$	70.0	67.3	2	−1	1	2	4.0*
248	PbI_2	70.0	76.9	2	−1	1	2	3.8*
249	$Pb(NO_3)_2$	70.0	71.42	2	−1	1	2	3.8*
250	$PrBr_3$	65.4	78.4	3	−1	1	3	6.0*
251	$PrCl_3$	65.4	76.3	3	−1	1	3	6.0*
252	$Pr(ClO_4)_3$	65.4	67.3	3	−1	1	3	6.3*
253	$Pr(NO_3)_3$	65.4	71.42	3	−1	1	3	6.0*
254	$RbBr$	77.5	78.4	1	−1	1	1	3.4**
255	$RbBrO_3$	77.5	55.8	1	−1	1	1	2.7**
256	$RbC_2H_3O_2$	77.5	41.0	1	−1	1	1	3.5*
257	$RbCl$	77.5	76.3	1	−1	1	1	3.6***
258	$RbClO_3$	77.5	64.6	1	−1	1	1	2.48**
259	$RbClO_4$	77.5	67.3	1	−1	1	1	1.86**
260	RbF	77.5	55.4	1	−1	1	1	4.01**
261	RbI	77.5	76.9	1	−1	1	1	3.51**
262	$RbIO_3$	77.5	41.0	1	−1	1	1	2.61**
263	$RbNO_2$	77.5	72.0	1	−1	1	1	7.95**
264	$RbNO_3$	77.5	71.42	1	−1	1	1	2.37**
265	$RbOH$	77.5	197.6	1	−1	1	1	3.0*
266	Rb_2SO_4	77.5	79.8	1	−2	2	1	3.3*
267	$SmCl_3$	65.8	76.3	3	−1	1	3	6.0*
268	$Sm(ClO_4)_3$	65.8	67.3	3	−1	1	3	6.3*

TABLE 15.2 *(Continued)*

No.	Electrolyte	λ_1^0	λ_2^0	z_1	z_2	v_1	v_2	a
269	Sm(NO$_3$)$_3$	65.8	71.42	3	−1	1	3	6.0*
270	SrBr$_2$	59.5	78.4	2	−1	1	2	4.0*
271	Sr(BrO$_3$)$_2$	59.5	55.8	2	−1	1	2	4.3*
272	Sr(CHO$_2$)$_2$	59.5	55.1#	2	−1	1	2	4.3*
273	Sr(C$_2$H$_3$O$_2$)$_2$	59.5	41.0	2	−1	1	2	4.8*
274	SrCl$_2$	59.5	76.3	2	−1	1	2	4.89***
275	Sr(ClO$_3$)$_2$	59.5	64.6	2	−1	1	2	4.3*
276	Sr(ClO$_4$)$_2$	59.5	67.3	2	−1	1	2	4.3*
277	Sr$_2$Fe(CN)$_6$	59.5	108.1#	2	−4	2	1	5.0*
278	SrI$_2$	59.5	76.9	2	−1	1	2	4.0*
279	Sr(NO$_2$)$_2$	59.5	72.0	2	−1	1	2	4.0*
280	Sr(NO$_3$)$_2$	59.5	71.42	2	−1	1	2	4.0*
281	SrSO$_4$	59.5	79.8	2	−2	1	1	4.5*
282	SrS$_2$O$_3$	59.5	87.4	2	−2	1	1	4.5*
283	TlC$_2$H$_3$O$_2$	74.9	41.0	1	−1	1	1	3.5*
284	TlCl	74.9	76.3	1	−1	1	1	2.8*
285	TlClO$_4$	74.9	67.3	1	−1	1	1	3.0*
286	TlF	74.9	55.4	1	−1	1	1	3.0*
287	TlNO$_2$	74.9	72.0	1	−1	1	1	2.8*
288	TlNO$_3$	74.9	71.42	1	−1	1	1	2.8*
289	Tl$_2$SO$_4$	74.9	79.8	1	−2	2	1	3.3*
290	YCl$_3$	65.2	76.3	3	−1	1	3	6.0*
291	Y(NO$_3$)$_3$	65.2	71.42	3	−1	1	3	6.0*
292	ZnBr$_2$	53.5	78.4	2	−1	1	2	4.5*
293	ZnCl$_2$	53.5	76.3	2	−1	1	2	4.5*
294	Zn(ClO$_3$)$_2$	53.5	64.6	2	−1	1	2	4.8*
295	Zn(ClO$_4$)$_2$	53.5	67.3	2	−1	1	2	4.8*
296	ZnF$_2$	53.5	55.4	2	−1	1	2	4.8*
297	ZnI$_2$	53.5	76.9	2	−1	1	2	4.5*
298	Zn(NO$_3$)$_2$	53.5	71.42	2	−1	1	2	4.5*
299	Zn(SCN)$_2$	53.5	66.5	2	−1	1	2	4.8*
300	ZnSO$_4$	53.5	79.8	2	−2	1	1	3.64***
301	ZnSeO$_4$	53.5	75.7	2	−2	1	1	5.0*

Source: From the Handbook of V.M.M. Lobo [7] with the Parameters Used to Calculate the Diffusion Coefficients.

The units of the equivalent ionic conductivities at infinitesimal concentration, λ_i^0, are cm^{-2} $^{-1}$ equiv^{-1}; the mean distance of closest approach of ions, a, by 10^{-8} cm. The equivalent ionic conductivities were mostly taken from Dobos [11], not being specially marked; the exceptions are marked with the symbol "#" which is an average of the values presented by Justice [13]. The number of asterisks in each value of a represents its reference. *Kielland [12]; **Justice [13]; and ***Harned [2].

15.4   CONCLUSION

Despite the limitations of the Onsager–Fuoss and Pikal theories, they can be used as simple and useful models for the estimation of diffusion coefficients of electrolytes in aqueous solutions. Doing so, we can, in many circumstances, overtake the experimental difficulties related with the techniques and related phenomena (e.g., hydrolysis), which affect a reliable and accurate analysis of experimental data, and, consequently, to get a better understanding of the structure of those systems.

To the best of our knowledge, no theory on diffusion of electrolyte solutions is able to estimate in general, reliable diffusion coefficients for a large set of electrolytes. However, stating that, we may conclude that the Onsager–Fuoss and Pikal equations can be successfully applied symmetrical uni-valent (1:1) and polyvalent (basically 2:2) electrolytes. Concerning nonsymmetrical polyvalent, we suggest the use of both Onsager–Fuoss and Pikal, assuming that the actual value of D should lie between them.

We can also conclude that, for all systems, the variation in D is mainly due to the variation of F_T (attributed to the nonideality in thermodynamic behavior), and, to a lesser extent, the electrophoretic effect in the mobility factor, F_M.

ACKNOWLEDGMENTS

The authors in Coimbra are grateful for funding from "The Coimbra Chemistry Centre" which is supported by the Fundaçãopara a Ciência e a Tecnologia (FCT), Portuguese Agency for Scientific Research, through the programs UID/QUI/UI0313/2019 and COMPETE.

The authors acknowledge the authorization given by the editor of Portugaliae Electrochimica Acta, Prof. Victor Lobo, to reproduce the theoretical diffusion coefficients for electrolytes.

KEYWORDS

- **diffusion coefficient**
- **electrolyte**
- **electrophoretic effect**
- **Onsager–Fuoss theory**
- **Pikal theory, solution**

REFERENCES

1. Robinson, R. A., & Stokes, R. H. (1959). *Electrolyte Solutions* (2nd edn.). Butterworths, London.
2. Harned, H. S., & Owen, B. B. (1964). *The Physical Chemistry of Electrolytic Solutions* (3rd edn.). Reinhold Pub. Corp.: New York.
3. Lobo, V. M. M., (1990). *Handbook of Electrolyte Solutions*. Elsevier, Amsterdam.
4. Tyrrell, H. J. V., & Harris, K. R. (1984). *Diffusion in Liquids: A Theoretical and Experimental Study*. Butterworths, London.
5. Cussler, E. L. (1984). *Diffusion: Mass Transfer in Fluid Systems*. Cambridge University Press, Cambridge.
6. Horvath, A. L. (1985). *Handbook of Aqueous Electrolyte Solutions: Physical Properties, Estimation and Correlation Methods*. John Wiley and Sons, New York.
7. Agar, J. N., & Lobo, V. M. M. (1971). *J. Chem. Soc. Faraday Trans. I 71*, 1659–1666.
8. Onsager, L., & Fuoss, R. M. (1932). *J. Phys. Chem., 36*, 2689.
9. Pikal, M. J. (1971). *J. Phys. Chem., 75*, 663.
10. Lobo, V. M. M., Ribeiro, A. C. F., & Andrade, S. G. C. S. (1996). *Diffusion coefficients of 301 electrolytes in aqueous solutions from Onsager–Fuoss and Pikal theories* (Vol. 14, pp. 45–124). Port. Electrochim. Acta.
11. Dobos, D. (1975). *Electrochemical Data*. Elsevier Scientific Publishing Company, Nova Iorque.
12. Kielland, (1937). Individual activity coefficients of ions in aqueous solutions. *J. Am. Chem. Soc.*, 1675.
13. Justice, Private Communication to V. M. M. Lobo.

CHAPTER 16

Thermodiffusion on the Fontainebleau Benchmark Mixture: An Overview

MARISA C. F. BARROS*, ANA C. F. RIBEIRO, and CECÍLIA I. A. V. SANTOS

Department of Chemistry, University of Coimbra, 3004535 Coimbra, Portugal

Corresponding author. E-mail: marisa.barros@gmail.com

ABSTRACT

In this chapter, a detailed overview of the present literature results for Soret coefficients is discussed using different methods, with emphasis on the agreement between them in the thermodiffusion process.

16.1 INTRODUCTION

Thermodiffusion is the effect observed as the result of the application of a temperature gradient in a multicomponent mixture, leading to a mass diffusion flow in the components of the mixture. This phenomenon was first observed by Ludwig, in 1856, that described his experiment on one single page [1]. Basically, the experiment was composed of a U-shaped tube filled with a sodium sulfate solution (8.98% concentration), heated on one side with boiling water and cooled the other one with melting ice. After 7 days, he observed a decrease of the salt concentration on the warm side compared to the cold side of the solution. Two decades later Charles Soret has independently discovered this phenomenon and subsequently studied it in greater detail, formulated equations for it, and wrote several papers on the subject [2–4]. Thermodiffusion is usually called by *Soret-effect* or *Ludwig-Soret effect* and has been objected of study in a wide range of systems: electrolytes [5], surfactant micelles [6], alcohols [7, 8], liquid hydrocarbons [9, 10], gases [11, 12], polymers [13–16], proteins [17], combustion [18], and many others.

This phenomenon is also present in several natural systems like the microstructure of the ocean [19] where we can find salty water and temperature gradients, convection instars [20] or even in biological systems, especially in mass transport through biological membranes as a result of small thermal gradients in living matter, and where thermodiffusion effect extent can play a big role if considered at major levels like organs or a tumor [21].

Thermodiffusion processes are quantified by mass and thermodiffusion coefficients, D, and D_T, respectively. Usually, the mass flux, J_x, for the x component on a binary mixture, is given by

$$J_x = -\rho D \frac{\partial c}{\partial x} - \rho D_T c (1-c) \frac{\partial T}{\partial x} \qquad (16.1)$$

where ρ is the density of the mixture, c is the concentration of the reference component, D is the Fickian diffusion coefficient and D_T is the thermodiffusion coefficient. The first term on the equation represents Fick's law of diffusion and the second term on the equation represents the mass separation due to a temperature gradient.

In steady-state conditions, when the two terms of the equation present equal intensity, there is no mass flux ($J_x = 0$), and

$$\frac{\partial c}{\partial x} = -\frac{D_T}{D} c (1-c) \frac{\partial T}{\partial x} \qquad (16.2)$$

Designating S_T by Soret coefficient and defined by

$$S_T = \frac{D_T}{D} \qquad (16.3)$$

And replacing this parameter in equation (16.2), we obtain (16.4)

$$\frac{\partial c}{\partial x} = -S_T c (1-c) \frac{\partial T}{\partial x} \qquad (16.4)$$

Being S_T, a measure of the concentration gradient that can be sustained by a given temperature gradient in the steady state, it may present positive or negative values depending on the sign of D_T or on the sense of migration of the reference component (toward the cold or the hot).

The thermal diffusion factor, α_T, which is related to the Soret coefficient, as a quantity used in the study of the thermodiffusion process, can be written by

$$\alpha_T = TS_T \qquad (16.5)$$

The thermodiffusion process is generally represented in the literature using D_T, S_T, or $_T$. Sometimes, the experimental data is translated to S_T or $\alpha_{T,}$ being necessary obtains additional information about D to find D_T.

One of the most important applications of the Soret effect is in the natural hydrocarbon reservoirs [22, 23]. The exploration of oil reservoirs requires an excellent optimization, and one of the biggest tasks is to keep the perfect knowledge of the fluid physics inside crude oil reservoirs. Currently, the modeling methods rely on the knowledge of pressure–temperature equilibrium diagrams and on gravity isolation of the different components. Yet there is a demand from the oil industry of improved models capable of a more accurate prediction on the concentrations of the different components. The concentration gradient of the different components in hydrocarbon mixtures is mostly given by diffusion and phase separation.

The accurate estimation of the gas–oil contact in an oil reservoir requires the local composition to be well known. The molecular segregation in the gravitational field is one of the principal reasons for the local variation on the composition. On the other side, the geothermal gradient may also induce local differences in the composition due to the thermodiffusion effect.

For a long time, there was no simple answer to the question as to whether published Soret and thermodiffusion coefficients are reliable and which employed techniques are capable of measuring correct values. In 1999, the EGRT European Group of Research in Thermodiffusion joined in Fontainebleau researchers from five different European investigation groups and launched a project with the aim of study the ternary model compounds used by the oil industry for numerical modeling of petroleum reservoirs, and object of great interest, composed by dodecane ($C_{12}H_{26}–nC_{12}$), isobutyl benzene (IBB) and 1,2,3,4-tetrahydronaphthalene (THN), and respectively binary mixtures. The different pairs of the components of the mixture later became known as the Fontainebleau benchmark systems. The obtained results for the three binaries systems were investigated by five independent groups and at different laboratories, and different techniques, and the results were published in a series of papers in 2003 [24], and later have also been used to validate other experimental techniques [25–27]. Also, the ternary mixtures with the three components have later been selected to prosecute investigation on the ground and under microgravity condition in the International Space Station (ISS) within DCMIX1 project [28–31].

This chapter compiles a broad overview of the present literature results for Soret coefficients, obtained with different methods, with emphasis on the agreement between them.

16.2 PHENOMENOLOGY CONCEPTS

The use of theoretical models to analyze thermodiffusion phenomenon was applied many years after Ludwig's and Soret's first experimental work. Models using nonequilibrium or irreversible thermodynamics were considered more helpful than others. Nonequilibrium thermodynamic models are based on the hypothesis that even in irreversible process; small elementary volumes of the system are in local equilibrium condition. In this point, a brief revision of theory and concepts from the literature [32–36] about nonequilibrium thermodynamics is presented.

In a barycentric reference system, the phenomenological equations are the primary topics to establish a linear relation between the flows and the general thermodynamic forces. Analyzing the binary isobaric case in absence of chemical reaction and $n = 2$ components, the reduced heat flow $J_q' = J_q - (h_1 - h_2)J_1$ and the mass flow J_1 of the independent component 1 are given by

$$J_q' = -L_{qq}\frac{\nabla T}{T^2} - L_{q1}\frac{(\partial \mu_1 / \partial c_1)_{p,T}}{c_2 T}\nabla c_1 \tag{16.6}$$

$$J_1 = -L_{1q}\frac{\nabla T}{T^2} - L_{11}\frac{(\partial \mu_1 / \partial c_1)_{p,T}}{c_2 T}\nabla c_1 \tag{16.7}$$

where J_q is the total heat flow and h_k is the partial specific enthalpy of species k, c_1 and $c_2 = 1 - c_1$ are the concentrations of the two components (in mass fractions), and μ_1 is the chemical potential per unit mass of component 1. L_{qq}, L_{11}, and $L_{1q} = L_{q1}$ are Onsager's phenomenological coefficients.

Experiments are preferably described in terms of the directly available concentration variables. The L_{q1}-term, corresponding to the Dufour effect, can usually be neglected in liquids. The diffusion coefficient D, the thermodiffusion coefficient D_T, and the thermal conductivity κ are defined as

$$D = \frac{L_{11}}{\rho c_2 T}\left(\frac{\partial \mu_1}{\partial c_1}\right)_{p,T}, D_T = \frac{L_{1q}}{c_2 1 c_2 \rho T^2}, = \frac{1}{T^2}\left(L_{qq} - \frac{L_{1q}^2}{L_{11}}\right) \tag{16.8}$$

In the limit, the thermophoretic drift velocity of the minority component is given by

$$v_T = -D_T \nabla T \tag{16.9}$$

After combining the flow equations (16.6) and (16.7) with conservation laws for energy and mass, the heat and the extended diffusion equation are obtained:

$$\rho c_p \frac{\partial T}{\partial t} = \nabla \cdot (\kappa \nabla T) + Q \tag{16.10}$$

$$\frac{\partial c}{\partial t} = \nabla \cdot \left[D \nabla c_1 + c_1 (1 - c_1) D_T \nabla T \right] \tag{16.11}$$

where ρ is the density, c_p the specific heat of the fluid, and Q is the heat production rate per unit volume. Absolute and reduced heats of transport $Q_{k,\text{abs}}^{'*}$ s and $Q_k^{'*} = Q_{k,\text{abs}}^{'*} - Q_{n,\text{abs}}^{'*}$, respectively are defined as the heats transported by the mass flows in an n-component isothermal system with n being the dependent component and $1 \ldots (n-1)$ the independent ones:

$$J_q' = \sum_{k=1}^{n-1} Q_k^{'*} J_k = \sum_{k=1}^{n} Q_{k,\text{abs}}^{'*} J_k^{\text{abs}} \tag{16.12}$$

with $J_k^{\text{abs}} = \rho_k v_k = J_k + \rho_k v$ is the absolute flow, v_k the velocity, and ρ_k the density of component k in the laboratory system; $v = \sum_{k=1}^{n} (\rho_k v_k)/\rho$ is the barycentric velocity.

After switching to molar instead of specific quantities, the Soret coefficient S_T can eventually be written in the form as follows:

$$S_T = \frac{\tilde{Q}_{1,\text{abs}}^{'*} - \tilde{Q}_{2,\text{abs}}^{'*}}{RT^2 \left[1 + (\partial \ln \gamma_1 / \partial \ln x_1)_{p,T} \right]} \tag{16.13}$$

where x_k is the mole fraction of the kth component, M_k its molar mass, γ_k is activity coefficient, and $\tilde{Q}_{k,\text{abs}}^{'*} = Q_{k,\text{abs}}^{'*} M_k$ is molar absolute reduced heat of transport. R is the gas constant. Equation (16.13) signifies that the Soret coefficient of a binary mixture is defined only by the difference $\tilde{Q}_{1,\text{abs}}^{'*} - \tilde{Q}_{2,\text{abs}}^{'*}$ of the heats of transport. At the same time, these quantities satisfy an additional condition following from the Gibbs–Duhem equation $x_1 \tilde{Q}_{1,\text{abs}}^{'*} + x_2 \tilde{Q}_{2,\text{abs}}^{'*} = 0$ (62).

Under an appropriate change of the reference value for the heats of transport, the latter condition can be eliminated [35]. Indeed, in terms of new variables,

$$\tilde{Q}^{*}_{1,abs} = Q_1 + q, Q^{*}_{2,abs} = Q_2 + q \tag{16.14}$$

where q is some reference value characteristic for the peculiar mixture; the equation for the Soret coefficient takes the final form:

$$S_T = \frac{Q_1 - Q_2}{RT^2 \left[1 + \left(\partial ln\gamma_1 / \partial lnx_1\right)_{p,T}\right]} \tag{16.15}$$

The new coefficients Q_k satisfy the more general relation $x_1 Q_1 + x_2 Q_2 = -q$. If Q_1 and Q_2 were known, this relation would provide a way to determine q. Similarly to $\tilde{Q}^{*}_{k,abs}$, we call these variables Q_k also heats of transport. In contrast, $\tilde{Q}^{*}_{1,abs}$ and $\tilde{Q}^{*}_{k,abs}$ the quantities Q_1 and Q_2 are independent variables.

The notation we have introduced is nowadays widely adopted but also alternative descriptions exist. Some authors prefer to include the term $c_1(1-c_1)$ in the definition of the thermodiffusion and the Soret coefficient [37] and some use the thermodiffusion factor $\alpha_T = TS$ [38] or the thermodiffusion ratio $k_T = Tc_1(1-c_1)S_T$ [39].

16.3   EXPERIMENTAL TECHNIQUES

The Soret coefficient is usually measured making use of the thermogravitational method or using the so-called *pure Soret effect* method. It may also be calculated from the electromotive force (e.m.f.) measurements in nonisothermal cells.

16.3.1   CLASSIC SORET CELL

In the pure Soret effect, method the solution to be analyzed is contained in a sandwich-type cell between two horizontal metallic plates and a constant temperature gradient is set up across the solution. Because of the temperature gradient, a concentration gradient is developed. The concentration distribution in the cell is then determined at appropriate intervals and the Soret coefficient is calculated either from observations on the final steady-state or from measurements of the rate of change of concentration after the application of temperature gradient. This concentration difference is related to the Soret coefficient, S_T, as

$$\Delta c = -S_T c_0 \left(1 - c_0\right) \Delta T \qquad (16.16)$$

The advantages of this technique include both simple principles and easy construction of the equipment. On the downside, the occurrence of convection is one of the most serious problems in these experiments. When present, convection prevents the attainment of the true steady-state concentration gradient and hence the measurements of the Soret coefficient may be erroneous. To avoid convention effects usually, small values of cell height are used.

16.3.2   SORET CELL WITH BEAM DEFLECTION TECHNIQUE

One of the traditional methods used to measure the Soret effect is the optical beam deflection technique that is used by different investigation groups [26, 40–42]. In this technique, in the absence of a thermal or the concentration gradient, a laser beam will pass through the glass walls containing the experimental mixture without any deflection. On the other hand, in the presence of a thermal or concentration gradient the laser beam will bend due to changes in the refractive index along the length of the liquid.

This method is widely used for binary mixtures using a single laser [26, 43], and for ternary mixtures [31] and has been tested against reference data for water–ethanol and the Fontainebleau benchmark mixtures [44]. Among its main advantages are the short duration of the typical experiment, the relative simplicity of the setup, and being supported by a well-established theory.

Departing from this method, a new experimental system was designed to study heat and mass transfer in liquids with Soret effect. This apparatus contains two principal parts: the optical cell [41] and an interferometric system in combination with equipment for digital recording and processing the phase information. This technique has the advantage to allow performing and reproducing experiments without disturbing the media. It is also possible measuring not only Δc between the hot and cold plates but also the concentration and temperature distributions along the diffusion path, giving the possibility to adopt this technique for some other purposes like the study hydrodynamic fluctuations.

In this technique, the experiments are performed in a transparent cubic cell filled initially by a homogenous mixture. The thermal gradient is imposed by heating and cooling the top and bottom walls of the cell, respectively, as long as the three-dimensional temperature variation induces mass transfer through the Soret effect. The entire cell is crossed by a laser beam perpendicular to the temperature gradient. Temperature and composition variations contribute to the spatial distribution of the refractive index that modulates the wavefront of the emerging optical beam. This gradient of refractive index, $\nabla\eta$, is related to the concentration and temperature gradient by the relation

$$\nabla\eta = \frac{\partial\eta}{\partial c}\nabla c + \frac{\partial\eta}{\partial T}\nabla T \qquad (16.17)$$

In the thermodiffusion, measurement is very important to have in mind that the relaxation time for the establishment of the thermal gradient is much smaller than that for the establishment of the concentration gradient. So, with the application of the thermal gradient, it is possible to verify the presence of an initial deflection $(\Delta\theta_{th})$ when the thermal gradient is established. Consequently, as the concentration gradient is increased in the system, the beam is subjected to additional deflection $(\Delta\theta_s)$. After recording the deflection of the beam during the experiment, Soret coefficient is evaluated using the relation

$$S_T = -\frac{1}{c}\frac{\partial\eta/\partial T}{\partial\eta/\partial c}\frac{\Delta\theta_s}{\Delta\theta_{th}} \qquad (16.18)$$

So, when the steady state is attained, the temperature gradient can be switched off.

In this way, the two walls will quickly return to the equal temperature, when the temperature gradient is zero. The present will cause a fast initial decrease in the deflection of the laser beam on the beam detection unit. Then again, the concentration gradient will disappear much slowly when compared to the thermal gradient. Therefore through a constant temperature between the two plates, the diffusion coefficient may similarly be determined by the beam deflection and the time necessary for it to return to the initial horizontal path when the concentration gradient reduces to zero.

The beam deflection method has been employed to study the Soret effect in binary mixtures since 1920 [5, 44–48].

The main advantages of the beam deflection technique reside on the simplicity of the apparatus. Another big advantage is that since it is an optical technique, the measurements can be made without disturbing the system.

16.3.3 THERMOGRAVITATIONAL COLUMN (TGC)

The thermogravitational column (TGC) method consists in imposing a temperature gradient in a sample contained between two closely spaced walls. The temperature gradient is activated by keeping one wall at a higher temperature than the other. This way the Soret effect will origin the movement of the components, resulting in a lateral density gradient and creating a resistance driven convection roll in the opposite region. The component that separates near the cold side will move to the bottom of the column while the other component will go up to the top of the column. Consequently, the separation process is governed by a combination of thermal, gravitational, and buoyancy forces.

Clusius and Dickel had first proposed a thermogravitaional column method in 1938 [49–51]. The validity limits of the theory have been very well presented by Valencia et al. [10].

The TGC apparatus can measure the Soret effect by two different methodologies: in the first methodology, at steady state, samples from the column at different heights can be drawn and evaluated to obtain the concentration distribution along with the column height. This information can be translated to the thermodiffusion coefficient by using the Furry–Jones–Onsager (FJO) theory [52] modified later by Majumdar [53]. The concentration gradient is related to D_T as

$$\Delta c = 504 \frac{v}{\alpha g} c_0 (1 - c_0) \frac{h}{\delta^4} D_T \qquad (16.19)$$

where α is the thermal expansion coefficient of the mixture, g is the acceleration due to the gravity, v is the kinematic viscosity, δ is the gap width between the plates in the x-direction, and h is the height of the column.

In the second methodology, implemented by Dutrieux et al. [24], the time history of the velocity amplitudes in the mixture are documented by means of a laser Doppler velocimetry. As a consequence of the measurements, the velocities will initially increase from 0 to v_{max} m/s and then decrease in amplitude to a steady-state value of v_{steady} m/s. The separation ratio of the mixture, φ, can be related to the velocities and consequently S_T can be determined by

$$\phi = \frac{\beta}{\alpha} S_T c_0 (1 - c_0) = \frac{V_{max}}{V_{steady}} - 1 \qquad (16.20)$$

TGC has not only been used for binary mixtures of small molecules at normal and high [55] pressures but also for polymer solutions [56] and critical mixtures [57] and, more recently, for ternary mixtures [58, 59]. Its principal disadvantage resides in the fact that the determination of D_T involves numerous simplifications and this can raise experimental errors. Also, small errors such as a marginal tilting of the column can increase/decrease the separation, that is, changing the value of S_T. Variations in temperature gradients will not reflect precision on the measured D_T values (cf. Ref. [10]). Another time, this is attributed to the oversimplification of the theoretical equations in the FJO theory. Finally, by keeping the Grashof numbers larger than a certain critical value, TGC can also be employed for mixtures with a negative S_T [60].

16.4   RESULTS AND CONCLUSIONS

In the previous section, different techniques to measure the Soret coefficients were described. It is clear that there is no universal method to study this phenomenon and perform to these measurements, each one with its own advantages and limitations.

Most of the existent results on Soret coefficients are for two-component mixtures. The fact that most of the relevant natural and technological systems are multicomponent made the focus shift toward ternary mixtures, which are seen as the simplest multicomponent mixtures. Nevertheless, available data in ternary mixtures is scarce and scattered, leading to a very difficult comparison of results. With the purpose to prove that precise values of the Soret coefficient can be attained, five European labs joined and decided to investigate independently the same systems, that is, same chemical compounds from the same batch with the same purity, composition, and at the same temperature (ternary mixture of THN, IBB, and n-dodecane (nC12), also known as the Fontainebleau benchmark mixture, at mass fractions 0.8/0.1/0.1 and at 298.15 K). This is a quasi-ideal mixture that can mimic representative compounds of crude oil. Results for Soret coefficient were obtained in ground laboratories and later validated in microgravity conditions, on board the ISS where Soret effect can be studied measuring the changes in heat and mass in fluids, while avoiding the effects of gravity.

Some years later, the comparison of results produced by the different laboratories showed that this benchmark campaign was a success. The highest difference from the mean was only 7% for one particular system [61]. This group of scientists proposed benchmark values to ground and underground conditions [61]. These results are presented in Tables 16.1 and 16.2.

TABLE 16.1 Soret Coefficients Measured in Ground Conditions for the Mixture THN-IBB-$nC12$ at the Mass Fraction of 0.8–0.1–0.1 and at 298.15 K

	Technique	$S'_{T1} \times 10^{-3}$ (K^{-1})	$S'_{T3} \times 10^{-3}$ (K^{-1})
	ODI[a]	1.04 ± 0.15	-0.94 ± 0.10
	OBD[b]	1.20 ± 0.09	-0.86 ± 0.06
Ground conditions	TG[c] + SST[d]	1.19 ± 0.09	-0.91 ± 0.15
	Average	1.17 ± 0.06	-0.88 ± 0.05

[a]ODI, optical digital interferometry;
[b]OBD, optical beam deflection;
[c]TG, thermogravitational column;
[d]SST, sliding symmetric tubes.

TABLE 16.2 Soret Coefficients Measured in Microgravity Conditions for the Mixture THN-IBB-$nC12$ at the Mass Fraction of 0.8–0.1–0.1 and at 298.15 K

	Team	$S_{T1} \times 10^{-3}$ (K^{-1})	$S_{T3} \times 10^{-3}$ (K^{-1})
	RAS[a]	1.40 ± 0.16	-0.83 ± 0.10
	ZS[b]	1.37 ± 0.06	-0.57 ± 0.05
Microgravity conditions	VS[c]	1.43 ± 0.21	-0.66 ± 0.07
	SVV[d]	1.39 ± 0.25	-0.49 ± 0.08
	Average	1.17 ± 0.06	-0.88 ± 0.05

[a]Team of Russian Academy of Sciences;
[b]Team of Ryerson University;
[c]Team of Université Libre de Bruxelles;
[d]Team of Université Libre de Bruxelles.

The authors chose to present only the values of S_{T1} and S_{T3}, for THN and nC_{12} (components 1 and 3, respectively), because they are the components chosen to determine the coefficients in most of the techniques. Results for component 2 have more experimental errors added and present higher dispersion. Also, one can easily calculate results for component 2 using $S_{T1} + S_{T2} + S_{T3} = 0$ [61].

Authors had considered to have achieved good agreement among the values measured in ground conditions. The weighted average of the three independent results has been proposed as the reference. It is interesting to note that consistent results have been obtained by three different methods: by two direct techniques and by the combination of the other two independent techniques [61].

In 2017, Mousavi et al. [62] joined the community and published the following results, obtained with optical interferometry technique:

	$S_{T1} \times 10^{-3}$ (K^{-1})	$S_{T3} \times 10^{-3}$ (K^{-1})
Mousavi et al.	1.24 ± 0.09	-0.84 ± 0.08

Again, this result presents a very good agreement with the previous ones (deviation inferior to 7%).

According to the authors' microgravity results, for the Soret coefficients presented in Table 16.2 [61] show acceptable agreement. Affording to these, the differences are larger when compared to the coefficients measured in ground conditions. This discrepancy between the measured results, in ground and at microgravity conditions, arises from apparatus differences (cell) and from the condition number used, that resulted in deviations of the visible separation developing in the cell from its analytical model. Simplified processing based on a straightforward fit to the analytical model results in a systematic error of 10%–15% (overvalued separation for THN and under-valued separation for nC_{12}). Authors call the attention to the fact that the variations of experimental data for the Soret coefficients are not independent and show a linear correlation due to the specific properties of the contrast factor matrix.

16.5   FINAL REMARKS

Results on the Soret (S_T) coefficients for quasi-ideal as hydrocarbon ternary mixtures with high relevance in research and industry were reviewed under the scope of this work. Soret coefficients for 1,2,3,4-tetrahydronaphthalene, IBB, and dodecane binary mixtures (also known as the Fontainebleau benchmark mixture) at mass fractions 0.8/0.1/0.1 and at 298 K, in-ground laboratories and microgravity conditions measured by several international teams and by means of different experimental methods are in fair agreement.

ACKNOWLEDGMENTS

CIAVS is grateful for the funding granted by FEDER-European Regional Development Fund through the COMPETE Programme and FCT-Fundaçãopara a Ciencia e a Tecnologia, for the KIDIMIX project POCI-01-0145-FEDER-030271.

KEYWORDS

- **Fontainebleau benchmark mixture**
- **thermodiffusion**
- **Soret coefficients**

REFERENCES

1. Ludwig C. (1856). Diffusion zwischen ungleich erwwärmten orten gleichzusammengestzter lösungen, *Sitz. Ber. Akad. Wiss. Wien Math-Naturw. Kl, 20,* 539.
2. Soret C. (1979). Sur l'état d'équilibre que prend au point de vue de sa concentration une dissolution saline primitivement homohéne dont deux parties sont portées à des températures différentes, *Archives des Sciences Physiques et Naturelles, 2,* 48.
3. Soret C. (1880). Influence de la température sur la distribution des sels dans leurs solutions, *Comptes rendus de l'Académie des Sciences, 91*(5), 289.
4. Soret, C. (1881). *Sur l'état d'équilibre que prend au point de vue de sa concentration une dissolution saline primitivement homohéne dont deux parties sont portées à des températures différentes. Annales de chimie et de physique, 22,* 293.
5. Tanner, C. C. (1927). The Soret effect. Part 1. *Transactions of the Faraday Society, 23,* 75.
6. Piazza, R., & Guarino, A. (2002). Soret effect in interacting micellar solutions. *Physical Review Letters, 88,* 208.
7. Kolodner, P., Williams, H., & Moe, C. (1988). Optical measurement of the Soret coefficient of ethanol/water solutions. *Journal of Chemical Physics, 88,* 6512.
8. Legros, J. C., Goemaere, P., & Platten, J. K. (1985). Soret coefficient and the two-component Benard convection in the benzene methanol system. *Physical Review E, 32*(3), 1903.
9. Costeséque, P., & Loubet, J. C. (2003). Measuring the Soret coefficient of binary hydro-carbon mixtures in packed thermogravitational columns (contribution of Toulouse University to the benchmark test). *Philosophical Magazine, 83*(17/18), 2017.
10. Bou-Ali, M. M., Valencia, J. J., Madariaga, J. A., Santamaria, C. M., Ecennaro, O., & Dutrieux, J. F. (2003). Determination of the thermodiffusion coefficient in three binary organic liquid mixtures by the thermogravitational method (contribution of the

Univesidad Del Pais Vasco Bilbao, to the benchmark test). *Philosophical Magazine, 83*(17e18), 2011.

11. Symons, M., Martin, M. L., & Dunlop, P. J. (1979). Thermal diffusion in mixtures of helium with argon, neon, nitrogen and carbon dioxide and of neon with argon. *Journal of the Chemical Society, Faraday Transactions, 1*(75), 621.

12. Davarzani, H., Marcoux, M., Costeseque, P., & Quintard, M. (2010). Experimental measurement of the effective diffusion and thermodiffusion coefficients for binary gas mixture in porous media. *Chemical Engineering Science, 65*(18), 5092.

13. Rauch, J., & Köhler, W., (2002). Diffusion and thermal diffusion of semi dilute to concentrated solutions of polystyrene in toluene in the vicinity of the glass transition. *Physical Review Letters, 88*(18), 185.

14. Kita, R., Wiegand, S., & Luettmer-Strathmann, J. (2004). Sign change of the Soret coefficient of poly(ethyleneoxide) in water/ethanol mixtures observed by thermal diffusion forced Rayleigh scattering. *Journal of Chemical Physics, 121*(8), 3874.

15. Rauch, J., & Köhler, W. (2003). Collective and thermal diffusion in dilute, semidilute, and concentrated solutions of polystyrene in toluene. *Journal of Chemical Physics, 119,* 11977.

16. Rauch, J., & Köhler, W. (2005). On the molar mass dependence of the thermal diffusion coefficient of polymer solutions. *Macromolecules, 38*(9), 3571.

17. Iacopini, S., & Piazza, R. (2003). Thermophoresis in protein solutions. *Europhysics Letters, 63*(2), 247.

18. Rosner, D. E., Israel, R. S., & La Mantia, B. (2000). Heavy species Ludwige Soret transport effects in air-breathing combustion. *Combustion and Flame, 123*(14), 547.

19. Gregg, M. (1973). The microstructure of the ocean. *Scientific American, 228,* 65.

20. Spiegel, E. A. (1972). Convection in stars-II: Special effects. *The Annual Review of Astronomy and Astrophysic, 10,* 261.

21. Bonner, F. J., & Sundelöf, L. O. (1984). Thermal diffusion as a mechanism for biological transport. *Zeitschriftfür Naturforschung C, 39,* 656.

22. Montel, F. (1994). Importance de la thermodiffusion en exploitation et production pétrolières. *Entropie, 184–185,* 86.

23. Montel, F. (1998). La place de la thermodynamique dans une modélisation des répartitions des espéces d'hydrocarbures dans les réservoirs pétroliers: Incidence sur les problémes de production. *Entropie, 214,* 7.

24. Platten, K., Bou-Ali, M. M., Costesèque, P., Dutrieux, J. F., & Köhler, W. (2003). Benchmark values for the Soret, thermal diffusion, and diffusion coefficients of three binary organic liquid mixtures. *Philosophical Magazine, 83*(17/18), 1965.

25. Mialdun, A., & Shevtsova, V. (2011). Measurement of the Soret and diffusion coefficients for benchmark binary mixtures by means of digital interferometry. *Journal of Chemical Physics, 134,* 044524.

26. Königer, A., Meier, B., & Köhler, W. (2009). Measurement of the Soret, diffusion, and thermal diffusion coefficients of three binary organic benchmark mixtures and of ethanol/water mixtures using a beam deflection technique. *Philosophical Magazine, 89,* 907.

27. Naumann, P., Martin, A., Kriegs, H., Larrañaga, M., Bou-Ali, M. M., & Wiegand, S. (2012). Development of a thermogravitational microcolumn with an interferometric contactless detection system. *Journal of Physical Chemistry B, 116,* 13889.

28. Blanco, P., Bou-Ali, M. M., Platten, J. K., De Mezquia, D. A., Madariaga, J. A., & Santamaria, C. (2010). Thermodiffusion coefficients of binary and ternary hydrocarbon mixtures. *The Journal of Chemical Physics, 132*, 114506.

29. Van Vaerenbergh, S., Srinivasan, S., & Saghir, M. Z. (2009). Thermodiffusion in multicomponent hydrocarbon mixtures: Experimental investigations and computational analysis. *The Journal of Chemical Physics, 131*, 114505.

30. Leahy-Dios, A., Bou-Ali, M., Platten, J. K., & Firoozabadi, A. (2005). Measurements of molecular and thermal diffusion coefficients in ternary mixtures. *The Journal of Chemical Physics, 122*, 234502.

31. Königer, A., Wunderlich, H., & Köhler W. (2010). Measurement of diffusion and thermal diffusion in ternary fluid mixtures using a two-color optical beam deflection technique. *The Journal of Chemical Physics, 132*, 174506.

32. Köhler, W., & Morozov, K. I. (2016). The Soret effect in liquid mixtures. *Journal of Non-Equilibrium Thermodynamics, 41*(3), 151.

33. De Groot, S. R., & Mazur, P. (1984). *Non-equilibrium thermodynamics*. Dover: New York.

34. Hartmann, S., Wittko, G., Köhler, W., Morozov, K. I., Albers, K., & Sadowski, G., (2012). Thermophobicity of liquids: Heats of transport in mixtures as pure component properties. *Physical Review Letters, 109*, 065901.

35. Hartmann, S., Wittko, G., Schock, F., Grob, W., Lindner, F., Köhler, W., & Morozov, K. I. (2014). Thermophobicity of liquids: Heats of transport in mixtures as pure component properties—the case of arbitrary concentration. *The Journal of Chemical Physics, 141*, 134503.

36. Denbigh, K. G., (1952). The heat of transport in regular solutions. *Transactions of the Faraday Society Home, 48*, 1.

37. Haugen, K. B., & Firoozabadi, A. (2007). Transient separation of multicomponent liquid mixtures in thermogravitational columns. *The Journal of Chemical Physics, 127*, 154507.

38. Abbasi, A., Saghir, M. Z., & Kawaji, M. A. (2009). New approach to evaluate the thermodiffusion factor for associating mixtures. *The Journal of Chemical Physics, 130*, 06450.

39. Lucas, P., & Tyler, A. (1977). Thermal diffusion ratio of a 3He/4He mixture Nearits λ transition: The onset of heat flush. *Journal of Low Temperature Physics, 27*, 281.

40. Mialdun, A., Yasnou, V., Shevtsova, V., Königer, A., Köhler, W., Alonso de Mezquia, D., & Bou-Ali, M. M. (2012). A comprehensive study of diffusion, thermodiffusion, and Soret coefficients of water-isopropanol. *The Journal of Chemical Physics, 136*, 244512.

41. Mialdun, A., & Shevtsova, V. (2008). Development of optical digital interferometry technique for measurement of thermodiffusion coefficients. *International Journal of Heat and Mass Transfer, 51*, 3164.

42. Vigolo, D., Brambilla, G., & Piazza, R. (2007). Thermophoresis of microemulsion droplets: Size dependence of the Soret effect. *Physical Review E, 75*, 040401.

43. Zhang, K. J., Briggs, M. E., Gammon, R. W., & Sengers, J. V. (1996). Optical measurement of the Soret effect and the diffusion coefficient of liquid mixture. *The Journal of Chemical Physics, 104*, 6881.

44. Tanner, C. C. (1952). The Soret effect. *Letters to Nature, 170*, 34.

45. Thomaes, G. (1951). Recherches Sur La thermodiffusion en phase liquide. *Physica, 17*(10), 885.

46. Anderson, T. G., & Horne, F. H. (1971). Pure thermal diffusion. II. Experimental thermal diffusion factors and mutual diffusion coefficients for CCl_4 and C_6H_{12}. *Journal of Chemical Physics, 55*(6), 2831.

47. Giglio, M., & Vendramini, A. (1974). Thermal lens effect in a binary liquid mixture: A new effect. *Applied Physics Letters, 25*(10), 555.

48. Agar, N., & Turner, J. C. R. (1960). Thermal diffusion in solutions of electrolytes. *Proceedings of the Royal Society of London, Series A, Mathematical and Physical Sciences, 255*, 307.

49. Clusius, G. D. (1938). Neues verfahrenzur gasentmischungund isotopentrennung. *Naturwissenschaften, 26*(33) 546.

50. Clusius, G. Dickel (1939). Das Trennrohrverfahren bei Flüssigkeiten. *Naturwissenschaften 27*(9), 148.

51. Janca J., Kaspárková V., Halabalová V., Simek L., J. Ruzicka, E. (2007). Barothermal field-flow fractionation of bacteria. *Journal of Chromatography B, 852*, 512.

52. Fury, W. H., Jones, R. C., & Onsager, L. (1939). On the theory of isotope separation by thermal diffusion. *Physical Review, 55*(11), 1083.

53. Majumdar, S. D. (1951). The theory of the separation of isotope by thermal diffusion. *Physical Review, 81*(5) 844.

54. Dutrieux, J. F., Platten, J. K., Chavepeyer, G., & Bou-Ali, M. M. (2002). On the measurements of positive Soret coefficients. *Journal of Physical Chemistry, 106*(23), 6104.

55. Urteaga, P., Bou-Ali, M. M., De Mezquia, D. A., Santamaria, J., Santamaria, C., & Madariaga, J. A. (2012). Measurement of thermodiffusion coefficient of hydrocarbon binary mixtures under pressure with the thermogravitational technique. *Review of Scientific Instruments, 83*, 074903.

56. Montel, F. (1998). *New tools for oil and gas reservoir fluid management* (Vol. 53, p. 9). Revue de l'Institut Francais du Petrole.

57. Ecenarro, O., Madariaga, J. A., Navarro, J. L., Santamaria, C. M., Carrion, J. A., & Saviron, J. M. (1993). Thermogravitational separation and the thermal diffusion factor nearcritical points in binary liquid mixtures. *Journal of Physics: Condensed Matter, 5*, 2289.

58. Platt, G., Vongvanich, T., & Rowley, R. L. (1982). The diffusion thermoeffect in ternary nonelectrolyte liquid mixtures. *The Journal of Chemical Physics, 77*, 2113.

59. Leaist, D. G., & Lu, H. (1990). Conductometric determination of the Soret coefficients of a ternary mixed electrolyte. Reversed thermal diffusion of sodium chloride in aqueous sodium hydroxide solutions. *The Journal of Physical Chemistry, 94*, 447.

60. Bou-Ali, M. M., Ecennaro, O., Madariaga, J. A., Santamaria, C. M., & Valencia, J. J. (1999). Stability of convection in a vertical binary fluid layer with an adverse density gradient. *Physical Review E, 59*, 1250.

61. Bou-Ali, M. M., Ahadi, A., De Mezquia, D. A., Galand, Q., Gebhardt, M., Khlybov, O., et al. (2015). Benchmark values for the Soret, thermodiffusion and molecular diffusion coefficients of the ternary mixture tetralin+isobutyl benzene+n-dodecane with 0.8-0.1-0.1 mass fraction. *The European Physical Journal E, 38*, 30.

62. Mousavil, S. A., Yousefi, T., & Saghir, Z. (2017). Experimental investigation on thermal diffusion in ternary hydrocarbon mixtures. *Fluid Dynamics and Materials Processing, 13*(4), 213.

CHAPTER 17

Hydrodynamic Radii for Sodium Salts in Aqueous Solutions: Contribution from Electrical Conductance and Diffusion

ANA C. F. RIBEIRO,[1*] LUIS M. P. VERISSIMO,[1] ARTUR J. M. VALENTE,[1] A. M. T. D. P. V. CABRAL,[2] and M. A. ESTESO[3,4]

[1]*Department of Chemistry, University of Coimbra, 3004-535 Coimbra, Portugal*

[2]*Faculty of Pharmacy, University of Coimbra, 3000-295 Coimbra, Portugal*

[3]*U.D. Physical Chemistry, University of Alcala, 28805 Alcala de Henares, Madrid, Spain*

[4]*Catholic University of "Santa Teresa de Jesus de Avila," Los Canteros street, 05005 Avila, Spain*

Corresponding author. E-mail: anacfrib@ci.uc.pt

ABSTRACT

In this chapter, we present the values of hydrodynamic radii for different anions of sodium simple salts in aqueous solutions estimated from their conductivities and by using the Stokes equation. In addition, we give the estimations of the tracer diffusion coefficient of each anion in these electrolytes. The aim of this work is to have a better understanding of the structure and transport behavior of these systems.

17.1 INTRODUCTION

17.1.1 OBJECTIVES

Aqueous solutions of sodium salt compounds are relevant systems in applied and fundamental research due to their wide range of applications related to different areas such as medicine, pharmaceutical, and food industries (e.g., [1–4]). However, despite the many reasons justifying the importance of these salts, the understanding of these complex systems has not yet been well established. We have been particularly interested in the study of some transport properties, highlighting the diffusion behavior of these salts and the corresponding hydrodynamic radii of the respective ions. These data help us a better understanding of the transport behavior and the structure of these aqueous systems.

The estimation of limiting diffusion coefficients of anions and the corresponding hydrodynamic radii can be made by using Nernst and Stokes equations [1–5], taking into account the previously reported electrical conductivity values.

In conclusion, this chapter reports theoretical data for hydrodynamic radii and limiting diffusion coefficients of anions (D_i) estimated from the Nernst equation for 31 sodium salt-containing systems. In addition, we also report differential binary mutual diffusion coefficients (D^0), theoretical, and experimental ones, for electrolytes (i.e., 27) and polyelectrolytes, respectively.

17.1.2 PHENOMENOLOGY OF TRANSPORT PROPERTIES

The transport processes in the aqueous solutions result from a situation in which these systems are far from their thermodynamic equilibrium states originated, for example, by heterogeneities of temperature, concentration (strictly, chemical potential), or electrical potential. In each case, a flow of a system variable, J, is associated with the corresponding macroscopic quantity gradient, X,

$$J = -K \operatorname{grad} X \tag{17.1}$$

where K represents the transport coefficient. In particular, the empirical relationships between the matter or the electric current flows and the correspondent gradients, described by linear phenomenological relationships (Fick's law and Ohm's law, respectively) [3], can be written as

$$J_x = -D \operatorname{grad} C \tag{17.2}$$

$$I = -kA \operatorname{grad} V \tag{17.3}$$

where J_x is the flow of component x in units of mass or mole, per unit area per unit time across defined reference plane perpendicular to the direction of flow, I is the total electric current flow, across of the same plane, D is the mutual diffusion coefficient and k is electric conductivity.

The measurements of transport properties provide data of outmost importance for different areas, such as corrosion and drug delivery [1–3]. However, information on many systems, of practical importance, is still lacking in the scientific literature. One of the reasons for that is because those measurements require laborious attention and attention to possible sources of error. In the last years, some authors (e.g., Lobo et al. [4]) have dedicated their time to reveal the main sources and the main methods of measurement of these properties as well as to discussing the present situation of the availability of data specifying the respective precision and accuracy. In this sense, they are contributing for the pure and applied chemistry research.

Among transport properties, we can mention the self-diffusion, D_i (or intradiffusion, tracer diffusion, single-ion diffusion, limiting ionic diffusion) that is distinguished from the mutual diffusion above indicated (Equation (17.2)) (or interdiffusion, concentration diffusion, salt diffusion) [1–4]. Methods such as those based in polarography, NMR, and capillary-tube techniques with radioactive isotopes measure self-diffusion coefficients, not mutual diffusion. However, for bulk substance transport, the appropriate parameter is the mutual diffusion coefficient, D. In the present work, the values for both parameters (D and D_i) for 31 sodium salts will be summarized.

17.2 HYDRODYNAMIC RADII

The Stokes–Einstein equation (Equation (17.4)) [1–4] can be used to estimate the size of solute molecules treated as Brownian particles immersed in a continuum fluid. Equation (17.4) (where k_B is the Boltzmann constant) establishes a relationship between the hydrodynamic radius of an equivalent spherical particle, R_H, and its limiting diffusion coefficient at infinitesimal concentration, D_i^0, also known as tracer diffusion coefficient, and estimated through the Nernst equation (Equation (17.5)). That is,

$$D_i^0 = \frac{k_B T}{6\pi\eta_0 R_H} \tag{17.4}$$

$$\lambda_i^0 = \frac{D_i^0 Z_a F^2}{RT} \tag{17.5}$$

where R is the ideal gas constant in J mol^{-1} K^{-1}, T is the absolute temperature, F is the Faraday constant, Z_a, and λ^0_i (in S m^2 mol^{-1}) represent the algebraic valence and the equivalent conductance at infinitesimal concentration of the anion, respectively. Table 17.1 shows the values for mutual diffusion coefficient, D, limiting diffusion coefficient, D_i, and hydrodynamic radius, R_H, for each anion for 24 sodium salts.

For some electrolytes containing polyanions (as indicated in Table 17.2), as far as the authors know, no data on conductivities have been published. So, it was necessary to estimate them. For that we take the experimental diffusion coefficients as a function of concentration and, from the extrapolation to c 0, D^0 can be computed; the D^0 is then used in the Nernst–Hartley equation (Equation (17.6)) for determination of the anion equivalent conductance at infinitesimal concentration, λ^0_a. That is

$$D^0 = \frac{RT}{F^2}\left(\frac{|Z_c|+|Z_a|}{|Z_c Z_a|}\right)\frac{\lambda^0_c \lambda^0_a}{\lambda^0_c + \lambda^0_a} \tag{17.6}$$

where D^0 is the mutual diffusion coefficient at infinitesimal ionic strength, R is the gas constant in J mol^{-1} K^{-1}, T is the absolute temperature, F is the Faraday constant, Z_c, and Z_a, and λ^0_c and λ^0_a (in S m^2 mol^{-1}) represent the algebraic valences and the equivalent conductances at infinitesimal concentration of Na$^+$ and anions, respectively, and taking the limiting ionic conductivity of sodium ion as equal to 50.100×10^{-4} S m^2 mol^{-1} [1]. The values found for the parameters λ^0_c, λ^0_a, D^0, D^0_a, and R_H are summarized in Table 17.2.

17.3   RESULTS AND DISCUSSION

The analysis of data reported in Table 17.1 shows that, for sodium salts, in general, the values of R_H, as obtained by Equation (2.1), are similar to the values of the hydrated ions given by Marcus [10, 11]. Considering that in aqueous solution the ions are generally hydrated, R_H may be greater than the sum of the crystallographic radii of the ions, and less than the sum of the radii of the hydrated ions. These differences may be justified if we consider the limitations of the Stokes equation (Equation (17.1)). Although this relationship is only approximated (arising from the acceptance that both the solute kinetic species and the solvent are not structured, along with the assumption that the viscosity is the only responsible for the slowest diffusivity), it can be used to estimate the radii of the moving species. Thus, this relation can only be considered as an approximated one (mainly arising from the fact that the structure of both the solute kinetic species and the solvent are disregarded).

On the other hand, for polyelectrolytes (Table 17.2), since the anions are relatively large compared to water molecules, the Stokes equation can be used in a more reliable way to get information on the correlation between size, shape, and limiting diffusion coefficient at infinitesimal concentration. For these cases, we believe that those estimations are closer to the corresponding real values (Table 17.2). Furthermore, the R_H values are higher when compared with those for other electrolytes (Table 17.1); however, it is observed the achievement of lower values for the tracer diffusion of anions, D^0_a (Equation (17.2)).

In all cases (Tables 17.1 and 17.2), however, all D^0_a values are always smaller than the correspondent mutual diffusion coefficients, D. The differences between the mutual diffusion coefficient of these electrolytes and the corresponding tracer diffusion coefficients for the respective anions characterize the electrostatic dragging effect of the sodium ions on the different anions.

Despite the limitations of these estimates for R_H, which limits their accurate determination, we have a rough idea of the corresponding values range; besides, we consider the method here presented as reasonable and acceptable for such purpose.

TABLE 17.1 Estimation[a] of Limiting Diffusion Coefficients of Anions, D_a^0, the Mutual Diffusion Coefficients at Infinitesimal Concentration, D^0, and the Hydrodynamics Radius of the Respective ions, R_H

Electrolyte	λ^0_c	λ^0_a	Z_c	Z_a	$D_a^0/(10^{-9} \text{ m}^2 \text{ s}^{-1})$	$D^0/(10^{-9} \text{ m}^2 \text{ s}^{-1})$	$R_{H(anion)}/(10^{-9}\text{m})$
NaBr	50.10	78.40	1	−1	2.085	1.627	0.116
NaBrO$_3$	50.10	55.80	1	−1	1.484	1.405	0.163
NaCHO$_2$	50.10	55.10	1	−1	1.465	1.397	0.165
Na$_2$CO$_3$	50.10	69.30	1	−2	0.922	1.161	0.265
Na$_2$C$_2$O$_4$	50.10	48.80	1	−2	0.649	0.987	0.373
NaCl	50.10	76.0	1	−1	2.029	1.611	0.119
NaClO$_3$	50.10	64.60	1	−1	1.718	1.502	0.141
NaClO$_4$	50.10	67.30	1	−1	1.790	1.529	0.158
Na$_2$CrO$_4$	50.10	85.00	1	−2	1.130	1.259	0.192
NaF	50.10	55.40	1	−1	1.473	1.401	0.173
Na$_4$Fe(CN)$_6$	50.10	108.10	1	−4	0.719	1.139	0.337

TABLE 17.1   *(Continued)*

Electrolyte	λ^0_c	λ^0_a	Z_c	Z_a	$D_a^0/(10^{-9}$ m^2 s$^{-1})$	$D^0/(10^{-9}$ m^2 s$^{-1}))$	$R_{H(anion)}{}^d/(10^{-9}$ m$)$
NaH$_2$AsO$_4$	50.10	34.00	1	−1	0.304	1.078	0.796
NaHCO$_3$	50.10	44.5'	1	−1	1.184	1.255	0.204
NaH$_2$PO$_4$	50.10	36.00	1	−1	0.357	1.115	0.678
Na$_2$HPO$_4$	50.10	57.00	1	−2	0.758	1.065	0.319
NaI	50.10	76.90	1	−1	2.045	1.615	0.118
NaIO$_3$	50.10	41.00	1	−1	1.090	1.200	0.220
NaMnO$_4$	50.10	62.80	1	−1	1.670	1.484	0.145
Na$_2$MoO$_4$	50.10	74.50	1	−2	0.991	1.196	0.244
NaNO$_2$	50.10	72.00	1	−1	1.915	1.573	0.126
NaNO$_3$	50.10	71.42	1	−1	1.900	1.568	0.127
NaOH	50.10	197.60	1	−1	5.255	2.128	0.046
NaSCN	50.10	66.50	1	−1	1.769	1.521	0.137
Na$_2$SO$_3$	50.10	72.0'	1	−2	0.357	1.180	0.678
Na$_2$SO$_4$	50.10	79.80	1	−2	1.061	1.229	0.228
Na$_2$S$_2$O$_3$	50.10	87.40	1	−2	1.162	1.272	0.208
Na$_2$WO$_4$	50.10	69.40	1	−2	0.323	1.162	0.749

[a]Z_c and Z_a, and λ^0_c and λ^0_a (in 10^{-4} S m^2 mol^{-1}) represent the algebraic valences and the equivalent conductances at infinitesimal concentration of Na$^+$ and of the anion in study, respectively [1].

TABLE 17.2A   Estimation[a] of Limiting Diffusion Coefficients of Some Polyanions, D^0_a, the Mutual Diffusion Coefficients at Infinitesimal Concentration, D^0, and the Hydrodynamics Radius of the Respective ions, R_H

Polyelectrolyte	λ^0_c	λ^0_a	Z_c	Z_a	$D^0_a/(10^{-9}$m^2 s$^{-1})$	$D^0/(10^{-9}$m^2 s$^{-1}))$	$R_{H(anion)}{}^d/ (10^{-9}m)$
Sodium hyaluronate (NaHA) [8]	50.10	40.05	1	−1	1.066	1.318	0.184

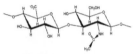

[a]Z_c and Z_a, and λ^0_c and λ^0_a (in 10^{-4} S m^2 mol^{-1}) represent the algebraic valences and the equivalent conductances at infinitesimal concentration of Na$^+$ and of the hyaluronate anion, respectively [8].

TABLE 17.2B Estimation[a] of Limiting Diffusion Coefficients of Some Polyanions, D^0_a, the Mutual Diffusion Coefficients at Infinitesimal Concentration, D^0, and the Hydrodynamics Radius of the Respective ions, R_H

Polyelectrolyte	λ^0_c	λ^0_a	Z_c	Z_a	$D^0_a/(10^{-9}m^2 s^{-1})$	$D^0/(10^{-9}m^2 s^{-1}))$	$R_{H(anion)}{}^d/(10^{-9}m)$
Sodium alginate ($NaC_6H_7O_6$) [7]	50.10	14.72	1	−1	0.551	0.606	0.445

[a]Z_c and Z_a, and λ^0_c and λ^0_a (in 10^{-4} S m^2 mol^{-1}) represent the algebraic valences and the equivalent conductances at infinitesimal concentration of Na$^+$ and of the alginate anion, respectively [7].

TABLE 17.2C Estimation[a] of Limiting Diffusion Coefficients of Some Polyanions, D^0_a, the Mutual Diffusion Coefficients at Infinitesimal Concentration, D^0, and the Hydrodynamics Radius of the Respective ions, R_H

Polyelectrolyte	λ^0_c	λ^0_a	Z_c	Z_a	$D^0_a/(10^{-9}m^2 s^{-1})$	$D^0/(10^{-9} m^2 s^{-1}))$	$R_{H(anion)}{}^d/(10^{-9}m)$
Tetrasodium-Tetraphenyl-Porphinesulfonate (Na_4TPPS) [6]	50.10	46.58	1	−4	0.310	0.800	0.790

[a]Z_c and Z_a, and λ^0_c and λ^0_a (in 10^{-4} S m^2 mol^{-1}) represent the algebraic valences and the equivalent conductances at infinitesimal concentration of Na$^+$ and of the TPPS^{4-} anion, respectively [6].

TABLE 17.2D Estimation[a] of Limiting Diffusion Coefficients of Some Polyanions, D^0_a, the Mutual Diffusion Coefficients at Infinitesimal Concentration, D^0, and the Hydrodynamics Radius of the Respective ions, R_H

Polyelectrolyte	λ^0_c	λ^0_a	Z_c	Z_a	$D^0_a/(10^{-9} m^2 s^{-1})$	$D^0/(10^{-9} m^2 s^{-1}))$	$R_{H(anion)}{}^d/(10^{-9}m)$
Ethyl sulfonated resorcinarene (Na_4ETRA)[9]	50.10	229.30	1	−4	0.384	0.881	0.638

[a]Z_c and Z_a, and λ^0_c and λ^0_a (in S m^2 mol^{-1}) represent the algebraic valences and the equivalent conductances at infinitesimal concentration of Na$^+$ and of the ERTA^{4-} anion, respectively [9].

17.4   CONCLUSION

We have estimated the limiting diffusion coefficients (tracer and mutual), and the hydrodynamic radii for different anions obtained by Nernst and Stokes models, respectively, and by using the limiting ionic conductivities. These data permit us a better understanding of the structure of aqueous electrolytes and polyelectrolyte systems.

ACKNOWLEDGMENTS

The authors in Coimbra are grateful for funding from "The Coimbra Chemistry Center" which is supported by the Fundaçãopara a Ciência e a Tecnologia (FCT), Portuguese Agency for Scientific Research, through the programs UID/QUI/UI0313/2019 and COMPETE.

KEYWORDS

- conductances
- diffusion
- hydrodynamic radius
- sodium salts
- solutions
- Taylor technique

REFERENCES

1. Robinson, R. A., & Stokes, R. H. (1959). *Electrolyte Solutions* (2nd edn.). Butterworths, London.
2. Harned, H. S., & Owen, B. B. (1964). *The Physical Chemistry of Electrolytic Solutions* (3rd edn.). Reinhold Pub. Corp., New York.
3. Tyrrell, H. J. V., & Harris, K. R. (1984). *Diffusion in Liquids: A Theoretical and Experimental Study*. Butterworths, London.
4. Lobo, V. M. M. (1990). *Handbook of Electrolyte Solutions*. Elsevier, Amsterdam.
5. Onsager, L., & Fuoss, R. M. (1932). Irreversible processes in electrolytes. Diffusion, conductance and viscous flow in arbitrary mixtures of strong electrolytes. *J. Phys. Chem.*, *36*, 2689–2778.

6. Da Costa, V. C. P., Ribeiro, A. C. F., Sobral, A. J. F. N., Lobo, V. M. M., Annunziata, O., Santos, C. I. A. V., Willis, S. A., Price, W. S., & Esteso, M. A. (2012). Mutual and self-diffusion of charged porphyrins in aqueous. *J. Chem. Thermodyn.*, *47*, 312–319.

7. Ribeiro, A. C. F., Fabela, I., Sobral, A. J. F. N., Verissimo, L. M. P., Barros, M. C. F., Melia, R. M., & Esteso, M. A. (2014). Diffusion of sodium alginate in aqueous solutions at T = 298.15 K. *J. Chem. Thermodyn.*, *74*, 263–268.

8. Verissimo, L. M. P., Valada, T. I. C., Sobral, A. J. F. N., Azevedo, E. E. F. G., Azevedo, M. L. G., & Ribeiro, A. C. F. (2014). Mutual diffusion of sodium hyaluronate in aqueous solutions. *J. Chem. Thermodyn.*, *71*, 14–18.

9. Galindres, D. M., Ribeiro, A. C. F., Valente, A. J. M., Esteso, M. A., Español, E. S., Vargas, E. F., et al. (2019). Ionic conductivities and diffusion coefficients of alkyl-substituted sulfonated resorcinarenes in aqueous solutions. *J. Chem. Thermodyn.*, *133*, 222–228.

10. Marcus, Y., (1988). Ionic radii in aqueous solutions. *Chem. Rev.*, *88*, 1475–1498.

11. Ribeiro, A. C. F., Esteso, M. A., Lobo, V. M. M., Burrows, H. D., Amado, A. M., Amorim Da Costa, A. M., et al. (2006). Mean distance of closest approach of ions: sodium salts in aqueous solutions. *J. Mol. Liq.*, *128*, 134–139.

PART V
Macromolecular Physics

CHAPTER 18

Pesticides and Their Environment and Health Impact: An Approach to Remediation Using Hydrogels

GIANLUCA UTZERI, LUIS M. P. VERISSIMO, ANA C. F. RIBEIRO, and ARTUR J. M. VALENTE*

Coimbra Chemistry Centre, Department of Chemistry, University of Coimbra, 3004-535 Coimbra, Portugal

Corresponding author. E-mail: avalente@ci.uc.pt

ABSTRACT

Over 2 million tons per year of pesticides are used in agriculture worldwide, and their presence in surface water bodies is reaching alarming concentration values. The use of low cost and environmentally friendly sorbents in the remediation of contaminated environments, especially those consisting of natural polymers, has been receiving significant interest in the last decades. In this work, a broad view of the pesticides, EU legislation, and remediation using biohydrogels is reported.

18.1 INTRODUCTION

Pesticides play an important role in the agricultural, household, and forestry areas by preventing or controlling pests and so increasing crops. In the European Union (EU), agrochemicals annual sales reach almost 400,000 tons with ca. 500 approved active substances (Figure 18.1) [1]. However, their persistence, toxicity, bioaccumulation, and nonspecificity as well as the capacity to permeate the soil, inquinate surface, and groundwater after leaching, run-off, and volatilize made them an issue of increasing concern

[2]. They have direct and indirect effects in biodiversity loss, ecosystem deterioration, and human diseases [3]. Moreover, commercial pesticides usually include more than one phytopharmaceutical, over inert ingredient and metabolites, leading to multiresidual contamination and resistant populations issues [4].

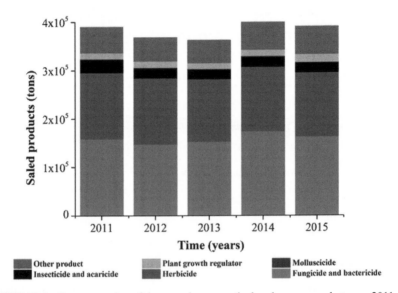

FIGURE 18.1   Representation of the agropharmaceuticals sales progress between 2011 and 2015.

Ecosystem protective actions are essential to safeguard essential food, clean water–soil–air, and health benefits and services. Currently, the policies concerned by the data needs to this issue are the EU Biodiversity Strategy 2020, the Common Agricultural Policy, the Water Framework Directive, and the Thematic Strategy on Soils [3].

The industrial production of pesticides began in 1940, switching from the use of organic fertilizer to synthetic products. The introduction of synthetic pesticides allowed increasing the agricultural production, by three times, in 50 years, attending to the increased need of requirements [4].

In the EU, the pesticide term defines two substance classes:

1. Plant production products (PPPs), mainly applied in agricultural fields. Whose entry in the marketplace is regulated by Reg.

110772009/CE with the goal to protect human health, animals, and environment.

2. Biocides, whose entry into the marketplace and handling are governed by Reg. 528/2012/UE.

Generally, pesticide formulations used in agricultural fields are made up of two sorts of compounds:

1. One or more active ingredients (phytopharmaceuticals) used to pest control or as a plant growth hormone for crop protection (approved by the European Commission, whose classification and labeling are controlled by Reg. 1272/2008/CE).
2. Inert ingredients that influence permeation and diffusion properties of pesticides in the plants and soils improve the product stability and increase the lifetime and solubility in the aqueous medium, among others. Labeling these ingredients is not compulsory [5–7].

The use of pesticides has positive effects in plants protection and maximization of the production process, a critical issue to deal with the growth of the global population (Figure 18.2) [8]. However, active and inert compounds both can produce negative effects for environment, human health, and biodiversity [7–9].

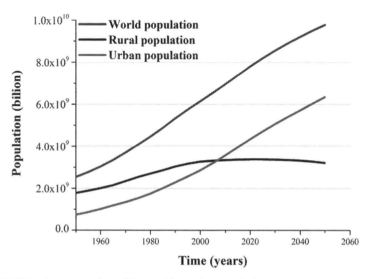

FIGURE 18.2 Representation of the world growing population.

The potential effect of pesticides is related with their own properties such as bioaccumulation, persistence, and toxicity, which significantly change toward the surrounding system conditions (temperature, pH, aqueous medium, soil, and air).

Different pathways of direct pesticide action are possible depending on its handling and application; indirect pathways include their presence in the food chain and drinking water (both superficial and groundwater sources).

Moreover, pesticides persistence is the key factor in the formation of resistant species, which consequently demands of higher amounts of pesticides or to the development of new PPPs [10].

Previous studies proved that ca. 0.5% of the pesticides reach the target, whereas 95% is dispersed in the environment [11].

With the purpose of minimizing the use and harmful effects of pesticides and their metabolites as well as to protect the environment, human, and animal health, organizations like Food and Agriculture Organization of the United Nations (FAO) and World Health Organization (WHO), among others, regulate and establish the residual maximum levels and issue good practices for use and handling of agropharmaceuticals, developing policies, and protocols like Integrated Pest Management [1, 9, 12, 13].

18.2   PESTICIDES

Chronologically agropharmaceuticals can be divided into the first and second generation. The products from the first generation were used before 1940 including both organic, characterized by active ingredients extracted from plants, for example, nicotine in this regard called botanicals, and inorganic pesticides as arsenic or sulfur salts [1, 10–14].

The synthetic pesticides used after 1940 are considered of the second generation, in which active ingredients are chemical derivatives of natural substances. Such synthetic functionalization allows improving the specificity of active ingredients, but, concomitantly, also led to an increase of toxicity, persistency, and less biodegradability. Subsequently, the PPPs can be classified in narrow or broad action spectrum pesticides with several action mechanisms [15].

Numerous studies relate the pesticide use and properties with endangered biodiversity and human diseases increase due to direct and indirect exposure [16, 17].

Based on the chemical structures, pesticides can be divided in four families: (1) organochlorine pesticides (OCP), (2) organophosphate pesticide (OPP), (3) carbamates, and (4) pyrethroids [18].

Finally, PPPs can be classified by (1) target pest, as herbicide, insecticide, pest growth regulator, nematicide (Taxonomically, nematodes (*phylum Nematode*) refer to organisms with bilateral symmetry, not metameric, with cylindrical, fusiform, or fusiform shapes.), termiticide, molluscicide, piscicide, avicide, rodenticide, bactericide, antimicrobial (or disinfectant), fungicide, or sanitizing agent; (2) source, as natural, synthetic, or biological; or (3) physical state, as liquid, solid, or gaseous [5].

18.2.1 ORGANOCHLORINE PESTICIDES

OCPs were widely applied in agricultural and industrial fields between 1950 and 1970. Subsequent studies demonstrated related high toxicity and persistence leading to their prohibition in EU and their classification as persistent organic pollutants (POPs) and biocumulable, as defined in the Stockholm Convention (*cf.* 18.4) [14–19].

Organochlorine-based pesticides were generally used as broad action spectrum insecticides. However, subsequent studies have demonstrated that OCPs have neurotoxic, carcinogenic, and mutagenic effects, as well as hepatotoxic and reprotoxic effects, also may acting as an endocrine disruptor, influencing the human hormonal levels [20].

The main common physical–chemical characteristics of OCP are the low polarity, high liposolubility, high volatility, and stability, which are due to similar chemical structures (aliphatic or aromatic) with one or more chlorides substituents [21].

The widespread use of OCPs was mainly due to the low cost, high efficiency, and simple application. Some examples of this class of pesticides are DDT (CAS: 50-29-3), alachlor (CAS: 15972-60-8), atrazine (CAS: 1912-24-9), and lindane (CAS: 319-86-8) (Figure 18.3); these OCPs have a half lifetime of 15, 2.5, 2.0, and 1.0 years, respectively [21, 22].

18.2.2 ORGANOPHOSPHATE PESTICIDES (OPP)

OPPs are ester derivatives of phosphoric acid, generally used as insecticides in substitution of OCPs [14]. The wide use of OPPs was due to the lower toxicity and persistence in the environment and higher biodegradability, when compared with OCP [10].

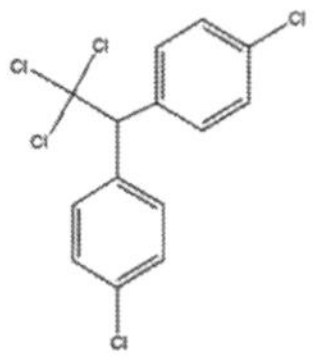

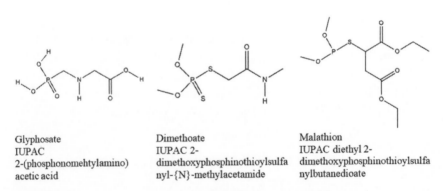

DDT
IUPAC: 1-chloro-4-[2,2,2-trichloro-1-(4-chlorophenyl)ethyl] benzene

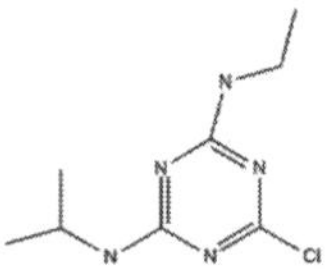

Atrazine
IUPAC: 6-chloro-4-{N}-ethyl-2-{N}-propan-2-yl-1,3,5-triazine-2,4-diamine

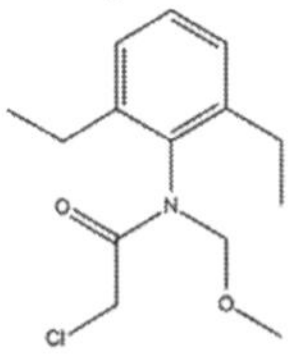

Alachlor
IUPAC: 2-chloro-{N}-(2,6-diethylphenyl)-{N}-(methoxymethyl)acetamide

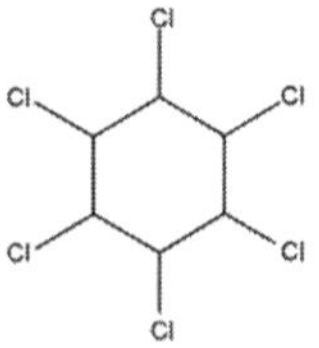

Lindane
IUPAC: 1,2,3,4,5,6-hexachlorocyclohexane

FIGURE 18.3   Molecular structures of some OCPs.

However, negative effects of OPPs for human and animals health are related to the residual amounts in food and drinking water because of their irreversible effect on the synthesis inhibition of enzyme acetylcholinesterase (acetylcholinesterase is a critical enzyme for human nervous function), affecting cerebral functions [14, 18, 19].

Examples of approved OPPs (Figure 18.4), in UE, are glyphosate (CAS: 1071-83-6), dimethoate (CAS: 60-51-5), and malathion (CAS: 121-75-5)-the last one, used exclusively as insecticide [22].

Glyphosate
IUPAC
2-(phosphonomehtylamino) acetic acid

Dimethoate
IUPAC 2-dimethoxyphosphinothioylsulfanyl-{N}-methylacetamide

Malathion
IUPAC diethyl 2-dimethoxyphosphinothioylsulfanylbutanedioate

FIGURE 18.4   Structures of some organophosphate pesticides (OPP).

18.2.3 CARBAMATES

The carbamates are carbamic acid derivatives used as insecticides, fungicides, nematicides, and herbicides. They show a lower toxicity than the organophosphates and faster biodegradation on the soil, with an average lifetime of less than 1 year [19].

The action mechanism is similar to that of OPPs, that is, by reversible inactivation of the acetylcholinesterase enzyme synthesis. Carbamates are dangerous for human and insects health due to its toxicity level [14, 18, 23].

Examples of carbamates (Figure 18.5) approved as active ingredient in the EU are oxamyl (CAS: 23135-22-097502–85-7), methomyl (CAS: 16752-77-5), and fenoxycarb (CAS: 72490-01-8), the latter one acting as a growth regulator on insects [10, 22, 24].

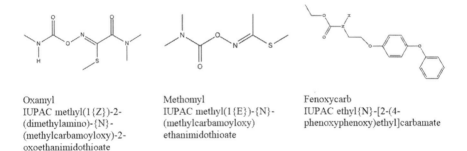

Oxamyl
IUPAC methyl(1{Z})-2-(dimethylamino)-{N}-(methylcarbamoyloxy)-2-oxoethanimidothioate

Methomyl
IUPAC methyl(1{E})-{N}-(methylcarbamoyloxy) ethanimidothioate

Fenoxycarb
IUPAC ethyl{N}-[2-(4-phenoxyphenoxy)ethyl]carbamate

FIGURE 18.5 Molecular structures of some carbamate pesticides.

18.2.4 PYRETHROID

Pyrethroid pesticides (Figure 18.6) are classified into two classes: natural and synthetic pyrethroid. The former ones have active ingredients obtained from *Chrysanthemum Coccineum* and *C. Cinerariaefolium* flowers [20]; six different compounds can be isolated, pyrethrin I and II (CAS 121-21-1 and 121-29-9), jasmolin I and II (CAS 4466-14-2 and 1172-63-0) and cinerin I and II (CAS 25402-06-6 and 121-20-0) [10]. In the case of synthetic pyrethroids, they are divided in the first generation, which are esters and alcohols derivatives of chrysanthemum acid that present a furan ring in the molecular structure and lateral chain, and the second generation, which generally are structural derivatives of the 3-phenoxybenzil alcohol [25]. The second-generation products are better insecticides than the first generation ones but, on contrary, show lower biodegradability.

Because of the less toxicity of pyrethroids than OPPs and carbamates, they are widely used both as household and agricultural insecticides.

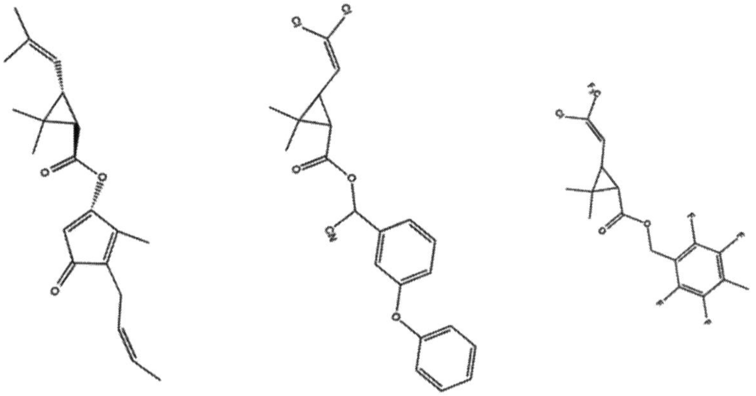

Cinerin I
IUPAC [(1{S})-3-[(~{Z})-but-2-enyl]-2-methyl-4-oxocyclopent-2-en-1-yl]-(1{R},3{R})-2,2-dimethyl-3-(2-methylprop-1-enyl)cyclopropane-1-carboxylate

Cypermethrin (CAS: 67375-30-8)
IUPAC [cyano-(3-phenoxyphenyl)methyl]-3-(2,2-dichloroehtnyl)-2,2-dimethylcyclopropane-1-carboxylate

Tefluthrin (CAS: 79538-32-3)
IUPAC (2,3,5,6-tetrafluor-4-methylphenyl)methyl-(1S,3S)-3-[(Z)-chloro-3,3,3-trifluoroprop-1-enyl]-2,2-dimethylcyclopropane-a-carboxylate

FIGURE 18.6 Molecular structures of some pyrethroid pesticides. Cinerin is a natural pyrethroid, whereas cypermethrin and tefluthrin are synthetic pyrethroid pesticides.

18.3 EUROPEAN UNION REGULATION

In the EU, since 2009, a complex legal system control and limit the entry on the marketplace and regulates the use of pesticides to protect the environment, human health, and biodiversity [26]. The legislative structure defines the correct use of pesticides related to the toxicity and environmental impact determining the maximum residual limits (MRLs) in the environment and foods [1]. The legislative system is based on the precautionary principle, which forces data harvesting about pesticides in order to grant the product security prior to the approval [4, 27, 28].

18.3.1 APPROVAL AND MARKETING

Firstly, it is necessary to distinguish between plant protection products and biocides, which are to be approved before entry in the marketplace by similar processes, but through different agencies and by different regulatory legislations.

Considering the PPPs, the first phase of the process for the approval of an agropharmaceutical consists in a technical and scientific evaluation of the phytopharmaceutical, inert ingredient, and metabolites by one or more member-states of the EU is the so-called Rapporteur Member State. The collected data are then submitted to the European Commission and European Food Safety Agency (EFSA) that develop a risk analysis for human health, environment, and specific pollution of superficial and groundwater sources [29, 30]. More-over, the residual pesticide risks in foods for humans and animals consumption are assessed with the emphasis in the reduction of agropharmaceuticals and establish and improve the permitted maximum residual values.

In the second step, the commercial product is evaluated to entry in the marketplace regulated by the Reg. 1107/2009/CE, which also establishes the rules for correct use and control of the pesticides, granting a high profile of protection [31].

Just after these preentry processes, the pesticide is approved for use at the national level. The process takes 3 years long for approval and is valid for 10 years, after which can be renewed for a maximum of 15 years [32].

The technical and scientific validation of biocides is done by the Euro-pean Chemical Agency (ECHA, ECHA is an agency established in 2007, based in the Reg. 1907/2006/CE (REACH) that standardize the registration, evaluation, authorization, and restriction of chemical substances.) and the security level and entry in the marketplace is ruled by Reg. 528/2012/UE [32].

In both cases, the regulation splits the EU geographical space into three different areas, North, Center, and South. The authorization for use of a given product, in each area, needs the approval from at least one member state in that area.

18.3.2 SUSTAINABLE USE OF AGROPHARMACEUTICALS

The directive EC/128/2009 regulates the sustainable use of pesticides, promoting the integrated pest control.

Beyond the phased process explained in Section 18.3.1, the member states must approve a National action plan as established in the directive EC/128/2009 with the aim of monitoring and reduce as much as possible the use of dangerous substances [33].

Moreover, based on this directive, handling, and application of plant protection products can only be done by qualified staff and merchants; however, when possible alternative pest control methods should be used. Finally, the directive regulates and enforces the member states to protect superficial and groundwater resources.

18.3.3   MAXIMUM RESIDUAL LIMIT

With the purpose to maintain the efficiency and capacity of the production process, plants, and production process protection is important, so, chemical and nonchemical methods can be used such as pesticides or crop rotation respectively.

Usually, chemical products are used for plants and crop protection; however, pesticides could have negative effects for the environment and living organisms, allowing the development of organism resistance issue.

All of that is regulated by Reg. 396/2005/EC through MRLs for active ingredients and their metabolites, after EFSA examination. By definition, the MRL refers to the highest pesticide residual concentration allowed on food for human and animal consumption.

Principles, requirements, and operational procedures for food safety are defined by Reg. 178/2002/EC.

The standard MRL value, based on the European regulation, generally is 0.01 mg kg^{-1} and is applied when no other specification exists about the examined pesticide [34].

More specific MRL values are defined considering the technical and scientific data, such as (1) good agricultural practices concerning amount and application periods; (2) toxicological references, based on both the chronic toxicity as acceptable daily intake and acute toxicity as acute reference dose; and (3) free international trade agreements [35].

18.3.4   WATER PROTECTION

Water natural resources, open-access, or groundwater are of critical importance and must be always protected. With this in mind, the directive 60/2000/

EC establishes the rules for protection and quality improvement of the water, always considering the international treaties objective son the environmental security.

The drainage basins management includes monitoring programs of the European member state, measuring the water quality with the evaluation of the ecological and chemical status and impacts of the anthropological and economic activities.

The water quality and ecological status of water resources are defined by physicochemical parameters such as pH, temperature, salinity, oxygen concentration, turbidity, and conductivity.

Compounds like OCPs, OPPs, hydrocarbons, organic persistent and biocumulative compounds, heavy metals, and biocides are just some examples of the possible water contaminants.

18.3.5 *PACKAGING, LABELING, AND PESTICIDE CLASSIFICATION*

The Reg. 1272/2008/EC establishes rules and principles related with the packaging, labeling, and classification process of substances and mixtures that may represent hazardous for the environment and human health, guaranteeing the free and correct commercialization and product register in the marketplace. The identification of a substance or mixture has to be included in the labeling together with the security record, in agreement with the Reg. 1907/2006/EC [Registration, Evaluation, and restriction of Chemicals (REACH)].

The substance, mixture, or impurity classification, in agreement with the regulations, is based on efficiency, target, and potential negative effects information obtained from the producer.

In accord to the REACH regulation, unsafety substances can be classified based on the toxicity level and effects on the environment and human health (Table 18.1) [29].

TABLE 18.1 Classification of Hazardous Substances

Chemicals		**Characteristics and Negative Effects**
Carcinogenic, mutagenic and/or reprotoxic (CMR)[a]	Carcinogenic	Induce cancer or increase its incidence
	Mutagenic	Increases occurrence of mutation
	Reproductive toxins	Effects on sexual function, male, and female fertility, toxicity to the fetus and effects by or via lactation

TABLE 18.1   *(Continued)*

Chemicals		Characteristics and Negative Effects
Persistent, biocumulative and toxic (PBT)	Persistent	Resistance to decomposition in the environment (bacteria, light, temperature among other)
	Biocumulative	Accumulation in living organisms, with concentration increasing along the food chain
	Toxic	Hazardous effects (e.g., DNA, neurologic system, reproduction or immunity system)
High persistent and high bio-cumulative (hPhB)		Resistance to environmental degradation and high degree of accumulation in living organisms
Hazardous chemicals equivalent to CMR substances	Endocrine disruptor	Functional alteration of the hormonal system
	Respiratory sensitizers	Induce respiratory allergies
	Specific organ toxicity	Toxicity to an organ or target system in the human body

[a]CMR, legislated by Reg. 1272/2008/EC.

18.4   INTERNATIONAL LEGISLATION

EU and member states, nongovernmental organizations, for example, Pesticide Action Network Europe, EFSA, Greenpeace, and international organizations, for example (*World Trade Organization* controls the market rules and act as a forum to negotiate international market agreements), WHO, FAO—set agreements with the aim of safeguard human health and the environment, improve the application of the good practices on the use of the pesticides, prevent, and control the negative effects resulting from the occurrence of pesticides traces in foods for human and animals consumption, as well as favoring the free and harmonized market of agropharmaceuticals. Some of those agreements are described as follows:

1. International Plant Protection Agreement, to define effective common actions and grant the correct use and appropriate chemicals and no-chemicals methods for plants and pests control protection [36];

2. *Codex Alimentarius* develop and adopt standards of reference in international food trade [37];

3. Rotterdam Convention promotes the accountability and cooperation in the international market of hazardous chemical substances to protect the environment and human health [38];

4. Stockholm convention establishes the international agreement on protection of the environment and human health from the POPs by reduction or elimination [39].

18.5 PESTICIDES ENVIRONMENTAL PATHWAYS

The contaminants introduced in the environment through the anthropogenic activities, depending on the inherent physicochemical characteristics, the environmental and application conditions may have shorter or longer persistence times in the environment as well as different degradation mechanisms.

The term persistent includes all organic compounds that not undergo structural modification over time in the considering environmental conditions [40]. The pesticides could be adsorbed and transported based on the potential persistence/mobility on the medium. On the other hand, the degradation pesticides process leads to the formation of degradation products or metabolites with different toxicity, persistence, and action mechanism when compared to the precursor.

The pesticides interactions with soil, superficial, and groundwater are complex and controlled by multiple chemical, physical, and biological factors. Aqueous solubility, adsorption trend in the soil, and persistence are fundamental features for the determination of the pesticides behavior in the media. For example, pesticides with high water solubility and persistence but less adsorption rate present a high-risk potential.

The pesticide adsorption rate in the soil highly depends on soil characteristics because, generally, dry soils present less pesticides retention capacity than wet soils, clay soils, or with high organic matter contents.

Moreover, the affinity of the pesticides at the soil as well as their water solubility influences the mobility and consequentially the leaching and the bioaccumulation phenomena.

The pesticide soil adsorption can be quantified by the partition coefficient, K_{oc} (Equation (18.1))

$$K_{oc} = \frac{\frac{\text{adsorbed conc}}{\text{dissolved conc}}}{\% \text{ organic carbon contents oil}} \tag{18.1}$$

where K_{oc} values above 1000 indicates a significant pesticide affinity for the soil, whereas values below 500 indicate high pesticide mobility [41].

However, substances with solubility lower than 1 mg L^{-1} are considered of low mobility, tendentially persisting in the surface or being moved through superficial transport phenomenon. The pesticides transport involves different processes as sorption/desorption, diffusion, leaching, phytosorption, degradation, and volatilization [14, 42].

18.6   REMEDIATION OF PESTICIDES

The active ingredients, metabolites, inert ingredients, and degradation products in commercial preparations represent a class of substances with high interest in the environmental field because of their toxicity and persistence. In the last decade, with the purpose of prevention and control the contaminant levels have been developed different technologies to remove or facilitate the concentration reduction of the pollutant in soil, water, and air. The methods actually most used for contaminant removal include chlorination, ozonization [41, 43–45] and advanced oxidation processes (AOPs). In the latter one, it is possible applied technologies to combine the action of hydrogen peroxide with ultraviolet radiation, which lead to the faster and higher formation of hydroxyl radicals, improved the oxidation efficiency and a decrease of the time reaction [46–48]. Different, but nonetheless important technologies consider absorbent and super-adsorbent materials application for environmental remediation.

Examples of some commonly used adsorbent materials include activated carbon, silica, chitosan, cellulose, and pectin, among others. However, the natural adsorbents usually require extensive treatment to reach their most effective form [49].

18.6.1   ADSORBENT MATERIALS

18.6.1.1   ACTIVATED CARBON

The activated carbons are produced from various type of carbonaceous materials [49–51]. They are the most common raw adsorbent material used because of microscale pores resulting in a great exposed area for a wide range of pollutants (e.g., dyes, heavy metals, and pesticides). Removal efficiency over 80% in a series of 11 pesticides has been observed for active carbon obtained from processed starch [50]. Specifically, the carbon nanotubes divided in single-walled carbon nanotube and multiwalled carbon nanotubes (MWCNTs). The MWCNTs are demonstrated high adsorption capacity in OPPs with almost 100% of removal efficiency for diazinon and malathion, respectively [52, 53].

Moreover, carbon-silica nanofibers were used for pesticides analysis. Specific surface area and fibers morphology highly influence the adsorption performance with a range of 80%–100% of removal efficiency [54].

18.6.1.2   SILICA

Silica is one of the most abundant materials on the earth, existing in several natural minerals and in synthetic products. It shows an amorphous structure with formula $SiO_2 \cdot nH_2O$, the high exposed surface area. The presence of SiOH functional groups in the surface lead at high adsorption capacity promoting hydrogen bonding formations [49]. These characteristics as well as mechanical, thermal, and chemical stability make silica an interesting material to be applied on environmental remediation processes. In addition, mesoporous silica can be easily modified. Recent studies show that mesoporous silica and its composite with polymers such as polyaniline present high chloridazon efficiency removal in aqueous media [55]. Silica mesoporous and silica composites doped with Ti and Fe as well as immobilized cyclodextrins that act as pesticide complexant [56]. These materials were tested for organophosphorus pesticides in wastewater samples and reached 80% as the highest efficiency removal [57].

18.6.2   HYDROGELS

The hydrogels are generally defined as crosslinked polymers with a three-dimensional structure, constituting for homo- or heteropolymers, with high water uptake capacity without dissolving itself in water as a consequence of the physical or chemical interpolymeric interactions.

The hydrophilicity is due to the presence of hydroxyl groups (–OH), carboxylic groups (–COOH), amide groups (–CONH–) or amine groups (–CNH$_x$), among others, along the central or lateral polymeric chain. On the other hand, the hydrophobic character of a hydrogel can be modified by using hydrophobically modified polymers [58].

The uptake water capacity and response to the external stimuli (pH, temperature, or ionic strength) make hydrogel materials of great interest with application in different fields such as in the production of contact lenses and other ophthalmic products, treatment of industrial effluents, biomedical engineering for cartilage repair or nerve regeneration, cosmetics, and as vectors for controlled drug delivery [59, 60].

The hydrogel structure and properties highly depend on the water content that, on the other side, depends on the following parameters:

1. Volume fraction of polymer in the swollen state, which is a measure of the swelling degree of the hydrogel;
2. Effective molecular weight of the polymeric chain between cross-linking points, related with the reticulation degree;
3. Dimensions of the crosslinking mesh (ξ), which indicate the distance between sequential crosslinking points.

Thus the solute diffusion is limited by the crosslinking mesh, which depends from different factors such as the gel crosslinking degree, polymers molecular structure, and capacity to respond at external stimuli [58, 59].

Moreover, the dimension of the crosslinking mesh influences the physical hydrogel properties as mechanical strength, biodegradability, and diffusivity in the release system.

However, by experimental design is possible to obtain a crosslinked structure with defined dimensions of crosslinking mesh and, accordingly, controls the analytes' mobility (Figure 18.7)

Finally, based on the polymer source is possible to classify the hydrogel as natural or synthetic and their synthesis can be gained for both chemical methods (covalent interchain linkage) and physical methods (hydrogen bonds, hydrophobic, or coacervation) [61].

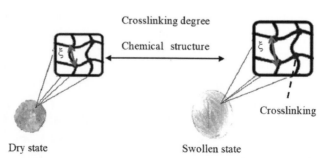

FIGURE 18.7   Representation of some factors that influence crosslinking mesh dimension.

18.6.2.1 NATURAL HYDROGELS

Polymeric natural-based hydrogels are widely studied and applied as much in environmental remediation as in biomedical engineering and regenerative medicine, because they are biodegradable, biocompatible with low toxicity and undergo enzymatic degradation.

Nevertheless, natural hydrogels usually show limited mechanical properties. Polymers obtained from natural sources commonly used in hydrogels synthesis include proteins or fibrous polysaccharides (Table 18.2) [62].

TABLE 18.2 Polymers Used in Natural Hydrogels Synthesis

Natural Polymer	Source	Synthesis Methods	Application
Collagen	Sheepskin, ligaments, cartilage, fish skin, and marine sponges	Neutralization	Biomedical engineering, drug delivery system
Gelatin	Acidic/alkaline treatment of collagen	Thermal treatment	Drug delivery system
Silk fibroin	Arthropods	Sol–gel transition in acidic medium, sonication, and lyophilization	Biomedical engineering, drug delivery system
Alginate	Brown seaweed and bacteria	Thermal treatment, electrostatic interactions, gamma beam	Biomedical engineering, drug delivery system, and cellular encapsulation
Agarose	Red seaweed	Thermal treatment	Cellular encapsulation, biomedical engineering
Hyaluronic acid	Synovial fluid, skin, and bacteria	Chemical treatment	Cosmetic, biomedical engineering, ophthalmology, and drug delivery systems
Cellulose and derivates	Plants	Physical or chemical treatment and radical polymerization	Controlled drug delivery system, cosmetic, biomedical engineering, and environmental remediation
Chitosan	Chitin derivative (crustaceans shell and insects)	Chemical or physical treatment	Biomedical engineering, drug delivery systems, and cellular encapsulation

18.6.2.1.1 Cellulose

The cellulose is the most abundant biorenewable polysaccharide presents on the earth, extracted from a broad range of plants and animals [63]. Its physicochemical and biological properties make it an interesting and outstanding material. Moreover, cellulose is easily modifiable due to the presence of the hydroxyl groups along the chain [64]. Because of that, cellulose and its derivatives were extensively explored and applied in several fields such as biomedical engineering and environmental remediation process [65, 66]. Cellulose-based membrane filtration has been applied in wastewater treatment for nitroaromatic pesticides retention [67]. Moreover, the use of cellulose-based nanofiltration membranes leads to high pesticide rejection efficiency [68]. Improved efficiency of pesticide adsorption of cellulose fiber was obtained by grafting of glycidyl methacrylate [69]. Removal efficiencies higher than ≥80%, for organophosphorus pesticides have been obtained by activated carbon cellulose and cellulose–graphene composite [70].

18.6.2.1.2 Chitosan

Chitosan is semicrystalline, heteropolysaccharide derived from chitin, which is mainly obtained from the shell of crustaceous [71]. Chitosan is produced by *N*-deacetylation of acetamido group in C2 position. Like cellulose, its physicochemical and biological properties made chitosan an interesting polymer applicable in several fields [72]. In particular, the presence of amino and hydroxyl groups allow for chemical modification through chemical or physical interactions to produce chitosan composites. Chitosan–silica blends and polyethylene glycol membrane was applied in wastewater treatment [73]. A recent study has demonstrated the removal efficiency of herbicide of chitosan-TiO_2 nanocomposite [74] as well as the improvement of the OPPs removal efficiency (>90%) by incorporation of Cu nanoparticles on chitosan [75].

18.6.1.2 PHYSICAL CROSS-LINKING METHODS

Physical synthesis methods of hydrogels actually are of great interest due to the easy methodologies of synthesis, without the need of crosslinker (e.g., glutaraldehyde, epichlorohydrin), and for the reversibility of the polyelectrolyte obtained complexes.

Based on the used polymers and controlling concentrations, pH, and temperature, among other parameters, it is possible to modify the physico-chemical characteristics of the hydrogel [71]. Number physical methods will be presented.

18.6.1.2.1 Thermal Treatment

The thermal treatment permits the synthesis of hydrogels by rapid temperature changes and resulting structural modification. This method is applied in the hydrogels synthesis of gelatin and carrageenan modifying their not ordinary conformation at high temperature in helix conformation at low temperature [61].

18.6.1.2.2 Ionic Interaction

The polymeric crosslinking by ionic interaction is based on electrostatic interaction between charged functional groups and species of opposite charge (usually with high-density charge) added in the polymeric solution; commonly are used di- or trivalent counter ions (e.g., Ca^{2+}, Ba^{2+}, Sr^{2+}, Al^{3+} or Fe^{3+}) or polymers with chargeable functional groups, for example, chitosan-polylysine or quitosano-dextran [76].

18.6.1.2.3 Complex Coacervation

Complex coacervation process permits to gain hydrogels by electrostatic interpolymeric interactions and hydrogen bonds, soluble or not as pH function. Chitosan-xanthan gum or chitosan-carboxymethyl cellulose is only two hydrogel examples obtained by complex coacervation method. However, the mechanisms and kinetics of the coacervation process are still not completely known [77].

18.6.1.2.4 Hydrogen Bonding

In this method, the pH is a fundamental parameter for the hydrogen bonding formation in the hydrogels synthesis process. As an example, the sodium carboxymethyl cellulose salt at low pH is protonated favoring intrachain hydrogen bonding. Moreover, it is possible to obtain hydrogels by hydrogen interpolymeric bonding, such as between xanthan and alginate [71].

18.6.1.2.5   *Freezing/Thawing Cycles*

Freezing cycles permit to obtain polymeric crosslinked structures by hydrogen bondings, which occur in the microcrystals formation areas. The most common example of hydrogel prepared by this method is the poly(vinyl alcohol) [61].

18.6.1.2.6   *Emulsion*

The emulsion methods are widely used to prepare micro- and nanohydrogels by stirring single-double- or triple-emulsion (oil/water, or O/W) systems.

The principal purpose of this technique is the formation of precursor crystals of hydrogels with the hydrophobic medium. Successively, the starter can be crosslinked by different mechanisms (e.g., temperature variation, pH, or UV radiation).

This method permits to control the hydrogel dimension changing the experimental parameters such as mechanical stirring, viscosity, and/or surfactant presence.

However, the emulsion system permit to control the hydrogel dimensions compared with other synthesis methods and usually it is applied in the preparation of spherical hydrogels.

Examples are the alginate hydrogels, collagen, and gelatin hydrogels applied in biomedical engineering or as drug delivery systems [78].

18.7   CONCLUSION

In this chapter, a broad and short view on the available pesticides, EU legislation, and remediation processes using natural hydrogels are highlighted. Ecosystem protective actions are essential to safeguard essential food production, recover water–soil–air, and for the development of sustainable projections of health benefits and services. Being highly flexible and adaptable, hydrogels surely are critical for the goals in this field.

ACKNOWLEDGMENTS

This work was financed by Portuguese funds through FCT-Fundaçãopara a Ciência e a Tecnologia (FCT) in the framework of the projects WaterJPI/0006/2016 and UID/QUI/00313/2019.

KEYWORDS

- collagen
- EU legislation
- hydrogels
- legislation
- pesticides
- remediation

REFERENCES

1. European Environment Agency (UE) (2017). *Annual Indicator Report Series (AIRS)*, *13*.
2. Demir, A. E. A., Dilek, F. B., & Yetis, U. (2019). A new screening index for pesticides leachability to groundwater. *J. Environ. Manage.*, *231*, 1193–1202. https://doi.org/10.1016/j.jenvman.2018.11.007 (accessed on 19 February 2020).
3. European Commission (EC) (2017). *Report from the Commission to the European Parliament and the Council, 18.*
4. Silva, V., Mol, H. G. J., Zomer, P., Tienstra, M., Ritsema, C. J., & Giessen, V. (2019). Pesticide residues in European agricultural soils—A hidden reality unfolded. *Sci. Total Environ.*, *653*, 1532–1545. https://doi.org/10.1016/j.scitotenv.2018.10.441 (accessed on 19 February 2020).
5. Erbach, G. (2012). Pesticide legislation in the EU: Towards sustainable use of plant protection products. *European Parliament: EP Library*, 6.
6. Bozzini, E. (2017). EU pesticide regulation: Principles and procedures. In: *Pesticide Policy and Politics in the European Union* (pp. 27–56). Palgrave Macmillan. doi: 10.1007/978-3-319-52736-9.
7. Khan, M. A., & Brown, C. D. (2016). Influence of commercial formulation on leaching of four pesticides through soil. *Sci. Total Environ.*, *573*, 1573–1579. https://doi.org/10.1016/j.scitotenv.2016.09.076 (accessed on 19 February 2020).
8. Vryzas, Z. (2018). Pesticide fate in soil-sediment-water environment in relation to contamination preventing actions. *Curr. Opin. Environ. Sci. Health*, *4*, 5–9. https://doi.org/10.1016/j.coesh.2018.03.001 (accessed on 19 February 2020).

9. Food and Agriculture Organization of the United Nation (2009). *Feeding the World in 2050* (p. 4). World Summit on Food Security manifest, Rome.

10. Metcalf, R. L. (2012). *Insect Control* (Vol. 19, pp. 264–322). Wiley-VCH Verlag GmbH & Co. KGaA. doi: 10.1002/143560018.a14_263.

11. Pimentel, D. (1995). Amounts of pesticides reaching target pests: Environmental impacts and ethics. *J. Agric. Environ. Ethic., 8*, 17–29.

12. Food and Agriculture Organization of the United Nations (2010). *International Code and Conduct on the Distribution and Use of Pesticides* (p. 39). Rome.

13. Food and Agriculture Organization of the United Nations and World Health Organization (2016). *Manual on Development and Use of FAO and WHO Specifications for Pesticides* (p. 306). Rome.

14. Handford, C. E., Elliott, C. T., & Campbell, K. (2015). A review of the global pesticide legislation and the scale of challenge in reaching the global harmonization of food safety standards. *SETAC, 11*, 525–536. doi: 10.1002/ieam.1635.

15. Burgeois, A., Klinkhamer, E., & Price, J. (2012). *Pesticide Removal from Water* (Thesis). Massachusetts: Polytechnic Institute Worcester.

16. Casida, J. E., & Bryant, R. J. (2017). The ABCs of pesticide toxicology: Amounts, biology, and chemistry. *Toxicology Res., 5*, 755–763. doi: 10.1039/c7tx00198c.

17. European Commission (EC) (2003). *Presence of Persistent Chemicals in the Human Body: Results of Commissioner Wallstrom's Blood Test* (p. 5). Brussels.

18. Wibawa, W., Rosli, B. M., Puteh, A. B., Omar, D., Jurami, A. S., & Abdullah, S. A. (2009). Residual phytotoxicity effects of paraquat, glyphosate and glufosinate-ammonium herbicides in soils from field-treated plots. *Int. J. Agric. Biol., 11*, 214–216.

19. Garcia, F. P., Ascencio, S. Y. C., Oyarzun, J. C. G., Hernandez, A. C., & Alavarado, P. V. (2012). Pesticide: Classification, uses, and toxicity. Measures of exposure and genotoxic risks. *J. Res. Environ. Sci. Toxicol., 1*, 279–293.

20. Jayaraj, R., Megha, P., & Sreedev, P. (2016). Organochlorine pesticides, their toxic effects on living organisms, and their fate in the environment. *Interdiscip. Toxicol., 9*, 90–100 doi: 10.1515/intox-2016-0012.

21. Pleština, R. (2003). Pesticides and herbicides. In: Trugo, L., & Finglas, P. M. (eds.), *Encyclopaedia of Food Science, Food Technology and Nutrition* (2nd edn., pp. 4473–4483). Academic Press.

22. Sparling, D. W. (2016). *Organochlorine Pesticides: Ecotoxicology Essentials* (Vol. 4, pp. 69–107). Academic Press. http://dx.doi.org/10.1016/B978-0-12-801947-4.00004-4 (accessed on 19 February 2020).

23. Gupta, R. C. (2014). Carbamate pesticides. In: Wexler, P. (ed.), *Encyclopedia of Toxicology* (3rd edn., Vol. 1, pp. 661–664). Academic Press. http://dx.doi.org/10.1016/B978-0-12-386454-3.00106-8 (accessed on 19 February 2020).

24. Fishel, F. M. (2005). *Toxicity Profile: Carbamate Pesticides* (p. 3). University of Florida Institute of Food and Agricultural Science.

25. Krysan, J. L., & Dunley, J. (1993). *Insect Growth Regulators*. Wenatchee: Washington State University, Tree Fruit Research and Extension Center.

26. Kaneko, H. (2010). Pyrethroid chemistry and metabolism. *Hayes' Handbook of Pesticide Toxicology, 76*, 1635–1663.

27. European Food Safety Authority (EU) (2011). The 2009 European Union report on pesticide residues in food. *EFSA J. 9*, 2430–2655.

28. European Food Safety Authority (EU) (2017). The 2015 European Union report on pesticide residues in food. *EFSA J. 15*, 4791–4925.

29. Bourguignon, D. (2016). *EU Policy and Legislation on Chemicals* (p. 37). European Parliamentary Research Service (EU).

30. Ramos-Peralonso, M. J. (2014). European Food Safety Authority (EFSA). In: Wexler, P., (ed.), *Encyclopedia of Toxicology* (3rd edn., Vol. 2, pp. 554–556). Academic Press. http://dx.doi.org/10.1016/B978-0-12-386454-3.00563-7 (accessed on 19 February 2020).

31. Storck, V., Karpouzas, D. G., & Martin-Laurent, F. (2017). Towards a better pesticide policy for the European Union. *Sci. Total Environ., 575*, 1027–1033. https://doi.org/10.1016/j.scitotenv.2016.09.167 (accessed on 19 February 2020).

32. European Crop Protection Authority (B) (2013). *Registering Plant Protection Products in the EU* (p. 20). Brussels.

33. European Food Safety Authority (EU) (2018). *How Pesticides are Regulated in the EU*, 10.

34. Pesticide Action Network International (D) (2016). *PAN International List of Highly Hazardous Pesticides* (p. 35). Hamburg.

35. Yeung, M. T., Kerr, W. A., Coomber, B., Lantz, M., & McConnell, A. (2017). The economics of International Harmonization of MRLs. In: Barrett, C. (ed.), *Declining International Cooperation on Pesticide Regulation* (pp. 27–43). Palgrave Macmillan. https://doi.org/10.1007/978-3-319-60552-4_4 (accessed on 19 February 2020).

36. Yeung, M. T., Kerr, W. A., Coomber, B., Lantz, M., & McConnell, A. (2017). Why maximum residue limits for pesticides are an important international issue. In: Barrett, C., (ed.), *Declining International Cooperation on Pesticide Regulation* (pp. 1–9). Palgrave Macmillan. https://doi.org/10.1007/978-3-319-60552-4_1 (accessed on 19 February 2020).

37. Cheng, C. (2019). Codex Alimentarius Commission. In: Ferranti, P., Berry, E. M., & Anderson, J. R. (eds.), *Encyclopedia of Food Security and Sustainability* (Vol. 1, pp. 50–55). Elsevier. https://doi.org/10.1016/B978-0-08-100596-5.22376-7 (accessed on 19 February 2020).

38. Secretariat of Rotterdam Convention (2015). *Rotterdam Convention, on the Prior Informed Consent Procedure for Certain Hazardous Chemicals and Pesticides in International Trade* (p. 51). Rome.

39. Secretariat of Stockholm Convention (2009). *Stockholm Convention on Persistent Organic Pollutants (POPs)* (p. 56). Stockholm.

40. International Union of Pure and Applied Chemistry (1980). Definition of persistence in pesticide chemistry. *Pure Appl. Chem., 52*, 2563–2566.

41. Gavrilescu, M. (2005). Fate of pesticides in the environment and its bioremediation. *Eng. Life Sci., 5*, 497–526. doi: 10.1002/elsc.200520098.

42. Álvarez-Martín, A. A., Rodríguez-Cruz, M. S., Soledad, M., Andrades, M. J., & Sánchez-Martín (2016). Application of a biosorbent to soil: A potential method for controlling water pollution by pesticides. *Environ. Sci. Pollut. Res., 23*, 9192–9203. doi: 10.1007/s11356-016-6132-4.

43. Cruz-Alcalde, A., Sans, C., & Esplugas, S. (2017). Priority pesticides abatement by advanced water technologies: The case of acetamiprid removal by ozonation. *Sci. Total Environ., 599–600*, 1454–1461. http://dx.doi.org/10.1016/j.scitotenv.20118.05.065 (accessed on 19 February 2020).

44. Gligorovski, S., Strekowski, R., Barbati, S., & Vione, D. (2015). Environmental implications of hydroxyl radicals (•OH). *Chem. Rev.*, *115*, 13051–13092. doi: 10.1021/cr500310b.

45. Wert, E. C., Rosario-Ortiz, F. L., & Snyder, S. A. (2009). Using ultraviolet absorbance and color to assess pharmaceutical oxidation during ozonation of wastewater. *Environ. Sci. Technol.*, *43*, 4858–4863. doi: 10.1021/es803524a CCC: $40.75.

46. Borowska, E., Bourgin, M., Hollender, J., Kienle, C., McArdell, C. S., & Von Gunten, U., (2016). Oxidation of cetirizine, fexofenadine and hydrochlorothiazide during ozonation: Kinetics and formation of transformation products. *Water Res.* 94, 350–362, http://dx.doi.org/10.1016/j.watres.2016.02.020 (accessed on 19 February 2020).

47. Gerrity, D., Gamage, S., Holady, J. C., Mawhinney, D. B., Quiñones, O., Trenholm, R. A., & Snyder, S. A. (2011). Pilot-scale evaluation of ozone and biological activated carbon for trace organic contaminant mitigation and disinfection. *Water Res.*, *45*, 2155–2165. doi: 10.1016/j.watres.2010.12.031.

48. Rosenfeldt, E. J., Linden, K. G., Canonica, S., & Von Guntan, U. (2006). Comparison of the efficiency of •OH radical formation during ozonation and the advanced oxidation processes O_3/H_2O_2 and UV/H_2O_2. *Water Res.*, *40*, 3695–3704. doi: 10.1016/j.watres.2006.09.008.

49. Morillo, E., & Villaverde, J. (2017). Advanced technologies for the remediation of pesticide-contaminated soils. *Sci. Total Environ.*, *586*, 576–5918. https://doi.org/10.1016/j.scitotenv.20118.02.020 (accessed on 19 February 2020).

50. Tareq, R., Akter, N., & Md. Azam, S. (2019). Biochars and biochar composites: Low-cost adsorbents for environmental remediation. In: Ok, Y. S., Tsang, D. C. W., Bolan, N., & Novak, J. M. (eds.), *Biochar from Biomass and Waste* (pp. 169–209). Amsterdam. https://doi.org/10.1016/B978-0-12-811729-3.00010-8 (accessed on 19 February 2020).

51. Suo, F., Liu, X., Li, C., Li, C., Yuan, M., Zhang, B., Wang, J., Ma, Y., Lai, Z., & Ji, M. (2019). *Mesoporous Activated Carbon from Starch for Superior Rapid Pesticides Removal*, *121*, 806–813. https://doi.org/10.1016/j.ijbiomac.2018.10.132 (accessed on 19 February 2020).

52. Sartova, K., Omurzak, E., Kambarova, G., Dzhumaev, I., Borkoev, B., & Abdullaeva, Z. (2019). Activated carbon obtained from the cotton processing wastes. *Diam. Relt. Mater.*, *91*, 90–918. https://doi.org/10.1016/j.diamond.2018.11.011 (accessed on 19 February 2020).

53. Dehghani, M. H., Kamalian, S., Shayeghi, M., Yousefi, M., Heidarinejad, Z., Agarwal, S., & Gupta, V. K. (2019). High-performance removal of diazinon pesticide from water using multi-walled carbon nanotubes. *Microchem. J.*, *145*, 486–491. https://doi.org/10.1016/j.microc.2018.10.053 (accessed on 19 February 2020).

54. Dehghani, M. H., Niasar, Z. S., Mehrnia, M. R., Shayeghi, M., Al-Ghouti, M. A., Heibati, B., McKay, G., & Yetilmezsoy, K. (2017). Optimizing the removal of organophosphorus pesticide malathion from water using multi-walled carbon nanotubes. *Chem. Eng. J.*, *310*, 22–32. http://dx.doi.org/10.1016/j.cej.2016.10.057 (accessed on 19 February 2020).

55. Jafari, M. T., Saraji, M., & Kermani, M. (2018). Sol-gel electrospinning preparation of hybrid carbon silica nanofibers for extracting organophosphorus pesticides prior to analyzing them by gas chromatography-ion mobility spectrometry. *J. Chromatogr. A*, *13*, 1–13. https://doi.org/10.1016/j.chroma.2018.05.014 (accessed on 19 February 2020).

56. El-Said, W. A., El-Khouly, M. E., Ali, M. H., Rashad, R. T., Elshehy, E. A., & Al-Bogami, A. S. (2018). Synthesis of mesoporous silica-polymer composite for the chloridazon pesticide removal from aqueous media. *J. Environ. Chem. Eng.*, *6*, 2214–2221. https://doi.org/10.1016/j.jece.2018.03.027 (accessed on 19 February 2020).

57. Pellicer-Castell, E., Belenguer-Sapiña, C., Amorós, P., El Haskouri, J., Herrero-Martínez, J. M., & Mauri-Aucejo, A. (2018). Study of silica-structured materials as sorbents for organophosphorus pesticides determination in environmental water samples. *Talanta*, *189*, 560–5618. https://doi.org/10.1016/j.talanta.2018.018.044 (accessed on 19 February 2020).

58. Sillanpää, M., Ncibi, M. C., & Matilainen, A. (2018). Advanced oxidation processes for the removal of natural organic matter from drinking water sources: A comprehensive review. *J. Environ. Manage.*, *208*, 56–76. https://doi.org/10.1016/j.jenvman.20118.12.009 (accessed on 19 February 2020).

59. Ganji, F., Farahani, S. V., & Farahani, E. V. (2010). Theoretical description of hydrogel selling: A review. *Iran. Polym. J.*, *19*, 375–398.

60. Dilaver, M. (2011). *Preparation and Characterization of Carboxymethylcellulose Based Hydrogels* (p. 62). DokuzEylul University (Thesis).

61. Caló, E., Vitaliy, V., & Khutoryanskiy, V. (2015). Biomedical applications of hydrogels: A review of patents and commercial products. *Eur. Polym. J.*, *65*, 252–2618. http://dx.doi.org/10.1016/j.eurpolymj.2014.11.024 (accessed on 19 February 2020).

62. Gulrez, S. K. H., Al-Assaf, S., Phillips, G. O. (2011). Hydrogels: Methods of preparation, characterization and applications. *INTECH*, *5*, 117–150.

63. Klemm, D., Heublein, B., Fink, H. P., & Bohn, A. (2005). Cellulose: Fascinating biopolymer and sustainable raw material. *Angew. Chem. Int. Ed.*, *44*, 3358–3393. doi: 10.1002/anie.2004605818.

64. Ma, J., Li, X., & Bao, Y. (2015). Advances in cellulose-based superabsorbent hydrogels. *RSC Adv.*, *5*, 59745–597518. doi: 10.1039/c5ra08522e.

65. Sapna, & Kumar, D. (2018). Biodegradable polymer-based nanoadsorbent for environmental remediation. In: Hussain, C. M., & Mishra, A. K. (eds.), *New Polymer Nanocomposites for Environmental Remediation* (pp. 261–278). Amsterdam. https://doi.org/10.1016/B978-0-12-811033-1.00012-3 (accessed on 19 February 2020).

66. BeMiller, J. N. (2019). Cellulose and cellulose-based hydrocolloids. In: BeMiller, J. N. (ed.), *Carbohydrate Chemistry for Food Scientists* (3rd edn., pp. 223–240). Duxford. https://doi.org/10.1016/B978-0-12-812069-9.00008-X (accessed on 19 February 2020).

67. Ghaemi, N., Madaeni, S. S., Alizadeh, A., Rajabi, H., Daraei, P., & Falsafi, M. (2012). Effect of fatty acids on the structure and performance of cellulose acetate nanofiltration membrane in retention of nitroaromatic pesticides. *Desalination*, *301*, 26–41. https://doi.org/10.1016/j.desal.2012.06.008.

68. Ghaemi, N., Madaeni, S. S., Alizadeh, A., Daraei, P., Vatanpour, V., & Falsafi, M. (2012). Fabrication of cellulose acetate/sodium dodecyl sulfatenanofiltration membrane: Characterization and performance in rejection of pesticides. *Desalination*, *290*, 99–106. https://doi.org/10.1016/j.desal.2012.01.013 (accessed on 19 February 2020).

69. Takács, E., Wojnárovits, L., Horváth, E. K., Fekete, T., & Borsa, J. (2012). Improvement of pesticide adsorption capacity of cellulose fibre by high-energy irradiation-initiated grafting of glycidyl methacrylate. *Rad. Phys. Chem.*, *81* 1389–1392. https://doi.org/10.1016/j.radphyschem.2011.11.016 (accessed on 19 February 2020).

70. Suo, F., Xie, G., Zhang, J., Li, J., Li, C., Liu, X., Zhang, Y., Ma, Y., & Ji, M. (2018). A carbonised sieve-like corn straw cellulose-graphene oxide composite for organophosphorus pesticide removal. *RSC Adv.*, *8*, 7735–7743. 10.1039/c7ra12898c.
71. Singh, M. R., Patel, S., & Singh, D. (2016). Natural polymer-based hydrogels as scaffolds for tissue engineering. *Nanobiomater. Soft Tissue Eng.*, *5*(9), 231–260. http://dx.doi.org/10.1016/B978-0-323-42865-1.00009-X.
72. Cui, S. W., & Wang, Q. (2006). Functional properties of carbohydrates: Polysaccharide gums. In: Hui, Y. H., (ed.), *Handbook of Food Science, Technology, and Engineering* (pp. 1–18). Boca Raton.
73. Mahatmanti, F. W., & Nuryono, N. (2016). Adsorption of Ca (II), Mg (II), Zn (II), and Cd (II) on chitosan membrane blended with rice hull ash silica and polyethylene glycol. *Indonesia J. Chem.*, *16*, 45–52. http://dx.doi.org/10.22146/ijc.1043 (accessed on 19 February 2020).
74. Le Cunff, J., Tomašić, V., & Wittine, O. (2015). Photocatalytic degradation of the herbicide terbuthylazine: Preparation, characterization and photoactivity of the immobilized thin layer of TiO_2/chitosan. *J. Photochem. Photobiol. A*, *309*, 22–29. https://doi.org/10.1016/j.jphotochem.2015.04.021 (accessed on 19 February 2020).
75. Jaiswal, M., Chauhan, D., & Sankararamakrishnan, N. (2012). Copper chitosan nanocomposite: Synthesis, characterization, and application in removal of organophosphorous pesticide from agricultural runoff. *Environ. Sci. Pollut. R*, *19*, 2055–2062. https://doi.org/10.1007/s11356-011-0699-6 (accessed on 19 February 2020).
76. Akhtar, M. F., Hanif, M., & Ranjha, N. M. (2016). Methods of synthesis of hydrogels: A review. *Saudi Pharm. J.*, *24*, 554–559. http://dx.doi.org/10.1016/j.jsps.2015.03.022 (accessed on 19 February 2020).
77. Khan, S., Ullah, A., Ullah, K., & Rehman, N. (2016). Insight into hydrogels. *Des. Monomers Polym.*, *19*, 456–478. http://dx.doi.org/10.1080/15685551.2016.1169380 (accessed on 19 February 2020).
78. Dubin, P., & Stewart, R. J. (2018). Complex coacervation. *Roy. Soc. Chem.*, *14*, 329–330. doi: 10.1039/C7SM90206A.

CHAPTER 19

A New Equation for the Computation of the Lattice Energies of Inorganic Ionic Crystals Based on the Parr's Electrophilicity Index

SAVAŞ KAYA,[1] ROBSON FERNANDES DE FARIAS,[2] and
NAZMUL ISLAM[3,*]

[1]*Department of Chemistry, Faculty of Science, Cumhuriyet University, Sivas 58140, Turkey*

[2]*The Federal University of Rio Grande do Norte, 59078-970 Natal, RN, Brazil*

[3]*Ramgarh Engineering College, Murubanda, Ramgarh, Jharkhand 825101, India*

Corresponding author. E-mail: nazmul.islam786@gmail.com

ABSTRACT

In the present work, based on the Parr's electrophilicity index a new equation for the computation of the lattice energies of inorganic ionic crystals is derived. The new equation is used to compute the lattice energies of some sets of inorganic ionic crystals. As for the validity test of the proposed equation of lattice energy, a comparative study of the results is done using the computed lattice energy with their experimental counterparts and some sets of theoretical lattice energy data. The comparative study shows excellent correlation with theoretical lattice energy data computed using our equation and experimental lattice energy data. The study revealed that the new equation is valid and reliable. The minimum electrophilicity Principle is one of the well-known electronic structure principles. We have also discussed the validity of minimum electrophilicity principle in inorganic ionic crystals and the study showed that this electronic structure principle contradicts with the lattice energies of inorganic solids.

19.1   INTRODUCTION

In the prediction of chemical properties of atoms, ions, and molecules, conceptual density functional theory [1] is one of the useful tools. In the mentioned theory, quantum chemical descriptors like electronegativity (or chemical potential) and hardness are defined as first and second derivatives of electronic energy with respect to the number of electrons at a constant external potential, respectively. Absolute hardness (η) and electronegativity (χ) are calculated depending on ground state ionization energy (IE) and electron affinity (EA) values of chemical species with the help of the following equations [2]:

$$\eta = IE - EA \tag{19.1}$$

$$\chi = (IE + EA) / 2 \tag{19.2}$$

Electrophilicity index (ω) proposed by Parr and [3] coworkers is calculated via the following equation depending on electronegativity and hardness values of chemical species.

$$\omega = \chi^2 / 2\eta \tag{19.3}$$

Hardness equalization principle and electronegativity equalization principle are some of the well-known electronic structure principles [4, 5]. It is important to note that hardness and electronegativity are charge-dependent chemical properties of atoms and molecules. Electronegativity equalization principle and hardness equalization principle have been proposed based on charge equalization during a chemical reaction. Electronegativity equalization principles state that "when two or more atoms initially different in electronegativity combine chemically, the electronegativities of said atoms have become equalized as a consequence of electron transfers among them." Chemical hardness is known as the resistance of the chemical species against the polarization or deformation of the electron cloud. It should be noted that chemical hardness also gets equalized like electronegativity during molecule formation. In recent years, considering hardness equalization and electronegativity equalization principles Kaya and coworkers [6, 7] derived the following equations to compute the molecular hardness (η_M) and molecular electronegativity (χ_M) of the molecules, respectively.

$$\eta_{M=}(2\sum_{i=1}^{N}\frac{b_i}{a_i})+q_M / \sum_{i=1}^{N}\frac{1}{a_i} \tag{19.4}$$

$$\chi_M = \sum_{i=1}^{N}\left(\frac{a_i}{b_i}\right)+2q_M / \sum_{i=1}^{N}\left(\frac{1}{b_i}\right) \tag{19.5}$$

In these equations, N and q_M are the total numbers of atoms in the molecule and charge of molecule. a_i and b_i parameters appearing in these equations are based on first IE and EA values of atoms and they are described as $a_i = (IE+EA)/2$ and $b_i = (IE-EA)/2$, respectively.

Lattice energy is a measure of the stability of inorganic ionic solids and it is defined as the energy required decomposing a solid into its constituent independent gaseous ions [8]. This energy is also used for the prediction of the feasibility of a reaction. Many theoretical methods are proposed to compute the lattice energies of simple and complex inorganic ionic solids. Born–Lande, Born–Mayer [9] and Kapustinskii [10] equations are some of the lattice energy calculation techniques introduced a long time ago. In the 1950s, Kudriavtsev [11] calculated the lattice energies of many inorganic ionic materials based on sound velocity data. In the 1990s, Reddy [12] explained the relationships between interionic separation, plasma energy, and lattice energy for alkali metal halides. One of the most comprehensive and useful equations used in the prediction of lattice energies of complex and simple inorganic ionic systems is Jenkins/Glasser equation. With this approach, known as volume-based thermodynamics (VBT), Jenkins [13] showed that there is a remarkable correlation between lattice energy (U) and molar volume (V_m). The details of this equation are given below

$$U(\text{kJ/mol}) = 2I[\alpha(V_m)^{-1/3} + \beta] \tag{19.6}$$

In the given equation, V_m is the molar volume of an ionic compound. α and β parameters appearing in the equation are coefficients depending on the stoichiometry in terms of the charge of the ionic compounds. I is the ionic strength of the lattice and it can be calculated as

$$I = 1/2\sum n_i z_i^2$$

where n_i is the number of ions of type i in the formula unit with charge z_i.

It should be noted that lattice energy values obtained by Born–Haber–Fajans thermochemical cycle for inorganic ionic materials are considered as experimental values. In addition to above-mentioned works, Petrov and Angelov [14, 15] calculated the lattice energies of complex ionic systems like lanthanide monosulfides, lanthanide monoaluminates, lanthanide orthophosphates ($LnPO_4$), lanthanide orthovanadates ($LnVO_4$), lanthanide orthoferrites ($LnFeO_3$), lanthanide gallium garnets ($Ln_3Ga_5O_{12}$), lanthanide iron garnets ($Ln_3Fe_5O_{12}$) with the help of Born–Haber–Fajans thermochemical cycle.

Leslie Glasser [16] showed a simple correlation between Madelung energy, E_M' and lattice energy and calculated the lattice energies of many simple and complex ionic systems with the equation $U(\text{kJ/mol}) = 0.963E_M'$.

The aim of this chapter is to derive a new equation for the computation of the lattice energies of inorganic ionic crystals based on Parr's electrophilicity index and molar volume concept and is to discuss whether minimum electrophilicity principle is valid in the formation of inorganic ionic solids.

19.2   THEORETICAL MODEL

Theoretical chemists present new electronic structure principles and equations to provide the backbone for the analysis of the stability or reactivity of chemical species. In Table 19.1, electronegativity, chemical hardness, electrophilicity, molar volume, and experimental lattice energy values of various inorganic ionic compounds are given. Chemical hardness and electronegativity values of the given compounds have been calculated with the help of Equations (19.4) and (19.5), respectively. Electrophilicity values of the compounds have been calculated using Parr's electrophilicity index formula and putting the Kaya molecular hardness (Equation (19.4)) and molecular electronegativity (Equation (19.5)) values of the corresponding systems. The VBT approach proposed by Jenkins is based on an equation introduced by Mallouk and coworkers [17]. As mentioned in Section 19.1, Jenkins equation includes the use of the molar volume concept. The convenience of the use of molar volume in the prediction of the lattice energies of inorganic ionic crystals is that the molar volumes of the ionic compounds can be easily determined with the help of their density and molar mass data without any structural detail.

To reveal the relationship between lattice energy and electrophilicity index and to discuss the validity of the minimum electrophilicity principle [18] in inorganic ionic solids, we firstly plotted Figures 19.1–19.7. All graphs (Figures 19.1–19.7) show that there is a remarkable correlation between experimental lattice energies and $\omega/V_m^{1/3}$ for many sets of inorganic ionic solids. It can be seen from these graphs that lattice energy increases linearly as $\omega/V_m^{1/3}$ ratio increases for simple inorganic ionic compounds.

TABLE 19.1 Electronegativity, Chemical Hardness, Electrophilicity, Molar Volume, and Experimental Lattice Energy Values of Various Inorganic Ionic Compounds

Inorganic Crystal	χ_M (eV)	η_M (eV)	ω (eV)	V_m (nm³)	U_{BFH} (kJ/mol)
LiH	4.139	7.15	1.197994	0.01692	918
NaH	3.984	6.93	1.145184	0.02853	807
KH	3.514	6.12	1.008840	0.04656	713
RbH	3.414	5.94	0.981094	0.05522	684
CsH	3.231	5.61	0.930424	0.06502	653
CuCl	6.042	7.49	2.436967	0.03965	996
CuBr	5.831	7.21	2.357875	0.05057	978
CuI	5.545	6.85	2.244308	0.05577	966
LiCl	4.79	5.99	1.915200	0.03403	853
LiBr	4.65	5.85	1.848077	0.04163	807
LiI	4.48	5.56	1.804892	0.05452	757
NaCl	4.63	5.81	1.844828	0.04482	787
NaBr	4.51	5.63	1.806403	0.05322	747
NaI	4.34	5.43	1.734401	0.06782	704
KCl	4.12	5.08	1.670709	0.06239	715
KBr	4.03	4.95	1.640495	0.07212	682
KI	3.90	4.77	1.594340	0.08826	649
RbCl	4.01	4.94	1.627540	0.07170	689
RbBr	3.93	4.82	1.602168	0.08197	660
RbI	3.80	4.65	1.552688	0.11338	630
CsCl	3.81	4.66	1.557521	0.07006	659
CsBr	3.73	4.54	1.532258	0.07959	631

TABLE 19.1 *(Continued)*

Inorganic Crystal	χ_M (eV)	η_M (eV)	ω (eV)	V_m (nm³)	U_{BFH} (kJ/mol)
CsI	3.62	4.39	1.492528	0.09565	604
CaO	4.55	7.85	1.318631	0.02688	3401
CaS	4.40	6.81	1.421439	0.04644	2966
SrS	4.22	6.47	1.376229	0.05372	2779
BaS	4.02	6.03	1.340000	0.06619	2643
PbO	5.23	8.79	1.555910	0.03889	3520
PbSe	4.84	7.32	1.600109	0.05866	3144
TiO	4.91	8.45	1.426515	0.02142	3811
MnO	5.17	9.25	1.444805	0.02168	3745
ZnO	5.83	10.72	1.585303	0.02413	3971
CdSe	5.24	8.61	1.594518	0.05463	3310
GeO	5.62	8.74	1.806888	0.03344	3919
VO	4.95	8.23	1.488609	0.01929	3863
MnS	4.81	8.05	1.437025	0.03620	3415
MnSe	4.69	7.85	1.401025	0.03977	3310
ZnS	5.50	8.91	1.697531	0.03950	3674
MgF_2	7.27	10.14	2.606	0.03922	2978
$MgCl_2$	6.62	8.23	2.660	0.06984	2540
$MgBr_2$	6.26	7.80	2.512	0.08100	2451
CaF_2	6.48	9.02	2.327	0.10101	2651
$CaCl_2$	6.02	7.47	2.425	0.03284	2271
$CaBr_2$	5.73	7.14	2.299	0.06814	2134
CaI_2	5.36	6.70	2.144	0.08210	2087
SrF_2	6.22	8.62	2.244	0.04076	2513
$SrCl_2$	5.83	7.16	2.373	0.08571	2170
$SrBr_2$	5.56	6.85	2.256	0.09900	2040
SrI_2	5.21	6.44	2.107	0.12320	1976
BaF_2	5.93	8.10	2.169	0.04919	2373

TABLE 19.1 *(Continued)*

Inorganic Crystal	χ_M (eV)	η_M (eV)	ω (eV)	V_m (nm³)	U_{BFH} (kJ/mol)
BaCl$_2$	5.60	6.74	2.326	0.08631	2069
BaBr$_2$	5.35	6.47	2.221	0.09730	1995
BaI$_2$	5.03	6.09	2.077	0.12460	1890
Li$_2$O	3.749	6.00	1.171	0.02458	2814
Na$_2$O	3.591	5.79	1.113	0.04533	2478
K$_2$O	3.119	4.99	0.974	0.06742	2232
Rb$_2$O	3.020	4.82	0.946	0.07760	2161
Cs$_2$O	2.839	4.52	0.891	0.10063	2063
Li$_2$S	3.725	5.45	1.272	0.04596	2472
Na$_2$S	3.575	5.27	1.212	0.06981	2203
K$_2$S	3.135	4.56	1.077	0.10521	1979
Rb$_2$S	3.042	4.41	1.049	0.11571	1949

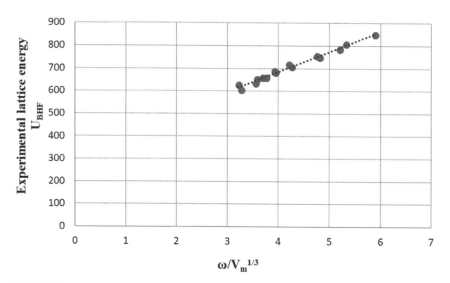

FIGURE 19.1 Plot of correlation between experimental lattice energies and $\omega/V_m^{1/3}$ values for alkali halides, $R^2 = 0.9841$.

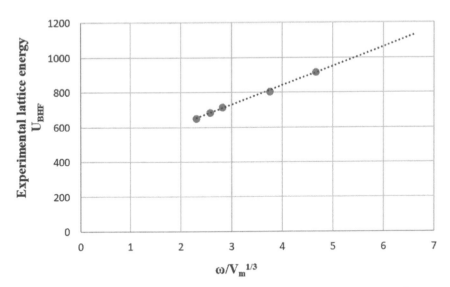

FIGURE 19.2   Plot of correlation between experimental lattice energies and $\omega/V_m^{1/3}$ values for alkali hydrides, $R^2 = 0.9985$.

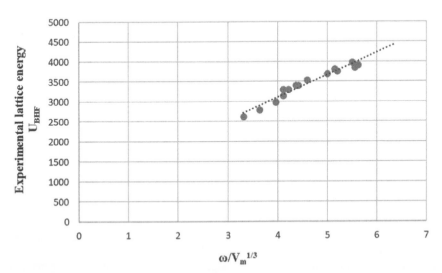

FIGURE 19.3   Plot of correlation between experimental lattice energies and $\omega/V_m^{1/3}$ values for MX (charge ratio 2:2) type in Table 19.1, $R^2 = 0.9519$.

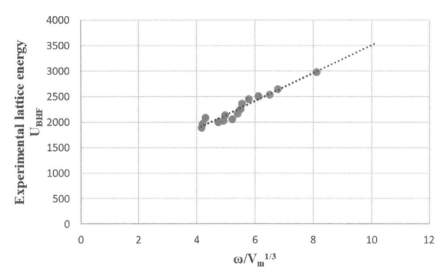

FIGURE 19.4   Plot of correlation between experimental lattice energies and $\omega/V_m^{1/3}$ values for MX_2 type in Table 19.1, $R^2 = 0.9421$.

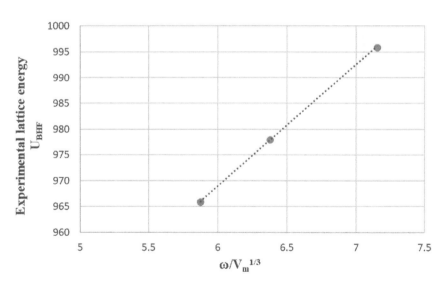

FIGURE 19.5   Plot of correlation between experimental lattice energies and $\omega/V_m^{1/3}$ values for copper (I) halides, $R^2 = 1.00$.

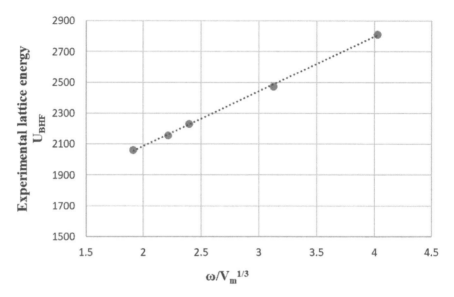

FIGURE 19.6   Plot of correlation between experimental lattice energies and $\omega/V_m^{1/3}$ values for alkali metal oxides, $R^2 = 1.9996$.

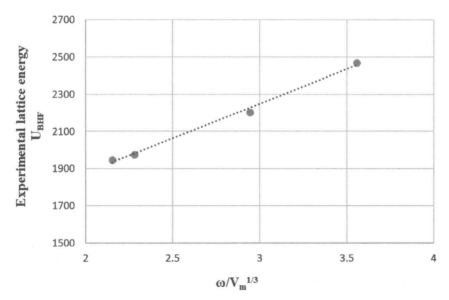

FIGURE 19.7   Plot of correlation between experimental lattice energies and $\omega/V_m^{1/3}$ values for alkali metal sulfides, $R^2 = 1.9933$.

By carefully analyzing the slope and cut values of the graphs drawn for the various simple inorganic ionic compounds, we presented the following equation that shows the correlation between electrophilicity and lattice energy to compute the lattice energies of simple ionic compounds. In this equation, we also used ionic strength (I) of the lattice instead of charge of cation and anion, Z^+, Z^- like Jenkins.

$$U(\text{kJ/mol}) = 2I[k\omega / V_m^{1/3} + l] \tag{19.7}$$

In this equation, U and V_m are lattice energy and molar volume, respectively. k and l parameters appearing in the equation are constants and their numerical values and units for various molecule groups are given in Table 19.2. The numerical values of these constants have been determined using Excel's regression analysis.

TABLE 19.2 The Numerical Values of k and l Constants to Use in Equation (19.7)

Molecules	Ionic Strength (I)	k (kJ nm/mol eV)	l (kJ/mol)
Alkali halides	1	44.41	164.73
Alkali hydrides	1	55.41	198.97
Copper(I) halides	1	11.78	413.84
Alkali metal oxides	3	59.27	229.46
Alkali metal sulfides	3	62.13	188.65
MX$_2$ type compounds in Table 19.1	3	45.418	130.37
MX (2:2) type compounds in Table 19.1	4	70.535	106.158

19.3 RESULTS AND DISCUSSION

This work presents a new correlation between lattice energy and Parr's electrophilicity index and shows the validity of the minimum electrophilicity principle for inorganic ionic solids. Here we found a linear correlation between their $\omega/V_m^{1/3}$ and lattice energy of inorganic ionic crystals. With the help of the mentioned correlation, we introduced a new equation that can be used for the prediction of lattice energies of various groups of inorganic ionic crystals. It is well known that lattice energies of inorganic ionic crystals cannot be experimentally measured directly and the data obtained using

Born–Fajans–Haber thermochemical cycle is accepted as experimental results. To prove the validity and reliability of the electrophilicity index-based lattice energy equations, in Table 19.3, we calculated the lattice energy values of some selected alkali metal halide molecules and then compared our results with well-known lattice energy calculation techniques like Born–Lande, Born–Mayer, Kapustinskii, Reddy, Kudriavtsev, and Jenkins equations. It is apparent from Table 19.3 that the present method provides more close results to experimental data compared to other equations. It should be noted that Born–Lande, Born–Mayer, Kapustinskii equations have been derived ignoring the covalent characters of inorganic ionic solids. For that reason, these equations are not useful in the prediction of lattice energies of the compounds having a high covalent character. In contrast to this situation, our electrophilicity index-based lattice energy equation is useful for both inorganic compounds having high covalent character and inorganic compounds having high ionic character (Table 19.4 and Figure 19.8).

TABLE 19.3 Comparison of Lattice Energies Determined with Various Theoretical Methods for Alkali Halides

Alkali Halide Crystal	Born–Fajans–Haber	Born–Lande	Born–Mayer	Kapuscinski	Reddy	Kudriavtsev	Jenkins	Present Work
LiCl	853	810	818	803.9	851.7	880.5	827	854
LiBr	807	765	772	792.6	813.9	844.5	780	803
LiI	757	713	710	713.1	755.8	–	721	752
NaCl	787	753	756	752.9	799.9	879.3	764	790
NaBr	747	717	719	713.5	765.0	844.1	727	756
NaI	704	671	670	673.4	712.8	768.8	678	707
KCl	715	686	687	680.9	732.0	799.3	695	703
KBr	682	658	659	675	699.8	772.1	667	679
KI	649	622	620	613.9	651.1	699.3	630	647
RbCl	689	659	661	662.1	701.5	740.7	668	677
RbBr	660	634	635	626.5	670.4	685.9	644	657
RbI	630	601	600	589.7	622.5	680.1	610	614
CsCl	659	621	621	625.2	644.0	672.1	672	665
CsBr	631	598	598	602.2	612.1	669.2	648	645
CsI	604	568	565	563.7	562.6	659.1	616	619

TABLE 19.4 Comparison of Experimental and Calculated Lattice Energy Values for Various Inorganic Ionic Crystals

Inorganic Crystal	U_{BFH}	U_{cal}	%SD
LiH	918	915	0.32
NaH	807	813	0.74
KH	713	708	0.70
RbH	684	683	0.14
CsH	653	654	0.15
CuCl	996	996	0.00
CuBr	978	978	0.00
CuI	966	966	0.00
CaO	3401	3333	1.99
CaS	2966	3080	3.84
SrS	2779	2907	4.60
BaS	2643	2718	2.83
PbO	3520	3440	2.27
PbSe	3144	3173	0.92
TiO	3811	3747	1.67
MnO	3745	3773	0.74
ZnO	3971	3944	0.67
CdSe	3310	3220	2.71
GeO	3919	4014	2.42
VO	3863	3981	3.05
MnS	3415	3306	3.19
MnSe	3310	3169	4.25
ZnS	3674	3660	0.38
MgF_2	2978	2999	0.70
$MgCl_2$	2540	2556	0.62
$MgBr_2$	2451	2357	3.83
CaF_2	2651	2624	1.01
$CaCl_2$	2271	2281	0.44
$CaBr_2$	2134	2136	0.09
CaI_2	2087	1956	6.27
SrF_2	2513	2451	2.46
$SrCl_2$	2170	2245	3.45
$SrBr_2$	2040	2118	3.82

TABLE 19.4   *(Continued)*

Inorganic Crystal	U_{BFH}	U_{cal}	%SD
SrI_2	1976	1931	2.27
BaF_2	2373	2295	3.28
$BaCl_2$	2069	2198	6.23
$BaBr_2$	1995	2072	3.85
BaI_2	1890	1910	1.05
Li_2O	2814	2809	0.17
Na_2O	2478	2486	0.32
K_2O	2232	2228	0.17
Rb_2O	2161	2165	0.18
Cs_2O	2063	2058	0.24
Li_2S	2472	2456	0.64
Na_2S	2203	2229	1.18
K_2S	1979	1982	0.15
Rb_2S	1949	1934	0.76

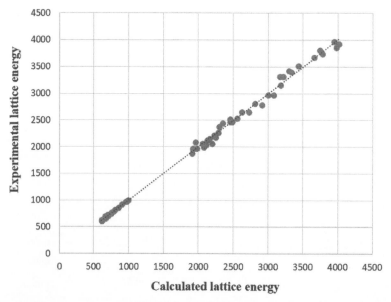

FIGURE 19.8 Comparison with the experimental lattice energies of calculated lattice energies for all molecules considered in this study, $R^2 = 0.9974$.

Table 19.4 contains experimental lattice energy values determined via Born–Fajans–Haber thermochemical cycle and calculated lattice energy values obtained using Equation (19.7). Percentage deviation values (%SD) between calculated (U_{cal}) and experimental lattice energy (U_{BFH}) values were determined considering the equation $\%SD = \left[|U_{BFH} - U_{cal}| / U_{BFH} \right] \times 100$. It is clear from standard deviation values that in the majority of case, the standard deviation is in between 0% and 1%. To see more clearly the agreement between experimental data and lattice energy values calculated via our method for all molecules considered in this study, we plotted and presented Figure 19.8. As it can be understood from the R^2 values of the mentioned figure that the present approach is quite useful. In 1995, Leslie Glasser [19] generalized the Kapustinskii equation and presented an equation that can be used for inorganic crystals containing multiple ions. Jenkins equation also can be used for the prediction of lattice energies of inorganic crystals containing complex ions. Here, it should be noted that our new method is valid for only simple inorganic ionic crystals.

The maximum hardness principle, minimum polarizability principle, and minimum electrophilicity principle are three important electronic structure principles that can be considered in the analysis of the stability or reactivity of chemical systems [20]. According to the maximum hardness principle introduced by Pearson [21], "There seems to be a rule of nature that molecules arrange themselves so as to be as hard as possible." A formal proof regarding the validity of the maximum hardness principle has been provided by Parr and coworkers [22]. The minimum polarizability principle states that polarizability is a stability measure and the preferred direction of a chemical reaction is toward lesser polarizability. Electrophilicity is one of the useful quantities used to discuss the stabilities of chemical compounds. Many authors demonstrated that electrophilicity can be considered a stability descriptor. Noorizadeh [23] explained considering Diels–Alder reaction types that major products in these reactions has always less electrophilicity than the minor product. In a study including 25 simple reactions, Noorizadeh [24] showed that the natural direction of a chemical reaction is toward a state of minimum electrophilicity. In short, if the minimum electrophilicity principle was valid in the formation of inorganic crystals, as electrophilicity decreases, lattice energy should increase. But it can be seen from Equation (19.7), lattice energy that is a stability criterion increases as electrophilicity increases. The graphs are drawn and the equation derived in this study show that the minimum electrophilicity principle is not valid for inorganic ionic crystals.

19.4   CONCLUSION

In the present chapter, a new equation depending on Parr's electrophilicity index and molar volume concept to compute the lattice energies of simple inorganic ionic crystals are presented. It is apparent from the results obtained that the new lattice energy equation provides more close results to experimental data compared to other theoretical approaches proposed for the calculation of the lattice energy. In addition, we discussed the validity of the minimum electrophilicity principle in inorganic ionic crystals and showed that this principle is not successful in the explanation of the stabilities of inorganic ionic crystals.

KEYWORDS

- **new equation**
- **computation**
- **lattice energies**
- **inorganic ionic crystals**
- **Parr's electrophilicity index**

REFERENCES

1. Geerlings, P., De Proft, F., & Langenaeker, W. (2003). Conceptual density functional theory. *Chem. Rev., 103*, 1793–1874.
2. Chattaraj, P. K., Giri, S., & Duley, S. (2010). Electrophilicity equalization principle. *J. Phys. Chem. Lett., 1*, 1064–1067.
3. Parr, R. G., Szentpaly, L. V., & Liu, S. (1999). Electrophilicity index. *J. Am. Chem. Soc., 121*, 1922–1924.
4. Ghosh, D. C., & Islam, N. (2011). Whether there is a hardness equalization principle analogous to the electronegativity equalization principle—A quest. *Int. J. Quant. Chem., 111*, 1961–1969.
5. Kaya, S., Kaya, C., & Islam, N. (2016). The nucleophilicity equalization principle and new algorithms for the evaluation of molecular nucleophilicity. *Computational and Theoretical Chemistry, 1080*, 72–78.
6. Kaya, S., & Kaya, C. (2015). A new equation for calculation of chemical hardness of groups and molecules. Mol. Phys., 113, 1311–1319.
7. Kaya, S., & Kaya, C. (2015). A new equation based on ionization energies and electron affinities of atoms for calculating of group electronegativity. *Comput. Theor. Chem., 1054*, 42–46.

8. Leslie, G., & Sheppard, D. A. (2016). Cohesive energies and enthalpies: complexities, confusions, and corrections. *Inorg. Chem., 55*, 7103–7110.

9. Born, M., & Mayer, J. (1932). Zur gittertheorie der ionenkristalle. *Z. Phys., 75*, 1–18.

10. Kapustinskii, A. F. (1956). Lattice energy of ionic crystals. *Q. Rev. Chem. Soc., 10*, 283–294.

11. Kudriavtsev, B. B. (1956). A relation connecting between the ultrasonic velocity in an electrolytic solution and the lattice energy, *Sov. Phys. Acoust., 2*, 36–45.

12. Reddy, R. R., Kumar, M. R., & Rao, T. V. R. (1993). Interrelations between interionic separation, lattice energy, and plasma energy for alkali halide crystals. *Cryst. Res. Technol., 28*, 973–977.

13. Jenkins, H. D. B., Roobottom, H. K., Passmore, J., & Glasser, L. (1999). Relationships among ionic lattice energies, molecular (formula unit) volumes, and thermochemical radii. *Inorg. Chem., 38*, 3609–3620.

14. Petrov, D., & Angelov, B. (2010). Lattice energies and crystal-field parameters of lanthanide monosulphides. *Phys. B., 405*, 4051–4053.

15. Petrov, D., & Angelov, B. (2012). Lattice energies and polarizability volumes of lanthanide monoaluminates. *Phys. B, 407*, 3394–3397.

16. Leslie, G. (2012). Simple route to lattice energies in the presence of complex ions. *Inorg. Chem., 51*, 10306–10310.

17. Mallouk, T. E., Rosenthal, G. L., Mueller, G., Brusasco, R., & Barlett, N. (1984). Fluoride ion affinities of germanium tetrafluoride and boron trifluoride from thermodynamic and structural data for $(SF3)_2GeF_6$, ClO_2GeF_5, and ClO_2BF_4. *Inorg. Chem., 23*, 3167–3173.

18. Pan, S., Sola, M., & Chattaraj, P. K., (2013). On the validity of the maximum hardness principle and the minimum electrophilicity principle during chemical reactions. *J. Phys. Chem. A., 117*, 1843–1852.

19. Glasser, L. (1995). Lattice energies of crystals with multiple ions: a generalized Kapustinskii equation. *Inorg. Chem., 34*, 4935–4936.

20. Kaya, S., Kaya, C., & Islam, N., (2016). Maximum hardness and minimum polarizability principles through lattice energies of ionic compounds. *Phys. B Condens. Matter, 485*, 60–66.

21. Pearson, R. G. (1993). The principle of maximum hardness. *Acc. Chem. Res., 26*, 250–255.

22. Parr, R. G., & Chattaraj, P. K. (1991). Principle of maximum hardness. *J. Am. Chem. Soc., 113*, 1854–1855.

23. Noorizadeh, S., & Maihami, H. (2006). A theoretical study on the regioselectivity of Diels-Alder reactions using electrophilicity index. *J. Mol. Struct. Theochem, 763*, 133–144.

24. Noorizadeh, S. (2007). Is there a minimum electrophilicity principle in chemical reactions? *Chin. J. Chem., 25*, 1439–1444.

Index

Printed and bound by CPI Group (UK) Ltd, Croydon, CR0 4YY

23/10/2024

01777675-0014